Zooplankton Challenges in a Changing World

Volume 1
A Worldwide Perspective and Research Approach

Editors

Luis R. Vieira

Interdisciplinary Centre of Marine
and Environmental Research (CIIMAR)
University of Porto
Matosinhos, Porto
Portugal

Fernando Morgado

The Centre for Environmental and Marine
Studies (CESAM) and Department of Biology
University of Aveiro
Aveiro, Portugal

CRC Press
Taylor & Francis Group
Boca Raton London New York

CRC Press is an imprint of the
Taylor & Francis Group, an **informa** business

A SCIENCE PUBLISHERS BOOK

First edition published 2025
by CRC Press
2385 NW Executive Center Drive, Suite 320, Boca Raton FL 33431

and by CRC Press
4 Park Square, Milton Park, Abingdon, Oxon, OX14 4RN

Library of Congress Cataloging-in-Publication Data (applied for)

ISBN: 978-1-032-64976-4 (hbk)
ISBN: 978-1-032-66226-8 (pbk)
ISBN: 978-1-032-66227-5 (ebk)

DOI: 10.1201/9781032662275

Typeset in Times New Roman
by Prime Publishing Services

Preface

Ocean science and marine biology highlight the dynamic interplay between marine organisms and their environments. Covering over 70% of the Earth, oceans are vital for sustaining life and climate systems. The present book delves into the multifaceted nature of oceanography and marine biology, emphasizing the critical role of interdisciplinary approaches in addressing the myriad stressors that oceans face today.

One of the foundational aspects of marine ecosystem exploration is the role of plankton, which constitutes a staggering 95% of marine biomass. Both phytoplankton and zooplankton are not only fundamental to marine food webs but also play a crucial role in carbon cycling, making them key indicators of aquatic ecosystem health. The technological revolution in marine observations has significantly enhanced researchers' capacity to monitor plankton diversity, ecosystem services, and the overall health of marine environments in real-time.

Global partnerships and new marine technologies are vital for comprehensive plankton research and effective ecosystem management. Research institutions worldwide act as hubs for interdisciplinary collaboration, addressing complex global challenges in ocean science. Collaborative works guide ocean research and conservation efforts.

The impact of climate change on coastal environments is multifaceted, influencing wind patterns, temperature, precipitation, upwelling patterns, and river discharge regimes. These changes, in turn, affect primary production and the entire food web. There is a dedicated chapter on the intricate connection between climate-level changes and zooplankton in coastal environments, with a representative study from the Chilean coast. Zooplanktonic organisms, due to their short life cycles are particularly sensitive to temperature changes, making them excellent sentinels for climate change.

The NW Iberian waters, part of the Canary Current Upwelling System, provide a fascinating case study for examining zooplankton dynamics concerning environmental variability. Long-term studies and updated zooplankton time series offer insights into the suitability of zooplankton-based metrics as indicators of ecosystem status. This research highlights the importance of database management and data analysis tools to understand long-term variations in zooplankton communities.

In the Mediterranean Sea, biomarkers like aminoacyl-tRNA synthetases (AARS) activity, electron transport system (ETS) activity, stable isotopes (SI), and fatty acids (FA) provide valuable insights into zooplankton physiology and their responses to environmental changes. These studies enhance our understanding of mesozooplankton ecology, especially in climate change scenarios.

The Portuguese coast's dynamic marine ecosystems, influenced by upwelling regimes and climate variability, are explored, emphasizing the importance of understanding zooplankton community dynamics for effective marine conservation strategies.

A chapter overviews zooplankton biodiversity over the last 20 years in the Southwest Atlantic Ocean, focusing on Brazilian estuaries, coastal, and oceanic regions.

The book also addresses the challenges faced by Mozambique's biodiverse yet vulnerable marine ecosystems, emphasizing the need for research infrastructure, scientific collaboration, and educational initiatives.

In the East Asian Seas, particularly the Sea of Japan and the Western North Pacific, zooplankton biomass and assemblages are undergoing accelerated changes due to global warming, industrial development, and fisheries activities, highlighting the need for enhanced monitoring and methodological advancements.

Zooplankton communities are crucial for ecosystem assessments, serving as reliable indicators of environmental changes. One of the chapters explores a 20-year dataset on the Bulgarian Black Sea coast, analysing spatio-temporal variations in zooplankton communities and the impacts of anthropogenic pressures.

Finally, the book explores the challenges and opportunities in plankton research across African coastal and marine ecosystems. With the continent's population expected to nearly double by 2050, intensified anthropogenic pressures on marine ecosystems are inevitable.

By exploring the complex interaction between environmental stressors, technological advancements, and collaborative research, this book represents a contribution from a team of researchers, from several regions of the world, with a common mission to enhance readers' understanding of current ocean science and marine biology while pointing towards future directions.

The Editors
Luis R. Vieira and Fernando Morgado

Contents

CHAPTER 1
Plankton Pressures in the Framework of the Marine Strategy

1.1

Global Ocean Science Infrastructures, Technological Development and Plankton Research

Fernando Morgado[1],* and *Luis R. Vieira*[2]

1. Introduction

Oceans occupy about 71% of the planet's surface area, corresponding approximately to a 361×10^6 km^2 area, and are also the largest repository of organisms on the planet (Edwards et al. 2016, Innes et al. 2016, United Nations 2016, 2017a, b, c, IOC-UNESCO 2017a, b, 2020). Marine and coastal biodiversity support 80% of the world's biodiversity, supply almost half the oxygen we breathe, and strongly influence the world's climate. Marine ecosystems regulate Earth's climate by absorbing and sequestering atmospheric CO_2, and its biodiversity is critical to providing resilience against future climate change and extremes (Edwards et al. 2016, Innes et al. 2016, Pearlman et al. 2019, Ocean Health and Human Well-Being 2020). They also provide ecosystem goods and services, such as nutrient cycling, raw materials, food, and tourism (Chiba et al. 2018, Moltmann et al. 2019, Batten et al. 2019, Bax et al. 2019, Beaugrand et al. 2019). Although the ocean was once thought to be a vast and indefinitely resilient compartment of the Earth system, able to absorb practically all pressures of the human population, from resource exploitation to fisheries and aquaculture development to marine transport, as a corollary of the increasing multi-stressors pressures, have shown enormous variations in their dynamics in different ocean basins and regions of the world (Bopp et al. 2013, Cloern et al. 2016, IOC-UNESCO 2017, 2020, Fennel et al. 2019, Bindoff et al. 2019). The interdependency of ocean-based industries and marine ecosystems, combined

[1] The Center for Environmental and Marine Studies (CESAM) and Department of Biology University of Aveiro, Campus Universitário de Santiago, Aveiro, Portugal.

[2] Interdisciplinary Center of Marine and Environmental Research (CIIMAR). University of Porto, Terminal de Cruzeiros do Porto de Leixões, Av. General Norton de Matos s/n, 2250–208 Matosinhos, Portugal.

* Corresponding author: fmorgado@ua.pt

with increasingly severe threats to the health of the ocean, have led to a growing recognition of the need for an integrated approach to ocean management (United Nations 2016, 2017a, b, c, Gleckler et al. 2016, OECD 2016, 2019, UNESCO 2019, Rudolph et al. 2020). Together with the anthropogenic pressures, climate variability and extreme events are also affecting the planet in many ways, manifesting in temperature increases and erratic and erosive rainfall that have led to soil and coastal erosion, and affecting urban and rural areas, cities, and coastal locations, with evident damage in ecosystems, endogenous resources (land and sea) and negatively influencing the services of reference ecosystems (Talley et al. 2016, IPPC 2019). The expected resulting changes in marine biodiversity will affect fundamental ecosystem functioning processes, such as biomass production and water quality, as well as the entire marine ecosystem structure (United Nations 2016, 2017a, b, c, O'Brien et al. 2017, UNESCO 2017, Benway et al. 2019) and required adequate ocean monitoring and management approaches focused on the marine geophysical structures and oceanographic and biological processes (Edwards et al. 2016, Innes et al. 2016, United Nations 2017, Fennel et al. 2019, Rudolph et al. 2020). In order to respond to the enormous environmental, scientific, and societal challenges, global ocean science and applied marine biology have shown increasing developments and scientific capacity and built progress in relation to knowledge, technology, and innovation development advancements (United Nations 2016, 2017a, b, c, OECD 2016, 2017, 2019) and also in societal and research governance at a local, regional, and international scales (IOC-UNESCO 2017a, b, 2020, Chiba et al. 2018, Moltmann et al. 2019, Batten et al. 2019, Bax et al. 2019, Beaugrand et al. 2019, Rudolph et al. 2020). In recent years, ocean science has been greatly facilitated by increasing the availability of new marine infrastructures, facilities, and technologies for *in situ* and remote sensing data for ocean mapping and classification of marine ecosystems and habitats, a study of large-scale spatial and temporal variability of several detailed environmental and biological variables (Bastos et al. 2016, Chiba et al. 2018, Batten et al. 2019, Bax et al. 2019, Beaugrand et al. 2019, Amani et al. 2021). In the past decades, we have witnessed a growing global economy and technological progress; however, despite this, there are still many societal challenges that need to be overcome to enhance human development (United Nations 2016, 2017a, b, c, 2018, UNESCO 2019, Balvanera et al. 2019, Ocean Health and Human Well-Being 2020, Borja et al. 2020a, b).

2. Global Marine Biology Science, Oceanography, and Plankton Research

Ocean science seeks to understand complex, multi-scale social-ecological systems and services, which requires observations and multidisciplinary and collaborative research (United Nations 2016, 2017a, b, c, IOC-UNESCO 2017, 2020). Because of the broad scope and multifaceted nature of the stressors, there has been growing recognition of the importance of an integrated ecosystem approach to the research and management of the ocean and coastal areas (Maxwell et al. 2015, Wunsch 2016, Wilkinson et al. 2016, Danovaro et al. 2017). Ocean science has a broader approach and includes interfaces of research disciplines related to the study of the ocean, such as physical, biological, chemical, geological, hydrographic, health, and social sciences, as well as engineering, the humanities, and multidisciplinary research on the relationship between science-policy, humans, ocean literacy, and the ocean (Santoro et al. 2016, Evans et al. 2019, UNESCO 2019, Borja et al. 2020a, b, Ocean Health and Human Well-Being 2020). The predictable variability over daily to interannual scales in ocean physical and biogeochemical processes, marine community's composition, functioning and processes, and the unpredictability of future climate change and other anthropogenic stressors (Edwards et al. 2010, Wilkinson et al. 2016, Pearlman et al. 2019, Tanhua et al. 2019a, b) requires a future marine planning and governance and a global integrated marine assessment (United Nations 2017a, b) to build ocean science-policy partnerships for the next decades (Government Office for Science 2018, European Commission 2019, Evans et al. 2019, Balvanera et al. 2019, UNESCO 2019, IOC-UNESCO 2019, 2020, Borja et al. 2020a, b). Global ocean science and applied marine biology try to obtain reliable predictions with the scales of space (on the order of hundreds of meters to tens of kilometers, horizontally and from meters to tens of meters, vertically) and of time (on the order of hours to days) in which the processes take place, in order to predict meteorological and oceanographic conditions in advance, obtaining real-time data on winds and currents so as to make predictions for the future (Mackas and Beaugrand 2010, O'Brien et al. 2017, UNESCO 2017, Benway et al. 2019). The state of the sea, the intensity, and the direction of surface currents, waves, and wind can also be predicted in real time by radar technology (remote sensing). However, radars do not provide water column data. These are obtained only by *in situ* instruments, of which currently the most used is the acoustic Doppler current profiler (ADCP). These instruments can be installed on the bottom or buoys and can measure the speed and direction of currents at different levels of the water column simultaneously (Borja et al. 2019, IOC-UNESCO 2017a, b, 2020, Pearlman et al. 2019, Dañobeitia et al. 2020, IOC-UNESCO 2020, World Bank 2020a). With the help of numerical modeling, it is possible to predict the winds, waves, and currents, which, together with the water level, constitute the main variables of operational interest in oceanography (Borja et al. 2019, IOC-UNESCO 2017a, b, 2020, Pearlman et al. 2019, Dañobeitia et al. 2020, IOC-UNESCO 2020, World Bank 2020a). The use of buoys with wind and solar energy capture allows data transmission in an efficient way by telephone, satellite, radio, or submarine cables.

Oceanography is essentially a multidisciplinary science (Maxwell et al. 2015, Wunsch 2016, Danovaro et al. 2017) that presents two different types of approaches: Descriptive Oceanography and Operational Oceanography. The first concerns a more theoretical study of the oceans, while the second concerns research and work produced in the field or simulated environments. Descriptive Oceanography may be divided into four main areas of study: physical oceanography, chemical oceanography, biological oceanography, and geological oceanography. Operational Oceanography can be defined as the activity of systematic and long-term routine measurements of the seas and oceans and atmosphere and their rapid interpretation and dissemination. For its pursuit, Operational Oceanography needs sophisticated monitoring infrastructures and new technologies for marine sampling, habitat mapping and classification, analysis, and modeling (Edwards et al. 2016, Innes et al. 2016, United Nations 2017, Everett et al. 2017, Capotondi et al. 2019, Fennel et al. 2019). During the last decades, an increase in synergetic investments was observed in strategic facilities and infrastructures for ocean research innovation and technology transfer in order to respond to future challenges of changes in marine ecosystems (OECD 2017, US Subcommittee on Ocean Science and Technology 2018, US Ocean Policy Committee 2018, 2019, Government Office for Science 2018, European Commission 2019, OceanObs'19 2019, European Marine Board 2019, Dañobeitia et al. 2020, IOC-UNESCO 2020, World Bank 2020a, b). Systematic and regular monitoring of the oceans has been largely improved in the last decades, with observations and models from small and medium-sized research institutions to large oceanographic and meteorological services (IOC-UNESCO 2017a, b, 2020, Chiba et al. 2018, Moltmann et al. 2019, Batten et al. 2019, Bax et al. 2019, Beaugrand et al. 2019). The marine observation and sampling in the context of oceanography include planning oceanographic campaigns (gathering information on bathymetry, oceanography, climatology, and weather forecast of the study areas, data acquisition, recording, metadata, and sample storage), physical oceanography (sampling equipment for *in situ* physical profiling of the water column), chemical oceanography (water sampling and preservation, analytical equipment: types, installation, and reagents, analytical determination of pH, alkalinity, oxygen, total inorganic carbon, total organic carbon, and nutrients), and data statistical methods (multivariate techniques used to analyze long-term series in oceanography), biological sampling (for planktonic organisms, categories, and sizes). Plankton samplers, such as neuston nets, plankton nets, pumps, plankton recorders, and accessories (Chiba et al. 2018, Moltmann et al. 2019, Batten et al. 2019, Bax et al. 2019, Beaugrand et al. 2019, OceanObs'19 2019, Dañobeitia et al. 2020). In the context of oceanography and marine biology, the approach has included the classic and the new modern and sophisticated methods and technologies for marine sampling for habitat classification and habitat mapping, such as *in situ* sampling, echosounder, Light Detection and Ranging (LiDAR) surveys, satellites, remote sensing, and small manned aerial vehicles (MAVs) and unmanned aerial vehicles (UAVs, also called drones) and Video Camera Systems (Bastos et al. 2016, Déniz-González et al. 2017, World Bank 2020a, b, Amani et al. 2021). Findable, accessible, interoperable, and reusable data services and principles have been proposed for the ocean (Tanhua et al. 2019a), and platforms for sharing best practices in ocean observing, data sharing, and community dialogue have also been established with the aim of improving the effective use of ocean data for the benefit of society (United Nations 2018, Balvanera et al. 2019, Pearlman et al. 2019, Ocean Health and Human Well-Being 2020, Borja et al. 2020). Results from the more recent marine scientific research and the new marine technologies and infrastructures provide input to a better understanding of the wide range of marine organisms and their interactions with the marine environment and for a prediction of the large-scale impact of pollution and climate change on the marine ecosystem (Cloern et al. 2016, Bindoff et al. 2019, IPCC 2019).

Planktonic ecological processes of the different taxonomic groups are part of a network of relationships with the environmental drivers and stressors that regulates marine food webs and the global carbon cycle that are fundamental for understanding marine ecological processes and networks and global environmental changes (Lombard et al. 2019, Batten et al. 2019, Borja et al. 2020). Plankton research is currently framed in the global axes of the key priority issues concerning future marine ecosystems governance and to face societal and research challenges at local, regional, and international scales (Chiba et al. 2018, Moltmann et al. 2019, Batten et al. 2019, Bax et al. 2019, Beaugrand et al. 2019). In addition, the plankton research allows interfaces with other fields of knowledge, contributing to the creation of new interdisciplinary and transdisciplinary fields for studies of biological and non-biological factors that structure biodiversity and the patterns of communities and populations from the mesoscale to the microscale, making scientific interfaces with different technological areas for solving biological issues (Lombard et al. 2019, Batten et al. 2019, Borja et al. 2019, 2020a, b). Global observations of planktonic ecosystems and trends in marine plankton diversity are essential for a global integrated marine assessment and contribute to understanding the global climate change and pollution impacts toward addressing global marine biodiversity conservation challenges and global ocean management (Edwards et al. 2016, Innes et al. 2016, Kavanaugh et al. 2016, United Nations 2017, Chiba et al. 2019, Batten et al. 2019, Ibarbalz et al. 2019, Lombard et al. 2019).

3. Plankton, an Ocean Essential Variable, and Plankton Research a Good Indicator of the State of the Marine Environment

The pelagic zone is the largest biome on Earth, where the plankton account for 95% of marine biomass, which sustain almost all marine food webs in marine ecosystems and provide the link between the physical environment and the fish, marine birds, and mammals (Lombard et al. 2019, Bar-On et al. 2019). The plankton is responsible for almost 50% of global primary production, and its structure and functions are the basis of a series of complex biogeochemical processes in the ocean that make up the biological pump, which involves the export of carbon and other elements from the atmosphere via surface waters to the oceans (Lombard et al. 2019, Batten et al. 2020). In addition, planktonic organisms are extremely sensitive to changing oceanic conditions and multiple pressures, such as climate-associated temperature changes and nutrient and pollutant dynamics that may affect the structure and function of the planktonic marine ecosystems (Boye et al. 2013, Beaugrand 2014, Maxwell et al. 2015, Kavanaugh et al. 2016, Wunsch 2016, Wilkinson et al. 2016, Danovaro et al. 2017, Chiba et al. 2018, Zhang et al. 2018, Righetti et al. 2019). Plankton is considered an ocean essential variable to understand actual and model predictions for marine ecology and dynamics, including population dynamics, nutrition, behavior and trophic and biophysical interactions, mechanisms that influence the abundance and distribution of marine plankton and species that have a planktonic larval stage in their life cycle (Chiba et al. 2018, Miloslavich et al. 2018, Muller-Karger et al. 2018, Batten et al. 2019, Lombard et al. 2019). In this framework, plankton research is crucial for ocean science and applied marine biology and also critical for marine biodiversity conservation strategies and ecosystem functions as a good indicator of the state of the marine environment (Chiba et al. 2018, Miloslavich et al. 2018, Muller-Karger et al.

2018, Batten et al. 2019, Lombard et al. 2019, Fennel et al. 2019). Plankton research includes the global patterns of phytoplankton and zooplankton diversity driven by temperature and environmental variability (Everett et al. 2017, Righetti et al. 2019, Salazar et al. 2019), the evolving paradigms in biological carbon cycling in the ocean and the ecosystem services sustainability, such as fisheries, nutrient cycling, and human health (Bedford et al. 2018, Zhang et al. 2018, Bax et al. 2019, Batten et al. 2019). The need to scale up research on plankton is gaining more attention at the national and international levels, and the recent plankton oceanic monitoring programs put in evidence the need for global partnerships that evolve governments, scientific international institutions, private sector stakeholders, and civil society (Chiba et al. 2018, Lombard et al. 2019, Batten et al. 2019, Ibarbalz et al. 2019). The plankton research has recently gained a more holistic sampling approach that, besides collecting organisms with simple or sophisticated nets and water collectors at different depths, use also a vast number of the new marine infrastructures and equipment from research vessels or in the water, such as cameras and acoustic sensors, robots, submarines or satellites in order to obtain a variety of data from the environment, and then send their data to dataloggers to be processed and stored (Batten et al. 2019, Von Schuckmann et al. 2016, Baumann-Pickering et al. 2019, Bax et al. 2019, Weller et al. 2019). The new marine facilities, infrastructures, and technologies offer opportunities for continuous obstinance of plankton data and more global and diversified use of advanced equipment, high resolution, and automatized processing data methods, complementary to more traditional techniques, that could be applied for long-term programs, to calibrate protocols and equipment, improve spatial and temporal resolution and standardization across countries or within the same geographic area of plankton data (Bopp et al. 2013, Talley et al. 2016, IOC-UNESCO 2017, 2020, OECD 2016, 2017, 2019a, b, Cloern et al. 2016, Bindoff et al. 2019, IPCC 2019).

An increasing effort has been observed in global marine biology science, oceanography, and plankton research in the advancement of knowledge, technological development, and innovation (OECD 2016, 2017, 2019a, b) concerning the conservation and protection of marine resources from anthropogenic destruction activities, including pollution, and from the effects of climate changes (JPI Oceans 2015, 2018, Sala et al. 2018, Frölicher et al. 2018, Wenhai et al. 2019, Bindoff et al. 2019, Hoegh-Guldberg et al. 2019, IPPC 2019). Responses for current marine, coastal ecosystems, and biological resources degradation must include interdisciplinary approaches regarding the Integrated ocean and coastal management and stakeholder participation in decision-making and for policymakers (OECD 2016, 2017, 2019a, b, 2021, Pearlman et al. 2019, Tanhua et al. 2019a, b). Several international governmental marine laboratories and agencies conduct marine observations sampling programs, model operations, and provide services to inform the scientific institutions, industry, the public, and other end users in order to obtain short and long-term studies of environmental and biological data (IOC-UNESCO 2017a, 2020, US Subcommittee on Ocean Science and Technology 2018, US Ocean Policy Committee 2018, 2019, Government Office for Science 2018, EU 2019, World Bank 2020a, b). In order to make integrated monitoring and evaluation of the ocean status, a great diversified oceanographic and marine biology activities and research programs around the world have been planned and taken (UNESCO 2015, JPI Oceans 2015, 2018), and criteria, guidelines, and capacity development strategies were defined in order to promote a sustainable and inclusive oceans management (Ryabinin et al. 2019, IOC-UNESCO 2016, 2017a, b, 2018, 2019, 2020).

4. Global Partnership for Ocean Science and Collaboration Networks

Ocean science and applied marine biology research were crucial for the increase of ocean knowledge for marine scientific research and the development of global and regional partnerships (United Nations 2016, 2017a, b, c, UNESCO 2017, 2019, IOC-UNESCO 2017, UNESCO 2017, 2019, 2020, Evans et al. 2019, Dañobeitia et al. 2020, Ocean Health and Human Well-Being 2020). Some pioneers' international institutions should be highlighted, responsible for the beginning of ocean science and for collaboration networks fundamental for the development of global and regional partnerships, such as the International Council for the Exploration of the Sea (ICES), created in 1902, the Scientific Committee of Ocean Research (SCOR), created in 1957 under the International Council of Scientific Unions (ICSU), later in 1960, the IOC/UNESCO, and finally in 1970, The National Aeronautics and Space Administration (NOAA) (IOC-UNESCO 2017, 2020, Evans et al. 2019, Dañobeitia et al. 2020, Ocean Health and Human Well-Being 2020).

The International Council for the Exploration of the Sea (ICES) is an intergovernmental marine science organization meeting societal needs for impartial evidence on the state and sustainable use of our seas and oceans. ICES was formed in 1902, and activities initially were focused on fisheries, although current ICES activities are much broader and include aquaculture, ecosystem observations, ecosystem processes and dynamics, fisheries, human interactions, and also support strategic initiatives on climate change effects on marine ecosystems and better understanding human dimensions of marine ecosystems. The ICES Strategic Plan defines the direction and priorities relating to science, data, and advice and develops the capacity to fulfill this commitment. ICES is a network of more than 5,000 scientists belonging to 20 different countries from over 700 marine institutes. Through strategic partnerships, ICES works in the Atlantic Ocean, the Arctic, the Mediterranean Sea, the Black Sea, and the North Pacific Ocean. It is important to highlight some relevant ICES groups and programs, such as the "Working Group on Phytoplankton and Microbial Ecology (WGPME)" and the "Working Group on Zooplankton Ecology (WGZE)" and ICES important projects, such as the "Harmful Algal Blooms" and the online global database "ICES Historical Plankton Dataset and ICES Data Portal" (https://www.ices.dk).

The SCOR is an international non-governmental, non-profit organization that includes approximately 600 scientists from nearly 60 countries who have formed national SCOR committees as a foundation for international. SCOR activities focus on promoting international cooperation in planning and conducting oceanographic research and solving methodological and conceptual problems that hinder research. SCOR covers all areas of ocean science and cooperates with other organizations with common interests to conduct several different activities and also to build the capacity for ocean science in developing countries. It is important to highlight some relevant SCOR global projects, such as the "Global Ocean Ecosystem Dynamics (GLOBEC)" and the "Global Ecology and Oceanography of Harmful Algal Blooms (GEOHAB)," and some relevant groups and programs, such as the "Microbial Carbon Pump in the Ocean," "Patterns of Phytoplankton Dynamics in Coastal Ecosystems," "Comparative Analysis of Time Series Observation," and the "Modern Planktonic Foraminifera and Ocean Changes" (https://scor-int.org).

The IOC/UNESCO, the UNESCO's Intergovernmental Oceanographic Commission (IOC), coordinates and promotes international cooperation and coordinates programs in marine research, services, observation systems, hazard mitigation, and capacity development in order to understand and effectively manage the resources of the ocean and coastal areas (Global Ocean Science Report 2017, 2020). The purpose of the IOC is to promote international cooperation and to coordinate programs in research, services, and capacity-building in order to learn more about the nature and resources of the ocean and coastal areas and to apply that knowledge for the improvement of management, sustainable development, the protection of the marine environment, and the decision-making processes of its member states. By applying this knowledge, the Commission aims to improve the governance, management, institutional capacity, and decision-making processes of its member states with respect to marine resources and climate variability and to foster sustainable development of the marine environment, particularly in developing countries. IOC is a body with functional autonomy within UNESCO that promotes international cooperation and coordinates programs in marine research, services, observation systems, hazard mitigation, and capacity development in order to manage better the nature and resources of the ocean and coastal areas (Global Ocean Science Report 2017, 2020). In addition to many other projects, it is important to highlight some relevant IOC/UNESCO projects such as "Harmful Algal Blooms" and "Time Series of Ocean Measurements" (https://ioc.unesco.org).

The NOAA was established in 1970 (NOOA 1970) and holds key leadership roles in shaping international ocean, fisheries, climate, space, and weather policies. Since then, NOAA's specific roles include (i) supplying environmental information products, (ii) providing environmental stewardship services, and (iii) conducting applied scientific research, with diversified activities, such as (1) monitoring and observing Earth systems with instruments and data collection networks, (2) understanding and describing Earth systems through research and analysis of data, (3) assessing and predicting the changes in these systems over time, (4) engaging, advising, and informing the public and partner organizations with relevant information, and (5) custodianship of environmental resources. Concerning marine and plankton research, some important and relevant NOOA projects could be highlighted, such as "Gulf of Mexico Plankton Investigations (Gulf of Mexico and Caribbean 1951–1953), *SKIPJACK* (South Pacific, 1970–1973), North Pacific and Bering Sea Oceanography (North Pacific, 1958), ORCAWALE (Oregon California and Washington Line-transect Expedition), NMFS Marine Mammals Surveys (North Pacific, Atlantic North, Gulf of Mexico, and Caribbean, 1965–1968 and 2001), NMFS-SWFSC Surveys (North Pacific, 1979–2000), International Ocean Atlas

Series, Harmful Algal Blooms Observation System (HABSOS), Ocean Archive System, World Ocean Plankton Database" (https://www.noaa.gov, 2022).

These pioneer institutions were responsible for the creation of many others that contributed to a major development of a global partnership for marine science through investment in infrastructures, such as oceanographic research vessels, the design of international research programs and expeditions, and also the creation of international committees to study global or regional subjects all over the world (IOC-UNESCO 2017, 2020, Evans et al. 2019, Dañobeitia et al. 2020, Ocean Health and Human Well-Being 2020). Some important were the Committee on Climatic Changes and the World Meteorological Organization (WMO), created in 1980, that were responsible for the Ocean and the World Climate Research Program (WCRP), and for the first global marine physical research projects, such as the World Ocean Circulation Experiment (WOCE) and the Tropical Ocean-Global Atmosphere Study (TOGA), the International Geosphere-Biosphere Program (IGBP), the Joint Global Ocean Flux Study (JGOFS), focusing on the role of the ocean in the global carbon cycle, the Global Ocean Ecosystem Dynamics (GLOBEC) program, that developed seven regional comparative studies to understand marine ecosystem responses to global changes, including both environmental and human pressures, the Integrated Marine Biogeochemistry and Ecosystem Research (IMBER) program, the Global Ecology and Oceanography of Harmful Algal Blooms (GEOHAB) program with a focus on obtaining an understanding of the ecological and oceanographic conditions that cause harmful algal blooms and promote their development (IOC-UNESCO 2017, 2020, Evans et al. 2019, Dañobeitia et al. 2020, Ocean Health and Human Well-Being 2020). The North Pacific Marine Science Organization (PICES) is also an intergovernmental scientific organization created by treaty in 1992 to advance scientific understanding of the North Pacific and adjacent seas. Present members include Canada, the People's Republic of China, Japan, the Republic of Korea, the Russian Federation, and the United States of America. PICES promotes and coordinates multidisciplinary marine research efforts and facilitates the exchange of scientific and technical information on a broad range of scientific disciplines. The organization provides an international forum to promote a greater understanding of the biological and oceanographic processes of the North Pacific Ocean and its role in a global environment. The Southern Ocean Observing System (SOOS), launched in 2011, is a joint initiative of the Scientific Committee on Oceanic Research (SCOR) and the Scientific Committee on Antarctic Research (SCAR), with the opening of the International Project Office (IPO), hosted by the Institute for Marine and Antarctic Studies (IMAS), University of Tasmania, Australia. Also noteworthy are other extremely relevant international programs related to the biology of the ocean and the description of new species for science, the Global Coral Reef Monitoring Network and the Census of Marine Life, which led to the creation of the Ocean Biodiversity Information System (OBIS), the largest repository of marine biodiversity to date.

Marine monitoring networks are becoming increasingly important all over the world for the collection, dissemination, and sharing of data for improved scientific understanding, assessment of the health of marine ecosystems, and forecasting the likely impacts of environmental change and human activities (Duarte et al. 2018, Buck et al. 2019, Smith et al. 2019a, 2019b, International Oceanographic Data and

Information Exchange (IODE) 2020). Currently, the Global Partnership for Oceans is an alliance of more than 100 governments, international organizations, civil society groups, and private sector interests committed to addressing the threats to the health, productivity, and resilience of the world's oceans, acting as a global network (Joint Programming Initiatives (JPIs) 2020, International Oceanographic Data and Information Exchange (IODE) 2020), that brought attention to current problems of overfishing, pollution, habitat loss ocean health (IOC-UNESCO 2017a, b, 2020, Chiba et al. 2018, Moltmann et al. 2019, Borga et al. 2019, 2020, Batten et al. 2019, Bax et al. 2019, Beaugrand et al. 2019). The global partnerships, besides the improvement of marine research, also help policymakers in the pursuit of development options and benefit society (National Research Council 2015, Rudd 2015, Joint Programming Initiatives (JPIs) 2020, International Oceanographic Data and Information Exchange (IODE) 2020) in terms of weather forecasting and natural disaster prevention (Bopp et al. 2013, Talley et al. 2016, Global Ocean Science Report 2017, OECD 2019, Cloern et al. 2016, Bindoff et al. 2019, IPCC 2019).

5. Global Marine Research Institutions and Resources

Research institutions and laboratory units represent a fundamental pillar in the consolidation of a modern and competitive scientific system that gathers critical mass appropriate to their mission and promotes creative environments where new ideas can emerge and where researchers find the right conditions for the realization of their scientific projects and the development of their careers (Government Office for Science 2018, EU 2019). They should bring together interdisciplinary and multidisciplinary resources that enhance the approach to complex problems and new social challenges (IOC-UNESCO 2017a, b, 2020, World Bank 2020a). The IOC-UNESCO Global Ocean Science Report (GOSR), as the first consolidated assessment of global ocean science, provides a status report on ocean science research and identifies and quantifies the elements that drive the productivity and performance of ocean science, including workforce, infrastructure, resources, networks, and outputs, assisting the science-policy interface and supports managers, policymakers, governments, and donors, as well as scientists beyond the ocean community (IOC-UNESCO 2017, 2020). The GOSR offers decision-makers an unprecedented tool to identify gaps and opportunities to advance international collaboration in ocean science and technology and harness its potential to meet societal needs, address global challenges, and drive sustainable development for all (IOC-UNESCO 2017, 2020). According to GOSR, almost 39% of ocean science research institutions worldwide work across a broad range of issues, whereas 35% are specialized in more limited themes, such as observations, and 26% in fisheries (IOC-UNESCO 2017a, b, 2020). At a global level, the USA has the highest number of research institutions (315) (US Subcommittee on Ocean Science and Technology 2018, US Ocean Policy Committee 2018, 2019), roughly equal to the total number of research institutions in Europe combined and greatly exceeding the number of institutions operating in Asia and Africa (UN General Assembly 2018, Government Office for Science 2018, UN ESCAP 2018, Claassen et al. 2019, European Commission 2019, IOC-UNESCO 2017, 2020, IIOE-2 2020). Marine field stations and laboratories provide access to a range of environments, including coral reefs, estuaries, kelp forests, marshes, mangroves, and urban coastlines. Globally, 784 marine stations are maintained by 98 countries, and the majority are located in Asia (23%), followed by Europe (22%), North America (21%), Antarctica (11%), South America (10%), Africa (8%) and Oceania (5%) (United Nations 2016, 2017a, b, c, IOC-UNESCO 2017a, b, 2020, World Bank 2020a, b).

IOC coordinates ocean observation and monitoring through the Global Ocean Observing System (GOOS), which aims to develop a unified network providing information and data exchange on the physical, chemical, and biological aspects of the ocean. Governments, industry, scientists, and the public use this information to act on marine issues (IOC-UNESCO 2017, 2020). IOC's work in ocean observation and science contributes to building the knowledge base of the science of climate change. IOC sponsors the World Climate Research Programme (WCRP), and the IOC's GOOS serves as the ocean component of the Global Climate Observing System (GCOS), which supports the Intergovernmental Panel on Climate Change (IPCC). UNESCO-IOC is co-convener with the WMO of the World Climate Change Conference, which aims to systematically make the existing knowledge on climate science available to a wide variety of potential users. IOC also coordinates and fosters the establishment of regional intergovernmental coordinating tsunami warning and mitigation systems in the Pacific and Indian Oceans, in the North East Atlantic, Mediterranean, and Caribbean seas (IOC-UNESCO 2017a, b, 2020, Tassa et al. 2022).

The European Marine Research Infrastructures is an EU and EU-Associate countries consortium of international relevance designed to further fundamental and applied marine biology and ecology research, interconnecting science, industry, and policy. It provides the necessary and relevant services, facilities and technology platforms to study marine organisms and ecosystems, and are also contributing to sustained *in situ* global ocean observation, notably some European Research Infrastructure Consortia constituted at ministry or institute level, such as the DANUBIUS-RI (International Center for Advanced Studies on River-Sea Systems), EMBRC (European Marine Biological Resource Centre), EMSO (European Multidisciplinary Seafloor and water column Observatory), Euro-Argo Research Infrastructure Sustainability and Enhancement, Eurofleets+ (an alliance of European marine research infrastructure to meet the evolving needs of the research), ICOS (Integrated Carbon Observation System), JERICO-RI (Joint European Research Infrastructure of Coastal Observatories), and LifeWatch ERIC Research infrastructures covers all research areas from biology to social sciences to physics, designed to accelerate the pace of scientific discovery and innovation from marine Bio-Resources (IOC-UNESCO 2017a, b, 2020, Tassa et al. 2022).

The Global Alliance of Continuous Plankton Recorder Surveys studies oceanic changes in planktonic biodiversity through the Continuous Plankton Recorder (CPR) surveys that are collecting data from the North Atlantic and the North Sea on biogeography and ecology of plankton since 1931. It is a consortium promoted by Sir Alister Hardy Foundation for Ocean Science (SAHFOS) and has as partners the National Institute of Polar Research (NIPR) from Japan, the National Institute of Water and Atmospheric Research (NIWA) from New Zealand, the North Pacific

Marine Science Organization (PICES) from Australia, the NOAA and the Integrated Marine Observing System (IMOS) from USA, and the Southern Ocean Observing System (SOOS).

The Coastal and Oceanic Plankton Ecology, Production and Observation Database, promoted by the National Marine Fisheries Services (NFMS-NOAA), have as partners, besides other, the ICES and SCOR, which has an online global database, the copepod's interactive time series explorer "COPEPOD" and the "NAUPLIUS," an interactive ecosystem data aggregation platform (NMFS Office of Science and Technology 2022).

6. Research Infrastructures, Facilities and Equipment

Global ocean science, plankton research, and applied marine biology studies depend on a lot of research infrastructures, including facilities, services, and resources that are used by the scientific community (IOC-UNESCO 2017a, b, 2020, Pearlman et al. 2019, Dañobeitia et al. 2020, IOC-UNESCO 2020, World Bank 2020a). In addition, the operational oceanography research for systematic, long-term, sustainable, and multidisciplinary measurements of the oceans and atmosphere, as well as the rapid interpretation of information and its dissemination, requires a set of facilities, monitoring, analysis, and modeling infrastructures (IOC-UNESCO 2017a, 2020, European Strategy Forum on Research Infrastructures (ESFRI) 2018, World bank 2020a) and the combination of remote and *in situ* ocean observations, including advances in ocean robotics (European Marine Board 2019). Oceanic and coastal research has evolved into the development of observation and monitoring activities with modern and robust tools and infrastructure based on new technologies for greater efficiency in ocean management and marine resource surveillance (IOC-UNESCO 2017, 2020, United Nations 2018, OECD 2016, 2019, Duarte et al. 2018, Buck et al. 2019, Grand et al. 2019, Smith et al. 2019a, Smith et al. 2019b). The main components of infrastructure for the monitoring and investigation of the oceans are:

i) **Facilities** – Land structures, laboratories and monitoring stations, remote platforms for the development of research, observation, monitoring, and surveillance that include ships, planes, satellites and submersibles, early warning systems, broadband telecommunications, and pollution control devices;

ii) **Hardware** – Research equipment, instrumentation, sensors, and new information technologies;

iii) **Technical Support** – Human resources specialized in monitoring, research, modeling, observation, and education. These infrastructures and technological innovations have contributed to protecting marine biodiversity and sustainable exploitation and protection of marine resources (IOC-UNESCO 2020, World Bank 2020a).

The great challenges for oceanographic research in the future require global cooperation in scientific and engineering investment in order to share infrastructure and new technologies that are very complex and expensive. New technologies have allowed the creation of more powerful sensors, robotic platforms, and new ocean observation systems for the development of interdisciplinary oceanographic research programs (IOC-UNESCO 2017a, b, 2020, World Bank 2020a). This framework promotes multi-stakeholder cooperation and collaborations between governments, the private sector, and research units and plays an increasingly important role in the advancement of knowledge and technology worldwide (Eurofleets + 2020, IOC-UNESCO 2017a, b, 2020, Pearlman et al. 2019, European Marine Board 2019, Dañobeitia et al. 2020). The infrastructures and equipment capacity of institutions and laboratories in terms of research vessels, autonomous underwater vehicles, acoustic systems, and other equipment availability are fundamental for the design and operation of ocean monitoring systems (Pearlman et al. 2019, IOC-UNESCO 2017a, b, 2020, Pearlman et al. 2019, Dañobeitia et al. 2020), are useful to increment long-term operation planning, the deployment of new observing and monitoring systems and offer new opportunities of cost-efficient maintenance and operational costs reduction (European Strategy Forum on Research Infrastructures (ESFIR), IOC-UNESCO 2017a, b, 2020, European Marine Board 2019, Pearlman et al. 2019, Dañobeitia et al. 2020).

6.1 Research Vessels

Research vessels are a key research infrastructure offering vital access to seas and ocean for conducting marine science and ocean observation, which due to their onboard equipment to collect weather and ocean information to divers, submersibles, and other observations deployed from ships, are arguably the most critical tool for scientists when it comes to exploring the ocean (European Strategy Forum on Research Infrastructures (ESFRI) 2018, European Marine Board 2019, Eurofleets + 2020, IOC-UNESCO 2017a, b, 2020, European Research Vessel Operators (ERVO) 2020) and according to the work they perform research vessels may be grouped as:

i) **Hydrographic vessels:** Built to serve as a basis for bathymetry and water body structure;

ii) **Fishery vessels:** Adapted for fisheries research;

iii) **Oceanographic vessels:** Designed to carry out almost all types of work of oceanography research.

According to their area of activity, four main types of research vessels are described:

1) **Special vessels:** Primarily designed to carry out certain specific types of work.

2) **Local research vessels:** Generally 10 to 35 meters in size and generally short in range, operating predominantly in coastal and estuarine areas.

3) **Regional research vessels:** They are generally between 35 meters and 55 meters in size and operate predominantly on a regional scale, for example in the Mediterranean or the North Sea. There are 46 regional research vessels in the world.

4) **Ocean research vessels:** Usually operate on an ocean scale meaning that they operate in an ocean, for example in the Atlantic or Pacific. These ships have sizes ranging from 55 to 65 meters.

5) **Global research vessels:** They are larger than 65 meters and can operate on a multi-oceanic scale (IRVSI 2014, United Nations 2016, 2017a, b, c, UNESCO 2017, 2019, IOC-UNESCO 2017a, European Strategy Forum on Research Infrastructures (ESFRI) 2018, European Marine Board 2019, Eurofleets + 2020, European Research Vessel Operators (ERVO) 2020, IOC-UNESCO 2017, 2020). The OCEANIC, MarineTraffic and Eurofleets international databases currently reported a total number of 1,081 vessels that are being used in ocean science-related and for ocean and coastal operations and activities (924 research vessels mainly used for ocean science and 157 ships of opportunity) (IRVSI 2014, United Nations 2016, 2017a, b, c, UNESCO 2017, 2019, IOC-UNESCO 2017a, European Strategy Forum on Research Infrastructures (ESFRI) 2018, European Marine Board 2019, European Research Vessel Operators (ERVO) 2020, IOC-UNESCO 2017, 2020, Eurofleets + 2020). Some of the current research vessels were built more than 60 years ago, while others have been in operation for less than five years, but, on average, the age of national fleets varies between < 25 years (Norway, Bahamas, Japan, and Spain) and > 45 years (Canada, Australia, and Mexico). More than 40% of research vessels focus primarily on coastal research, while 20% engage in global research (IOC-UNESCO 2017a, b, 2020, European Strategy Forum on Research Infrastructures (ESFRI) 2018, European Research Vessel Operators (ERVO) 2020, International Research Ship Operators (IRSO) 2020). Globally, they mainly belong to eight countries, such as the USA (441), Japan (50), Sweden (42), Canada (40), the Republic of Korea (26), the UK (26), Germany (25), and Turkey (24) (IRVSI 2014, United Nations 2016, IOC-UNESCO 2017a, European Strategy Forum on Research Infrastructures (ESFRI) 2018, European Research Vessel Operators (ERVO) 2020, International Research Ship Operators (IRSO) 2020, IOC-UNESCO 2017, 2020, Eurofleets + 2020). Although the different vessels may vary in the type of equipment they carry, 129 of all the reported fleet have icebreaker capabilities, 103 can berth for the ocean and coastal operations and deploy remotely operated vehicles, autonomous underwater vehicles, or submersibles, and all have echo-sounding capabilities, while 187 are equipped with Conductivity, Temperature and Depth probes (CTDs), 124 with ADCP, 116 with multibeam mapping systems, and 57 with dynamic positioning systems (IRVSI 2014, United Nations 2016, 2017a, b, c, UNESCO 2017, 2019, European Strategy Forum on Research Infrastructures (ESFRI) 2018, European Marine Board 2019, European Research Vessel Operators (ERVO) 2020, International Research Ship Operators (IRSO) 2020, Eurofleets + 2020, IOC-UNESCO 2017a, b, 2020).

6.2 Autonomous Underwater Vehicles (AUVs), Acoustic Systems and Other Equipment

Currently, there are sophisticated equipment and advanced technologies for ocean monitoring that use advanced technologies in information, communication, automatic control, system optimization, and management to examine, record, and analyze the ocean. The contribution of engineering knowledge to the logistic systems and management for the new marine research structures allows the coordination of systematic and scientific ocean monitoring. Some technologies include innovative platforms, such as vessels and submersibles, observing systems and sensors, communication technologies, and diving technologies. They are designed to collect critical data regarding the health and welfare of oceans and marine resources. Collected data can be stored or transmitted acoustically via acoustic modems or any other standard or custom interconnect and cable solutions. Some examples are satellites, AUVs, remotely operated vehicles (ROVs), SeaGliders, human-occupied vehicles (HOVs), drifters, buoy systems, CTDs, clod cards, telepresence technologies, submersibles, XBTs (expendable bathythermographs), magnetometers, geographic information systems (GIS), photogrammeters, submersible collectors, trawls, technical diving and environmental DNA samples (eDNA) (Colomina and Molina 2014, Hardin et al. 2018, Chirayath et al. 2019, Greenwood et al. 2019, Bedford et al. 2020, IOC-UNESCO 2017a, b, 2020, Pearlman et al. 2019, Dañobeitia et al. 2020, IOC-UNESCO 2020, World Bank 2020a, Angnuureng et al. 2020, IOC-UNESCO 2017, 2020, Pitois et al. 2021, Amani et al. 2022). Radars and acoustic mapping, high-frequency radars, sonars, side-scan sonars, eco-acoustic multibeam, acoustic ground discrimination systems (AGDS), ADCPs, multibeam sonars, side-scan sonars, split-beam sonars, synthetic aperture sonars (SAS), technologies for ocean acoustic monitoring, underwater hydrophones, and multispectral analysis of time series images from videos and drones (Colefax et al. 2018, Hensel et al. 2018, Provost et al. 2019, Joyce et al. 2019, Lana and Castello 2020, Geraeds et al. 2017, Hardin et al. 2018, Raoult et al. 2018, Provost et al. 2019, Bergsma et al. 2019, IOC-UNESCO 2017a, b, 2020, Pearlman et al. 2019, Dañobeitia et al. 2020, IOC-UNESCO 2020, World Bank 2020a, IOC-UNESCO 2017a, b, 2020, Amani et al. 2022).

6.3 Earth Observation Satellites (EO) and Global Navigation Satellite Systems (GNSS)

Since the development of infrastructures and technologies orbiting in space in the 1960s, an increasing number of satellites equipped with specialized sensors are covering vast and remote areas of the Earth, that creates the "Earth observation system" (Group on Earth Observations 2016, GSA 2017, 2020, United Nations 2018, UNOOSA 2018, Global Earth Observation System of Systems (GEOSS) 2020, ESA (European Space Agency) 2019, 2021). So far, more than 300 Earth Observation satellites have been launched worldwide, improving knowledge of the various components of this system, including its atmosphere, land, oceans, and ice coverage

(Group on Earth Observations 2016, GSA 2017, 2020, United Nations 2018, UNOOSA 2018, Global Earth Observation System of Systems (GEOSS) 2020, ESA (European Space Agency) 2019, 2021). Earth observations (EO) refer to monitoring the planet using sensors in, on, or around the Earth, providing information on the Earth's surface, its atmosphere, and marine systems, while GNSS is the term used for a satellite navigation system with worldwide coverage to determine a precise position anytime, anywhere on the globe (Group on Earth Observations 2016, GSA 2017, 2020, United Nations 2018, UNOOSA 2018, Global Earth Observation System of Systems (GEOSS) 2020, ESA (European Space Agency) 2019, 2021). The role of EO and geolocation (provided by global navigation satellite systems [GNSS]) is currently being widely employed; besides pinpointing the user's position, velocity, and time on the globe with great accuracy, GNSS systems also provide information for all modes of transport (road, aviation, and maritime), fleet management, high-precision, and consumer applications, provision of time information in critical national infrastructures as well as scientific applications, such as measuring the impact of space weather on the Earth, of earthquakes, and climate change on human communities (Group on Earth Observations 2016, GSA 2017, 2020, United Nations 2018, UNOOSA 2018, ESA (European Space Agency) 2019, 2021). At the global level stand out the GNSS and EO, with a special focus on European GNSS and Copernicus (COPERNICUS 2020, Copernicus Climate Change Service 2020). The Group on Earth Observations (GEO) is an intergovernmental partnership that provides open access to more than 400 million open EO data and information resources that are relevant for research, policy, and decision-making (Group on Earth Observations 2016, GSA 2017, 2020, UNOOSA 2018, ESA (European Space Agency) 2019, 2021).

At the international level, there are several actors that play an important role in the promotion of EO:

i) **The Group on Earth Observations (GEO):** It aims to connect the demand for consistent and timely environmental information with the supply of data and information about the Earth, advocate for broad, open data policies in order to ensure that the data collected through national, regional, and global observing systems are both made available and applied to decision-making;

ii) **The Committee on Earth Observation Satellites (CEOS):** It coordinates civil space-borne observations of the Earth, enhances international coordination and data exchange, and optimizes societal benefit;

iii) **UNOOSA:** UNOOSA is actively engaged in the preparation of UNISPACE + 50, Global Partnership for the Coordination of the Development, Operation, and Utilization of Space-Related Infrastructure, Data, Information and Services, including Earth Observation (Group on Earth Observations 2016, GSA 2017, 2020, United Nations 2018, UNOOSA 2018, Global Earth Observation System of Systems (GEOSS) 2020, ESA (European Space Agency) 2019, 2021).

Currently, there are four operational GNSS:

i) **The United States Global Positioning System (GPS):** It was created in 1995 and is the first GNSS fully operational and managed by the United States Department of Defense;

ii) **The Russian Federation Global Navigation Satellite System (GLONASS):** The Russian GNSS has been fully operational since 2011 and is managed by the Russian Aerospace Defence Forces;

iii) **Galileo; The European GNSS:** It became fully operational in 2016, managed by the European Union (the only GNSS under civilian control);

iv) **The BeiDou Navigation Satellite Systems (BDS):** The Chinese GNSS is managed by the governmental China Satellite Navigation Office (Group on Earth Observations 2016, GSA 2017, 2020, United Nations 2018, UNOOSA 2018, Global Earth Observation System of Systems (GEOSS) 2020, ESA (European Space Agency) 2019, 2021).

For the European GNSS, the European Union, the European Commission, the European Space Agency, and the European GNSS Agency have responsibilities, respectively, with regard to the political oversight, the infrastructure deployment, and the service provision and commercialization of the services (NEREUS (European Space Agency and European Commission) 2018, NEXTSPACE 2019, Hanan et al. 2020, COPERNICUS 2020, Copernicus Climate Change Service 2020, Baumgart et al. 2021, Von Schuckmann et al. 2021, Tassa et al. 2022). UNOOSA, through its different programs, works on building the necessary capacity to enable the effective use of space by all nations. Among the space technologies, the two European flagship programs—the European GNSS European Geostationary Navigation Overlay Service (EGNOS) and Galileo and the European EO program Copernicus—could be used not only in Europe but worldwide (UNOOSA 2018, COPERNICUS 2020, ESA European Space Agency 2019, 2021). The European Union's efforts in infrastructure deployment and market uptake have resulted in the launch of EGNOS, Galileo, and Copernicus operational services (NEREUS (European Space Agency and European Commission) 2018, NEXTSPACE 2019, Hanan et al. 2020, Copernicus Climate Change Service 2020, Baumgart et al. 2021, Von Schuckmann et al. 2021, Tassa et al. 2022). These services support a continuously increasing number of users in many different market application domains: from transport-related services (for example, aviation, road, maritime, and rail) and consumer solutions to professional applications (for example, agriculture, construction, and infrastructure monitoring) (NEREUS (European Space Agency and European Commission) 2018, NEXTSPACE 2019, Hanan et al. 2020, Copernicus Climate Change Service 2020, Baumgart et al. 2021, Von Schuckmann et al. 2021, Tassa et al. 2022).

Some global observatories and Earth observation satellites can be listed (Group on Earth Observations 2016, Déniz-González et al. 2017, GSA 2017, 2020, United

Nations 2018, UNOOSA 2018, Global Ocean Science Report 2017, 2020, ESA (European Space Agency) 2019, 2021):

QuikSCAT (Quick Scatterometer) – NASA, USA: An Earth observation satellite carrying the SeaWinds scatterometer (a wide array of applications, measures the surface wind speed and direction over the ice-free global oceans and has contributed to climatological studies, weather forecasting, meteorology, oceanographic research, marine safety, commercial fishing, tracking large icebergs, and studies land and sea ice).

SSM/I (Special Sensor Microwave Imager) and SSMIS (Special Sensor Microwave Imager Sounder), NASA: SSM/I and SSMIS are satellite passive microwave radiometers that measure atmospheric, ocean, and terrain microwave brightness temperatures incident upon a seven-port horn antenna (support Department of Defense operations, onboard Defense Meteorological Satellite Program (DMSP) satellites).

WindSat – NOAA, USA: A satellite-based polarimetric microwave radiometer developed by the Naval Research Laboratory Remote Sensing Division and the Naval Center for Space Technology for the US Navy and the National Polar Orbiting Operational Environmental Satellite System (NPOESS) Integrated Program Office (IPO) (designed to demonstrate the capability of polarimetric microwave radiometry to measure the ocean surface wind vector from space).

ASCAT (Advanced Scatterometer), European Space Agency (ESA), AND European Organization for the Exploitation Of Meteorological Satellites (EUMETSAT): A real aperture radar (measures wind speed and direction over the oceans, but also provides useful data in a variety of studies, including polar ice and tropical vegetation).

TOPEX/Poseidon – Centre National d'Etudes Spatiales (CNES), France and NASA, USA: Jointly conducted by the USA's NASA and the French Space Agency, Centre National d'Etudes Spatiales (CNES) for studying the global circulation from space (used the technique of satellite altimetry to make precise and accurate observations of sea level for several years).

JASON – Centre National d'Etudes Spatiales (CNES), France, NASA and NOAA, USA, and EUMETSAT: A group of three altimetry satellites (monitors global ocean circulation, studies the ties between the ocean and the atmosphere, improves global climate forecasts and predictions, and monitors events, such as El Niño Southern Oscillation (ENSO) and ocean eddies).

AVHRR (Advanced Very High Resolution Radiometer) NOAA, USA: This sensor is carried on NOAA's Polar Orbiting Environmental Satellites (POES) (acquired High Resolution Picture Transmission (HRPT), Local Area Coverage (LAC), and Global Area Coverage [GAC]).

AATSR (Advanced Along-Track Scanning Radiometer) and European Space Agency (ESA): A multi-channel imaging radiometer on board (measure global Sea Surface Temperature [SST]).

AMSR-E (Advanced Microwave Scanning Radiometer for NASA's Earth Observing System) NASA, USA: A multifrequency, dual-polarized microwave radiometer that detects faint microwave emissions from the Earth's surface and atmosphere (dedicated to the observation of climate and hydrology).

SMOS (Soil Moisture and Ocean Salinity) and European Space Agency (ESA): A radio telescope in orbit that uses a Microwave Imaging Radiometer (map levels of land soil moisture for hydrology studies and ocean salinity for enhanced understanding of ocean circulation).

AQUARIUS NASA, USA: A NASA instrument in the board (measure global sea surface salinity to predict future climate conditions better and to provide insight into observations of variations in salinity and creating global ocean salinity distribution maps).

CZCS (Coastal Zone Color Scanner) NASA, USA: The first instrument flown on a spacecraft was devoted to measuring ocean color (used primarily for ocean color).

SeaWiFS (Sea-Viewing Wide Field-of-View Sensor) NASA, USA: Satellite designed to obtain global high-precision, moderate resolution, multispectral visible observations of ocean radiance for research in biogeochemical processes, climate change, and oceanography.

VIIRS (Visible Infrared Imaging Radiometer Suite) NOAA, USA: A scanning radiometer collects visible and infrared imagery and radiometric measurements of the land, atmosphere, cryosphere, and oceans (used to measure cloud and aerosol properties, ocean color, sea and land surface temperature, ice motion and temperature, fires, and Earth's albedo).

MODIS (Moderate Resolution Imaging Spectroradiometer) NASA, USA: A radiometer on board (study global dynamics of the Earth's atmosphere, land, ice, and oceans).

MERIS (Medium Resolution Imaging Spectrometer) European Space Agency (ESA): A programmable, medium-spectral resolution imaging spectrometer operating in the solar reflective spectral range for observing the color of the ocean.

7. Ocean Observatories and Remote Sensing for Environmental and Biological Datasets

The ocean observing systems have been developed and maintained for managing marine biodiversity and for sustainable exploitation of marine resources, preserving healthy marine ecosystems, ensuring public health and safe and efficient navigation (Halpern et al. 2017, Audzijonyte et al. 2019, Moltmann et al. 2019, Bastos et al. 2016, Déniz-González et al. 2017, Amani et al. 2022). Ocean observatories provide long-term monitoring to understand ocean and coastal variability and integrate all process steps from data to information. Stringed and towed instruments have been the basics of ocean monitoring, especially for monitoring their physical parameters (Wang et al. 2019, Zolich et al. 2019, Camus et al. 2019, Moore et al. 2019, Amani et al. 2022) to collect, process, analyze and supply large amounts of oceanic data, such as physical, meteorological, biological and chemical parameters, data on sediments, ocean noise, bottom topography, seismology, or coastline morphology (Bastos et al. 2016, Le Quéré et al. 2018, Canonico et al. 2019, Toonen and Bush 2020, Koerich et al. 2020). Coastal and Oceanic observatories are able to collect data in a regular and consistent way at adequate temporal and spatial scales (Bastos et al. 2016, Déniz-González et al. 2017, Amani et al. 2022). With the emergence of a wide range of instrumentation in observatories, high bandwidth wired instruments, and the rapid development of autonomous vehicles, *in situ* monitoring of ocean ecosystems is revolutionizing (Déniz-González et al. 2017, Jiang et al. 2019, Simoniello et al. 2019, Amani et al. 2022). Video sampling, along with real-time image analysis, is effective for surveying some components of marine ecosystems, and *in situ* DNA analysis leads to species-specific sampling (Déniz-González et al. 2017, Luther et al. 2017, Stelzenmüller et al. 2018, Amani et al. 2022). These coastal and oceanic observatories provide real-time information for operational oceanography at regional and local scales about oceanic patterns and dynamics, crucial for the sustainability of marine and coastal ecosystems and services and to a broad range of stakeholders and decision-makers integrated coastal zone management (Bastos et al. 2016, Déniz-González et al. 2017, Amani et al. 2022).

The ocean environment monitoring using advanced technologies, such as remote data detection, which provides various consistent and frequent datasets from different spatial scales for later various applications over oceans and coastal areas (Wang et al. 2019, Zolich et al. 2019, Camus et al. 2019), which, in conjunction of *in situ* measurements, enables the development of more advanced methods (Moore et al. 2019, Amani et al. 2022). Remote sensing is organized considering the most important variable obtained for oceanographical and atmospherically studies: atmospheric data, hydrographic data, biological data, and new satellites offering a wide variety of variables (Le Quéré et al. 2018, Canonico et al. 2019, Toonen and Bush 2020, Koerich et al. 2020). Remote sensing system has two primary components, the platform and the sensor, where the platform is a vehicle utilized to deploy the sensor, and the sensor is the device that records data (Déniz-González et al. 2017, Jiang et al. 2019, Simoniello et al. 2019, Amani et al. 2022). Satellites and global ocean observations constitute, integrate and coordinate networks of globally comparable data systems (Bojinski et al. 2014, Miloslavich et al. 2018, Moltmann et al. 2019) that can be used for a wide variety of functionalities in the ocean, such as evaluation of cumulative effects of multiple pressures, biodiversity study and management, marine ecosystems, and habitats conservation (Déniz-González et al. 2017, Luther et al. 2017, Stelzenmüller et al. 2018, Amani et al. 2022), and also for human development and sustainability (Halpern et al. 2017, Audzijonyte et al. 2019, Moltmann et al. 2019).

At global level several ocean observation platforms are also known the Regional Coastal Ocean Observing Systems (RCOOS) (Seim et al. 2009) of the IOOS that includes the Alaska Ocean Observing System (AOOS), the European Marine Observation and Data Network (EMODNET) 2020, the European Ocean Observing System (EOOS) 2020, the Central California Ocean Observing System (CeNCOOS), the Great Lakes Observing System (GLOS), the Gulf of Maine Ocean Observing

System (GoMOOS), the Gulf of Mexico Coastal Ocean Observing System (GCOOS), the Pacific Islands Ocean Observing System (PacIOOS), the Mid-Atlantic Coastal Ocean Observing Regional Association (MACOORA), the Northwest Association of Networked Ocean Observing Systems (NANOOS), the Southern California Ocean Observing System (SCCOOS), the Southeast Coastal Ocean Observing Regional Association (SECOORA), and the Caribbean Integrated Ocean Observing System (CarICOOS), Regional Programmes for the National Network of Regional Coastal Monitoring Programmes of England (Bastos et al. 2016) that include the North West Strategic Regional Coastal Monitoring Program, the North East Coastal Observatory (NECO), the East Riding of Yorkshire Council's Coastal Team (ERYC), Anglian Coastal Monitoring, Southeast Strategic Regional Coastal Monitoring Program, and Southwest Strategic Regional Coastal Monitoring Program (Bastos et al. 2016).

At the regional level, there are also several regional Coastal Observatories: NEPTUNE in the NE Pacific and VENUS in the Salish Sea (both from the Ocean Networks Canada) (Taylor 2009), Martha's Vineyard Coastal Observatory (US), the U.S. Cabled Array Fiber-Optic Ocean Observatory across the Juan de Fuca tectonic plate in the National Science Foundation's OOI, the Mayotte Island coastal observatory a French island in the Indian Ocean (Jeanson et al. 2014), the Liverpool Bay Coastal Observatory (England) (Howarth and Palmer 2011), the Observatorio Medio ambiental del Estrecho de Gibraltar (OME, South Spain), OBSEA (Barcelona, Spain) (Aguzzi et al. 2011), and the RAIA Observatory (NW Iberian Peninsula, Spain, and Portugal) (Bastos et al. 2016).

7.1 The Global Ocean Observing System (GOOS) and Integrate Ocean Databases Network

The creation of GOOS, in operation since 2005, established the first global ocean observing system that deals with environmental, biological, and pollution aspects of the ocean and coastal seas, that creates a network to obtain effectively and integrate ocean data, allowing management capacity and understanding as a large and distributed scientific facility or infrastructure, and one of the mechanisms required to support sustainable development (IOC/GOOS 2015, IOC-UNESCO 2017, 2020, European Global Ocean Observing System (EuroGOOS) 2020, GOOS 2020). The IOC/UNESCO and WMO gave priority to the implementation of the physical oceanographic component of GOOS, as the ocean component of the climate observing system GCOS that includes some operating systems (IOC/GOOS 2015, TAO 2015, IOC-UNESCO 2017, 2020, European Global Ocean Observing System (EuroGOOS) 2020, GOOS 2020), such as;

i) **Surface moorings:** Large fixed buoys moored to the bottom of the ocean to measure daily data broadcasted to shore through satellite links of surface winds, air temperature, relative humidity, sea surface temperature, and subsurface temperatures (IOC/GOOS 2015, TAO 2015, IOC-UNESCO 2017, 2020).

ii) **Argo Profiling Floats Program:** Autonomous observation systems which drift with ocean currents making detailed physical measurements of the upper

2 kilometers of the water column with records of seawater conductivity (salinity), temperature and pressure, information of geographical position via GPS and satellite data transmits to Argo data centers. Argo data have transformed ocean circulation studies. Currently, Argo data are routinely assimilated into global circulation models, giving accurate and timely global views of the circulation patterns and heat distribution of the ocean. This product has become an essential element of atmospheric forecast models and greatly improves seasonal climate, monsoon, and El Niño forecasts, as well as tropical cyclone simulations. The value of subsurface heat content measurements to the study of global warming and climate change has made the Argo an invaluable component of twenty-first-century environmental observation systems (IOC/GOOS 2015, TAO 2015, IOC-UNESCO 2017, 2020, European Global Ocean Observing System (EuroGOOS) 2020, GOOS 2020).

iii) **Ship-of-Opportunity Program (SOOP):** Utilization of ordinary cargo ships on regular routes, whose owners and crew agree to carry and, where necessary, operate oceanographic equipment during their regular voyages to obtain upper-ocean data. This constitutes a global network of Expendable Bathythermograph (XBT) (expendable (disposable) temperature- and depth-profiling system), ThermoSalinoGraphs (TSG) (continuous automated sea surface temperature and salinity measurements), CTDs (make precise conductivity, temperature, and depth measurements), XCTD (expendable (disposable) conductivity, temperature and depth-profiling system), ADCP (different depths acoustic current profiles) and pCO_2 (measurements of the "partial pressure of CO_2") systems on board of merchant ships, from which data are transmitted in real time and made available to the oceanographic and meteorological communities for operational use in ocean models and for other scientific purposes (IOC/GOOS 2015, TAO 2015, IOC-UNESCO 2017, 2020, European Global Ocean Observing System (EuroGOOS) 2020, GOOS 2020).

iv) **Direct sampling of ocean water by lowering bottles:** Use of CTD rosettes equipped with Niskin bottles water samples up onboard ship for sample CO_2, chlorophyll, microorganisms, biogeochemistry, and other wide variety of analysis to obtain databases of tens of thousands of hydrography profiles throughout the world's ocean. This method provides precise and accurate *in situ* measurements to evaluate the penetration of heat in the ocean, the changes in alkalinity, and monitoring the ocean's uptake of CO_2 that changes the ocean acidity levels (IOC/GOOS 2015, TAO 2015, IOC-UNESCO 2017, 2020).

v) **Surface drifting buoys:** Deployment of surface drifting buoys around the world to the measurement of seawater-surface temperature, salinity, and marine meteorological variables, telemetered in real time through the WMO's Global Telecommunications System (GTS) to support global meteorological services, climate research, and monitoring Together with the Argo profilers; these surface drifters has contributed to the success of a real-time monitoring system of the oceans, enabling much more accurate weather and climate forecasts (IOC/GOOS 2015, TAO 2015, IOC-UNESCO 2017, 2020, European Global Ocean Observing System (EuroGOOS) 2020, GOOS 2020).

vi) Continuous Plankton Recorder (CPR): In operation since 1946, on research and merchant vessels, captures plankton from the near-surface waters as the ship tows the instrument during its normal sailing (IOC/GOOS 2015, TAO 2015, IOC-UNESCO 2017, 2020, European Global Ocean Observing System (EuroGOOS) 2020).

vii) Global Sea Level Observing System (GLOSS): Monitoring gauges installed along the coasts of over 70 countries for oceanographic sea-level monitoring and real-time transmission of information via satellite links. The main component of GLOSS is the "Global Core Network" (GCN), comprising around 290 sea-level stations around the world for long-term climate change and oceanographic sea-level monitoring. Sea-level observations are global tsunami warning systems and are useful for local navigation and continuous refinement of tide-table predictions. Tide gauges measure rising water levels from storms and extreme tides (IOC/GOOS 2015, TAO 2015, IOC-UNESCO 2017, 2020, European Global Ocean Observing System (EuroGOOS) 2020, GOOS 2020).

The international cooperation models developed by international intergovernmental organizations like GOOS and the IOC and the World Meteorological Organization (WMO) have allowed new global datasets to be available to improve predicting models and forecasting capabilities (TAO 2015, IOC/GOOS 2015, Global Ocean Science Report 2017, 2020, European Global Ocean Observing System (EuroGOOS) 2020, GOOS 2020). Currently, global high-resolution operational systems and databases are available, and also for some regional or local regions and specific issues, in order to planning and development models and systems for operational oceanography, including modeling, observation, and data assimilation. Databases and repositories are fundamental to providing different types of marine data, global or for specific regions. There are **General databases**, databases of **Oceanographical data from specific regions** (North Atlantic and adjacent seas, Baltic Sea), **Climatological data, Geospatial data** (North East Atlantic and Baltic Sea), **General databases for Taxonomy, Traits and Biogeography, specific groups databases for Taxonomy, Traits and Biogeography** (fish, plankton, Corals, and Invasive Species), **Genetic databases, Fish stock assessments and Fishery data, Ecological indicators databases, Regime shifts databases** and some other type of Repositories (www.marinedatascience.co).
A resume of Global Marine Databases is presented in Table 1.

Table 1. Global Marine Databases.

General Databases	
PANGAEA - Data Publisher for Earth and Environmental Science	www.pangaea.de
ICES Data Center	http://ices.dk/marine-data/data-portals/
AODN Portal; Australian Ocean Data Network (AODN)	https://portal.aodn.org.au/
WHOI Data	www.whoi.edu/data/
Oceanographical Data	
World Ocean Database (WOD)	www.nodc.noaa.gov/OC5/WOD/pr_wod.html
IODE Ocean Data Portal (ODP) or **International Oceanographic Data and Information Exchange**	www.oceandataportal.org
OceanSITES	www.oceansites.org
NOAA Data Buoy Center (NDBC)	www.ndbc.noaa.gov
EOSDIS Ocean Data	https://earthdata.nasa.gov/discipline/ocean
OceanColor	https://oceancolor.gsfc.nasa.gov/data/overview/
USGS Earth Explorer	https://earthexplorer.usgs.gov/
Oceanographical Data for North Atlantic and Adjacent Seas	
ICES Data Center - Ocean	http://ices.dk/marine-data/data-portals/Pages/ocean.aspx
Oceanographical Data for Baltic Sea	
IOWDB – Oceanographic database of the IOW	https://odin2.io-warnemuende.de
BED – Baltic Environmental Database:	www.balticnest.org/bed
Climatological Data	
CCAFS – Climate	http://ccafs-climate.org
NCAR Climate Data Guide	https://climatedataguide.ucar.edu
NOAA Climate Indices	www.esrl.noaa.gov/psd/data/climateindices/
NOAA Gridded Climate Datasets	www.esrl.noaa.gov/psd/data/gridded/index.html
Geospatial Data	
Marine Regions	www.marineregions.org
Natural Earth	www.naturalearthdata.com
Geospatial Data for North East Atlantic	
ODIMS - OSPAR Data and Information Management System:	https://odims.ospar.org
Geospatial Data for Baltic Sea: HELCOM Map and Data Service	**HELCOM Metadata catalogue and HELCOM Map and Data Service**
General Taxonomy, Traits, and Biogeography	
WoRMS – World Register of Marine Species	www.marinespecies.org
Marine Species Identification Portal	http://species-identification.org
ITIS – Integrated Taxonomic Information System	www.itis.gov

Table 1 contd. ...

...Table 1 contd.

Catalogue of Life	www.catalogueoflife.org/
Encyclopedia of Life	http://eol.org/
Encyclopedia of Life TraitBank	http://eol.org/info/516
MARBEF Data system – ERMS	www.marbef.org/data/erms.php
Census of Marine Life	www.coml.org/
OBIS – Ocean Biogeographic Information System	www.iobis.org
GBIF – Global Biodiversity Information Facility	www.gbif.org
ReefBase – A global information system for coral reefs	www.reefbase.org
IUCN Red List	www.iucnredlist.org
Specific Groups Taxonomy, Traits, and Biogeography	
Fish. FishBase	www.fishbase.org
Corals. Coral Trait Database	https://coraltraits.org
Invasive Species	
GISD - Global Invasive Species Database	www.iucngisd.org/gisd/
GRIIS - Global Register of Introduced and Invasive Species	www.griis.org
Baltic Sea Alien Species Database	www.corpi.ku.lt/nemo/
Genetic Databases	
BOLD - Barcode of Life Data System	http://boldsystems.org
Tara Oceans Data Portal	www.taraoceans-dataportal.org
Fish Stock Assessments and Fishery Data	
Stock Assessments of ICES Fish Stocks	www.stockassessment.org/login.php
RAM Legacy Stock Assessment Database	http://ramlegacy.org/
FAO Global Fishery Database	www.fao.org/fishery/statistics/collections/en
ICES Data Center	www.ices.dk/marine-data/dataset-collections/Pages/Fish-catch-and-stock-assessment.aspx
Eurostat – EU Fishery statistics	https://ec.europa.eu/eurostat/web/fisheries/data
OpenFisheries	www.openfisheries.org
Sea around us	www.seaaroundus.org/
Ecological Indicators	
HELCOM Indicators	www.helcom.fi/baltic-sea-trends/indicators/
OSPAR Indicators	ODIMS Source Data files > IE2017
MARMONI Indicator Database	www.sea.ee/marmoni/marmoni_pulk/start_indicator_database.html
DEVOTES Catalogue of Indicators and DEVOTool	www.devotes-project.eu/devotool/
Regime Shifts	
Marine Regime Shifts Dataset	www.regimeshifts.org
Other Repositories Types	
KNB – Knowledge Network for Biocomplexity	https://knb.ecoinformatics.org
DataONE – Data Observation Network for Earth	www.dataone.org

Table 2. Global Ocean Mapping Databases.

GENERAL BATHYMETRIC CHART OF THE OCEAN (GEBCO)	International ocean mapping is operating under the joint auspices of the IOC of UNESCO and the International Hydrographic Organization (IHO). Development of a range of bathymetric datasets and data products (gridded bathymetric data sets, GEBCO Digital Atlas, GEBCO world map, and GEBCO Gazetteer of Undersea Feature Names). Provide the most authoritative publicly-available bathymetry of the world's oceans.
ONEGEOLOGY PORTAL	An international web-accessible Geological Survey initiative of global geoscience data at the best possible scales.
INTERNATIONAL COMPREHENSIVE OCEAN-ATMOSPHERE DATA SET (ICOADS)	A global ocean marine meteorological and oceanographic variables dataset (sea surface and air temperatures, wind, pressure, humidity, and cloudiness and surface ocean dataset). Contain measurements and visual observations from ships (commercial, navy, and research), moored and drifting buoys, coastal stations, and other marine platforms.
EXTENDED RECONSTRUCTED SEA SURFACE TEMPERATURE (ERSST) NOAA, USA	Analysis based on the International Comprehensive Ocean-Atmosphere Data Set (ICOADS) (long-term global and basin-wide studies, local and short-term variations).
OCEANCOLOR WEB NASA, USA	Open access information on ocean color data and products at various levels of information.
PERMANENT SERVICE FOR MEAN SEA LEVEL (PSMSL)	It is based at the National Oceanography Center (NOC), Liverpool (United Kingdom), which is a component of the UK Natural Environment Research Council (NERC). Funded by the Federation of Astronomical and Geophysical Data Analysis Services (FAGS), the Intergovernmental Oceanographic Commission (IOC-UNESCO), and the UK Natural Environment Research Council (NERC) Collection, publication, analysis, and interpretation of sea-level data from the global network of tide gauges (contains monthly and annual mean values of sea level from almost 2,000 tide gauge stations around the world.
SURFACE OCEAN CO2ATLAS (SOCAT) DIFFERENT DATA OWNERS. THE SURFACE OCEAN CO$_2$ ATLAS (SOCAT).	A global surface quality controlled fCO$_2$ (fugacity of carbon dioxide) dataset from the global oceans. It includes the Arctic and coastal seas (quantification of the ocean carbon sink and ocean acidification and the evaluation of ocean biogeochemical models). Under the auspices of The IOCCP (International Ocean Carbon Coordination Project), SOLAS (Surface Ocean Lower Atmosphere Study) and IMBE (Integrated Marine Biogeochemistry and Ecosystem Research), and function under the auspices of the IOC-UNESCO and SCOR (Scientific Committee on Oceanic Research).

GLOBAL MARINE INFORMATION SYSTEM (GMIS) INSTITUTE FOR ENVIRONMENT & SUSTAINABILITY, JOINT RESEARCH CENTER, EUROPEAN COMMISSION	An activity of the European Commission, DG Joint Research Center (JRC), developed within the Water Resources Unit of the Institute for Environment and Sustainability (IES) to provide the scientific community and other users with an appropriate set of biophysical information to monitor and conduct water quality assessment in the coastal and marine waters (Earth Observation data and the provision of continuous, detailed and accurate information on relevant marine biophysical parameters as derived from optical and infrared satellite sensors).
WORLD OCEAN DATABASE 2013 (WOD13) NATIONAL OCEANOGRAPHIC DATA CENTER (NODC), USA	A set of historical ocean profile data and plankton measurements in digital form, along with ancillary metadata and quality control flags. Study climate and the ocean environment, providing uniform, easy, and quality-assured access to nearly 19,000 datasets consisting of more than 200,000 oceanographic cruises from the National Oceanographic Data Center archive (contains nearly 13 million temperature profiles and almost 6 million salinity measurements). Integrates ocean profile data from approximately 90 countries around the world, collected from buoys, ships, gliders, and other instruments.
OCEAN BIOGEOGRAPHIC INFORMATION SYSTEM (OBIS)	An open-access database that allows users to search marine species datasets from the world's oceans and marginal seas (taxonomically and geographically resolved data on marine life and the ocean environment-interoperability with similar databases-software tools for data exploration and analysis).

8. Challenges for Ocean Science and Plankton Research Under the UN Agenda 2015–2030 Sustainable Development Goals

The marine ecosystem is currently under a wide range of challenges, pressures, and threats, such as climate change and anthropogenic multi-stressors that impact ocean biodiversity and functions and the physical and biogeochemical conditions of the ocean (Talley et al. 2016, United Nations 2016, 2017a, b, c, Ramírez et al. 2017, IOC-UNESCO 2017, 2020, Fennel et al. 2019, Balvanera et al. 2019, Bindoff et al. 2019, IPPC 2019). These drivers are also impacting all economic activities at sea that, include shipping, fishery, energy, land-ocean interactions, coastal protection, sustainable environmental and ecosystem management, tourism, and security (OECD 2016, 2019, UNESCO 2017, 2019, WMO-UNEP-GCP-IOC-Met Office 2020). The new marine technologies and operational oceanographic infrastructures and services have been fundamental for the development, in the short and medium term, to all marine and research activities through an integrated approach to face these challenges (O'Brien et al. 2017, UNESCO 2017, 2019, Edwards et al. 2016, Innes et al. 2016, United Nations 2016, 2017a, b, c, UNESCO 2017, 2019, IOC-UNESCO 2017a, b, 2020, Fennel et al. 2019, Rudolph et al. 2020, Leape et al. 2020). Although knowledge about the oceans and marine scientific research is still very limited, the development of new marine technologies for sustainable exploitation and protection of marine biodiversity often depends on the potential to exploit and preserve their marine resources (UN 2019, Independent Group of Scientists 2019, Österblom et al. 2020, European Marine Board 2020). The long history of global ocean monitoring programs which have been expanded, modified, and developed over time, together with methodological differences between nations, results in difficulties in the integration and holistic assessment of the data (United Nations 2016, 2017a, b, c, UNESCO 2017, 2019, IOC-UNESCO 2017a, b, 2020). In the coming years, in order to increase the current understanding of challenges in a changing world, the study of oceans, seas, and plankton research will increase significantly, in part encouraged by the defined qualitative elements of Aichi Biodiversity (Targets 11 and 15) and by the UN Agenda 2015–2030 Sustainable Development Goals, delivering the Essential Ocean Variables and Essential Biodiversity Variables to the Global Ocean Observing System and Group on Earth Observation's Biodiversity Observation Networks, which defined strategies to advancing understanding of the ocean and planktonic research plankton dynamics and informing policy and management decisions, in order to foster human health through ocean sustainability in the twenty-first century (UNOOSA (United Nations Office for Outer Space Affairs) 2018b, 2020, Hurd et al. 2018, Rees et al. 2018, Sala et al. 2018, Ryabinin et al. 2019, Fleming et al. 2019, European Marine Board 2020, Ocean Health and Human Well-Being 2020, Baumgart et al. 2021). The future of ocean and coastal research, management, surveillance, and observations will depend on the global partnership for ocean science and collaboration networks and the availability of infrastructure and a technological transition for operational use in research centers in oceanography and marine biology (United Nations 2016, 2017a, b, c, IOC-UNESCO 2017a, b, 2020, UNESCO 2017, 2019). Strategies need cooperation and communication between different actors from the sectors of marine science, industry, companies, and governments in order to create a critical scientific mass and an informed society for the development of adequate governance and policies that guarantee a future sustainable marine ecosystem based on in the knowledge (United Nations 2016, 2017a, b, c, IOC-UNESCO 2017a, b, 2020, UNESCO 2017, 2019, Österblom et al. 2020, European Marine Board 2020). The main challenge to respond to the enormous environmental, scientific, and societal challenges for a global and European marine research community is to increase the consistency and the sustainability of dispersed networks and infrastructures by integrating them within a shared pan-European framework for global ocean science policies and applied marine biology, incrementing scientific capacity and builds progress in relation to knowledge, technology, and innovation development advancements (United Nations 2016, 2017a, b, c, OECD 2016, 2017, 2019) and also in societal and research governance at a local, regional, and international scales (IOC-UNESCO 2017a, b, 2020, Chiba et al. 2018, Moltmann et al. 2019, Garcia-Garcia et al. 2019, Borga et al. 2019, 2020a, b, Batten et al. 2019, Bax et al. 2019, Beaugrand et al. 2019, Rudolph et al. 2020).

In the framework of the challenges for ocean science and plankton research under the context of the UN Agenda 2015–2030 Sustainable Development Goals, it is also important to highlight the UN Agenda 2015–2030 Goal 12 'Ensure Sustainable Consumption and Production Patterns' and Goal 14 'Protecting Sea Life: Conserving and Using the Oceans, Seas and Marine Resources' sustainably for sustainable development, that specifically calls for an increase to scientific knowledge research capacity and marine technology transfer to improve ocean health and to enhance the contribution of marine biodiversity to economic development (United Nations 2016, 2017a, b, c, UNESCO 2017, 2019, IOC-UNESCO 2017a, b, 2020). These goals set by this objective require constant and real time continuous monitoring of marine ecosystems to control pollution, fishing stocks, and marine biodiversity and thus achieve integrated ocean management. The interconnectedness of all SDGs requires the adoption of new integrated ways of cooperation across SDGs and countries to make use of the vast potential of plankton and marine resources for food security, the reduction of poverty, and better livelihoods. This is a particular challenge, but also an opportunity, for the implementation of the ocean-related SDGs allowing for the specific ecological, economic, and social transboundary characteristics and challenges of marine regions to be properly addressed:

i) Increasing the level of collective ambition and the diversity of solutions available;

ii) Partnering;

iii) Providing flexibility that can better ensure the participation of civil society stakeholders in adaptive decision-making processes;

iv) Allowing parties to cooperate by sharing expertise, developing joint processes, and coordinating and harmonizing their governance efforts (United Nations 2016, 2017a, b, c, IOC-UNESCO 2017a, b, 2020, OECD 2016, 2017, 2019, UNESCO 2017, 2019).

These needs are gaining more attention at national and international levels and efforts are been developed to provides de overarching legal framework for activities affecting the conservation and sustainable use of the ocean and its resources, such as the UN Convention on the Law of the Sea (UNCLOS), the UN Framework Convention on Climate Change (UNFCCC), and its related Paris Agreement, that have recognized the importance of ensuring the integrity of oceans (JPI Oceans 2015, 2018, United Nations 2016, 2017a, b, c, IOC-UNESCO 2017a, b, 2020, OECD 2016, 2017, 2019, UNESCO 2017, 2019). The new international legal policies are binding instrument for the conservation and sustainable use of marine biodiversity beyond nation jurisdiction (BBNJ) that provides a global platform for area-based management tools, including marine protected areas, supporting environmental impact and strategic assessments, enhancing capacity development, accessing appropriate technologies, and sharing benefits from the use of marine genetic resources (Santoro et al. 2016, Government Office for Science 2018, Stelzenmüller et al. 2018, Pearlman et al. 2019, International Oceanographic Data and Information Exchange (IODE) 2020, Simoniello et al. 2019, Wenhai et al. 2019, Borja et al. 2020a, b, Joint Programming Initiatives (JPIs) 2020, Hanan et al. 2020, World Bank 2020a).

References

Aguzzi, J., Mànuel, A., Condal, F., Guillén, J., Nogueras, M., del Rio, J. et al. 2011. The new seafloor observatory (OBSEA) for remote and long-term coastal ecosystem monitoring. Sensors 11: 5850–5872. doi: 10.3390/s110605850.

Amani, M., Mehravar, S., Asiyabi, R.M., Moghimi, A., Ghorbanian, A., Ahmadi, S.A. et al. 2022. Ocean remote sensing techniques and applications: a review (Part II). Water 14: 3401. https:// doi. org/10.3390/w14213401.

Angnuureng, D.B., Jayson-Quashigah, P-N., Almar, R., Stieglitz, T.C., Anthony, E.J., Aheto, D.W. et al. 2020. Application of shore-based video and unmanned aerial vehicles (drones): complementary tools for beach studies. Remote Sens 12: 394.

Audzijonyte, A., Pethybridge, H., Porobic, J., Gorton, R., Kaplan, I. and Fulton, E.A. 2019. Atlantis: a spatially explicit end-to-end marine ecosystem model with dynamically integrated physics, ecology and socio-economic modules. Methods in Ecology and Evolution 10(10): 1814–1819. https://doi. org/10.1111/2041-210X.13272.

Balvanera, P., Pfaff, A., Viña, A., Frapolli, E., Hussain, S., Merino, L. et al. 2019. Chapter 2: Status and trends; indirect and direct drivers of change. In IPBES Global Assessment on Biodiversity and Ecosystem Services, ed. IPBES. Bonn: IPBES Secretariat.

Bar-On, Y.M. and Milo, R. 2019. The biomass composition of the oceans: a blueprint of our blue planet. Cell 179(7): 1,451–1,454, https://doi.org/10.1016/j.cell.2019.11.018.

Batten, S., Abu-Alhaija, R., Chiba, S., Edwards, M., Graham, G., Jyothibabu, R. et al. 2019. A global plankton diversity monitoring program. Front. Mar. Sci. 6: 321.

Bastos, L., Bio, A. and Iglesias, I. 2016. The importance of marine observatories and of RAIA in particular. Front. Mar. Sci. 3: 140. doi: 10.3389/fmars.2016.00140.

Baumann-Pickering, S., Carreiro-Silva, M., Gjerde, K., Howe, B., Janssen, F., Katsumata, K. et al. 2019. Deep ocean observing strategy. Science and implementation guide. Deep Ocean Observing Strategy Project Paris, France, 68 pp.

Baumgart, A., Vlachopoulou, E.I., Vera, J.D.R. and Di Pippo, S. 2021. Space for the sustainable development goals: mapping the contributions of space-based projects and technologies to the achievement of the 2030 agenda for sustainable development. Sustainable Earth. 4: 6. https://doi. org/10.1186/s42055-021-00045-.

Bax, N., Miloslavich, P., Muller-Karger, F., Allain, V., Appeltans, W. and Batten, S. 2019. A response to scientific and societal capacity needs for marine biological observations. Front. Mar. Sci.

Beaugrand, G. 2014. Marine Biodiversity, Climatic Variability and Global Change. Routledge, London, 486 pp., https://doi.org/10.4324/9780203127483.

Beaugrand, G., Conversi, A., Atkinson, A., Cloern, J., Chiba, S., Fonda-Umani, S. et al. 2019. Prediction of unprecedented biological shifts in the global ocean. Nat. Clim. Change 9: 237–243.

Bedford, J., Johns, D., Greenstreet, S. and McQuatters-Gollop, A. 2018. Plankton as prevailing conditions: a surveillance role for plankton indicators within the marine strategy framework directive. Mar. Policy 89: 109–115. doi: 10.1016/j.marpol.2017.12.021.

Bedford, J., Ostle, C., Johns, D.G., Atkinson, A., Best, M. and Bresnan, E. 2020. Lifeform indicators reveal large-scale shifts in plankton across the north-west European shelf, in global change biology. https:// doi.org/10.1111/gcb.15066.

Benway, H., Lorenzoni, L., White, A., Fiedler, B., Levine, N., Nicholson, D. et al. 2019. Ocean time series observations of changing marine ecosystems: an era of integration, synthesis, and societal applications. Front. Mar. Sci.

Bergsma, E.W.J., Conley, D.C., Davidson, M.A., O'Hare, T.J. and Almar, R. 2019. Storm Event to Seasonal evolution of nearshore bathymetry derived from shore-based video imagery. Remote Sens. 11: 519.

Bindoff, N.L., Cheung, W.W.L., Kairo, J.G., Arístegui, J., Guinder, V.A., Hallberg, R. et al. 2019. Changing ocean, marine ecosystems, and dependent communities. pp. 447–587. *In*: Pörtner, H.-O., Roberts, D.C., Masson-Delmotte, V., Zhai, P., Tignor, M., Poloczanska, E. et al. (eds.). IPCC Special Report on the Ocean and Cryosphere in a Changing Climate. Cambridge University Press, Cambridge, UK and New York, NY, USA.

Bojinski, S., Verstraete, M., Peterson, T., Richter, C., Simmons, A. and Zemp, M. 2014. The concept of essential climate variables in support of climate research, applications, and policy. Bulletin of the American Meteorological Society 95(9): 1431–1443.

Borja, A., Santoro, F., Scowcroft, G., Fletcher, S. and Strosser, P. (eds.). 2020a. Connecting People to their Oceans: Issues and Options for Effective Ocean Literacy. Lausanne (Switzerland), Frontiers Media SA.

Borja, A., Andersen, J.H., Arvanitidis, C.D., Basset, A., Buhl-Mortensen, L., Carvalho, S. et al. 2020b. Past and future grand challenges in marine ecosystem ecology. Front. Mar. Sci. 7: 362. doi: 10.3389/fmars.2020.00362.

Bopp, L., Resplandy, L., Orr, J.C., Doney, S.C., Dunne, J.P., Gehlen, M. et al. 2013. Multiple stressors of ocean ecosystems in the 21st century: projections with CMIP5 models, Biogeosciences 10: 6225–6245. doi:10.5194/bg-10-6225-2013.

Bossier, S., Palacz, A.P., Nielsen, J.R., Christensen, A., Hoff, A. and Maar, M. 2018. The baltic sea atlantis: an integrated end-to-end modelling framework evaluating ecosystem-wide effects of human-induced pressures. PloS One 13(7).

Buck, J.J.H., Bainbridge, S.J., Burger, E.F., Kraberg, A.C., Casari, M., Casey, K.S. et al. 2019. Ocean data product integration through innovation-the next level of data interoperability, Front. Mar. Sci. 6. doi:10.3389/fmars.2019.00032.

Chiba, S., Batten, S., Martin, C.S., Ivory, S., Miloslavich, P. and Weatherdon, L.V. 2018. Zooplankton monitoring to contribute towards addressing global biodiversity conservation challenges. J. Plankton Res. 40: 509–518. doi: 10.1093/plankt/fby030.

Cloern, J.E., Abreu, P.C., Carstensen, J., Chauvaud, L., Elmgren, R., Grall, J. et al. 2016. Human activities and climate variability drive fast-paced change across the world's estuarine–coastal ecosystems. Global Change Biology 22(2): 513–529.

Camus, L., Pedersen, G., Falk-Petersen, S., Dunlop, K., Daase, M., Basedow, S.L. et al. 2019. Autonomous surface and underwater vehicles reveal new discoveries in the arctic ocean. In OCEANS 2019-Marseille, pp. 1–8. IEEE.

Canonico, G., Buttigieg, P.L., Montes, E., Muller-Karger, F.E., Stepien, C., Wright, D. et al. 2019. Global observational needs and resources for marine biodiversity. Frontiers in Marine Science 6: art. 367. https://doi.org/10.3389/fmars.2019.00367.

Capotondi, A., Jacox, M., Bowler, C., Kavanaugh, M., Lehodey, P. and Barrie, D. 2019. Observational needs supporting marine ecosystems modeling and forecasting. Front. Mar. Sci. 6. https://doi.org/10.3389/fmars.2019.00623.

Claassen, M., Zagalo-Pereira, G., Soares-Cordeiro, A.S., Funke, N. and Nortje, K. 2019. Research and innovation cooperation in the south atlantic ocean. South African Journal of Science 115(9/10).

Chirayath, V. and Li, A. 2019. Next-generation optical sensing technologies for exploring ocean worlds—NASA FluidCam, MiDAR, and NeMO-Net. Front. Mar. Sci. 6: 521.

Colefax, A.P., Butcher, P.A. and Kelaher, B.P. 2018. The potential for unmanned aerial vehicles (UAVs) to conduct marine fauna surveys in place of manned aircraft. ICES J. Mar. Sci. 75: 1–8.

Colomina, I. and Molina, P. 2014. Unmanned aerial systems for photogrammetry and remote sensing: a review. ISPRS J. Photogramm. Remote Sens. 92: 79–97.

Costanza, R., Groot, R., Sutton, P., Van der Ploeg, S., Anderson, S., Kubiszewski, I. et al. 2014. Changes in the global value of global ecosystems. Global Environmental Changes 26: 152–58.

Copernicus Climate Change Service. 2020. European state of the climate 2020. Available at: https://climate.copernicus.eu/esotc/2020.

COPERNICUS. 2020. European programme for the establishment of a European capacity for earth observation. Available at: https://www.copernicus.eu.

Dañobeitia, J.J., Pouliquen, S., Johannessen, T., Basset, A., Cannat, M., Pfeil, B.G. et al. 2020. Toward a comprehensive and integrated strategy of the european marine research infrastructures for ocean observations. Front. Mar. Sci., Sec. Ocean Observation, https://doi.org/10.3389/fmars.2020.00180.

Danovaro, R., Aguzzi, J., Fanelli, E., Billet, D., Gjerde, K., Jamieson, A. et al. 2017. An ecosystem-based deep-ocean strategy. Science 355: 452–454. doi: 10.1126/science.aah7178.

Déniz-González, I., Pascual-Alayón, P., Chioua, J., García-Santamaría, M.T. and Valdés, J.L. 2017. Directory of Atmospheric, Hydrographic and Biological Datasets for the Canary Current Large

Marine Ecosystem. 3rd Edition: Revised and Expanded. IOC-UNESCO, Paris. IOC Technical Series 110: 287.

Duarte, C.M., Poiner, I. and Gunn, J. 2018. Perspectives on a global observing system to assess ocean health. Front. Mar. Sci. 5: 265. doi:10.3389/fmars.00265.

Edwards, M., Helaouet, P., Alhaija, R.A., Batten, S., Beaugrand, G., Chiba, S. et al. 2016. Global marine ecological status report: results from the global CPR survey 2014/2015. SAHFOS Technical Report, 11: 1–32. Plymouth.

ESA (European Space Agency). 2019. Earth observations for SDG. Compendium of Earth Observation Contributions to the SDG Targets and Indicators. https://eo4society.esa.int/wp-content/uploads/2021/01/EO_Compendium-for-SDGs.pdf.

ESA (European Space Agency). 2021. Observing the earth/living planet. http://www.esa.int/Applications/Observing_the_Earth/ESA_for_Eart.

Eurofleets + 2020. EC Project. Available at: www.eurofleets.eu.

European Global Ocean Observing System [GOOS]. 2020. Available at: http://eurogoos.eu.

European Commission. 2019. International ocean governance: an agenda for the future of our oceans. Available at: https://ec.europa.eu/maritimeaffairs/sites/maritimeaffairs/files/list-of-actions_en.pdf.

European Marine Board. 2019. Navigating the future V: marine science for a sustainable future. Paper Presented at 24 of the European Marine Board (Ostend: European Marine Board).

European Marine Board. 2020. Policy needs for oceans and human health. EMB Policy briefs, 8. Ostend (Belgium), EMB. Available at: https://www.marineboard.eu/sites/marineboard.eu/files/public/publication/EMB_PB8_Policy_Needs_v3_web.pdf.

European Marine Observation and Data Network (EMODNET). 2020. Available at: http://www.emodnet.eu/.

European Ocean Observing System (EOOS). 2020. Available at: www.eoos-ocean.eu.

European Research Vessel Operators (ERVO). 2020. Available at: www.ervo-group.eu.

European Strategy Forum on Research Infrastructures (ESFRI). 2018. Strategy report on research infrastructures in europe, roadmap 2018. Available at: http://roadmap2018.esfri.eu.

Evans, K., Chiba, S., Bebianno, M.J., Garcia-Soto, C., Ojaveer, H., Park, C. et al. 2019. The global integrated world ocean assessment: linking observations to science and policy across multiple scales. Frontiers in Marine Science 6: art. 298.

Everett, Jason, D., Baird, Mark E., Buchanan, Pearse, Bulman, Cathy, Davies, Claire, Downie, Ryan et al. 2017. Modeling what we sample and sampling what we model: challenges for zooplankton model assessment. Frontiers in Marine Science 4: 77, 19 pp. DOI:10.3389/fmars.2017.00077.

Fennel, K., Gehlen, M., Brasseur, P., Brown, C., Ciavatta, S., Cossarini, G. et al. 2019. Advancing marine biogeochemical and ecosystem reanalyses and forecasts as tools for monitoring and managing ecosystem health. Front. Mar. Sci. 6: 89. doi: 10.3389/fmars.2019.00089.

Fleming, L.E., Maycock, B., White, M.P. and Depledge, M.H. 2019. Fostering human health through ocean sustainability in the 21st century. People and Nature 1: 276–83.

Frölicher, T.L., Fischer, E.M. and Gruber, N. 2018. Marine heatwaves under global warming. Nature 560(7718): 360–64.

García-García, L., van der Molen, J., Sivyer, D., Devlin, M., Paintin, S. and Collingridge, K. 2019. Optimizing monitoring programs: a case study based on the OSPAR eutrophication assessment for UK waters. Front. Mar. Sci. 19 pp. doi:10.3389/fmars.2018.00503.

Geraeds, M., van Emmerik, T., de Vries, R. and Ab Razak, M.S. 2019. Riverine plastic litter monitoring using unmanned aerial vehicles (UAVs). Remote Sens. 11: 2045.

Gleckler, P.J., Durack, P.J., Stouffer, R.J., Johnson, G.C. and Forest, C.E. 2016. Industrial-era global ocean heat uptake doubles in recent decades. Nat. Clim. Change 6: 394–398. doi: 10.1038/nclimate2915.

Global Earth Observation System of Systems (GEOSS). 2020. Available at: https://www.earthobservations.org/geoss.php.

Global Ocean Observing System [GOOS]. 2020. Available at: https://www.goosocean.org.

Government Office for Science. 2018. Foresight future of the sea: a report from the government chief scientific adviser. London, Government Office for Science. Available at: https://assets.publishing.service.gov.uk/government/uploads/system/uploads/attachment_data/file/706956/foresight-future-of-the-sea-report.pdf.

Greenwood, N., Devlin, M.J., Best, M., Fronkova, L., Graves, C.A., Milligan, A. et al. 2019. Utilizing eutrophication assessment directives from transitional to marine systems in the thames estuary and liverpool Bay, UK. In Frontiers in Marine Science. https://doi.org/10.3389/fmars.2019.00116.

Group on Earth Observations. 2016. Maximize the value of earth observation data in a big data world. [image online] https://www.slideshare.net/BYTE_EU/maximize-the-value-of-earth-observation -data-in-a-big-data-world.

GSA. 2017. GNSS market report. Issue 5. European Global Navigation Satellite Systems Agency. Available: https://www.gsa.europa.eu/system/files/ reports/gnss_mr_2017.pdf.

GSA. 2020. Making space for the oceans. European Global Navigation Satellite Systems Agency. Available: https://www.gsa.europa.eu/newsroom/ news/making-space-oceans.

Halpern, B.S., Frazier, M., Afflerbach, J., O'Hara, C., Katona, S., Stewart Lowndes, J.S. et al. 2017. Drivers and implications of change in global ocean health over the past five years. PLOS ONE 12(7): 1–23. https://doi.org/10.1371/journal.pone.0178267.

Hanan, N.P., Limaye, A.S. and Irwin, D.E. 2020. Editorial: use of earth observations for actionable decision making in the developing world. Front. Environ. Sci. 8: 601340. doi:10.3389/fenvs.2020.601340.

Hardin, P.J., Lulla, V., Jensen, R.R. and Jensen, J.R. 2018. Small unmanned aerial systems (sUAS) for environmental remote sensing: challenges and opportunities revisited. GISci. Remote Sens. 56: 309–322.

Hensel, E., Wenclawski, S. and Layman, C.A. 2018. Using a small, consumer-grade drone to identify and count marine megafauna in shallow habitats. Lat. Am. J. Aquat. Res. 46: 1025–1033.

Hoegh-Guldberg, O., Caldeira, K., Chopin, T., Gaines, S., Haugan, P., Hemer, M. et al. 2019. The Ocean as a Solution to Climate Change: Five Opportunities for Action. Washington D.C, World Resources Institute.

Howarth, J. and Palmer, M. 2011. The liverpool bay coastal observatory. Ocean Dynam. 61: 1917–1926. doi: 10.1007/s10236-011-0458-8.

Hurd, C.L., Lenton, A., Tilbrook, B. and Boyd, P.W. 2018. Current understanding and challenges for oceans in a higher-CO_2 world. Nature Climate Change 8: 686–94.

Ibarbalz, F.M., Henry, N., Brandão, M.C., Martini, S., Busseni, G., Byrne, H. et al. 2019. Global trends in marine plankton diversity across kingdoms of life. Cell 179(5, 14): 1084–1097.e21.

Innes, L., Simcock, A., Ajawin, A.Y., Alcala, A.C., Bernal, P., Calumpong, H.P. et al. 2016. The First Global Integrated Marine Assessment: World Ocean Assessment 1. New York, NY: United Nations.

IOC-UNESCO. 2017a. Global Ocean Science Report—The Current Status of Ocean Science Around the World. Valdés, L. et al. (eds.). Paris, UNESCO Publishing. Available at: https://en.unesco.org/gosr.

IOC-UNESCO. 2020. Global Ocean Science Report 2020–Charting Capacity for Ocean Sustainability. Isensee, K. (ed.). Paris, UNESCO Publishing. Available at: https://oceandecade.org/news/49/ Dialogue-mobilizes-over-twenty-Foundations-to-build-ocean-partnerships-for-the-next-decade.

IIOE-2. 2020. Second international indian ocean expedition (IIOE-2) (website). Available at: https:// incois.gov.in/IIOE-2/index.jsp.

International Oceanographic Data and Information Exchange [IODE]. 2020. Available at: www.iode.org.

International Research Ship Operators [IRSO]. 2020. Available at: www.irso.info.

IOC/GOOS (Intergovernmental Oceanographic Commission – Global Ocean Observation System). 2015. http://www.ioc-goos.org/.

IOC-UNESCO. 2016. IOC Capacity Development Strategy. 2015–2021. Paris, UNESCO Publishing. Available at: https://unesdoc.unesco.org/ark:/48223/pf0000244047.

IOC-UNESCO. 2017a. Global Ocean Science Report—The Current Status of Ocean Science Around the World. Valdés, L. et al. (eds.). Paris, UNESCO Publishing. Available at: https://en.unesco.org/gosr.

IOC-UNESCO. 2017b. Ocean Literacy for all: Toolkit, A, and Santoro, F. et al. (eds.). Paris, UNESCO Publishing. Available at: https://unesdoc.unesco.org/ark:/48223/pf0000260721.

IOC-UNESCO. 2018. Revised Roadmap for the UN Decade of ocean science for sustainable development. Paris, UNESCO Publishing. Available at: https://www.oceandecade.org/resource/44/REVISED-ROADMAP-FOR-THE-UN-DECADE--OF-OCEAN-SCIENCE-FOR-SUSTAINABLE-DEVELOPMENT.

IOC-UNESCO. 2019a. Summary report of the first global planning meeting: UN decade of ocean science for sustainable development. Gilleleje (Denmark), UNESCO Publishing. Available at: https://www.

oceandecade.org/resource/58/Summary-Report-of-the-First-Global-Planning-Meeting-UN-Decade-of-Ocean-Science-for-Sustainable-Development-Copenhagen-13-15-May-2019.

IOC-UNESCO. 2019b. The Science we need for the future we want. Paris, UNESCO. https://www.oceandecade.org/assets/The_Science_We_Need_For_The_Ocean_We_Want.pdf.

IOC-UNESCO. 2020. Dialogue mobilizes over twenty foundations to build ocean partnerships for the next decade (web article). https://oceandecade.org/news/49/Dialogue-mobilizes-over-twenty-Foundations-to-build-ocean-partnerships-for-the-next-decade.

IPCC. 2019: IPCC Special Report on the Ocean and Cryosphere in a Changing Climate [Pörtner, H.-O., Roberts, D.C., Masson-Delmotte, V., Zhai, P., Tignor, M. et al. (eds.)]. Cambridge University Press, Cambridge, UK and New York, NY, USA, 755 pp. https://doi.org/10.1017/9781009157964.

IRVSI. 2014. University of delaware, international research vessels schedules and information (http://www.researchvessels.org/.

Jeanson, M., Dolique, F. and Anthony, E.J. 2014. A GIS-based coastal monitoring and surveillance observatory on tropical islands exposed to climate change and extreme events: the example of Mayotte Island, Indian Ocean. J. Coast. Conserv. 18: 567–580. doi: 10.1007/s11852-013-0286-8.

Jiang, Z.-P., Yuan, J., Hartman, S.E. and Fan, W. 2019. Enhancing the observing capacity for the surface ocean by the use of volunteer observing ship. Acta Oceanologica Sinica 38(7): 114–120. https://doi.org/10.1007/s13131-019-1463-3.

Joint Programming Initiatives (JPIs). 2020. Available at: https://ec.europa.eu/programmes/horizon2020/en/h2020-section/joint-programming-initiatives.

Joyce, K., Duce, S., Leahy, S., Leon, J.X. and Maier, S. 2018 Principles and practice of acquiring drone-based image data in marine environments. Mar. Freshw. Res. 70: 952–963.

JPI Oceans. 2015. JPI oceans strategic research and innovation agenda. Available at: http://www.jpi-oceans.eu/news-events/news/jpi-oceans-sria-now-available.

JPI Oceans. 2018. Operational procedures, joint programming initiative on healthy and productive seas and oceans AISBL. Available at: http://www.jpi-oceans.eu/library?THES%5b%5d=169497.

Kavanaugh, M.T., Oliver, M.J., Chavez, F.P., Letelier, R.M., Muller-Karger, F.E. and Doney, S.C. 2016. Seascapes as a new vernacular for pelagic ocean monitoring, management and conservation. ICES J. Mar. Sci. 73: 1839–1850. doi: 10.1093/icesjms/fsw086.

Koerich, G., Assis, J., Costa, G.B., Sissini, M.N., Serrão, E.A., Rörig, L.R. et al. 2020. How experimental physiology and ecological niche modelling can inform the management of marine bioinvasions? Science of The Total Environment 700: 134692. https://doi.org/10.1016/j.scitotenv.2019.134692.

Le Quéré, C., Andrew, R.M., Friedlingstein, P., Sitch, S., Hauck, J., Pongratz, J. et al. 2018. Global carbon budget 2018. Earth System Science Data 10(4): 2141–2194. https://doi.org/10.5194/essd-10-2141-2018.

Leape, J., Abbott, M. and Sakaguchi. H. 2020. Technology, Data and New Models for Sustainably Managing Ocean Resources. Washington D.C., World Resources Institute.

Lombard, F., Boss, E., Waite, A.M., Vogt, M., Uitz, J., Stemmann, L. et al. 2019. Globally consistent quantitative observations of planktonic ecosystems. Front. Mar. Sci. 6: 196. doi: 10.3389/fmars.2019.00196.

Luther, J., Hainsworth, A., Tang, X., Harding, J., Torres, J. and Fanchiotti, M. 2017. World meteorological organization: concerted international efforts for advancing multi-hazard early warning systems. pp. 129–41. *In*: Kyoji Sassa, Matjaž Mikoš and Yueping Yin (eds.). Advancing Culture of Living with Landslides. Cham, Switzerland: Springer International Publishing.

Mackas, D.L. and Beaugrand, G. 2010. Comparisons of zooplankton time series. J. Mar. Syst. 79: 286–304. doi: 10.1016/j.jmarsys.2008.11.030.

Maxwell, S.M., Hazen, E.L., Lewison, R.L., Dunn, D.C., Bailey, H., Bograd, S.J. et al. 2015. Dynamic ocean management: defining and conceptualizing real-time management of the ocean. Mar. Policy 58: 42–50.

Miloslavich, P., Bax, N.J., Simmons, S.E., Klein, E., Appeltans, W., AburtoOropeza, O. et al. 2018. Essential ocean variables for global sustained observations of biodiversity and ecosystem changes. Glob. Change Biol. 24: 2416–2433. doi: 10.1111/gcb.14108.

Moltmann, T., Turton, J., Zhang, H.-M., Nolan, G., Gouldman, C., Griesbauer, L. et al. 2019. A global ocean observing system (GOOS), delivered through enhanced collaboration across regions, communities, and new technologies. Frontiers in Marine Science 6: art. 291. https://doi.org/10.3389/fmars.2019.00291.

Moore, A.M., Martin, M.J., Akella, S., Arango, H.G., Balmaseda, M., Bertino, L. et al. 2019. Synthesis of ocean observations using data assimilation for operational, real-time and reanalysis systems: a more complete picture of the state of the ocean. Frontiers in Marine Science 6: art. 90. https://doi.org/10.3389/fmars.2019.00090.

Muller-Karger, F.E., Miloslavich, P., Bax, N.J., Simmons, S., Costello, M.J., Sousa Pinto, I. et al. 2018. Advancing marine biological observations and data requirements of the complementary essential ocean variables (eovs) and essential biodiversity variables (ebvs) frameworks. Front. Mar. Sci. 5: 211. doi: 10.3389/fmars.2018.00211.

National Research Council. 2015. Sea change: 2015–2025 decadal survey of ocean sciences. Washington DC, National Academies Press. Available at: https://www.nap.edu/catalog/21655/sea-change-2015-2025-decadal-survey-of-ocean-sciences.

NEREUS, European Space Agency and European Commission. 2018. The ever-growing use of copernicus across europe's regions: a selection of 99 user stories by local and regional authorities, 277. https://www.nereus-regions.eu/copernicus4regions/publication/.

NEXTSPACE. 2019. Study on the copernicus data policy post-2020. Final Report. 93 pages. https://www.copernicus.eu/sites/default/files/2019-04/Study-on-the-Copernicus-data-policy 2019_0.pdf.

NMFS Office of Science and Technology. 2022. COPEPOD: The coastal and oceanic plankton ecology, production, and observation database from 2010-06-15 to 2010-08-15. NOAA National Centers for Environmental Information, https://www.fisheries.noaa.gov/inport/item/25610.

NOOA. 1970. Reorg. Plan No. 4 of 1970, F.R. Doc. 70-13375. US Federal Register 35(194).

O'Brien, T.D., Lorenzoni, L., Isensee, K. and Valdes, L. 2017. Chapter 6 southern ocean, in what are marine ecological time series telling us about the ocean? A Status Report, O'Brien, T.D., Lorenzoni, L., Isensee, K. and Valdés, L. (eds.). (IOCUNESCO, IOC Technical Series) 129: 1–297 (Available online at: https:// unesdoc.unesco.org/ark:/48223/pf0000247014).

Ocean Health and Human Well-Being. 2020. http://documents.worldbank.org/curated/en/111381468155703584/Indispensable-ocean-aligning-ocean-health-and-human-well-bei ng-guidance-from-the-blue-ribbon-panel-to-the-global-partnerships-for-ocean.

OceanObs'19. 2019. Conference statement. Available at: http://www.oceanobs19.net/statement/.

OECD. 2016. The ocean economy in 2030. Paris, OECD. Available at: https://read.oecd-ilibrary.org/economics/the-ocean-economy-in-2030_9789264251724-en#page1.

OECD. 2017. Strengthening the Effectiveness and Sustainability of International Research Infrastructures. Paris: OECD Publishing.

OECD. 2019a. Reference Framework for Assessing the Scientific and Socio-Economic Impact of Research Infrastructures. Paris: OECD Publishing.

OECD. 2019b. Rethinking Innovation for a Sustainable Ocean Economy, OECD Publishing, Paris, https://doi.org/10.1787/9789264311053-en.

Österblom, H., Wabnitz, C.C.C. and Tladi, D. (lead authors). 2020. Towards Ocean Equity. Washington D.C. World Resources Institute. https://www.oceanpanel.org/sites/default/files/2020-04/towards-ocean-equity.pdf.

Pearlman, J., Bushnell, M., Coppola, L., Karstensen, J., Buttigieg, P.L., Pearlman, F. et al. 2019. Evolving and sustaining ocean best practices and standards for the next decade. Front. Mar. Sci. 6: 277. doi: 10.3389/fmars.2019.00277.

Provost, E.J., Butcher, P.A., Colefax, A.P., Coleman, M.A., Curley, B.G. and Kelaher, B.P. 2019. Using drones to quantify beach users across a range of environmental conditions. J. Coast. Conserv. 23: 633–642.

Ramírez, F., Afán, I., Davis, L.S. and Chiaradia, A. 2017. Climate impacts on global hot spots of marine biodiversity. Science Advances 3(2).

Raoult, V., Tosetto, L. and Williamson, J.E. 2018. Drone-based high-resolution tracking of aquatic vertebrates. Drones 2: 37.

Rees, S.E., Foster, N.L., Langmead, O., Pittman, S. and Johnson, D.E. 2018. Defining the qualitative elements of aichi biodiversity target 11 with regard to the marine and coastal environment in order to strengthen global efforts for marine biodiversity conservation outlined in the united nations sustainable development goal 14. Marine Policy 93: 241–50.

Righetti, D., Vogt, M., Gruber, N., Psomas, A. and Zimmermann, N.E. 2019. Global pattern of phytoplankton diversity driven by temperature and environmental variability. Science Advances 5(5). https://doi.org/10.1126/sciadv.aau6253.

Rudolph, T.B., Ruckelshaus, M., Swilling, M., Allison, E.H., Österblom, H., Gelcich, S. et al. 2020. A transition to sustainable ocean governance. Nature Communications 11(1): 3600.

Rudd, M.A. 2015. Scientists' framing of the ocean science–policy interface. Global Environmental Change 33: 44–60.

Ryabinin, V., Barbière, J., Haugan, P., Kullenberg, G., Smith, N., McLean, C. et al. 2019. The UN decade of ocean science for sustainable development. Frontiers in Marine Science 6(470).

Sala, E., Lubchenco, J., Grorud-Colvert, K., Novelli, C., Roberts, C. and Sumaila, U.R. 2018. Assessing real progress towards effective ocean protection. Marine Policy 91: 11–13.

Salazar, G., Paoli, L., Alberti, A., Huerta-Cepas, J., Ruscheweyh, H-J., Cuenca, M.A. et al. 2019. Gene expression changes and community turnover differentially shape the global ocean metatranscriptome. Cell 179(5, 14): 1068–1083.e21.

Santoro, F., French, V., Lescrauwaet, A.K. and Seys, J. 2016. Science-Policy Interfaces in International and European Marine Policy. EU Sea Change Project.

Simoniello, C., Jencks, J., Lauro, F.M., Loftis, J.D., Weslawski, J.M., Deja, K. et al. 2019. Citizen-science for the future: advisory case studies from around the globe. Frontiers in Marine Science 6: art. 225. https://doi.org/10.3389/fmars.2019.00225.

Seim, H.E., Fletcher, M., Mooers, C.N.K., Nelson, J.R. and Weisberg, R.H. 2009. Towards a regional coastal ocean observing system: an initial design for the southeast coastal ocean observing regional association. J. Mar. Syst. 77: 261–277. doi: 10.1016/j.jmarsys.2007.12.016.

Smith, L.M., Yarincik, K., Vaccari, L., Kaplan, M.B., Barth, J.A., Cram, G.S. et al. 2019a. Lessons learned from the united states ocean observatories initiative. Frontiers in Marine Science 5: 1–7. doi:10.3389/fmars.2018.00494.

Smith, N., Kessler, W.S. et al. 2019b. Tropical pacific observing systems. Front. Mar. Sci. 839. doi:10.3389/fmars.2019.00031.

Stelzenmüller, V., Coll, M., Mazaris, A.D., Giakoumi, S., Katsanevakis, S., Portman, M.E. et al. 2018. A risk-based approach to cumulative effect assessments for marine management. Science of The Total Environment 612: 1132–1140. https://doi. org/10.1016/j.scitotenv.2017.08.289.

Talley, L.D., Feely, R.A., Sloyan, B.M., Wanninkhof, R., Baringer, M.O., Bullister, J.L. et al. 2016. Changes in ocean heat, carbon content, and ventilation: a review of the first decade of GO-SHIP global repeat hydrography. Ann. Rev. Mar. Sci. 8: 185–215.

Tanhua, T., Pouliquen, S., Hausman, J., O'Brien, K., Bricher, P., de Bruin, T. et al. 2019a. Ocean fair data services. Frontiers in Marine Science 6: art. 440.

Tanhua, T., McCurdy, A., Fischer, A., Appeltans, W., Bax, N., Currie, K. et al. 2019b. What we have learned from the framework for ocean observing: evolution of the global ocean observing system? Frontiers in Marine Science 6: art. 471. https://doi. org/10.3389/fmars.2019.00471.

TAO (Tropical Atmosphere Ocean project). 2015. http://www.pmel.noaa.gov/tao/.

Tassa, A., Willekens, S., Lahcen, A., Laurich, L. and Mathieu, C. 2022. On-going european space agency activities on measuring the benefits of earth observations to society: challenges, achievements and next steps. Front. Environ. Sci. 10: 788843. doi: 10.3389/fenvs.2022.788843.

Taylor, S.M. 2009. Transformative ocean science through the VENUS and NEPTUNE Canada ocean observing systems. Nucl. Instrum. Methods Phys. Res. A Accel. Spectrom. Detect. Assoc. Equip. 602: 63–67. Proceedings of the 3rd International Workshop on a Very Large Volume Neutrino Telescope for the Mediterranean Sea. doi: 10.1016/j.nima.2008.12.019.

Toonen, Hilde M. and Simon R. Bush. 2020. The digital frontiers of fisheries governance: fish attraction devices, drones and satellites. Journal of Environmental Policy and Planning 22(1): 125–137. https://doi.org/10.1080/1523908X.2018.1461084.

UN. 2019. Independent Group of scientists. 2019. The Future is Now—Science for Achieving Sustainable Development. Global Sustainable Development Report 2019. New York, United Nations.

UN ESCAP. 2018. Assessment of capacity development needs of the countries in asia and the pacific for the implementation of sustainable development goal 14. Bangkok, United Nations Economic and

Social Commission for Asia and the Pacific (ESCAP). Available at: https://www.unescap.org/sites/default/files/ESCAP%20Ocean%20Assessment_1.pdf.

United Nations. 2016. The First Global Integrated Marine Assessment: World Ocean Assessment I. New York, NY: United Nations.

United Nations. 2017a. The Conservation and Sustainable Use of Marine Biological Diversity of Areas Beyond National Jurisdiction: A Technical Abstract of The First Global Integrated Marine Assessment. New York, NY: United Nations.

United Nations. 2017b. The Impacts of Climate Change and Related Changes in the Atmosphere on the Oceans: A Technical Abstract of the First Global Integrated Marine Assessment. New York, NY: United Nations.

United Nations. 2017c. The Ocean and the Sustainable Development Goals Under The 2030 Agenda for Sustainable Development: A Technical Abstract of the First Global Integrated Marine Assessment. New York, NY: United Nations.

United Nations. 2018a. European global navigation satellite systems and copernicus: supporting the development goals, building blocks towards the 2030 Agenda [pdf]. www.unoosa.org/res/oosadoc/data/documents/2018/stspace/stspace71_0_html/ st_space_71E.pdf.

United Nations. 2018b. The "Space2030" agenda and the global governance of outer space activities. [online]. http://www.unoosa.org/oosa/oosadoc/data/documents/2018/aac.105/aac.1051166_0.htm.

United Nations. 2019. The "Space2030" Agenda: space as a driver of sustainable development. Committee on the Peaceful Uses of Outer Space Sixty-second session Vienna, 12–21 June 2019. A/AC.105/2019/CRP.15; 2019b. https://www.unoosa.org/res/oosadoc/data/documents/2019/aac_1052019crp/aac_1052019crp_15_0_html/AC105_2019_CRP15E.pdf.

UNESCO. 2015. UNESCO Science Report: Towards 2030. Paris, UNESCO Publishing. Available at: https://unesdoc.unesco.org/ark:/48223/pf0000235406.locale=f.

UNESCO. 2017. What are marine ecological time series telling us about the ocean? A status report, O'Brien, T.D., Lorenzoni, L., Isensee, K. and Valdés, L. (eds.). (Paris: IOC-UNESCO), 171–190.

UNESCO. 2019. UN decade of oceanic science for sustainable development (2021–2030). Available at: https://en.unesco.org/ocean-decade.

UN, General Assembly. 2018. Sustainable Development of the Caribbean Sea for Present and Future Generations, Report of The Secretary-General. United Nations General Assembly, Seventy-Third Session. Available at: https://digitallibrary.un.org/record/838645?ln=en.

UNOOSA, United Nations Office for Outer Space Affairs. 2018. European global navigation satellite system and copernicus: supporting the sustainable development goals building blocks towards the 2030 agenda. https://www.unoosa.org/res/oosadoc/data/documents/2018/stspace/stspace71_0_html/st_space_71E.pdf.

UNOOSA, United Nations Office for Outer Space Affairs. 2018a. Proposal by the bureau of the working group on the "Space2030" agenda on a draft structure of a "Space2030" agenda [pdf] http://www.unoosa.org/res/oosadoc/data/documents/2019/aac_105c_1l/aac_105c_1l_372_0_html/V1808639.pdf.

UNOOSA, United Nations Office for Outer Space Affairs. 2018b. European global navigation satellite system and copernicus: supporting the sustainable development goals. Building blocks towards a 2030 agenda. Vienna: Office for Outer Space Affairs, United Nations; 2018. Available: https://www.unoosa.org/oosa/en/oosadoc/data/documents/201 8/stspace/stspace71_0.Html.

UNOOSA, United Nations Office for Outer Space Affairs. 2018b. Annual report 2018. Vienna: United Nations Office for Outer Space Affairs, United Nations; 2019. Available: https://www.unoosa.org/documents/pdf/annualreport/UNOOSA_Annual_Report_2018.pdf.

UNOOSA, United Nations Office for Outer Space Affairs. 2019a. Fifty years since the first united nations conference on the exploration and peaceful uses of outer space (1968–2018): UNISPACE+50. Available at: https://unoosa.org/oosa/en/ourwork/unispaceplus/.

UNOOSA, United Nations Office for Outer Space Affairs. 2020. Space supporting the sustainable development goals. How space can be used in support of the 2030 Agenda for Sustainable Development. Vienna: United Nations Office for Outer Space Affairs, United Nations; 2020. Available: https://www.unoosa.org/oosa/en/ourwork/spaceplus50/index.html.

US, Ocean Policy Committee. 2019. Summary of the 2019 white house summit on partnerships in ocean science and technology. Washington DC, Office of Science and Technology Policy and Council on Environmental Quality. Available at: https://www.whitehouse.gov/wp-content/uploads/2019/12/Ocean-ST-Summit-Readout-Final.pdf.

US, Subcommittee on Ocean Science and Technology. 2018. Science and technology for america's oceans: a decadal vision. Washington DC, Office of Science and Technology Policy. Available at: https://www.whitehouse.gov/wp-content/uploads/2018/11/Science-and-Technology-for-Americas-Oceans-A-Decadal-Vision.pdf.

von Schuckmann, K., Le Traon, P.Y., Smith, N., Pascual, A., Djavidnia, S., Gattuso, J.P. et al. 2021. Copernicus marine service ocean state report, Issue 5. J. Oper. Oceanography 14: 1–185. doi:10.1080/1755876X.2021.1946240.

Wang, Z.A., Moustahfid, H., Mueller, A.V., Michel, A.P.M., Mowlem, M., Glazer, B.T. et al. 2019. Advancing observation of ocean biogeochemistry, biology, and ecosystems with cost-effective *in situ* sensing technologies. Frontiers in Marine Science 6: art. 519. https://doi.org/10.3389/fmars.2019.00519.

Weller, R., Baker, D., Glackin, M., Roberts, S., Schmitt, R. and Twigg, E. 2019. The challenge of sustaining ocean observations. Front. Mar. Sci. 6: 105. doi: 10.3389/fmars.2019.00105.

Wenhai, L., Cusack, C., Baker, M., Tao, W., Mingbao, C., Paige, K. et al. 2019. Successful blue economy examples with an emphasis on international perspectives. Frontiers in Marine Science 6.

Wilkinson, M.D., Dumontier, M., Aalbersberg, I.J., Appleton, G., Axton, M., Baak, A. et al. 2016. The fair guiding principles for scientific data management and stewardship. Sci. Data 3: 160018. doi: 10.1038/sdata. 2016.18.

WMO-UNEP-GCP-IOC-Met Office. 2020. United in science 2020: a multi-organization high-level compilation of the latest climate science. Available at: https://library.wmo.int/doc_num.php?explnum_id=10361.

World Bank. 2020a. Global environment facility projects. Available at: https://www.thegef.org/projects.

World Bank. 2020b. Overview of the caribbean. Available at: https://www.worldbank.org/en/country/caribbean/overview.

Wunsch, C. 2016. Global ocean integrals and means, with trend implications. Ann. Rev. Mar. Sci. 8: 1–33.

Zhang, C., Dang, H., Azam, F., Benner, R., Legendre, L., Passow, U. et al. 2018. Evolving paradigms in biological.

Zolich, A., Palma, D., Kansanen, K., Fjørtoft, K., Sousa, J., Johansson, K.H. et al. 2019. Survey on communication and networks for autonomous marine systems. Journal of Intelligent and Robotic Systems 95(3): 789–813. https://doi.org/10.1007/ s10846-018-0833-5.

1.2

Current and Forthcoming Threats to Coastal Zooplankton

Patterns Facing Increasing Freshwater Needs

Alvaro T. Palma,[1,*] *M. Loreto Torreblanca,*[2,*] *Luis R. Daza*[1,*]
and *Jessica A. Rayo*[1,*]

1. Introduction

Global water demand is projected to increase by 20% to 30% per year by 2050 (UNESCO 2018), and water scarcity will affect a vast proportion of humanity (Ihsanullah 2019). Human population growth, alongside climate change—although other factors could be added to this list—is responsible for increasing water scarcity (Teow and Mohammad 2019, UNESCO 2020, Ihsanullah et al. 2021). Hence, initiatives such as water reuse and desalination technologies are accelerating as cities worldwide come to terms with the vulnerability of their water supply (International Desalination Association 2021). While the utilization of seawater becomes part of the solution, it also represents a challenge for it to be a sustainable practice.

The above-described scenario highlights the apparent connection between climate-level changes and zooplankton inhabiting coastal environments from two parallel sources of impact; (i) at a global scale, the ever more accelerating upsetting of "normal" (or at least expected) environmental conditions affecting oceanographic processes and (ii) at a more local scale, the impact exerted by large volumes of water drawn by desalination practices.

Under current conditions, with the average global temperature 1.0 degrees above that of the pre-industrial period (1850–1900), it is forecasted that between 2030 and 2052, this value will climb to 1.5 degrees, while by 2100, the warming could be 3–4 degrees higher if the current level of emissions persists (IPCC 2018). Such a scenario could have profound consequences on the structure and taxonomic

---

[1] Fisioaqua SpA, Vitacura 2909 Las Condes, 7550024 Santiago, Chile.

[2] Caliptopis Ltda, San Pio X 2460 Of. 313 Providencia, Santiago, Chile.

* Corresponding author: apalma@fisioaqua.com; ltorreblanca@caliptopis.cl; rdaza@fisioaqua.com; jrayo@fisioaqua.com

composition of the zooplankton. The impact of climate change on zooplankton has received less attention in comparison with terrestrial organisms, despite the former playing a relevant role as a critical link in the carbon transfer between the atmosphere and deeper portions of the ocean (Steinberg and Landry 2017, Archibald et al. 2019, Kobari et al. 2003, Dam and Baumann 2018). Specific links between climatic warming and the decline of zooplankton have been reported since 1951 (i.e., Roemmich and McGowan 1995), whereas biomass of macrozooplankton in waters off Southern California has decreased by 80%. During the same period, the surface layer warmed, and the temperature difference across the thermocline increased. Likewise, another response of zooplankton to global warming is the broadening of its distribution into higher latitudes (Beaugrand et al. 2002, McKinnon et al. 2007, Richardson 2008). Another case shows that the warming trend of the Northeast Pacific registered over the past three decades has led to a mismatch between zooplankton life history events and the more-nearly-fixed seasonality of insolation, stratification, and food supply (Mackas et al. 2007).

Furthermore, these changes in zooplankton are not likely to occur synchronized, thus increasing the probability of temporal predator-prey mismatch. For example, during their larval transition, fish form part of the zooplankton and feed on copepods; therefore, variations in composition, time of occurrence, abundance, or size of zooplankton, could affect fish recruitment with likely regional repercussions on fisheries (Beaugrand et al. 2003). Aside from the direct consequences of temperature increase on the oceans, it is likely to observe variability in the responses to climate change at local and regional levels due to shifts in wind patterns affecting coastal upwelling, such as observed by Sydeman et al. (2014) and the intensification of upwelling winds in California, Benguela, and Humboldt systems.

Given its short life cycle and poikilothermal nature, zooplankton's physiological functions and life-history traits are susceptible to temperature and can quickly respond to environmental changes (Dam 2013, Dam and Baumann 2018). Hence, zooplankton can be considered a good sentinel for climate change (*sensu* Dam and Baumann 2018). This background information highlights the importance of studies on the composition and distribution of zooplankton in coastal areas at different spatial and temporal scales, more so when coastal ecosystems are likely to be threatened by local human activity and affected by global climate changes and the need for sustainable practices is no longer optional.

Alongside the former general climate-change-related issues affecting marine ecosystems, there is a foreseeable impact on plankton (both phyto and zooplankton) due to the intake of large volumes of seawater from coastal environments (Hogan 2015, Missimer and Maliva 2018, Ihsanullah et al. 2021), relevant quantitative information regarding the impact on the planktonic community inhabiting these environments requires further attention, such as highlighted by several authors (Roberts et al. 2010, Missimer 2016, Vega et al. 2020).

In this chapter, our main objective is to understand better the interaction between zooplankton and its environment in coastal areas of highly productive temperate latitudes on the Chilean coast. In turn, this will allow a better scaling of the threat level due to seawater intake generated by the operation of desalination facilities.

2. Zooplankton Susceptible to Desalination Activity

The northern coast of Chile is among the most productive ecosystems in the world, fueled by intense, upwelled, nutrient-rich cold waters (Escribano et al. 2007). Winds between 5–35°S are favorable to the occurrence of upwelling throughout most of the year (Thiel et al. 2007). The upwelling conditions provoke an ascent of equatorial sub-superficial water to the narrower and shallower continental shelf, decreasing temperature, increasing salinity, and producing occasional episodes of shallowing of the oxygen minimum zone (OMZ) (Escribano et al. 2009). These general conditions are favorable for an increment in the overall primary productivity in this zone during the spring-summer periods (Escribano and Hidalgo 2001, Thiel et al. 2007, Henríquez et al. 2007, Palma et al. 2009), which also tend to have a positive effect on the zooplankton inhabiting the coastal areas (Escribano et al. 2012). For instance, in a study that encompassed ten years (2002–2012), a community response of zooplankton was observed in an upwelling system in Chile in which the size structure and composition of the zooplankton responded to a water column that became less stratified, more saline, and colder; the mixed layer deepened, and the OMZ became shallower. The taxonomic and size diversity of the zooplankton community increased under these conditions, like small-sized copepods (< 1 mm) decreased in abundance, being replaced by larger-sized (> 1.5 mm) and medium-size copepods (1–1.5 mm), whereas euphausiids, decapod larvae, appendicularian, and ostracods increased their abundance (Medellín-Mora et al. 2016). Another issue related to the strengthening of upwelling intensity corresponds with the offshore advection of surface-dwelling larvae away from their favorable coastal habitat (Bakun et al. 2015). Although, upwelling can, in some cases, benefit certain meroplanktonic organisms, like competent larvae of the highly valuable mollusk *Concholepas concholepas* that occurs throughout the coast of Chile. Through their vertical migration capability, competent larvae can find themselves between an upwelling front and the coast during upwelling conditions, hence in the right environment for recruiting (Poulin et al. 2002a, 2002b). Moreover, the interplay of these generalized upwelling-favorable conditions together with geomorphological features of the coastline, like the presence of embayments facing north (protected from the prevailing south-west winds and the north-east swell), are favorable for the retention time of phytoplankton and zooplankton (Palma et al. 2006, Henríquez et al. 2007, Palma et al. 2009).

While zooplankton is composed of a wide variety of organisms, both vertebrates and invertebrates (Krautz et al. 2021), and where all existing phyla in the marine system exist within this community (Brierley 2017), it is vital to know the zooplankton that is available and enters a desalination facility in coastal settings. An essential part of the meroplankton in such an environment is represented by species that utilize these coastal settings as the habitat where they settle and recruit into their benthic/demersal or even water column adult habitat (Palma et al. 2006). The most common meroplankton groups present along the coast of Chile (particularly in the north) correspond to gastropods, mollusks, sea urchins, barnacle nauplii, and cyprids, different larval stages of decapod crustaceans, as well as eggs and larval stages of fishes (Torreblanca et al. 2016), many of which develop into important commercially harvested species (Cárcamo et al. 2021). The other fraction of the zooplanktonic

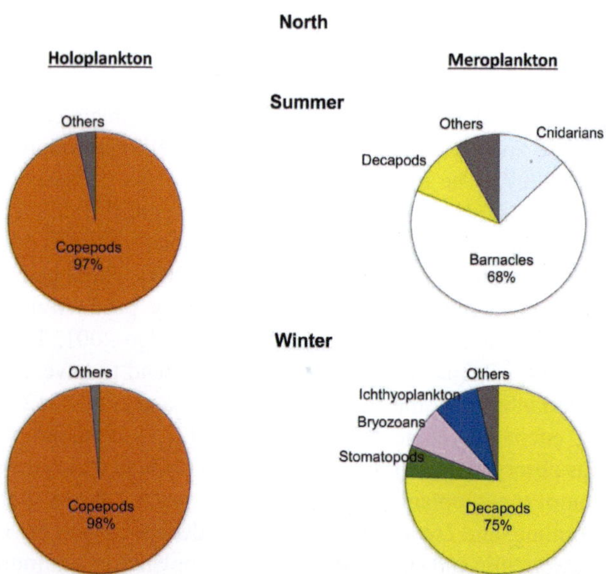

Figure 1. Composition (relative abundance) of several taxonomic groups of holo- and meroplankton in northern Chile during contrasting periods (summer 2022/winter 2020).

community, namely the holoplankton, spends their entire ontogeny in the water column near the shore. This latter group is mainly represented by copepods, the most abundant and diverse coastal zooplankton of the northern zone of the Humboldt current system (Torreblanca et al. 2016, Escribano and Hidalgo 2000).

In order to present the type of zooplanktonic community susceptible to being affected by human interventions along the coast, we present the following example based on actual field results corresponding to baseline surveys in the context of environmental regulations required in Chile for any initiative such as desalination projects. It informs about coastal zooplankton present in two regions of the coast of Chile susceptible to being affected by seawater intake operations (Fisioaqua 2022), highlighting only two of several sources of variability, namely, distance (study locations > 1,500 km apart) and time (contrasting seasons) that can affect diversity and abundance of zooplankton. Data resulted from plankton tows using a Bongo net (210 μm of mesh opening) that sampled the whole water column (25 m) in two locations, one in the north and the other on the central coast of Chile. Both systems correspond to bays facing north (semi-protected from the swell and the predominant southerly winds), and both are affected by coastal upwelling; however, in the north, upwelling tends to occur throughout the year (Fonseca and Farias 1987, Escribano et al. 2004, Thiel et al. 2007), while in Central Chile, this phenomenon tends to be more seasonal (Thiel et al. 2007, Letelier et al. 2009).

The zooplankton community in the north (Figure 1) shows differences between summer and winter, particularly for the meroplankton, where the dominant group in the summer (68%) were barnacle larvae (both cyprids and nauplii), followed by equivalent proportions of decapods and cnidarian larvae. Whereas in the winter, the meroplankton was dominated (75%) by decapod larvae (crab zoea and megalopae

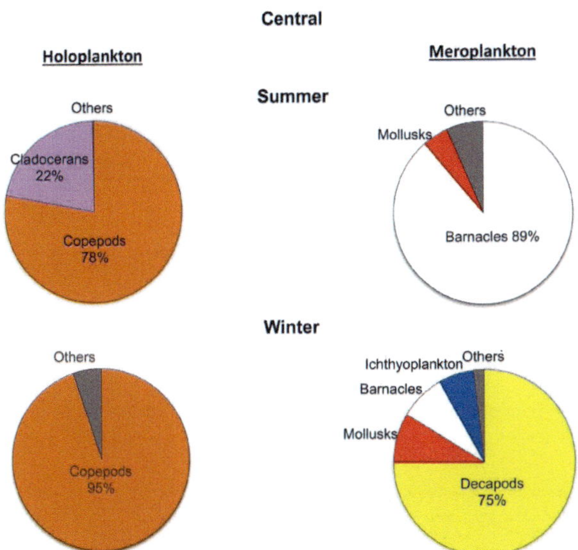

Figure 2. Composition (relative abundance) of several taxonomic groups of holo- and meroplankton in Central Chile during contrasting periods (summer 2022/winter 2021).

and shrimp Mysis), and the remaining groups corresponded to stomatopods, bryozoans, and fish larvae. In contrast, the holoplankton was very similar during both seasons and a predominance of copepods (> 97%), with *Acartia tonsa* and *Paracalanus indicus* being the most abundant species, described as common and abundant coastal copepods characteristic of the Humboldt current system (Hidalgo et al. 2010, Escribano et al. 2013, Torreblanca et al. 2016).

Although the pattern observed in the Central Chile location shares similarities with those observed in the north, the proportions vary (Figure 2). During summer, except for 22% of cladocerans, represented by *Evadne normanii*, a cosmopolitan neritic surface-dwelling species (Mujica and Espinoza 1994), the holoplankton was dominated by copepods in summer (78%) and winter (95%) with *A. tonsa* and *P. indicus* being the most abundant species. On the other hand, the meroplankton in the summer was dominated by barnacles (89%) and a minor presence of mollusk larvae, while during the winter, decapod larvae (zoea and megalopae) dominated, but the remaining abundance included mollusk, barnacle and fish larvae.

The species composition in these surveys (Table 1) shows that in the holoplankton, the highest richness corresponds to copepods, with 11 taxa shared by the two zones. On the other hand, some secondary groups, in terms of diversity, like Cnidaria, shared four taxa between zones.

The diversity of the meroplankton (expressed here as the number of taxa), however, exhibited more groups that contributed with different levels of richness. While decapods were the most diverse and shared 28 taxa between zones, other groups like Actinopterygii also exhibited moderate diversities and a high proportion of shared taxa.

Table 1. Composition (number of taxa) that conform the holo and the meroplankton in the northern and Central Chile study areas during contrasting seasons (summer and winter).

Holoplankton

| Group | North | | Central | | N° of shared taxa |
	Summer-22	Winter-20	Summer-22	Winter-21	Between zones
Cnidaria	6	1	6	8	4
Ctenophora	1	1	–	–	–
Polychaeta	–	–	1	1	–
Molluska	–	1	–	–	–
Copepoda	14	8	18	31	11
Anphipoda	3	1	2	1	2
Cladocera	1	–	1	1	–
Ostracoda	1	–	–	–	–
Euphausiacea	–	1	1	1	–
Chaetognatha	3	2	–	3	2
Chordata	–	1	1	3	–
Total N° of taxa	29	16	30	49	19

Meroplankton

| Group | North | | Central | | N° of shared taxa |
	Summer	Winter	Summer	Winter	Between zones
Platyhelminthes			1	–	
Cnidaria	1	–	–	–	
Polychaeta	2	–	3	3	1
Molluska	2	3	2	4	2
Equinodermata	2		1	–	1
Cirripedia	1	1	2	1	1
Isopoda	1	1	1	1	1
Mysida	–	–	1	1	–
Stomatopoda	–	1	–	1	–
Decapoda	23	19	31	28	28
Bryozoa	1	1	1	1	1
Chordata	2	–	–	1	1
Actinopterygii	11	14	15	16	10
Total N° of taxa	46	40	58	57	46

In the study above, performed similarly in the two zones, both characterized for being affected by upwelling events, the composition and seasonal dominance of zooplankton showed similarities. Such brief exercise highlights which groups of zooplanktonic organisms are more prone to be affected by activities in coastal settings that imply the use of seawater.

With a growing scarcity of freshwater, mainly in Northern-Central Chile, linked to climatic changes (Vargas et al. 2020), we provide insight that considers the potential effect on zooplanktonic communities of the increase in demand for seawater (both for mining-related and domestic activities); a legitimate type of concern where climate change at a global scale can have significant local-level repercussions.

3. Impact of Desalination Intake on Zooplankton

With a growing scarcity of freshwater, mainly in Northern-Central Chile, linked to climatic changes (Vargas et al. 2020), we provide insight that considers the potential effect on zooplanktonic communities of the increase in demand for seawater (both for mining-related and domestic activities), a legitimate type of concern where climate change at a global scale can have significant local-level repercussions. Chile's freshwater production through desalination is 8,600 L/s and is likely to double by 2026 and triple by 2032. The expansion of mining in Northern Chile has been the main incentive for desalination developments during the past decade (ACADES 2022). These numbers should be enough justification for considering knowing the zooplankton along the northern coast of Chile as a good subject of study.

It is important to recognize aspects such as the type of zooplankton present in coastal areas and how it will likely interact with desalination water intakes (as shown in the previous section). Thus, we aim to focus our contribution—on our own experience studying coastal communities along the central and northern coast of Chile for the past two decades (i.e., Poulin et al. 2002a, 2002b, Hernández et al. 2003, Palma et al. 2006, Torreblanca et al. 2016, Palma et al. 2021)—on a better understanding of the links between environmental conditions such as those highlighted above and zooplankton inhabiting coastal settings.

With technologies for producing water through desalination becoming increasingly accessible (Pinto and Marques 2017, Darre and Toor 2018), adverse environmental impacts (EIs) associated with this activity still represent an issue in need of better and sustainable solutions (Roberts et al. 2010, Dawoud and Al Mulla 2012). While brine production and its disposal have received the most attention when it comes to EIs (Hogan 2015, Roberts et al. 2010, Baten and Stummeyer 2013), the impact on aquatic biota that is susceptible to entering desalination facilities with open water intakes has received comparatively less attention (Missimer and Maliva 2018). Whit the co-occurrence of several technologies for seawater desalination, reverse osmosis (SWRO) is one of the most common (Jones et al. 2019), and like most others, it affects plankton (Maliva and Missimer 2012, Missimer et al. 2015). Regardless of the type of surface intake system that collects water from the open ocean above the seabed (e.g., deep-water intakes, offshore intake, or passive-screen intakes), it has potential EIs, of which that of most significant concern is the impingement and entrainment of marine organisms (Hogan 2015). This impact will also depend on the availability of organisms susceptible to entering and the design and function of the intake facility. For example, modern desalination plants utilize velocity caps and screens, which are effective ways of decreasing inlet velocities which in turn have been shown to minimize entrainment and impingement of marine organisms (Craig 2015).

Typically, a coastal water column harbors a high density of small-size fauna with none or little mobility (i.e., plankton), subject to natural or artificial currents along the coast. This plankton is composed of taxa, such as copepods, that spend their whole life in the water column (holoplankton), while others share a similar size range and occupy this habitat only during their dispersive larval phase (meroplankton). The latter category fits a wide variety of organisms, among which many represent harvestable species of commercial value. However, the mobility capabilities of certain zooplankters, such as larval stages of fish and invertebrates, can be significant and, in shallow coastal water columns, represent an effective mechanism in favor of remaining near their nursery (Hernández et al. 2003) or settling (Palma et al. 2006) habitat.

In surface water intakes such as desalination plants, it is fundamental to know the impact generated on these organisms by the intake flow of such operations since the elevated probability of passive entry. This aspect is relevant, both for holoplankton as well as for meroplankton. The latter is of particular importance given the high diversity of meroplaktonic species occurring in coastal settings, which eventually can cause a diminished availability of individuals of ecological and commercial interest (York and Foster 2005). Therefore, it is essential to evaluate such losses from the ecosystem due to industrial seawater uptakes (York and Foster 2005, Steinbeck et al. 2007) and determine how these losses can eventually be mitigated (Strange et al. 2004).

A life history strategy associated with an elevated per capita production of offspring, characteristic of organisms undergoing a dispersive larval stage like fish and invertebrates (meroplanktonic species), involves typically high levels of natural mortality (Rumrill 1990, Pineda 1994, Underwood and Keough 2001). It is expected that a critical percentage (even higher than 99.9%) of individuals that spawn from a typical female fish or invertebrate die before adulthood (Thorson 1946, Caddy and Sharp 1986, McConaugha 1992, López et al. 1998, Greer et al. 2016, Aranceta-Garza et al. 2016). The literature, however, stresses the need to generate information that would allow a better understanding of the natural rates of mortality in the early stages in meroplanktonic species—like in a study by Vaughn and Allen (2010) highlighting the importance of predation as a source of mortality—in order to perform better stock assessments of exploited species (Punt et al. 2021).

Currently, the primary trend among studies oriented to estimate the impacts on the plankton of operations that take up marine water for their functioning (i.e., desalination plants, LNG plants, thermic cooling plants) is to extrapolate the actual number of individuals affected in equivalents of adult numbers or the equivalent of biomass production available for predators. Such an approach estimates the number of individuals entrained or impinged who otherwise would have survived into a future age (age of equivalency). The Equivalent Adult (EA) model, initially described by Horst (1975) and Goodyear (1978), adjusts the losses so that the natural mortality that would have occurred between the age of entrainment or impingement and the age of equivalency can be quantified.

The age of equivalency refers then to the age at which losses can be extrapolated utilizing the EA model. The researcher defines this age and can vary depending on the study's objective, for example, from 1 year of age, age of sexual maturity, or the age of recruitment into the fishery. However, studies are still limited based on the available life history information for each species of interest (Barnthouse 2004). Therefore, this model is better suited in cases where species' life cycle and early developmental stages are well known, usually the case for highly valuable commercial species and some of the particular ecological relevance but never to the entire list of species in a community.

3.1 Estimates of Zooplankton Losses by Seawater Intakes

The estimation of losses due to water intakes can be evaluated from an ongoing project or prior to a facility being operational, thus helping in the decision-making for engineering improvements of the project and aiding with the environmental permitting process. In either case, it is crucial to recognize that when it comes to meroplankton, one must be aware of the larval stage or age of the organisms susceptible to being withdrawn by the water intake and how that translates into EA.

Barnthouse 2004 (EPRI technical report 1008471) identifies and compiles detailed information on the historical use of EA models, documenting relevant equations and explaining its methods of use in detail. These models include the "Forward Projection Approach," which utilizes estimations of fractions of survivorship—for different developmental stages—to extrapolate losses due to entrainment or impingement into the number of adults that survive to some future age (age of equivalency). Another model is the "Fecundity Hindcasting Approach," which is an estimation of survivorship fractions from the "egg" stage to the stage/age of mortality (due to intake) to calculate the number of females needed in order to produce the lost individuals.

In either case, it is worth reminding the relevance of being capable of providing the best possible taxonomy of the zooplanktonic community subject to these impacts, as well as recognizing the different life stages of individual species.

Considering the context and perspective of the present analysis (i.e., studying the impact of industrial seawater uptake on the diverse zooplanktonic community present in coastal settings), we decided to explore the Forward Projection Approach more in-depth. This model is one that requires less background information—such as detailed life history data or accurate data on the status of the adult population—in order to be executed. Additionally, this model expresses the losses of different developmental stages (due to water uptake) in standard EA units.

Given the high probability that different developmental stages of the same species would be affected by an industrial water intake, the number of EAs will be the sum of the surviving fraction of each stage. The number of adult equivalent losses based on the FP model will be:

$$EA = \sum_{i=1}^{n} S_{i,A} N_i$$

where:

EA: Impinged or entrained individuals that would have naturally survived to the age of equivalency.

$S_{i,A}$: Fraction of individuals expected to survive from age "*i*" to the age of equivalence.

N_i: number of individuals in ontogenetic stage "i" that is expected to be affected by the intake for a given period (i.e., days or months).

Furthermore, the fraction of individuals expected to survive is defined by the following equations:

$$S_{i,A} = \prod_{j=i}^{j_{max}} S_j$$

$$S_j = e^{-k_i t_i}$$

where:

Sj: Survival fraction from stage "j" to stage "j+1."

K_i: per day natural mortality rate of the ontogenetic stage "i."

t_i: Duration in days of the ontogenetic stage "i."

Since individuals affected by an intake are constantly developing and can be caught within days or weeks since hatching, their survival probability is greater (Barnthouse 2004). Because of this, Nielsen (2005) defines an adjusted surviving fraction as "S*i" that applies to the affected stage of individuals. Besides considering individuals' natural mortality, this adjustment also considers a mortality rate due to uptake "m", which corresponds to the quotient between the volume of water captured and the volume of water that is susceptible to being captured. Thus, the adjusted surviving fraction is:

$$S^* = \frac{e - kt_s \left(\dfrac{e - mt_s}{m} + t_s - \dfrac{1}{m} \right)}{e - (k+m)t_s \left(\dfrac{1}{k+m} - \dfrac{e^{mt_s}}{k} \right) - \left(\dfrac{1}{k+m} - \dfrac{1}{k} \right)}$$

where:

S*: Survival fraction of individuals based on their stage of development at the moment of entering the intake and until the next larval stage.

k: Natural mortality rate of individuals based on their stage of development at the time they are taken in.

t_s: Duration of developmental stages of individuals, based on the ontogenetic stage at the time they are taken in.

m: Mortality rate due to passive intake

Finally, $S_{i,A}$ can be corrected by:

$$S_{i,A} = S^* \prod_{j=i+1}^{j_{max}} S_j$$

where:

S_j: Survival fraction of individuals following ontogenetic stages from the one they were taken in until their remaining larval stages (eventually reaching the age of equivalency).

3.2 Caveats Regarding Adult Equivalent Losses

Several initiatives and authors have documented Adult Equivalents analyses (i.e., Lawler et al. 1981, Otto 1989, Tenera 2000, PSEG 1999, 2001, USEPA 2001). However, the under and over-estimation of individuals that will reach the age of equivalency remains a significant issue in evaluating Equivalent Adult Losses (EAL). Some factors that can produce such uncertainties are the following:

3.2.1 Wrongly Estimating the Number of Affected Individuals

Globally, most large-capacity SWRO plants use open-ocean intake systems, with the actual intake located either onshore or offshore. The most common offshore intake type uses a velocity cap at the top of the inverted pipe. Inshore or offshore passive-screen intakes are used to reduce the impacts of impingement and entrainment (Pankratz 2015). The design and depth of the intake towers are such that their influence reaches only up to a few meters above the structure, which leaves the rest of the water column unaffected (WateReuse Association 2011). Therefore, EAL zooplankton surveys should be stratified to cover the whole water column. This way, the vertical distribution of zooplankton can be recognized, and its density circumscribed to the volume of water susceptible to intake. However, this should also be understood, knowing that zooplankton can migrate vertically, which can happen differentially during a day-night period (Poulin et al. 2002, Lyczkowski-Shultz and Steen 1991, Holt and Holt 2000).

3.2.2 Lack of Early Life History Information

The scarcity or lack of early life history information represents an additional source of uncertainty for which further investigation is needed. However, assumptions of high natural mortality rates are funded (i.e., Underwood and Keough 2001), and estimates of those rates (i.e., McConaugha 1992, López et al. 1998, Greer et al. 2016, Aranceta-Garza et al. 2016) exist and are helpful for the models discussed above.

The absence or incomplete information on spawning periods of target species that coexist with facilities that utilize large volumes of water is also noticeable, stressing the need to obtain more information on the organisms that exist in the source water column at a reasonable (seasonal) time scale so natural variability can be incorporated (Vega et al. 2020).

Hence, the extrapolation of EALs based on isolated and discrete surveys falls into the risk of under or overestimating the impact of water intake. However, as depicted in Figure 3, natural temporal variability in larval abundance for two important invertebrates like the red sea urchin *Loxechinus albus* and the gastropod *Concholepas concholepas* (both highly valued resources along the coast of Chile) is such that it becomes clear the need to design a surveying schedule, frequent enough, so an estimation of EAL is possible. Moreover, the former does not even include oceanographic forcings like El Niño or La Niña, which can further affect the abundance variability of organisms like these.

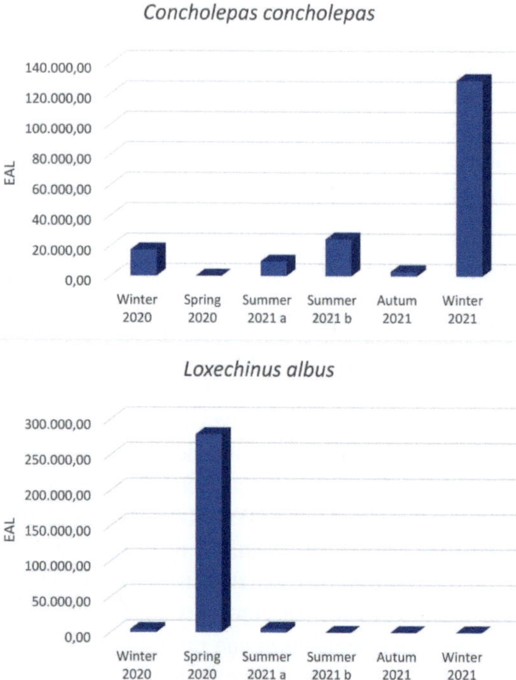

Figure 3. Estimates of EAL to the age of recruitment into the benthos for the edible gastropod *C. concholepas* and the red edible sea urchin *L. albus* based on abundances registered at 15 m above the bottom during six campaigns. Surveys were performed inside an embayment in northern Chile where a SWRO facility is projected.

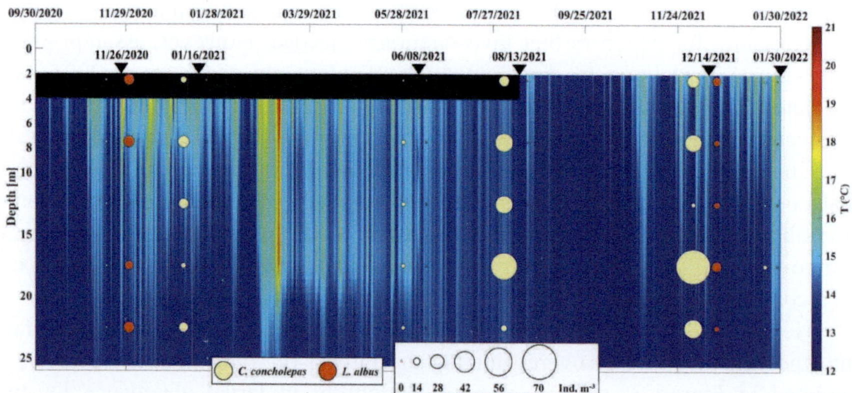

Figure 4. Abundance of the gastropod *C. concholepas* and the sea urchin *L. albus* obtained larvae at different depths (every 5 m depth) by means of a WP-2 plankton net of 210 μm mesh size on six different occasions. Black inverted triangles correspond to the sampling dates. The temperature section for the water column along time was constructed from temperature readings of loggers secured to a mooring line every 4 m (6 loggers until August 2021 and 7 loggers from then on with the shallowest positions at 2 m).

Moreover, considering that seawater intakes do not affect the whole water column equally, it becomes relevant to understand how this meroplankton is vertically distributed in an area of interest and through time. While long-temporal disparities in the abundance of zooplankton are expected for larval stages of species that do not necessarily spawn year-round, as depicted above from Bongo net tows (Figure 3), another scale of variability is spatial (depth, for instance). On surveys obtained at different depths (samples every 5 m intervals using a WP-2 net with closing mechanisms triggered from the surface) at the exact location in northern Chile, the abundance of competent larvae of the same two species (*C. concholepas* and *L. albus*) were quantified throughout the water column (Figure 4). At this point of interest, the thermal structure of the water column was recorded continuously utilizing a series of temperature loggers installed along a permanent mooring line. The sea urchin *L. albus* exhibited the highest abundance throughout the water column in the austral spring of 2020 and 2021 during non-stratified conditions and a minor abundance or absence the rest of the time. On the other hand, *C. concholepas* exhibited its highest abundance throughout the water column (except the deepest section) during the winter and spring of 2021 and minor the rest of the time. In both cases, the water column was cold and non-stratified.

4. Discussion and Conclusions

In the face of an increased need for freshwater via desalination, regions of the world like Chile, particularly its northern coast along the arid and home of some of the most extensive copper mining operations worldwide, should be considered with more attention. Not only are the coastal zooplankton observed in this region diverse and abundant, but their related magnitudes are also highly variable in space and time. The most notorious temporal change occurs seasonally (i.e., Palma et al. 2006, Gaymer et al. 2010, Torreblanca et al. 2016), mainly through changes in the diversity and abundance of species belonging to a highly productive coastal ecosystem (Thiel et al. 2007, Krautz et al. 2021, Palma et al. 2021). Furthermore, the dispersive larval phase of many coastal species is of particular concern in shallow settings since this environment is crucial for their settlement and subsequent recruitment (Palma et al. 1999, Palma et al. 2006, Henández et al. 2003). At a local scale, and as depicted in our results of two economically relevant invertebrates (among several others), variation in abundance with depth close to the coast is also a factor that becomes ever more relevant when water intakes can represent a primary source of disturbance.

Understanding and quantifying the potential impact on zooplankton generated by the intake of large volumes of water must consider several approaches. Firstly, it is fundamental to know the local patterns of variability in diversity, abundance, and distribution of the zooplankton to contrast the potential losses with its source. Secondly, an anticipation of the adult equivalent losses is needed, thus making it possible to design mitigation, remediation, or even reparation measures. Thirdly, more detailed information on the zooplankton behavior at different temporal and local spatial scales are in dire need to help develop strategies (both technological as well as better procedures) that will reduce the impact on coastal zooplankton. Finally, it is as relevant as ever to ensure the best possible taxonomic information about the zooplanktonic community being affected, as well as the recognition of the different life stages susceptible to entering the desalination plants.

5. Acknowledgments

We want to express our gratitude to José Miguel Bogdanovich for his critical review of an early draft of this chapter, as well as the participation of several cohorts of students and interns at FisioAqua and Caliptopis. They all participated in the studies that form part of the content of this chapter.

References

ACADES. 2022. https://www.acades.cl/los-proyectos-que-vienen-en-2022-para-avanzar-en-desalinizacion/. Access date March 20, 2022.

Aranceta-Garza, F., Arreguín-Sánchez, F., Ponce-Díaz, G. and Seijo, J.C. 2016. Natural mortality of three commercial penaeid shrimps (Litopenaeus vannamei, L. stylirostris and farfantepenaeus californiensis) of the gulf of california using gnomonic time divisions. Scientia Marina 80: 199–206. https://doi.org/10.3989/scimar.04326.29A.

Archibald, K.M., Siegel, D.A. and Doney, S.C. 2019. Modeling the impact of zooplankton diel vertical migration on the carbon export flux of the biological pump. Global Biogeochemical Cycles 33: 181–199. https://doi.org/10.1029/2018GB005983.

Bakun, A., Black, B.A., Bograd, S.J., Garcia-Reyes, M., Miller, A.J., Rykaczewski, R.R. et al. 2015. Anticipated effects of climate change on coastal upwelling ecosystems. Current Climate Change Reports 1: 85–93. https://doi.org/10.1007/s40641-015-0008-4.

Barnthouse, L.W. 2004. Extrapolating impingement and entrainment losses to equivalent adults and production foregone. Electrical Power Research Institute, Report, 1008471.

Baten, R. and Stummeyer, K. 2013. How sustainable can desalination be? Desalin. Water Treat. 51: 44–52. https://doi.org/10.1080/19443994.2012.705061.

Beaugrand, G., Reid, P.C., Ibanez, F., Lindley, J.A. and Edwards, M. 2002. Reorganization of North Atlantic marine copepod biodiversity and climate. Science 296: 1692–1694. https://doi.org/10.1126/science.1071329.

Beaugrand, G., Brander, K.M., Alistair Lindley, J., Souissi, S. and Reid, P.C. 2003. Plankton effect on cod recruitment in the North Sea. Nature 426: 661–664. https://doi.org/10.1038/nature02164.

Brierley, A. 2017. Plankton. Current Biology 27: 478–483. https://doi.org/10.1016/j.cub.2017.02.045.

Caddy, J.F. and Sharp, G.D. 1986. An Ecological Framework for Marine Fishery Investigations. F.A.O. Fish. Tech. Pap., Rome.

Cárcamo, P., Henríquez-Antipa, L., Galleguillos, F. and Figueroa-Fabrega, L. 2021. Marine stocking in Chile: a review of past progress and future opportunities for enhancing marine artisanal fisheries. Bulletin of Marine Science -Miami-. https://doi.org/10.5343/bms.2020.0052.

Craig, K. 2015. Sydney and gold coast desalination plant intake design, construction and operating experience, Chapter 3. pp. 39–56. *In*: Missimer, T.M., Jones, B. and Maliva, R.G. (eds.). Intakes and Outfalls for Seawater Reverse-osmosis Desalination Facilities: Innovations and Environmental Impacts. Springer International Publishing, Switzerland.

Dam, H.G. 2013. Evolutionary adaptation of marine zooplankton to global change. Annual Review of Marine Science 5: 349–370. https://doi.org/10.1146/annurev-marine-121211-172229.

Dam, H.G. and Baumann, H. 2018. Climate change, zooplankton and fisheries. pp. 851–874. *In*: Phillips, B.F. and Pérez-Ramírez, M. (eds.). Climate Change Impacts on Fisheries and Aquaculture: A Global Analysis. 2. https://doi.org/10.1002/9781119154051.ch25.

Darre, N.C. and Toor, G.S. 2018. Desalination of water: a review. Current Pollution Reports 4: 104–111. https://doi.org/10.1007/s40726-018-0085-9.

Dawoud, M.A. and Al Mulla, M.M. 2012. Environmental impacts of seawater desalination. Arabian Gulf case study, Int. J. Environ. Sustain. 1: 22–37. https://doi.org/10.24102/ijes.v1i3.96.

Escribano, R. and Hidalgo, P. 2000. Spatial distribution of copepods in the north of the humboldt current region off chile during coastal upwelling. Journal of the Marine Biological Association of the UK 80: 283–290. https://doi.org/10.1017/S002531549900185X.

Escribano, R. and Hidalgo, P. 2001. Circulación inducida por el viento en bahía de antofagasta, norte de chile (23°S). Revista de Biología Marina y Oceanografía 36: 43–60. http://dx.doi.org/10.4067/S0718-19572001000100005.

Escribano, R., Rosales, S. and Blanco, J. 2004. Understanding upwelling circulation off antofagasta (northern chile): a three-dimensional numerical-modeling approach. Continental Shelf Research 24: 37–53. https://doi.org/10.1016/j.csr.2003.09.005.

Escribano, R., Hidalgo, P., González, H., Giesecke, R., Riquelme-Bugueno, R. and Manríquez, K. 2007. Seasonal and inter-annual variation of mesozooplankton in the coastal upwelling zone off central-southern chile. Progress in Oceanography 75: 470–485. https://doi.org/10.1016/j.pocean.2007.08.027.

Escribano, R., Hidalgo, P. and Krautz, C. 2009. Zooplankton associated with the oxygen minimum zone system in the northern upwelling region of chile during march 2000. Deep Sea Research Part II: Topical Studies in Oceanography 56: 1083–1094. https://doi.org/10.1016/j.dsr2.2008.09.009.

Escribano, R., Hidalgo, P., Fuentes, M. and Donoso, K. 2012. Zooplankton time series in the coastal zone of chile: variation in upwelling and responses of the copepod community. Progress in Oceanography 97: 174–186. https://doi.org/10.1016/j.pocean.2011.11.006.

Escribano, R., Hidalgo, P., Valdés, V. and Frederick, L. 2013. Temperature effects on development and reproduction of copepods in the humboldt current: the advantage of rapid growth. Journal of Plankton Research 36: 104–116. https://doi.org/10.1093/plankt/fbt095.

Fonseca, T.R. and Farías, M. 1987. Estudio del proceso de surgencia en la costa chilena utilizando percepción remota. Investigaciones Pesqueras 34: 33–46.

Gaymer, C.F., Palma, A.T., Vega, J.M.A., Monaco, C.J. and Henríquez, L.A. 2010. Effects of la niña on recruitment and abundance of juveniles and adults of benthic community-structuring species in northern chile. Mar. Freshwater Res. 61: 1185–1196. https://doi.org/10.1071/MF09268.

Greer, A.T., Woodson, C.B., Guigand, C.M. and Cowen, R.K. 2016. Larval fishes utilize batesian mimicry as a survival strategy in the plankton, marine ecology progress series, 10.3354/meps11751. 551: 1–12. https://doi.org/10.3354/meps11751.

Henríquez, L.A., Daneri, G., Muñoz, C., Montero, P., Veas, R. and Palma, A.T. 2007. Primary production and phytoplanktonic biomass in shallow marine environments of central chile: effect of coastal geomorphology. Estuarine, Coastal and Shelf Sciences 73: 137–147. https://doi.org/10.1016/j.ecss.2006.12.013.

Hernández, E.H., Palma, A.T. and Ojeda F.P. 2003. Coastal assemblages of ichthyoplankton. the importance of near shore environments as nursery grounds for distinctive fish species. Estuarine, Coastal and Shelf Sciences 56: 1–18.

Hidalgo, P., Escribano, R., Vergara, O., Jorquera, E., Donoso, K. and Mendoza, P. 2010. Patterns of copepod diversity in the chilean coastal upwelling system. Deep Sea Research Part II: Topical Studies in Oceanography 57: 2089–2097. 10.1016/j.dsr2.2010.09.012.

Hogan, T.W. 2015. Impingement and entrainment at SWRO desalination facility intakes. pp. 57–78. *In*: Missimer, T., Jones, B. and Maliva, R. (eds.). Intakes and Outfalls for Seawater Reverse-Osmosis Desalination Facilities. Environmental Science and Engineering. Springer, Switzerland.

Holt, G.J. and Holt, S.A. 2000. Vertical distribution and the role of physical processes in the feeding dynamics of two larval sciaenids sciaenops ocellatus and cynoscion nebulosus. Mar. Ecol. Prog. Ser. 193: 181–190. https://doi.org/10.3354/meps193181.

Ihsanullah. 2019. Carbon nanotube membranes for water purification: developments, challenges, and prospects for the future. Sep. Purif. Technol. 209: 307–337. https://doi.org/10.1016/j.seppur.2018.07.043.

Ihsanullah, I., Atieh, M.A., Sajid, M. and Nazal, M.K. 2021. Desalination and environment: a critical analysis of impacts, mitigation strategies, and greener desalination technologies. Science of the Total Environment 780: 1–18. https://doi.org/10.1016/j.scitotenv.2021.146585.

International Desalination Association [WWW Document]. 2021. URL. https://idadesal.org/ (accessed 7.17.22).

IPCC. 2018. Global warming of 1.5°C, an IPCC special report on the impacts of global warming of 1.5°C above pre-industrial levels and related global greenhouse gas emission pathways, in the context of strengthening the global response to the threat of climate change, sustainable development, and efforts to eradicate poverty. Intergovernmental Panel on Climate Change.

Jones, E., Qadir, M., van Vliet, M.T.H., Smakhtin, V. and Kang, S. 2019. The state of desalination and brine production: a global outlook. Sci. Total Environ. 657: 1343–1356. https://doi.org/10.1016/j.scitotenv.2018.12.076.

Kobari, T., Shinada, A. and Tsuda, A. 2003. Functional roles of interzonal migrating mesozooplankton in the western subarctic Pacific. Progress in Oceanography 57: 279–298. https://doi.org/10.1016/S0079-6611(03)00102-2.

Krautz, M.C., Hernández-Miranda, E., Veas, R, Anabalón, V. and Quiñones, R. 2021. Fito y zooplancton en programas de monitoreo costero: la necesidad de vincular su diversidad y abundancia con estimaciones de su estado vital. pp. 289–316. En: Castilla, J.C., Fariña, J.M. and Camaño, A. (eds.). Programas de monitoreo del medio marino costero: diseños experimentales, muestreos, métodos de análisis y estadística asociada. Ediciones Universidad Católica. Santiago, Chile.

Lawler, J.P., Hogarth, W.T., Copeland, B.J., Weinstein, M.P., Hodson, R.G. and Chen, H.Y. 1981. Techniques for assessing the impact of entrainment and impingement as applied to the Brunswick Steam Electric Plant. pp. 159–198. *In*: Jensen, L.D. (ed.). Issues Associated with Impact Assessment. EA Communications, Inc., Sparks, MD.

Letelier, J., Pizarro, O. and Núñez, S. 2009. Seasonal variability of coastal upwelling and the upwelling front off central chile. Journal of Geophysical Research 114. https://doi.org/10.1029/2008JC005171.

López, S., Turon, X., Montero, E., Palacín, C., Duarte, C.M. and Tarjuelo, I. 1998. Larval abundance, recruitment and early mortality in paracentrotus lividus (Echinoidea). Interannual variability and plankton-benthos coupling. Mar. Ecol. Prog. Ser. 172: 239–251. https://doi.org/10.3354/meps172239.

Lyczkowski-Shultz, J. and Steen Jr, J.P. 1991. Diel vertical distribution of red drum *Sciaenops ocellatus* larvae in the northcentral Gulf of Mexico. Fish. Bull. 89: 631–641.

Mackas, D.L., Batten, S. and Trudel, M. 2007. Effects on zooplankton of a warmer ocean: recent evidence from the northeast pacific. Progress in Oceanography 75: 223–252. https://doi.org/10.1016/j.pocean.2007.08.010.

Maliva, R. and Missimer, T.M. 2012. Desalination: desalination in arid lands. *In*: Arid Lands Water Evaluation and Management. Environmental Science and Engineering. Springer, Berlin, Heidelberg. https://doi.org/10.1007/978-3-642-29104-3_27.

McConaugha, J.R. 1992. Decapod larvae: dispersal, mortality, and ecology: a working hypothesis. Amer. Zool. 32: 512–523. https://doi.org/10.1093/icb/32.3.512.

McKinnon, A.D., Richardson, A.J., Burford, M.A. and Furnas, M.J. 2007. Vulnerability of great barrier reef plankton to climate change. pp. 122–152. *In*: Johnson, J.E. and Marshall, P.A. (eds.). Climate Change and the Great Barrier Reef: A Vulnerability Assessment. Great Barrier Reef Marine Park Authority and the Australian Greenhouse Office, Australia.

Medellín-Mora, J., Escribano, R. and Schneider, W. 2016. Community response of zooplankton to oceanographic changes (2002–2012) in the central/southern upwelling system of chile. Progress in Oceanography 142: 17–29. https://doi.org/10.1016/j.pocean.2016.01.005.

Missimer, T.M., Jones, B. and Maliva, R.G. (eds.). 2015. Intakes and Outfalls for Seawater Reverse Osmosis Desalination Facilities: Innovations and Environmental Impacts. Springer, New York.

Missimer, T.M. 2016. The california ocean plan: does it eliminate the development of large-capacity SWRO plants in california? Proceedings of the American Membrane Technology Association Conference and Exposition, San Antonio, Texas 17: 1–5.

Missimer, T.M. and Maliva, R.G. 2018. Environmental issues in seawater reverse osmosis desalination: intakes and outfalls. Desalination 434: 198–215. https://doi.org/10.1016/j.desal.2017.07.012.

Mujica, A. and Espinoza, E. 1994. Cladóceros marinos chilenos (18°30′–37°30′S). Revista chilena de Historia Natural 67: 265–272.

Otto, R.G. 1989. Mitigative Alternatives for Entrainment Losses at the Potomac Electric Power Company Chalk Point Station. Prepared for Potomac Electric Power Company, Washington, D.C.

Palma, A.T., Steneck, R.S. and Wilson, C. 1999. Settlement-driven, multiscale demographic patterns of large benthic decapods in the gulf of maine. Journal of Experimental Marine Biology and Ecology 241: 107–136.

Palma, A.T., Pardo, L.M., Veas, R.I., Cartes, C., Silva, M., Manriquez, K. et al. 2006. Coastal brachyuran decapods: settlement and recruitment under contrasting coastal geometry conditions. Marine Ecology Progress Series 316: 139–153. https://doi:10.3354/meps316139.

Palma, A.T., Henríquez, L.A. and Ojeda, F.P. 2009. Phytoplanktonic primary production in a highly dynamic environment in central chile modulated by coastal geomorphology: a short-term effect. Revista de Biología Marina y Oceanografía 44: 325–334. https://dx.doi.org/10.4067/S0718-19572009000200006.

Palma, A.T., Torreblanca, M.L. and San Martín, B. 2021. Diseño de muestreo y análisis de información oceanográfica biológica para programas de monitoreo del medio marino costero. pp. 219–231. Castilla, J.C., Fariña, J.M. and Camaño, A. (eds.). Programas de Monitoreo del Medio Marino Costero: Diseños Experimentales, Muestreos, Métodos de Análisis y Estadística Asociada. Ediciones Universidad Católica. Santiago, Chile.

Pankratz, T. 2015. Overview of intake systems for seawater reverse osmosis facilities, chapter 1. pp. 3–17. *In*: Missimer, T.M., Jones, B. and Maliva, R.G. (eds.). Intakes and Outfalls for Seawater Reverse-osmosis Desalination Facilities: Innovations and Environmental Impacts. Springer International Publishing, Switzerland.

Pineda, J. 1994. Spatial and temporal patterns in barnacle settlement along a southern California rocky shore. Marine Ecology Progress Series 107: 125–138. https://doi.org/10.3354/meps107125.

Pinto, F.S. and Marques, R.C. 2017. Desalination projects economic feasibility: a standardization of cost determinants. Renew. Sust. Energ. Rev. 78: 904–915. https://doi.org/10.1016/j.rser.2017.05.024.

Poulin, E., Palma, A.T., Leiva, G., Hernández, E., Martinez, P., Navarrete, S. and Castilla, J.C. 2002. Temporal and spatial variation in the distribution of premetamorphic larvae of *concholepas concholepas* (gastropoda: muricidae) along the central coast of chile. Marine Ecology Progress Series 229: 95–104. https://doi.org/10.3354/MEPS229095.

Poulin, E., Palma, A.T., Leiva, G., Narvaez, D., Pacheco, R., Navarrete, S.A. et al. 2002. Avoiding offshore transport of competent larvae during upwelling events: the case of the gastropod concholepas concholepas in central chile. Limnology and Oceanography 47: 1248–1255. https://doi.org/10.4319/lo.2002.47.4.1248.

PSEG. 1999. Salem Generating station, permit renewal application NJPDES No. NJ0005622. Newark, NJ.

PSEG. 2001. Mercer generating station 316(b) demonstration. PSEG Fossil LLC, 80 Park Plaza, Newark, NJ.

Punt, A.E., Castillo-Jordán, C., Hamel, O., Cope, J.M. and Maunder, M. 2021. Consequences of error in natural mortality and its estimation in stock assessment models. Fisheries Research 233: 105759. https://doi.org/10.1016/j.fishres.2020.105759.

Richardson, A.J. 2008. In hot water: zooplankton and climate change. ICES Journal of Marine Science 65: 279–295. https://doi.org/10.1093/icesjms/fsn028.

Rivera, J., Guzman, G. and Palma, S. 2019. Cross-shelf distribution of decapod larvae in a coastal upwelling zone of northern chile: some oceanographic implications. Continental Shelf Research 181: 50–71. https://doi.org/10.1016/j.csr.2019.05.006.

Roberts, D.A., Johnston, E.L. and Knott, N.A. 2010. Impacts of desalination plant discharges on the marine environment: a critical review of published studies. Water Res. 44: 5117–5128. https://doi.org/10.1016/j.watres.2010.04.036.

Roemmich, D. and McGowan, J. 1995. Climatic warming and the decline of zooplankton in the California Current. Science 267: 1324–1326. https://doi.org/10.1126/science.267.5202.1324.

Rumrill, S.S. 1990. Natural mortality of marine invertebrate larvae. Ophelia 32: 163–198. https://doi.org/10.1080/00785236.1990.10422030.

Steinbeck, J., Hedgpeth, J., Raimondi, P., Cailliet, G. and Mayer, D. 2007. Assessing power plant cooling water intake system entrainment impacts. CEC Report 700-2007-010. California Energy Commission, Sacramento.

Steinberg, D.K. and Landry, M.R. 2017. Zooplankton and the ocean carbon cycle. Annual Review of Marine Science 9: 413–444. https://doi.org/10.1146/annurev-marine-010814-015924.

Strange, E., Allen, D., Mills, D. and Raimondi, P. 2004. Research on estimating the environmental benefits of restoration to mitigate or avoid environmental impacts caused by California power plant cooling water intake structures. CEC Report 500-04-092. California Energy Commission, Sacramento.

Sydeman, W.J., García-Reyes, M., Schoeman, D.S., Rykaczewski, R.R., Thompson, S.A., Black, B.A. et al. 2014. Climate change and wind intensification in coastal upwelling ecosystems. Science 345: 77–80. https://doi.org/10.1126/science.1251635.

Tenera. 2000. Diablo canyon power plant: 316(b) demonstration report. Prepared for Pacific Gas and Electric Company, 77 Beale St., San Francisco, CA.

Teow, Y.H. and Mohammad, A.W. 2019. New generation nanomaterials for water desalination: a review. Desalination 451: 2–17. https://doi.org/10.1016/j.desal.2017.11.041.

Thiel, M., Castilla, J.C., Fernández Bergia, M.E. and Navarrete, S. 2007. The humboldt current system of northern and central chile. pp. 195–344. *In*: Gibson, R.N., Atkinson R.J.A. and Gordon, J.D.M. (eds.). Oceanography and Marine Biology: An Annual Review. Florida.

Thorson, G. 1946. Reproduction and larval development of danish marine bottom invertebrates. Medd. Komm. Dan. Fisk. Havunders. Ser. Plankt. 4: 1–523.

Torreblanca, M.L., Pérez-Santos, I., San Martín, B., Varas, E., Zilleruelo, R., Riquelme-Bugueño, R. et al. 2016. Seasonal dynamics of zooplankton in a northern chile bay exposed to upwelling conditions. Revista de Biología Marina y Oceanografía 51: 273–291. http://dx.doi.org/10.4067/S0718-19572016000200006.

Underwood, A.J. and Keough, M.L. 2001. Supply-side ecology: the nature and consequences of variations in recruitment of intertidal organisms. pp. 183–200. *In*: Bertness, M.D., Gaines, S.D. and Hay, M.E. (eds.). Marine Community Ecology. Sinauer Associates, Sunderland MA, USA.

UNESCO. 2018. World Water Assessment Programme: The United Nations World Water Development Report 2018: Nature-Based Solutions for Water; Facts and Figures. Perugia, Italy.

UNESCO, UN-Water. 2020. United Nations World Water Development Report 2020: Water and Climate Change. Paris, France.

USEPA. 2002. Case study analysis for the proposed section 316(b) phase II existing facilities rule. EPA 821-R-02-002. Office of Water, U.S. Environmental Protection Agency, Washington, D.C. 20460.

Vargas, C.A., Garreaud, R., Barra, R., Vásquez-Lavin, F., Saldías, G.S. and Parra, O. 2020. Environmental costs of water transfers. Nature Sustainability 3: 408–409. https://doi.org/10.1038/s41893-020-0526-5.

Vaughn, D. and Allen, J.D. 2010. The peril of the plankton. Integrative and Comparative Biology 50: 1–19. https://doi.org/10.1093/icb/icq037.

Vega, J.M.A., Caillaux, L., Valdebenito, M. and Bravo, J. 2020. Composition and abundance of zooplankton in the bilge of the seawater intake of a desalination plant in caldera, region of atacama, chile. Gestión Ambiental 38: 5–18.

WateReuse Association. 2011. Desalination plant intakes: impingement and entrainment impacts and solutions. WateReuse Association White Paper. https://doi.org/10.13140/RG.2.1.2604.5848.

York, R. and Foster, M.S. 2005. Issues and environmental impacts associated with once through cooling at California's coastal power plants. CEC Report 700-2005-013 + Appendices (CEC 700-2005-013-AP-A). California Energy Commission, Sacramento.

1.3

Climate Change Challenges on Coastal Environments
Physical Processes in Upwelling and Estuarine Systems

*Carina Lurdes Lopes, Magda Sousa,
Ana Picado* and *João Miguel Dias**

1. Introduction

Climate change can affect coastal environments, mainly through changes in wind patterns, air temperature and precipitation, that consequently will impact the upwelling patterns, mean sea level and river's discharge regime. In inland coastal systems, physical processes are forced by tides, freshwater inflow, wind stress and atmosphere-ocean exchange, as well as by the interaction with the adjacent shelf where coastal phenomena, such as upwelling and river plumes dispersion occur, changing the present hydrodynamics and salinity patterns. Therefore, changes in upwelling system boundaries and estuarine systems in response to climate change will affect the primary production and therefore, the whole food web. The modifications expected for the future will influence the coastal circulation, salinity and water temperature patterns, which are important physical factors for the organisms living in these regions. Therefore, this chapter aims to describe a general review of the climate change impact on the physical and biogeochemical processes driving the main coastal regions, using as case studies the Iberian Upwelling System (IUS) and the Portuguese estuarine systems, which are critically endangered by the mean sea level (MSL) rise (Figure 1).

CESAM, Centre for Environmental and Marine Studies, Physics Department, University of Aveiro, Campus de Santiago, 3810–193, Aveiro, Portugal.
Emails: carinalopes@ua.pt; mcsousa@ua.pt; ana.picado@ua.pt
* Corresponding author: joao.dias@ua.pt

Figure 1. Iberian Peninsula coast with the location of the estuarine systems under study. The red line indicates the location of the cross-section used to compute the upwelling index and the Brunt-Väisälä frequency.

2. Upwelling Systems

The Eastern Boundary Upwelling Systems (EBUS) includes Benguela, Canary, Humboldt (Chile and Peru) and California current systems (Chavez and Messié 2009, García-Reyes et al. 2015, Sydeman et al. 2014). The EBUS are among the most productive of the world's ecosystems, covering a small ocean area (< 1%) but contributing to 20% of the global fish catches (Pauly and Christensen 1995), representing, therefore, areas with substantial ecological and socioeconomic impact to coastal communities. The atmospheric large-scale processes that induce equatorward winds force an offshore Ekman transport at the surface layer and, consequently, due to mass conservation, the rising of cold, nutrient-rich waters with low pH and dissolved oxygen concentrations and high dissolved CO_2 concentrations (Huyer 1983). At the surface layers, where light is available, these rich waters boost primary production by phytoplankton, generating rich areas in terms of biological abundance.

Consequently, owing to their ecological and economic importance, these systems have been extensively studied in terms of fisheries and biological productivity (Gutknecht et al. 2013, Lachkar and Gruber 2011, Pauly and Christensen 1995, Rossi et al. 2009), physical processes description (Brink et al. 1983, Gutknecht et al. 2013, Pitcher et al. 2010), upwelling trends (Narayan et al. 2010, Patti et al. 2008, Seo et al. 2012, Varela et al. 2015) or distribution of marine biodiversity (Kämpf and Chapman 2016).

In the past, Bakun (1990) hypothesized a strengthening of coastal upwelling intensity, as a global warming consequence (intensification of the land-sea thermal contrast), which would result in the ocean surface cooling. However, recently contradictory results were obtained by Bakun et al. (2015), Barton et al. (2013), García-Reyes and Largier (2010), Gutiérrez et al. (2011) and Varela et al. (2015).

According to Sydeman et al. (2014), who analyzed up to 20 studies, winds have intensified in the California, Benguela and Humboldt upwelling systems and weakened in the Iberian system over time scales ranging up to 60 years. This study emphasized a high dependence on the selected region, the season analyzed, the length of the data used and the database as well.

In both cases, impacts on marine ecosystems are expected. On the one hand, stronger upwelling-favorable winds will reinforce the input of nutrients at the surface; however, the strong winds will enhance surface offshore advection (Bakun et al. 2010, 2015, Cury and Roy 2011), upper water column mixing and light limitation (Cury and Roy 2011, Yáñez et al. 2001) that may affect phytoplankton productivity. Also, the increased thermal stratification may limit the nutrient surface supply. On the other hand, weaker upwelling-favorable winds may limit the rising of cold and nutrient-rich water to the euphotic zone (Chhak and Di Lorenzo 2007), which may impact marine ecosystems, inducing latitudinal displacements of phytoplankton and zooplankton populations toward cooler regions (Richardson and Richardson 2008, Richardson and Schoeman 2004).

Upwelling-favorable winds are sensitive to climate variability (Macias et al. 2012, Montecinos et al. 2003) and, therefore, are highly likely to be affected by global warming (Bakun et al. 2015). The comprehension of future upwelling trends in upwelling regions has become important considering its potential and possible changes in marine ecosystems and socioeconomic characteristics. The results from the last Intergovernmental Panel on Climate Change (IPCC) report, indicate that by the end of the 21st century, the global mean surface temperature will increase, resulting in global ocean warming worldwide (Lee et al. 2021). However, this warming is not homogeneous along the ocean (Ruela et al. 2020), and different rates may be found for offshore regions or coastal areas located in upwelling systems. For instance, upwelling intensification may result in lower warming rates in coastal locations compared to offshore. Several works based on historical data confirm the existence of different warming rates at oceanic and coastal locations in areas where coastal upwelling plays a key role. For instance, Gómez-Gesteira et al. (2008) analyzed coastal warming using satellite-derived sea surface temperature (SST) along the continental Atlantic Arc from 1985 to 2005, detecting an inhomogeneous warming trend, ranging from 3.5°C/century at latitudes close to 48°N to 1.2°C/century at latitudes close to 37°N. More recently, Varela et al. (2018) carried out a global analysis of temperature trends using the NOAA OISST ¼ high-resolution dataset between 1982 and 2015 for the coastline of the continents. These authors found significant warming over the majority of the locations, with differences between upwelling and non-upwelling areas. In upwelling regions, for 92% of the locations, the coastal warming trend is lower than the ocean, whereas in the non-upwelling regions, this percentage decreased to 58%. Seabra et al. (2019)

performed a similar study and observed a reduced nearshore warming associated with the EBUS.

Stratification is also key to understanding the link between climate change and biology since intensified stratification negatively correlates with net primary production (Behrenfeld et al. 2006). As a consequence of global warming, the upper ocean has warmed considerably over the last century and projections show that it will keep warming over the next one. This fact will contribute to an increase in ocean stratification, which can modify the behavior of the upper ocean in different ways. First, temperature anomalies in the upper ocean will not penetrate along the water column, remaining confined near the surface. Second, enhanced stratification will hinder, or at least decrease, nutrient exchange through vertical mixing. However, it should be kept in mind that coastal warming could increase thermal stratification and render upwelling less effective (García-Reyes et al. 2015, Gruber 2011). There is limited research focused on the impact of global warming on the stratification of coastal upwelling systems, which relies on good horizontal resolution close to the shore.

2.1 *Iberian Peninsula Upwelling System*

The western coast of the Iberian Peninsula (IP) is the northern boundary of the Canary Upwelling Ecosystem (CUE), which is one of the four major EBUS of the world (Pelegrí et al. 2005). This is a region of pronounced hydrologic and biogeochemical activity due to phytoplankton blooms associated with coastal upwelling (Fraga 1981, Peliz et al. 2002, Relvas et al. 2007, Tenore et al. 1995). Previous works show that coastal upwelling in the region shows a well-defined seasonality (Fraga 1981, Picado et al. 2016, Wooster et al. 1976), with active and persistent conditions prevailing from June to September. Nevertheless, upwelling events may change in strength (Alvarez et al. 2008, 2011, Gómez-Gesteira et al. 2006) under different atmospheric conditions (especially those changing the wind stress intensity), impacting the ecosystem through changes in SST and phytoplankton patterns. The upwelling events throughout the IUS have been characterized in terms of their physical processes (Alvarez-Salgado et al. 1993, Cordeiro et al. 2018, Fiuza et al. 1982, Peliz et al. 2002), and their associated biological consequences (Arístegui et al. 2009, Queiroga et al. 2007, Santos et al. 2005). Indeed, this region is one of the primary spawning and recruitment areas for sardine (Carrera and Porteiro 2003, Marques et al. 2005, Santos et al. 2001), horse mackerel (Murta et al. 2008, Santos et al. 2001) and several marine invertebrates (Dos Santos et al. 2008) and therefore has attracted large interest in the scientific community. The IP coast is also influenced by numerous estuarine outflows, which transport freshwater and sediment from rivers that sustain coastal ecosystems, and carry nutrients that boost biological productivity.

In the last decades, several studies focusing on chlorophyll-a variability throughout the IP coast have been carried out, either based on data collected from cruises (Castro et al. 2000, Varela et al. 2005), from remote sensing (Alvarez et al. 2011, Picado et al. 2013, 2016), or generated by numerical models (Reboreda et al. 2014). In most of these studies, coastal upwelling and continental runoff appeared as the main driving forces of pelagic primary productivity and phytoplankton composition. According

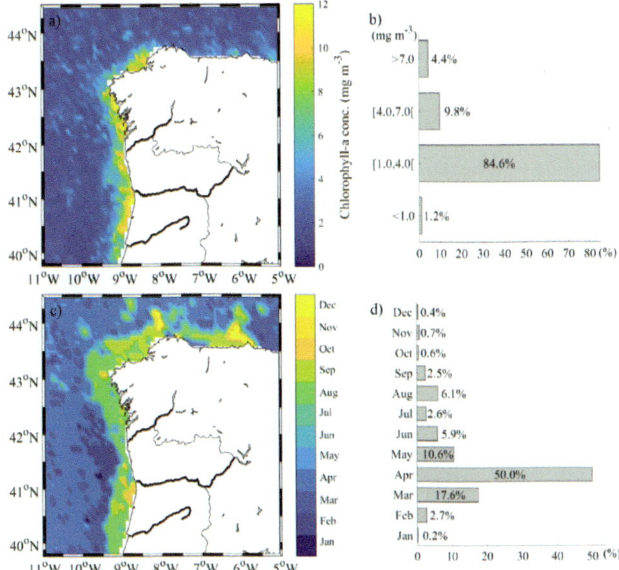

Figure 2. (a) Maximum chlorophyll concentration (mg m^{-3}) and (c) correspondent period of occurrence during 1998–2007. (b) and (d) represent the percentage of occurrence of (a) and (b), respectively.

to a recent study (Picado 2016) that used satellite-derived data (Sea-viewing Wide Field-of-View Sensor), the highest chlorophyll-a concentrations (> 4 mg m^{-3} Figure 2a and 2b) occur during spring-summer months in the western IP coastal region (Figure 2c and 2d) and during spring and autumn in the north (Figure 2c and 2d), triggered by the upwelling events. However, during the winter, the river discharges also play an important role in the delivery of nutrients to the coast, promoting high primary production (Figure 2c and 2d).

Future projections along the major coastal upwelling regions worldwide, predict significant changes over the next century, especially at high latitudes (DeCastro et al. 2016, Sousa et al. 2017, Wang et al. 2015). These are particularly noticeable in the northwestern IP (Sousa et al. 2017), where was found the intensification of the upwelling index from historical to future climates (Figure 3a). These changes are induced by the migration and intensification of the Azores High (Rykaczewski et al. 2015, Sousa et al. 2017).

Relvas et al. (2009) and Santos et al. (2005, 2012a) detected weaker warming trends at coastal locations along the CUE. This pattern has also been observed in other areas worldwide (Santos et al. 2012b, 2016, Varela et al. 2015). The relevance of coastal upwelling is twofold since apart from pumping nutrients to the surface, which means an increase in primary production by phytoplankton, upwelling can also attenuate global warming in coastal areas (Freitas et al. 2009, Santos et al. 2011, 2012a, Seabra et al. 2019, Varela et al. 2018). However, the increased supply of nutrient-rich and oxygen-poor deep waters will generate conditions to increase hypoxia that may affect the coastal ocean food web (Bakun et al. 2015, Grantham et al. 2004).

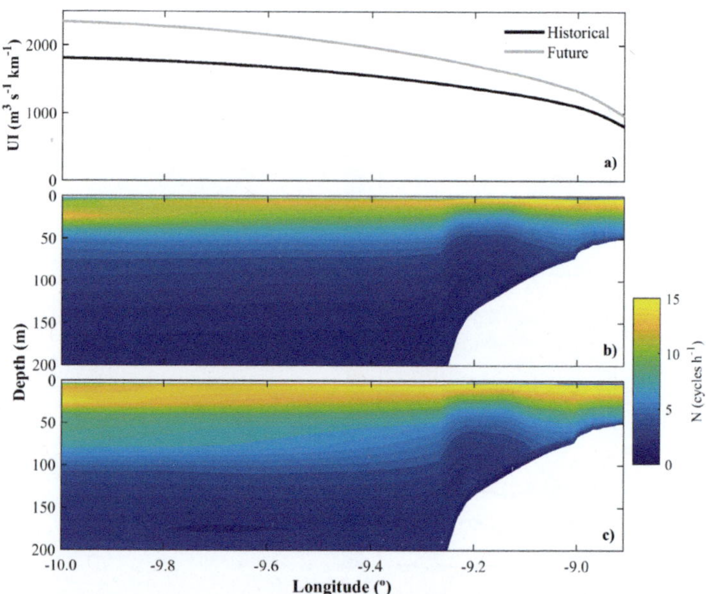

Figure 3. Upwelling index (a). Predicted water column stratification for the historical (b) and future periods (c) for the longitudinal section represented in Figure 1.

Previous studies have analyzed the impact of climate change along the IP coast using basin-scale or global circulation models (Miranda et al. 2013, Pires et al. 2013, 2016), although they were only focused on the shelf. They analyzed upwelling changes based on the IPCC A2 emission scenario and indicated an increase in coastal upwelling, supporting the idea initially proposed by Bakun (1990). However, the typical spatial resolution of climate models for these studies was on the order of 50–100 km, which could be insufficient for resolving upwelling patterns in localized areas. Recent results obtained through the development and exploitation of a more realistic model configuration, including the interaction between shelf and estuarine processes (Sousa et al. 2020) showed that coastal upwelling will be less effective due to the increase in the stratification of the upper layers (Figure 3b and 3c) caused by sea surface warming, reducing the surface input of nutrients. These findings are in agreement with Gutiérrez et al. (2011), who also observed a negative impact on productivity due to ocean warming. Similarly, Roemmich and McGowan (1995) observed the deepening of the thermocline, which resulted in the decline of zooplankton production in the California Current.

3. MSL Rise and General Implications for Coastal Regions

Global MSL rise is predominantly driven by thermal expansion and ocean mass gain due to melting ice from mountain glaciers and the Antarctic and Greenland ice sheet (Church et al. 2013, Fox-Kemper et al. 2021, Frederikse et al. 2020, Oppenheimer et al. 2019). Changes in global land-water storage (e.g., water stored in dams and reservoirs, river runoff, groundwater extraction, etc.), also affect global MSL, but to a lesser extent (Fox-Kemper et al. 2021, Oppenheimer et al. 2019). Analysis of sea level data recorded over the period 1901–1990, revealed that global MSL increased at a rate of 1.35 mm/year (Fox-Kemper et al. 2021). The thermal expansion and ocean mass gain contributed by 32% and 81%, respectively, to this increase, while the global land-water storage contributed to a decrease of 14% (Fox-Kemper et al. 2021). Analysis of more recent data, for the periods 1993–2018 and 2006–2018, revealed that the global MSL rise rate increased to 3.25 mm/year and 3.69 mm/year, respectively, and that this acceleration was in great part attributed to the melting of glaciers and ice sheets (Fox-Kemper et al. 2021, Oppenheimer et al. 2019). It is consensual that the global MSL will continue to rise throughout the 21st century; however, there is also a great uncertainty associated with the magnitude of global MSL rise. Global MSL is projected to increase between 0.18 and 0.23 m by 2050 under Shared Socioeconomic Pathway (SSP) scenarios SSP1-1.9 and SSP5-8.5, respectively (Fox-Kemper et al. 2021). The differences between the projections are accentuated for 2100, where increases of 0.38 and 0.77 m under scenarios SSP1-1.9 and SSP5-8.5 are forecasted. SSP1-1.9, assumes very strong mitigation and is aligned with the 1.5°C goal of the Paris Agreement, while SSP5-8.5 is characterized by low mitigation and high adaptation (Chen et al. 2021).

Although the generalized increase in MSL across the planet, it is important to note that the ocean surface varies spatially and therefore MSL is not changing at the same rate globally. Several factors influence the spatial variations of MSL, including changes in ocean dynamics. Climate changes and climate variability modify ocean currents and, consequently, the redistribution of mass, heat and salt, which may result in considerable sea level variability (Stammer et al. 2013). For example, the weakening of the Gulf Stream transport may have contributed to enhancing the MSL rise over the northeastern US coast (Boon 2012, Ezer 2013, Sallenger et al. 2012). Changes in the Earth's gravity field induced, for example by ice melting, affect MSLs locally. As the ice sheets melt the weight of ice is reduced and, in response, the land beneath them tends to uplift. For example, in response to the melting of the Greenland ice sheet, the MSL is expected to rise less than the global average in the Baltic Sea and the northern Coast of Norway (Slangen et al. 2014). Moreover, Glacial Isostatic Adjustment (GIA) is also changing the Earth's gravity field in response to the melting of glaciers formed during the last ice age (Peltier et al. 2015, Whitehouse 2018).

The rise in the MSL is one of the most well-known consequences of global warming, given its impacts on coastal regions, including (1) more frequent and severe coastal flooding, (2) accelerated coastal erosion, (3) permanent submergence of lands, (4) loss and change of coastal ecosystems and (5) saltwater intrusion and salinization of soils (Cooley et al. 2022). The impacts of rising sea levels highly vary spatially, given the spatial variations in both MSLs and the exposure of coastal regions. For example, it is estimated that coastal flooding affects 100,000 people and causes average damage of 1.4 billion € across Europe annually (Vousdoukas et al. 2020). Furthermore, projections by 2100 show that in the absence of adaptation measures, the number of people affected may increase to 1.6 and 3.9 million, and damage can reach between 210 billion € and 1.3 trillion € (Vousdoukas et al. 2020).

Besides the negative consequences on the socioeconomy, MSL rise will also have a major impact on the hydrodynamics and salinity of the worldwide estuarine systems, potentially affecting ecosystems' functioning and underlying services (Khojasteh et al. 2021, Lopes et al. 2022). Estuarine hydrodynamics are dictated by external forcing, including sea level variations (tides, storm surges, waves, MSL, etc.) and freshwater inflows (rivers, groundwater inflows, etc.); but at the same time, it is also highly dependent on the estuarine geomorphology. The shape of estuaries (e.g., depth, width, length, inlets, etc.) induces changes in the external forcing as it propagates throughout the estuaries (Khojasteh et al. 2021). For example, tidal waves can be amplified if the effect of shoaling (depth decrease) and funneling (width decrease) exceeds the dissipation caused by increased friction. Otherwise, in highly dissipative estuaries (shallow or constituted by extensive tidal flats), the tidal wave dampens as it propagates upstream.

Attending this, MSL rise affects not only the external oceanic forcing but also the estuarine hydrodynamics. This is why it is so important to integrate precise local MSL rise projections with high-resolution numerical models when assessing the impacts of MSL rise on estuarine hydrodynamics and salinity (Lopes et al. 2022). Moreover, it is important to note that the geomorphology of estuarine systems should continue to be altered by human actions. As the MSL rises, flood protection structures will have to be built to protect settlements and economic activities hosted in the adjacent low-lying margins. Given this complexity and considering that each estuary has singular features, the effects of MSL rise on estuarine circulation and salinity highly vary among estuaries.

Previous studies revealed that the influence of MSL on tidal range depends on the inundation extent and, therefore, it could increase or decrease depending on whether flooding of adjacent regions is prevented or allowed (Ross et al. 2017). Furthermore, Lopes et al. (2022) and Zhong et al. (2008) verified that MSL rises would modify the resonance characteristics of Tagus Estuary and Chesapeake Bay, respectively, suggesting that rising sea levels would increase tidal range. Increases in the tidal amplitude and currents due to MSL rise have been identified in Grand Bay (Passeri et al. 2015, 2016) and Gironde Estuary (van Maanen and Sottolichio 2018). Additionally, Chen et al. (2016) forecasted an increase in residual currents as the MSL rises in the Pearl River Estuary, while Valentim et al. (2013) found a considerable decrease in residual circulation in the Tagus Estuary. Concerning tidal asymmetry, Passeri et al. (2015) verified that MSLR would increase the ebb dominance in Grand Bay and otherwise, Palmer et al. (2019) showed that flood dominance would decrease in Tamar Estuary.

There is consensus in the literature that MSL rise would increase saltwater intrusion and modify stratification patterns and mixing processes, profoundly affecting the horizontal and vertical distribution of salinity and water temperature within estuaries. For instance, increases in saltwater intrusion with MSL rise have been reported in Tamsui River Estuary (Chen et al. 2015), Weser Estuary (Grabemann et al. 2001), Pearl River Estuary (Hong et al. 2020), James and Chickahominy Rivers (Rice et al. 2012). Moreover, saltwater intrusion tends to be more pronounced during the dry season when the river discharges are low (Chua and Xu 2014, Liu and Liu 2014). Prandle and Lane (2015) further verified that salinity intrusion is higher in shallow than in deeper estuaries of the UK, particularly in England and Wales. Other studies have also shown that the influence of MSL rise on vertical stratification varies among and sometimes within estuaries, highlighting its dependence on river flows. Hong et al. (2020) verified that MSL rise enhances vertical stratification in the Pearl River Estuary; however, the highest increases were found during high flow conditions in the upstream estuary and low flow conditions in the adjacent bay.

3.1 Impacts of MSL Rise in Portuguese Estuaries

Analysis of tidal gauge records on the Portuguese Coast shows that the MSL has risen during the last century (Antunes 2019), and climate model-based projections evidence that this rising trend will continue throughout the end of the 21st century (Lopes et al. 2011, 2022). The most recent projections indicate that mean sea levels would rise between 0.47 m and 0.79 m on the Portuguese Coast by the end of the 21st century, under scenarios SSP1-2.6 and SSP5-8.5, respectively (https://sealevel. nasa.gov/ipcc-ar6-sea-level-projection-tool). Moreover, as the MSL hardly varies along the Portuguese Coast (Lopes et al. 2022) all Portuguese estuaries should be subjected to the same rate of increase. However, as the estuaries have distinct and unique geomorphological features, the impacts of rising sea levels could vary among estuarine systems.

Previous studies used numerical modeling to evaluate the effect of MSL rise on the salinity of the Minho (Pereira et al. 2022), Lima (Pereira et al. 2022), Tagus (Rodrigues et al. 2019) and Guadiana (Mills et al. 2020) estuaries and of the Ria de Aveiro (Vargas et al. 2017) and Ria Formosa (Rodrigues et al. 2021) coastal lagoons. The conclusions retrieved by such studies are very similar and indicate that the salinity would increase in all estuarine systems, and the upper estuarine regions would be affected by saltwater intrusion, particularly under low river discharges. Vargas et al. (2017) went further and assessed changes in Ria de Aveiro haline zonation by applying the Venice System scheme, concluding that haline zones would be displaced landward as the MSL rises, consequently affecting most of the living resources that populate this lagoon.

Recently, Lopes et al. (2022) applied hydrodynamic models to the five Portuguese estuarine systems most endangered by MSL rise (Ria de Aveiro, Mondego, Tagus, Sado and Ria Formosa) to evaluate its influence on the inundation extent. Their results (Figure 4) evidence that all estuarine systems should experience high inundation extents under rising sea levels; however, the increase rate varies among estuaries and depends on the elevation of adjacent regions and changes in the long-wave (combining tidal and storm surge waves) as it propagates within each estuarine system. Ria de Aveiro, Mondego and Tagus were found to be the estuaries experiencing the highest inundation extents; however, it is important to highlight that this study disregards future adaptation measures aiming to prevent or reduce flooding. Besides changing maximum inundation extents, MSL rise would also affect the extent of subtidal and intertidal regions. Silveira et al. (2021) found for the Ria de Aveiro lagoon that the extent of subtidal regions should increase and the area of intertidal regions should maintain with MSL rise when considering the absence of flood protection structures. Although this analysis has not been done for other

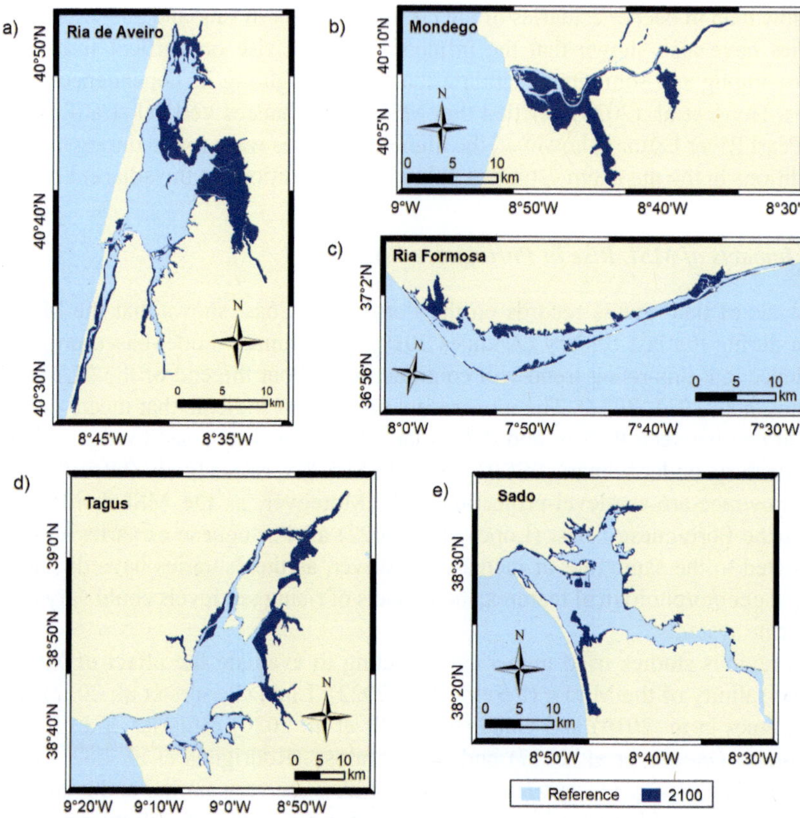

Figure 4. Inundation extents predicted by Lopes et al. (2022) for 2100 when considering MSLR and storm surges with return periods of 100 years in (a) Ria de Aveiro, (b) Mondego Estuary, (c) Ria Formosa, (d) Tagus Estuary and (e) Sado Estuary.

Portuguese estuaries, it is expected that other estuaries presenting extensive intertidal areas, such as Sado, Tagus, Mondego and Ria Formosa nowadays, should experience a similar behavior.

Previous studies based on numerical modeling have also reported changes in estuarine circulation patterns. Rising sea levels would reduce the current velocity in Douro and Guadiana estuaries (Mendes et al. 2013, Mills et al. 2020) and intensify tidal currents in Ria de Aveiro (Lopes and Dias 2015). In addition, Valentim et al. (2013) verified that residual circulation in Tagus Estuary should decrease considerably with MSLR, while Lopes and Dias (2015) found that residual currents could increase or decrease depending on the lagoon geometry.

Given that the abundance, diversity and zonation of zooplankton in estuarine environments are strictly linked with environmental factors (Appeltans et al. 2003, Chaalali et al. 2013, David et al. 2016, Marques et al. 2007, Modéran et al. 2010, Mouny and Dauvin 2002), the aforementioned environmental modifications induced by MSL rise would modify zooplankton communities in the Portuguese estuaries. For instance, given that zooplankton species are distributed according to the salinity gradient (Appeltans et al. 2003, David et al. 2016, Marques et al. 2007), the distribution of species throughout the Portuguese estuaries would modify in response to the expected increase in salinity and changes in haline zonation. Moreover, zooplankton species distribution could be changed in response to modifications in subtidal and intertidal extents following the conclusions achieved by David et al. (2016) for the Gironde Estuary, which found a clear structuration of the zooplankton community between subtidal and intertidal regions during high tide. Hydrodynamic changes induced by MSL rise could also modify estuarine zooplankton dynamics since organisms have limited locomotion and use tidal currents to move throughout estuarine regions (Kimmerer et al. 1998, 2014, Petrusevich et al. 2020, Villate 1997, Wooldridge and Erasmus 1980).

4. Final Considerations

This chapter contributed to a general view of the main biogeochemical and physical processes that influence primary production along the western IP coast at seasonal and annual scales, considering historical and future conditions. Taking advantage of the high-resolution numerical model applications and recent remote-sensing databases, more scientific work is needed to fully understand upwelling, estuarine dynamics and underlying ecosystem functioning and services. As the upwelling and estuarine circulation is highly dependent on climate dynamics and local morphological features, the impacts of climate changes on plankton communities and consequent food web are still uncertain and would vary worldwide.

5. Acknowledgements

Through national funds, we acknowledge financial support to CESAM by FCT/MCTES (UIDP/50017/2020 + UIDB/50017/2020 + LA/P/0094/2020). We thank the projection authors for developing and making the sea-level rise projections available, multiple funding agencies for supporting the development of the projections and the NASA Sea-Level Change Team for developing and hosting the IPCC AR6 Sea-Level Projection Tool.

References

Alvarez-Salgado, X.A., Rosón, G., PéRez, F.F., Pazos, Y., Alvarez-Salgado, X.A., Rosón, G. et al. 1993. Hydrographic variability off the rías baixas (NW spain) during the upwelling season. JGR 98: 14,447–14,455. DOI: 10.1029/93JC00458.

Alvarez, I., Gomez-Gesteira, M., DeCastro, M. and Dias, J.M. 2008. Spatiotemporal evolution of upwelling regime along the western coast of the iberian peninsula. Journal of Geophysical Research Oceans 113: C07020. DOI: 10.1029/2008JC004744.

Alvarez, I., Gomez-Gesteira, M., DeCastro, M., Lorenzo, M.N., Crespo, A.J.C. and Dias, J.M. 2011. Comparative analysis of upwelling influence between the western and northern coast of the iberian peninsula. Continental Shelf Research 31: 388–399. DOI: 10.1016/j.csr.2010.07.009.

Antunes, C. 2019. Assessment of sea level rise at west coast of portugal mainland and its projection for the 21st century. Journal of Marine Science and Engineering 7: 61. DOI: 10.3390/jmse7030061.

Appeltans, W., Hannouti, A., Van Damme, S., Soetaert, K., Vanthomme, R. and Tackx, M. 2003. Zooplankton in the schelde estuary (Belgium/The Netherlands). The distribution of Eurytemora affinis: Effect of oxygen? Journal of Plankton Research 25. DOI: 10.1093/plankt/fbg101.

Arístegui, J., Barton, E.D., Álvarez-Salgado, X.A., Santos, A.M.P., Figueiras, F.G., Kifani, S. et al. 2009. Sub-regional ecosystem variability in the canary current upwelling. Progress in Oceanography 83: 33–48. DOI: 10.1016/j.pocean.2009.07.031.

Bakun, A. 1990. Coastal ocean upwelling. Science 247: 198–201. DOI: 10.1126/science.247.4939.198.

Bakun, A., Field, D.B., Redondo-Rodriguez, A. and Weeks, S.J. 2010. Greenhouse gas, upwelling-favorable winds, and the future of coastal ocean upwelling ecosystems. Global Change Biology 16: 1213–1228. DOI: 10.1111/J.1365-2486.2009.02094.X.

Bakun, A., Black, B.A., Bograd, S.J., García-Reyes, M., Miller, A.J., Rykaczewski, R.R. et al. 2015. Anticipated effects of climate change on coastal upwelling ecosystems. Current Climate Change Reports 1: 85–93. DOI: 10.1007/S40641-015-0008-4/FIGURES/2.

Barton, E.D., Field, D.B. and Roy, C. 2013. Canary current upwelling: more or less? Progress in Oceanography 116: 167–178. DOI: 10.1016/J.POCEAN.2013.07.007.

Behrenfeld, M.J., O'Malley, R.T., Siegel, D.A., McClain, C.R., Sarmiento, J.L., Feldman, G.C. et al. 2006. Climate-driven trends in contemporary ocean productivity. Nature 444: 7120, 444: 752–755. DOI: 10.1038/nature05317.

Boon, J.D. 2012. Evidence of sea level acceleration at US and canadian tide stations, Atlantic coast, north America. Journal of Coastal Research 28. DOI: 10.2112/JCOASTRES-D-12-00102.1.

Brink, K.H., Halpern, D., Huyer, A. and Smith, R.L. 1983. The physical environment of the peruvian upwelling system. Progress in Oceanography 12: 285–305. DOI: 10.1016/0079-6611(83)90011-3.

Carrera, P. and Porteiro, C. 2003. Stock dynamic of the iberian sardine (Sardina pilchardus, W.) and its implication on the fishery off galicia (NW Spain). Scientia Marina 67: 245–258. DOI: 10.3989/SCIMAR.2003.67S1245.

Castro, C.G., Pérez, F.F., Álvarez-Salgado, X.A. and Fraga, F. 2000. Coupling between the thermohaline, chemical and biological fields during two contrasting upwelling events off the NW iberian peninsula. Continental Shelf Research 20: 189–210. DOI: 10.1016/S0278-4343(99)00071-0.

Chaalali, A., Chevillot, X., Beaugrand, G., David V., Luczak, C., Boët, P. et al. 2013. Changes in the distribution of copepods in the gironde estuary: a warming and marinization consequence? Estuarine, Coastal and Shelf Science 134. DOI: 10.1016/j.ecss.2012.12.004.

Chavez, F.P. and Messié, M. 2009. A comparison of eastern boundary upwelling ecosystems. Progress in Oceanography 83: 80–96. DOI: 10.1016/J.POCEAN.2009.07.032.

Chen, D., Rojas, M., Samset, B.H., Cobb, K., Diongue Niang, A., Edwards, P. et al. 2021. Framing, context, and methods. pp. 147–286. *In*: Masson-Delmotte, V., Zhai, P., Pirani, A., Connors, S.L., Péan, C., Caud, N. et al. (eds.). Climate Change 2021: The Physical Science Basis. Contribution of Working Group I to the Sixth Assessment Report of the Intergovernmental Panel on Climate Change. Cambridge University Press.

Chen, W.B., Liu, W.C. and Hsu, M.H. 2015. Modeling assessment of a saltwater intrusion and a transport time scale response to sea-level rise in a tidal estuary. Environmental Fluid Mechanics 15: 491–514. DOI: 10.1007/S10652-014-9367-Y/FIGURES/12.

Chen, Y., Zuo, J., Zou, H., Zhang, M. and Zhang, K. 2016. Responses of estuarine salinity and transport processes to sea level rise in the zhujiang (Pearl River) estuary. Acta Oceanologica Sinica 35(5): 38–48. DOI: 10.1007/S13131-016-0857-2.

Chhak, K. and Di Lorenzo, E. 2007. Decadal variations in the california current upwelling cells. Geophysical Research Letters 34. DOI: 10.1029/2007GL030203.

Chua, V.P. and Xu, M. 2014. Impacts of sea-level rise on estuarine circulation: an idealized estuary and san francisco bay. Journal of Marine Systems 139: 58–67. DOI: 10.1016/J.JMARSYS.2014.05.012.

Church, J.A., Gregory, J.M., Cazenave, A., Gregory, J.M., Jevrejeva, S., Levermann, A. et al. 2013. Sea level change. pp. 1137–1216. *In*: Stocker, T.F., Qin, D., Plattner, G-K., Tignor, M., Allen, S.K., Boschung, J. et al. (eds.). Climate Change 2013: The Physical Science Basis. Contribution of Working Group I to the Fifth Assessment Report of the Intergovernmental Panel on Climate Change. Cambridge University Press, Cambridge, United Kingdom and New York, NY, USA. DOI: 10.1017/CBO9781107415324.026.

Cooley, S., Schoeman, D., Bopp, L., Boyd, P., Donner, S., Ito, S. et al. 2022. Oceans and coastal ecosystems and their services. pp. 379–550. *In*: Pörtner, H-O., xRoberts, D.C., Tignor, M., Poloczanska, E.S., Mintenbeck, K., Alegría, A. et al. (eds.). Climate Change 2022: Impacts, Adaptation and Vulnerability.

Contribution of Working Group II to the Sixth Assessment Report of the Intergovernmental Panel on Climate Change. Cambridge University Press.

Cordeiro, N.G.F., Dubert, J., Nolasco, R. and Barton, E.D. 2018. Transient response of the northwestern Iberian upwelling regime. PLoS ONE. DOI: 10.1371/journal.pone.0197627.

Cury, P. and Roy, C. 2011. Optimal environmental window and pelagic fish recruitment success in upwelling areas. https://doi.org/10.1139/f89-086 46: 670–680. DOI: 10.1139/F89-086.

David, V., Selleslagh, J., Nowaczyk, A., Dubois, S., Bachelet, G., Blanchet, H. et al. 2016. Estuarine habitats structure zooplankton communities: implications for the pelagic trophic pathways. Estuarine, Coastal and Shelf Science 179. DOI: 10.1016/j.ecss.2016.01.022.

DeCastro, M., Sousa, M.C., Santos, F., Dias, J.M. and Gómez-Gesteira, M. 2016. How will somali coastal upwelling evolve under future warming scenarios? Scientific Reports 6(1): 1–9. DOI: 10.1038/srep30137.

Dos Santos, A., Santos, A.M.P., Conway, D.V.P., Bartilotti, C., Lourenço, P. and Queiroga, H. 2008. Diel vertical migration of decapod larvae in the portuguese coastal upwelling ecosystem: implications for offshore transport. Marine Ecology Progress Series 359: 171–183. DOI: 10.3354/MEPS07341.

Ezer, T. 2013. Sea level rise, spatially uneven and temporally unsteady: why the US East coast, the global tide gauge record, and the global altimeter data show different trends. Geophysical Research Letters 40. DOI: 10.1002/2013GL057952.

Fiuza, A., Demacedo, M. and Guerreiro, M. 1982. Climatological space and time-variation of the portuguese coastal upwelling. Oceanologica Acta 5: 31–40.

Fox-Kemper, B., Hewitt, H.T., Xiao, C., Aðalgeirsdóttir, G., Drijfhout, S.S., Edwards, T.L. et al. 2021. Ocean, cryosphere and sea level change. pp. 1211–1362. *In*: Masson-Delmotte, V., Zhai, P., Pirani, A., Connors, S.L., Péan, C., Caud, N. et al. (eds.). Climate Change 2021: The Physical Science Basis. Contribution of Working Group I to the Sixth Assessment Report of the Intergovernmental Panel on Climate Change. Cambridge University Press.

Fraga, F. 1981. Upwelling off the galician coast, northwest Spain. pp. 176–182. *In*: Richardson, F. (ed.). Coastal Upwelling. American Geophysical Union. DOI: 10.1029/CO001p0176.

Frederikse, T., Landerer, F., Caron, L., Adhikari, S., Parkes, D., Humphrey, V.W. et al. 2020. The causes of sea-level rise since 1900. Nature 584. DOI: 10.1038/s41586-020-2591-3.

Freitas, V., Costa-Dias, S., Campos, J., Bio, A., Santos, P. and Antunes, C. 2009. Patterns in abundance and distribution of juvenile flounder, platichthys flesus, in minho estuary (NW iberian peninsula). Aquatic Ecology 43: 1143–1153. DOI: 10.1007/s10452-009-9237-8.

García-Reyes, M. and Largier, J. 2010. Observations of increased wind-driven coastal upwelling off central California. Journal of Geophysical Research: Oceans 115. DOI: 10.1029/2009JC005576.

García-Reyes, M., Sydeman, W.J., Schoeman, D.S., Rykaczewski, R.R., Black, B.A., Smit, A.J. et al. 2015. Under pressure. climate change, upwelling, and eastern boundary upwelling ecosystems. Frontiers in Marine Science 2: 109. DOI: 10.3389/FMARS.2015.00109/BIBTEX.

Garner, G.G., Hermans, T., Kopp, R.E., Slangen, A.B.A., Edwards, T.L., Levermann, A. et al. 2021. IPCC AR6 Sea Level Projections. Version 20210809. https://doi.org/10.5281/zenodo.5914709.

Gomez-Gesteira, M., Moreira, C., Alvarez, I. and DeCastro, M. 2006. Ekman transport along the galician coast (northwest Spain) calculated from forecasted winds. Journal of Geophysical Research 111: C10005. DOI: 10.1029/2005JC003331.

Gómez-Gesteira, M., De Castro, M., Álvarez, I., Lorenzo, M.N., Gesteira, J.L.G. and Crespo, A.J.C. 2008. Spatio-temporal upwelling trends along the canary upwelling system (1967–2006). Annals of the New York Academy of Sciences 1146: 320–337. DOI: 10.1196/ANNALS.1446.004.

Grabemann, H.J., Grabemann, I., Herbers, D. and Müller, A. 2001. Effects of a specific climate scenario on the hydrography and transport of conservative substances in the Weser estuary, Germany: a case study. Climate Research 18: 77–87. DOI: 10.3354/CR018077.

Grantham, B.A., Chan, F., Nielsen, K.J., Fox, D.S., Barth, J.A., Huyer, A. et al. 2004. Upwelling-driven nearshore hypoxia signals ecosystem and oceanographic changes in the northeast pacific. Nature 429(6993): 749–754. DOI: 10.1038/nature02605.

Gruber, N. 2011. Warming up, turning sour, losing breath: ocean biogeochemistry under global change. Philosophical Transactions of the Royal Society A: Mathematical, Physical and Engineering Sciences 369: 1980–1996. DOI: 10.1098/RSTA.2011.0003.

Gutiérrez, D., Bouloubassi, I., Sifeddine, A., Purca, S., Goubanova, K., Graco, M. et al. 2011. Coastal cooling and increased productivity in the main upwelling zone off Peru since the mid-twentieth century. Geophysical Research Letters 38. DOI: 10.1029/2010GL046324.

Gutknecht, E., Dadou, I., Le Vu, B., Cambon, G., Sudre, J., Garçon, V. et al. 2013. Coupled physical/biogeochemical modeling including O_2-dependent processes in the eastern boundary upwelling systems: Application in the Benguela. Biogeosciences 10: 3559–3591. DOI: 10.5194/BG-10-3559-2013.

Hong, B., Liu, Z., Shen, J., Wu, H., Gong, W., Xu, H. et al. 2020. Potential physical impacts of sea-level rise on the pearl river estuary, China. Journal of Marine Systems 201: 103245. DOI: 10.1016/J. JMARSYS.2019.103245.

Huyer, A. 1983. Coastal upwelling in the California current system. Progress in Oceanography 12: 259–284. DOI: 10.1016/0079-6611(83)90010-1.

Kämpf, J. and Chapman, P. 2016. Upwelling systems of the world: a scientific journey to the most productive marine ecosystems. Upwelling Systems of the World: A Scientific Journey to the Most Productive Marine Ecosystems 1–433. DOI: 10.1007/978-3-319-42524-5/COVER.

Khojasteh, D., Glamore, W., Heimhuber, V. and Felder, S. 2021. Sea level rise impacts on estuarine dynamics: a review. Science of The Total Environment 780: 146470. DOI: 10.1016/J. SCITOTENV.2021.146470.

Kimmerer, W.J., Burau, J.R. and Bennett, W.A. 1998. Tidally oriented vertical migration and position maintenance of zooplankton in a temperate estuary. Limnology and Oceanography 43. DOI: 10.4319/lo.1998.43.7.1697.

Kimmerer, W.J., Gross, E.S. and MacWilliams, M.L. 2014. Tidal migration and retention of estuarine zooplankton investigated using a particle-tracking model. Limnology and Oceanography 59. DOI: 10.4319/lo.2014.59.3.0901.

Lachkar, Z. and Gruber, N. 2011. What controls biological production in coastal upwelling systems? insights from a comparative modeling study. Biogeosciences 8: 2961–2976. DOI: 10.5194/BG-8-2961–2011.

Lee, J.-Y., Marotzke, J., Bala, G., Cao, L., Corti, S., Dunne, J.P. et al. 2021. Future global climate: scenario-based projections and near- term information. pp. 553–672. *In*: Masson-Delmotte, V., Zhai, P., Pirani, A., Connors, S.L., Péan, C., Berger, S. et al. (eds.). Climate Change 2021: The Physical Science Basis. Contribution of Working Group I to the Sixth Assessment Report of the Intergovernmental Panel on Climate Change. DOI: 10.1017/9781009157896.006.

Liu, W.C. and Liu, H.M. 2014. Assessing the impacts of sea level rise on salinity intrusion and transport time scales in a tidal estuary, taiwan. Water 6: 324–344. DOI: 10.3390/W6020324.

Lopes, C.L., Silva, P.A., Dias, J.M., Rocha, A., Picado, A., Plecha, S. et al. 2011. Local sea level change scenarios for the end of the 21st century and potential physical impacts in the lower ria de aveiro (portugal). Continental Shelf Research 31: 1515–1526. DOI: 10.1016/j.csr.2011.06.015.

Lopes, C.L. and Dias, J.M. 2015. Tidal dynamics in a changing lagoon: flooding or not flooding the marginal regions. Estuarine, Coastal and Shelf Science 167: 14–24. DOI: 10.1016/J.ECSS.2015.05.043.

Lopes, C.L., Sousa, M.C., Ribeiro, A., Pereira, H., Pinheiro, P., Vaz, L. et al. 2022. Evaluation of future estuarine floods in a sea level rise context. Scientific Reports 12: 8083. DOI: 10.1038/s41598-022-12122-7.

Macias, D., Landry, M.R., Gershunov, A., Miller, A.J. and Franks, P.J.S. 2012. Climatic control of upwelling variability along the western north-american coast. PLOS ONE 7: e30436. DOI: 10.1371/ JOURNAL.PONE.0030436.

Marques, S.C., Pardal, M.A., Pereira, M.J., Gonçalves, F., Marques, J.C. and Azeiteiro, U.M. 2007. Zooplankton distribution and dynamics in a temperate shallow estuary. Hydrobiologia. DOI: 10.1007/s10750-007-0682-x.

Marques, V., Chaves, C., Morais, A., Cardador, F. and Stratoudakis, Y. 2005. Distribution and abundance of snipefish (*Macroramphosus* spp.) off portugal (1998–2003). Scientia Marina 69: 563–576. DOI: 10.3989/SCIMAR.2005.69N4563.

Mendes, R., Vaz, N. and Dias, J.M. 2013. Potential impacts of the mean sea level rise on the hydrodynamics of the douro river estuary. Journal of Coastal Research 165. DOI: 10.2112/si65-330.1.

Mills, L., Janeiro, J., Neves, A.A.S. and Martins, F. 2020. The impact of sea level rise in the guadiana estuary. Journal of Computational Science 44: 101169. DOI: 10.1016/J.JOCS.2020.101169.

Miranda, P.M.A., Alves, J.M.R. and Serra, N. 2013. Climate change and upwelling: response of Iberian upwelling to atmospheric forcing in a regional climate scenario. Climate Dynamics 40: 2813–2824. DOI: 10.1007/s00382-012-1442-9.

Modéran, J., Bouvais, P., David, V., Le Noc, S., Simon-Bouhet, B., Niquil, N. et al. 2010. Zooplankton community structure in a highly turbid environment (charente estuary, France): spatio-temporal patterns and environmental control. Estuarine, Coastal and Shelf Science 88. DOI: 10.1016/j. ecss.2010.04.002.

Montecinos, A., Purca, S. and Pizarro, O. 2003. Interannual-to-interdecadal sea surface temperature variability along the western coast of south America. Geophysical Research Letters 30. DOI: 10.1029/2003GL017345.

Mouny, P. and Dauvin, J.C. 2002. Environmental control of mesozooplankton community structure in the seine estuary (english channel). Oceanologica Acta 25. DOI: 10.1016/S0399-1784(01)01177-X.

Murta, A.G., Abaunza, P., Cardador, F. and Sánchez, F. 2008. Ontogenic migrations of horse mackerel along the iberian coast. Fisheries Research 89: 186–195. DOI: 10.1016/J.FISHRES.2007.09.016.

Narayan, N., Paul, A., Mulitza, S. and Schulz, M. 2010. Trends in coastal upwelling intensity during the late 20th century. Ocean Science 6: 815–823. DOI: 10.5194/OS-6-815-2010.

Oppenheimer, M., Glavovic, B.C., Hinkel, J., Wal, R. van de., Magnan. A.K., Abd-Elgawad. A. et al. 2019. Sea level rise and implications for low-lying islands, coasts and communities. *In*: IPCC Special Report on the Ocean and Cryosphere in a Changing Climate. IPCC 355.

Palmer, K., Watson, C. and Fischer, A. 2019. Non-linear interactions between sea-level rise, tides, and geomorphic change in the tamar estuary, Australia. Estuarine, Coastal and Shelf Science 225: 106247. DOI: 10.1016/J.ECSS.2019.106247.

Passeri, D.L., Hagen, S.C., Medeiros, S.C. and Bilskie, M.V. 2015. Impacts of historic morphology and sea level rise on tidal hydrodynamics in a microtidal estuary (grand bay, mississippi). Continental Shelf Research 111: 150–158. DOI: 10.1016/J.CSR.2015.08.001.

Passeri, D.L., Hagen, S.C., Plant, N.G., Bilskie, M.V., Medeiros, S.C. and Alizad, K. 2016. Tidal hydrodynamics under future sea level rise and coastal morphology in the northern gulf of Mexico. Earth's Future 4: 159–176. DOI: 10.1002/2015EF000332.

Patti, B., Guisande, C., Vergara, A.R., Riveiro, I., Maneiro, I., Barreiro, A. et al. 2008. Factors responsible for the differences in satellite-based chlorophyll a concentration between the major global upwelling areas. Estuarine, Coastal and Shelf Science 76: 775–786. DOI: 10.1016/J.ECSS.2007.08.005

Pauly, D. and Christensen, V. 1995. Primary production required to sustain global fisheries. Nature 374(6519): 255–257. DOI: 10.1038/374255a0.

Pelegrí, J.L., Arístegui, J., Cana, L., González-Dávila, M., Hernández-Guerra, A., Hernández-León, S. et al. 2005. Coupling between the open ocean and the coastal upwelling region off northwest Africa: water recirculation and offshore pumping of organic matter. Journal of Marine Systems 54: 3–37. DOI: 10.1016/j.jmarsys.2004.07.003.

Peliz, Á., Rosa, T.L., Santos, A.M.P. and Pissarra, J.L. 2002. Fronts, jets, and counter-flows in the western iberian upwelling system. Journal of Marine Systems 35: 61–77. DOI: 10.1016/S0924-7963(02)00076-3.

Peltier, W.R., Argus, D.F. and Drummond, R. 2015. Space geodesy constrains ice age terminal deglaciation: the global ICE-6G-C (VM5a) model. Journal of Geophysical Research: Solid Earth 120. DOI: 10.1002/2014JB011176.

Pereira, H., Sousa, M.C., Vieira, L.R., Morgado, F. and Dias, J.M. 2022. Modelling salt intrusion and estuarine plumes under climate change scenarios in two transitional ecosystems from the NW Atlantic coast. Journal of Marine Science and Engineering 10: 262. DOI: 10.3390/JMSE10020262.

Petrusevich, V.Y., Dmitrenko, I.A., Niemi, A., Kirillov, S.A., Kamula, C.M., Kuzyk, Z.Z.A. et al. 2020. Impact of tidal dynamics on diel vertical migration of zooplankton in hudson bay. Ocean Science 16: 337–353. DOI: 10.5194/OS-16-337-2020.

Picado, A., Alvarez, I., Vaz, N. and Dias, J.M. 2013. Chlorophyll concentration along the northwestern coast of the iberian peninsula vs. atmosphere-ocean- land conditions. Journal of Coastal Research 65: 2047–2052. DOI: 10.2112/SI65-346.1.

Picado, A. 2016. Influence of Physical Processes on the Primary Production along the Iberian Peninsula Northwestern Coast. Ph.D. Thesis, University of Aveiro, 191 pp.

Picado, A., Lorenzo, M.N., Alvarez, I., DeCastro, M., Vaz, N. and Dias, J.M. 2016. Upwelling and chl-a spatiotemporal variability along the galician coast: dependence on circulation weather types. International Journal of Climatology 36: 3280–3296. DOI: 10.1002/joc.4555.

Pires, A.C., Nolasco, R., Rocha, A. and Dubert, J. 2013. Assessing future climate change in the Iberian upwelling system. Journal of Coastal Research 165: 1909–1914. DOI: 10.2112/SI65-323.1

Pires, A.C., Nolasco, R., Rocha, A., Ramos, A.M. and Dubert, J. 2016. Climate change in the Iberian upwelling system: a numerical study using GCM downscaling. Climate Dynamics 47: 451–464. DOI: 10.1007/S00382-015-2848-Y/TABLES/1.

Pitcher, G.C., Figueiras, F.G., Hickey, B.M. and Moita, M.T. 2010. The physical oceanography of upwelling systems and the development of harmful algal blooms. Progress in Oceanography 85: 5–32. DOI: 10.1016/J.POCEAN.2010.02.002.

Prandle, D. and Lane, A. 2015. Sensitivity of estuaries to sea level rise: vulnerability indices. Estuarine, Coastal and Shelf Science 160: 60–68. DOI: 10.1016/J.ECSS.2015.04.001.

Queiroga, H., Cruz, T., dos Santos, A., Dubert, J., González-Gordillo, J.I., Paula, J. et al. 2007. Oceanographic and behavioural processes affecting invertebrate larval dispersal and supply in the western iberia upwelling ecosystem. Progress in Oceanography 74: 174–191. DOI: 10.1016/J. POCEAN.2007.04.007.

Reboreda, R., Cordeiro, N.G.F., Nolasco, R., Castro, C.G., Álvarez-Salgado, X.A., Queiroga, H. et al. 2014. Modeling the seasonal and interannual variability (2001–2010) of chlorophyll-a in the iberian margin. Journal of Sea Research 93: 133–149. DOI: 10.1016/j.seares.2014.04.003.

Relvas, P., Barton, E.D., Dubert, J., Oliveira, P.B., Peliz, Á., da Silva, J.C.B. et al. 2007. Physical oceanography of the western iberia ecosystem: latest views and challenges. Progress in Oceanography 74: 149–173. DOI: 10.1016/j.pocean.2007.04.021.

Relvas, P., Luís, J. and Santos, A.M.P. 2009. Importance of the mesoscale in the decadal changes observed in the northern canary upwelling system. Geophysical Research Letters 36: L22601. DOI: 10.1029/2009GL040504.

Rice, K.C., Hong, B. and Shen, J. 2012. Assessment of salinity intrusion in the james and chickahominy rivers as a result of simulated sea-level rise in chesapeake bay, east coast, USA. Journal of Environmental Management 111: 61–69. DOI: 10.1016/J.JENVMAN.2012.06.036.

Richardson, A.J. and Schoeman, D.S. 2004. Climate impact on plankton ecosystems in the northeast Atlantic. Science 305: 1609–1612. DOI: 10.1126/SCIENCE.1100958/SUPPL_FILE/RICHARDSON.SOM.PDF.

Richardson, A.J. and Richardson, A.J. 2008. In hot water: zooplankton and climate change. ICES Journal of Marine Science 65: 279–295. DOI: 10.1093/ICESJMS/FSN028.

Rodrigues, M., Fortunato, A.B. and Freire, P. 2019. Saltwater intrusion in the upper tagus estuary during droughts. Geosciences (Switzerland) 9. DOI: 10.3390/geosciences9090400.

Rodrigues, M., Rosa, A., Cravo, A., Jacob, J. and Fortunato, A.B. 2021. Effects of climate change and anthropogenic pressures in the water quality of a coastal lagoon (ria formosa, portugal). Science of the Total Environment 780. DOI: 10.1016/j.scitotenv.2021.146311.

Roemmich, D. and McGowan, J. 1995. Climatic warming and the decline of zooplankton in the California current. Science 267: 1324–1326. DOI: 10.1126/SCIENCE.267.5202.1324.

Ross, A.C., Najjar, R.G., Li, M., Lee, S.B., Zhang, F. and Liu, W. 2017. Fingerprints of sea level rise on changing tides in the chesapeake and delaware bays. Journal of Geophysical Research: Oceans 122: 8102–8125. DOI: 10.1002/2017JC012887.

Rossi, V., López, C., Hernández-García, E., Sudre, J., Garcon, V. and Morel, Y. 2009. Surface mixing and biological activity in the four Eastern boundary upwelling systems. Nonlinear Processes in Geophysics 16: 557–568. DOI: 10.5194/NPG-16-557-2009.

Ruela, R., Sousa, M.C., deCastro, M. and Dias, J.M. 2020. Global and regional evolution of sea surface temperature under climate change. Global and Planetary Change 190: 103190. DOI: 10.1016/J. GLOPLACHA.2020.103190.

Rykaczewski, R.R., Dunne, J.P., Sydeman, W.J., García-Reyes, M., Black, B.A. and Bograd, S.J. 2015. Poleward displacement of coastal upwelling-favorable winds in the ocean's eastern boundary currents through the 21st century. Geophysical Research Letters 42: 6424–6431. DOI: 10.1002/2015GL064694.

Sallenger, A.H., Doran, K.S. and Howd, P.A. 2012. Hotspot of accelerated sea-level rise on the Atlantic coast of north America. Nature Climate Change 2. DOI: 10.1038/nclimate1597.

Santos, A.M.P., Borges, M.D.F. and Groom, S. 2001. Sardine and horse mackerel recruitment and upwelling off portugal. ICES Journal of Marine Science 58: 589–596. DOI: 10.1006/JMSC.2001.1060.

Santos, A.M.P., Kazmin, A.S. and Peliz, Á. 2005. Decadal changes in the canary upwelling system as revealed by satellite observations: their impact on productivity. Journal of Marine Research 63: 359–379. DOI: 10.1357/0022240053693671.

Santos, F., Gomez Gesteira, M. and deCastro, M. 2011. Coastal and oceanic SST variability along the western Iberian peninsula. Continental Shelf Research 31: 2012–2017. DOI: 10.1016/J. CSR.2011.10.005.

Santos, F., DeCastro, M., Gómez-Gesteira, M. and Álvarez, I. 2012a. Differences in coastal and oceanic SST warming rates along the canary upwelling ecosystem from 1982 to 2010. Continental Shelf Research 47: 1–6. DOI: 10.1016/J.CSR.2012.07.023.

Santos, F., Gómez-Gesteira, M., deCastro, M. and Álvarez, I. 2012b. Variability of coastal and ocean water temperature in the upper 700 m along the western iberian peninsula from 1975 to 2006. PLOS ONE 7: e50666. DOI: 10.1371/JOURNAL.PONE.0050666.

Santos, F., Gómez-Gesteira, M., Varela, R., Ruiz-Ochoa, M. and Dias, J.M. 2016. Influence of upwelling on SST trends in la guajira system. Journal of Geophysical Research: Oceans 121: 2469–2480. DOI: 10.1002/2015JC011420.

Seabra, R., Varela, R., Santos, A.M., Gómez-Gesteira, M., Meneghesso, C., Wethey, D.S. et al. 2019. Reduced nearshore warming associated with eastern boundary upwelling systems. Frontiers in Marine Science 6: 104. DOI: 10.3389/FMARS.2019.00104/BIBTEX.

Seo, H., Brink, K.H., Dorman, C.E., Koracin, D. and Edwards, C.A. 2012. What determines the spatial pattern in summer upwelling trends on the US west coast? Journal of Geophysical Research: Oceans 117. DOI: 10.1029/2012JC008016.

Silveira, F., Lopes, C.L., Pinheiro, J.P., Pereira, H. and Dias, J.M. 2021. Coastal floods induced by mean sea level riseandmdash; ecological and socioeconomic impacts on a mesotidal lagoon. Journal of Marine Science and Engineering 9: 1430. DOI: 10.3390/JMSE9121430.

Slangen, A.B.A., Carson, M., Katsman, C.A., van de Wal, R.S.W., Köhl, A., Vermeersen, L.L.A. et al. 2014. Projecting twenty-first century regional sea-level changes. Climatic Change 124. DOI: 10.1007/s10584-014-1080-9.

Sousa, M.C., deCastro, M., Alvarez, I., Gomez-Gesteira, M. and Dias, J.M. 2017. Why coastal upwelling is expected to increase along the western Iberian peninsula over the next century? Science of The Total Environment 592: 243–251. DOI: 10.1016/j.scitotenv.2017.03.046.

Sousa, M.C., Ribeiro, A., Des, M., Gomez-Gesteira, M., deCastro, M. and Dias, J.M. 2020. NW Iberian peninsula coastal upwelling future weakening: competition between wind intensification and surface heating. Science of The Total Environment 703: 134808. DOI: 10.1016/J.SCITOTENV.2019.134808.

Stammer, D., Cazenave, A., Ponte, R.M. and Tamisiea, M.E. 2013. Causes for contemporary regional Sea level changes. Annual Review of Marine Science 5. DOI: 10.1146/annurev-marine-121211-172406.

Sydeman, W.J., García-Reyes, M., Schoeman, D.S., Rykaczewski, R.R., Thompson, S.A., Black, B.A. et al. 2014. Climate change. Climate Change and Wind Intensification in Coastal Upwelling Ecosystems. Science (New York, N.Y.) 345: 77–80. DOI: 10.1126/SCIENCE.1251635.

Tenore, K.R., Alonso-Noval, M., Alvarez-Ossorio, M., Atkinson, L.P., Cabanas, J.M., Cal, R.M. et al. 1995. Fisheries and oceanography off galicia, NW spain: mesoscale spatial and temporal changes in physical processes and resultant patterns of biological productivity. Journal of Geophysical Research 100: 10943. DOI: 10.1029/95JC00529.

Valentim, J.M., Vaz, N., Silva, H., Duarte, B., Caçador, I. and Dias, J.M. 2013. Tagus estuary and ria de aveiro salt marsh dynamics and the impact of sea level rise. Estuarine, Coastal and Shelf Science 130: 138–151. DOI: 10.1016/j.ecss.2013.04.005.

van Maanen, B. and Sottolichio, A. 2018. Hydro- and sediment dynamics in the gironde estuary (france): sensitivity to seasonal variations in river inflow and sea level rise. Continental Shelf Research 165: 37–50. DOI: 10.1016/J.CSR.2018.06.001.

Varela, R., Álvarez, I., Santos, F., DeCastro, M. and Gómez-Gesteira, M. 2015. Has upwelling strengthened along worldwide coasts over 1982–2010? Scientific Reports 5(1): 1–15. DOI: 10.1038/srep10016.

Varela, R., Lima, F.P., Seabra, R., Meneghesso, C. and Gómez-Gesteira, M. 2018. Coastal warming and wind-driven upwelling: a global analysis. Science of The Total Environment 639: 1501–1511. DOI: 10.1016/J.SCITOTENV.2018.05.273.

Varela, R.A., Rosón, G., Herrera, J.L., Torres-López, S. and Fernández-Romero, A. 2005. A general view of the hydrographic and dynamical patterns of the rías baixas adjacent Sea area. Journal of Marine Systems 54: 97–113. DOI: 10.1016/j.jmarsys.2004.07.006.

Vargas, C.I.C., Vaz, N. and Dias, J.M. 2017. An evaluation of climate change effects in estuarine salinity patterns: application to ria de aveiro shallow water system. Estuarine, Coastal and Shelf Science 189: 33–45. DOI: 10.1016/j.ecss.2017.03.001.

Villate, F. 1997. Tidal influence on zonation and occurrence of resident and temporary zooplankton in a shallow system (estuary of mundaka, bay of biscay). Scientia Marina 61.

Vousdoukas, M.I., Mentaschi, L., Mongelli, I., Carlos Ciscar, J., Hinkel, J., Ward, P. et al. 2020. Adapting to rising coastal flood risk in the EU under climate change. Luxembourg. DOI: 10.2760/456870.

Wang, D., Gouhier, T.C., Menge, B.A. and Ganguly, A.R. 2015. Intensification and spatial homogenization of coastal upwelling under climate change. Nature 518(7539): 390–394. DOI: 10.1038/nature14235.

Whitehouse, P.L. 2018. Glacial isostatic adjustment modelling: historical perspectives, recent advances, and future directions. Earth Surface Dynamics 6. DOI: 10.5194/esurf-6-401-2018.

Wooldridge, T. and Erasmus, T. 1980. Utilization of tidal currents by estuarine zooplankton. Estuarine and Coastal Marine Science 11. DOI: 10.1016/S0302-3524(80)80033-8.

Wooste, W.S., Bakun, A. and McLain, D.R. 1976. Seasonal upwelling cycle along the eastern boundary of the north Atlantic. Journal of Marine Research 34: 131–141.

Yáñez, E., Barbieri, M.A., Silva, C., Nieto, K. and Espíndola, F. 2001. Climate variability and pelagic fisheries in northern chile. Progress in Oceanography 49: 581–596. DOI: 10.1016/S0079-6611(01)00042-8.

Zhong, L., Li, M. and Foreman, M.G.G. 2008. Resonance and sea level variability in chesapeake bay. Continental Shelf Research 28: 2565–2573. DOI: 10.1016/J.CSR.2008.07.007.

CHAPTER 2
Integrated Approaches and Tools for Plankton Research

2.1

Zooplankton Dynamics at the Northern Boundary of the Canary Current Upwelling System (NW Iberian Peninsula). Prospects From Updated Time Series in the 'Ría de Vigo' and Adjacent Shelf

Enrique Nogueira, * *Gerardo Casas, Fernando Rayón-Viña,*
Esther Velasco-Senovilla and *Ana Miranda*

1. Introduction

The zooplankton community is a highly diverse component of the plankton, comprising both protist and metazoans, which inhabit all types of aquatic realms and cover an ample range of sizes, trophic modes, reproductive strategies or life-history traits (Harris et al. 2000, Kehayias 2014, Bucklin et al. 2021). It plays a pivotal role in the functioning of the food web, channeling matter and energy from primary producers to higher trophic levels and contributing to propagating environmental changes to the whole pelagic system (Chiba et al. 2006). It is also essential for the maintenance of exploitable fish stocks, especially those of foraging species such as sardines and anchovies (Robinson et al. 2014). Nowadays, it is also amply recognized its role in biogeochemical cycles and ocean carbon pump, for instance, through the production of large, fast-sinking fecal pellets that transport chemical elements downwards through different pelagic domains of the water column and eventually to the sea floor where they are sequestered. This downward transport is particularly relevant for carbon due to its impact on global warming and ocean acidification (Steinberg and Landry 2017).

As a component of the plankton ecosystem, zooplankton is highly responsive to environmental variability (Hays et al. 2005). Bottom-up processes, such as temperature and quantity/quality of nutritional resources, set the habitat conditions that modulate life-history traits, such as growth, fecundity and survival. This tight

Centro Oceanográfico de Vigo (IEO-CSIC), Subida a Radio Faro 50, 36390 Vigo, Spain.

* Corresponding author: enrique.nogueira@ieo.csic.es

coupling with environmental variability, added to short generation times and not commercial exploitation of the vast majority of species, make zooplankton good sentinels of ecosystem alterations, either promoted by natural variability or human-induced due to factors such as pollution, over-exploitation of resources and climate change, which may act synergistically (Taylor et al. 2002, Perry et al. 2004, Hays et al. 2005). Accordingly, several zooplankton-based metrics, built on bulk properties, functional groups or characteristic species, have been suggested as surveillance indicators to assess ecosystem state (Bedford et al. 2018). For instance, within the framework of the EU Marine Strategy Framework Directive (MSFD) (Directive 2008/56/EC), zooplankton-based indicators have been proposed to assess the environmental status of pelagic habitats in relation to ecosystem descriptors linked to biodiversity characteristics (descriptor D1) and food-web functioning (D4) (McQuatters-Gollop et al. 2019).

The western Iberian Peninsula is located at the northern stretch of the Canary Current Upwelling System (CanCUS), one of the four major Eastern Boundary Upwelling Systems (EBUS) of the world (Kämpf and Chapman 2016). The northernmost part corresponds to the Galician sub-region (Arístegui et al. 2009), where coastal upwelling driven by northerly winds is relatively feeble and shows a strong seasonality in comparison to other sub-regions along the CanCUS (Wooster et al. 1976). During the upwelling season, extending on average from March to October, upwelling takes place as a series of upwelling cycles occurring with a periodicity of 1–2 weeks (Buttay et al. 2022), showing characteristic spin-up and spin-down phases interspaced by shorter interludes of upwelling relaxation or downwelling (Smayda and Trainer 2010). Upwelling promotes the ascent of sub-surface cold and nutrient-rich Eastern North-Atlantic Central Water (ENACW) (Fraga 1981, Ríos et al. 1992). In autumn and winter, southerly and westerly winds predominate, causing downwelling conditions in the shelf and the piling up of surface waters toward the coast. The passage of low atmospheric pressure systems from the Atlantic brings abundant precipitations to the sub-region during the downwelling season. This period is also characterized by the development of a warm and saline (thus spicy, González-Nuevo and Nogueira 2005) poleward current flowing along the slope. This poleward flow, named Iberian Poleward Current (IPC) (Peliz et al. 2003) (Figure 1A), Portugal Coastal Counter Current (PCCC) (Ambar and Fiuza 1994, Álvarez-Salgado et al. 2003) or "Navidad" (Christmas) Current (García-Soto et al. 2002) is a common feature of the winter circulation of eastern ocean margins (Neshyba et al. 1989).

Capes Finisterre and Ortegal mark abrupt changes in the orientation of the Galician coast, dividing it into three domains: west (south of Cape Finisterre), north (east of Cape Ortegal) and the intermediate or northwest between these two capes (Figure 1B). Coastal relief is indented by characteristic physiographic features named 'rías' (von Richthofen 1886), drowned river valleys that present an estuarine part at the head and extend seawards conforming a coastal embayment connected with shelf waters (Gómez-Gesteira et al. 2011). Hydrodynamically, the rías behave on average as a partially mixed embayment with a positive residual circulation pattern, which is modulated by the influence of coastal upwelling and downwelling dynamics on the shelf and runoff inputs, particularly of river discharges at the inner reaches of

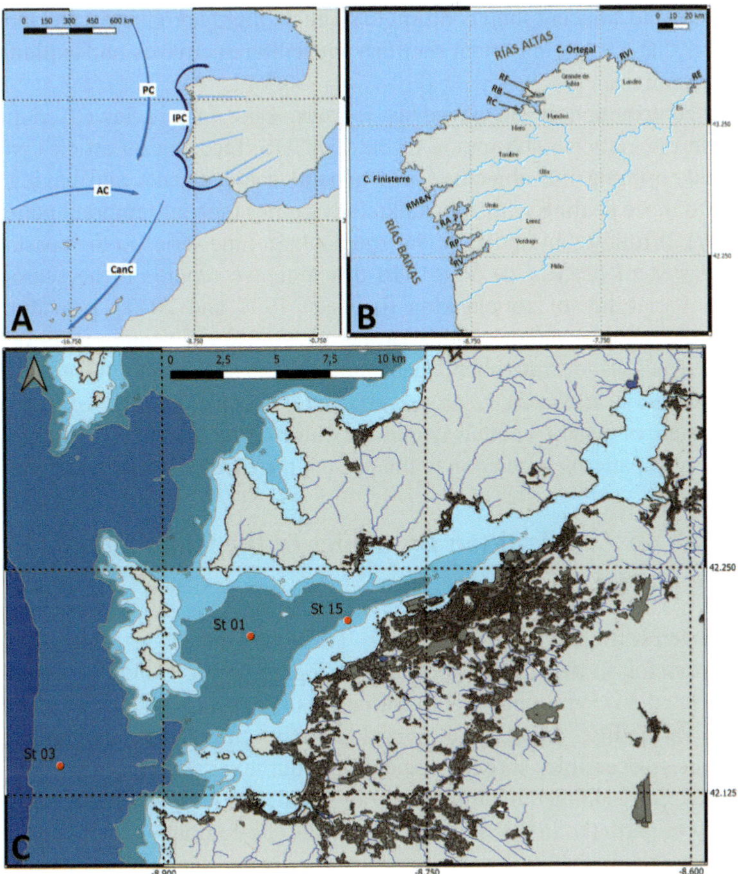

Figure 1. Map of the study area. (A) Main currents around the Iberian Peninsula: IPC, Iberian Poleward Current; PC, Portugal Current; AC, Azores Current; CanC, Canary Current. On the mainland, light blue arrows represent the main freshwater inputs (modified from Aristegui et al. 2009). (B) Galician Rías. Rías Baixas: RV, Ría de Vigo; RP, Ría de Pontevedra; RA, Ría de Arousa; RM&N, Ría de Muros-Noia; Rías Altas: RC, Ría de A Coruña; RB, Ría de Betanzos; RF, Ría de Ferrol; RVi, Ría de Viveiro; RE, Ría del Eo (Ribadeo). (C) Oceanographic stations in the running Coastal Ocean Observing System 'Radial de Vigo' (RV section), sampling within the frame of the RADIALES program (https://www.seriestemporales-ieo.net/).

the rías (Fraga and Margalef 1979, Gilcoto et al. 2001). The Galician sub-region can be further sub-divided into southern and northern areas, with the boundary around Cape Finisterre (42.9°N), according to distinctive climatic conditions, with increasing influences of oceanic climate type (C*fb* type) northwards and temperate warm-summer Mediterranean type climate southwards (C*sb* according to Köppen-Geiger classification) (Beck et al. 2018), and topographic features such as the change in the orientation of the coast and isobaths, shelf width and morphology of the rías (larger, deeper and more funnel-shaped in the southern than in the northern area of 'Rías Baixas' and 'Rías Altas', respectively (Figure 1B). These differences result in a higher impact of coastal upwelling, in terms of strength and frequency of occurrence, in the southern than in the northern areas (Álvarez et al. 2012, Picado et al. 2016). Due to the influence of upwelling and the characteristics of the coast, the

Galician rías and adjacent shelf conform to singular, highly productive and biodiverse ecosystems that sustain important shellfish aquaculture activities and exploitation of fisheries resources.

This contribution is organized as follows. First, we reviewed zooplankton research in the Galician sub-region of the CanCUS, focusing on articles published since 1990, priming the objectives and general achievements and paying spatial attention to those dealing with zooplankton dynamics (spatial-temporal patterns and variability). It follows an update of the zooplankton time series in the coastal-ocean observing system of 'Ría de Vigo'. To this aim, we report on the seasonal and long-term variability of zooplankton (between 1997 and 2017), providing some examples for total abundance (bulk property), functional groups (life-form pair: copepods versus cladocerans abundance) and indicator species (abundance of *Temora longicornis* and *T. stylifera*), as an example set of zooplankton-based indicators to assess the state of pelagic habitats within the frame of the MSFD. Finally, we discuss these observed patterns in relation to the present meteorological/climatic scenario.

2. Studies on Zooplankton Dynamics in the Galician Sub-Region of the CanCUS

We reviewed articles published since 1990 on metazoan holo-zooplankton (zooplankton for short) in the Galician sub-region of the CanCUS, dealing with aspects related to community composition (taxa or species), spatial distribution, temporal variability, process rates and abiotic and biotic interactions. The publications, accessible through scientific journal databases or institutional repositories, corresponded mainly (73 out of 99 articles reviewed) to studies carried out, totally or partially, in the southern area of the sub-region, a major contributor to the fisheries and aquaculture socio-economic sectors of Galicia and where several marine research facilities are established. Milestones and contributions that impulse zooplankton research in the sub-region initiated around 1885, stirred by a sardine stock crisis in French waters, are summarized in Annex 1.

Zooplankton investigations were mostly based on observations at sea (*in situ*), either from short-term ('quasi-synoptic') surveys or long-term time-series programs. Short-term surveys were conducted during several weeks or months of intensive (daily-weekly) sampling, which extended for 1–2 years when they were part of research projects (Isla et al. 2004) or annually repeated when they were associated with stock assessment programs (Vandromme et al. 2014). Long-term sustained (weekly-monthly) observations are framed on time series programs, such as RADIALES (https://www.seriestemporales-ieo.net/) (Valdés et al. 2021) and the Continuous Plankton Recorder (CPR, with Standard Area F4 containing the Galician sub-region; https://www.cprsurvey.org/) (Bode et al. 2009, Nogueira et al. 2012). These *in situ* studies followed mostly the Eulerian method, and samples are usually collected at fixed locations along a series of across-shelf sections covering mainly coastal-shelf domains (Barquero et al. 1998, Cabal et al. 2008). The Lagrangian method, in which sampling is conducted following the track of drifting water masses, was applied in process rates studies (Batten et al. 2001), in some cases in combination with the Eulerian method (Isla and Anadón 2004). Finally, some investigations were based on experimental approaches, focusing preferentially on the autecology of copepod species common in the area (*Euterpina acutifrons, Acartia clausi* or *Temora longicornis*) in relation to the dynamics of Harmful Algal Blooms (HABs) (Bagoien et al. 1996, Guisande et al. 2002, Barreiro et al. 2007).

In the present review, we focused on the investigations of zooplankton dynamics, considering temporal as well as spatial variability and its relationship with the abiotic and biotic environmental factors. Studies based on the observations of natural populations are fundamental to assessing the suitability of zooplankton-based metrics as indicators of ecosystem status.

2.1 Short-Term (Quasi-Synoptic) Surveys

Short-term studies focused mainly on discerning spatial distribution patterns of zooplankton at different organization (aggregation) levels and their underlying causes. The usual bulk variables analyzed were total or size-fractionated biomass and elemental (CNH) composition (Isla et al. 2004), size-fractionated composition of stable isotopes of carbon (^{13}C) and nitrogen (^{15}N) (Bode et al. 2003, 2004, 2007, 2020) and size-spectra features (slope and biomass by size/biomass classes) (Nogueira et al. 2004, Vandromme et al. 2014). Other studies focused on species composition and relative abundances of zooplankton groups, most of them including bulk variables such as total zooplankton abundance and biomass (dry weight). Special emphasis was given to the distribution of copepods due to their large diversity and major contribution to the community (Valdés et al. 1990, Fernández de Puelles et al. 1996, Poulet et al. 1996, Blanco-Bercial 2006, Cabal et al. 2008, Roura et al. 2013, López-López et al. 2014). Meroplankton also received considerable attention, with most studies targeting plankton stages of economically relevant taxa/species such as decapods (Fusté and Gili 1991), mussel *Mytilus galloprovincialis* (Cáceres-Martínez and Figueiras 1998), cephalopods (Rocha et al. 1999, González et al. 2005), barnacles (Macho et al. 2005) and fishes (Rodríguez et al. 2015, Uberos et al. 2021).

Estimations of zooplankton process rates were also based on short-term, multidisciplinary studies that usually embrace several trophic levels of the pelagic community. The work by Tenore et al. (1995) could be deemed the first comprehensive study of the plankton ecosystem in the NW Iberian shelf, including physical-biological interactions and multiple components of the pelagic biocenosis (bacteria, phytoplankton, zooplankton and ichthyoplankton) and their process rates (bacterial production, primary production, nitrate uptake by phytoplankton and zooplankton grazing). Several zooplankton processes rates studies, based on a Lagrangian sampling, conducted incubation experiments to estimate grazing rates of copepods on micro-zooplankton (Batten et al. 2001) or phytoplankton (herbivory) (Isla and Anadón 2004), grazing rates of micro-zooplankton on phytoplankton (Fileman and Burkill 2001) or sedimentation and production rates of meso-zooplankton fecal pellets (Riser et al. 2001). Herbivory and metabolism (respiration and excretion rates) of copepods (Isla et al. 2004) or egg production and gonad development of *Calanus helgolandicus* and *C. carinatus* (Ceballos et al. 2004) were also derived from studies applying an Eulerian sampling approach.

Zooplankton distribution was resolved at different scales depending on spatial coverage and temporal extent of the studies, ranging from local (ca. 10 km, days) (Varela et al. 2010, Roura et al. 2013), sub-regional (ca. 10^2 km, weeks) (Blanco-Bercial et al. 2006) or regional scales (ca. 10^3 km, months) (Vandromme et al. 2014). The observed patterns were related to the main physical processes and the resultant hydrographic structures. It was soon recognized the prominent role of upwelling and outflow from the rías as structuring factors of zooplankton in the Galician sub-region at the level of bulk properties (Tenore et al. 1995), taxonomic groups and species (Valdés 1992) and processes (Tenore et al 1995, Batten et al. 2001). Across-shelf circulation driven by upwelling dynamics, added to the rías' outflow, promotes across-shelf variability of the abundance and relative proportions of the species of the main zooplankton groups (copepods, appendicularians, doliolids and siphonophores), as well as an offshore decreasing contribution of meroplankton (mainly of lamellibranches, polychaeta and bryozoans larvae) (Roura et al. 2013, López-López et al. 2014). This resulted in differentiated communities in coastal, shelf (frontal area of the across-shelf upwelling cell) and oceanic domains (Bode et al. 1994, Cabal et al. 2008, Roura et al. 2013). Large herbivorous species (> 1,000 μm) predominated in coastal-shelf waters (Blanco-Bercial et al. 2006, Cabal et al. 2008) in response to blooms of large-sized phytoplankton (> 20 μm, mostly chain-forming diatoms) driven by shifting hydrographic conditions occurring seasonally, during the periods of transient thermoclines in spring and autumn and at sub-seasonal scales, coupled to upwelling cycles occurring during the upwelling season. In these conditions, phytoplankton biomass was enough to fulfill metabolic demands of the zooplankton community and herbivory predominated (Isla and Anadón 2004, Isla et al. 2004). Neritic holoplanktonic species such as *Temora longicornis, Acartia clausi, Pseudocalanus elongatus, Paracalanus parvus* and *Euterpina acutifrons*, coexist with meroplankton inside the rías (Roura et al. 2013). Selective feeding of copepod species on the phytoplankton community (composed mainly of diatoms, dinoflagellates chlorophytes and prymnesiophytes) was observed in coastal habitats ('Ría de Vigo'), where food supply is usually not a limiting factor for the zooplankton community (holo- and meroplankton); this food niche partitioning relaxes inter-specific competition, favoring coexistence (Guisande et al. 2002). In contrast, small-medium sized copepods (200–1,000 μm), appendicularians and thaliacea (*Salpa fusiformis* and *Thalia democratica*), adapted to feed on the small-sized plankton community (< 5 μm, pico- and nano-phytoplankton, micro-heterotrophic flagellates and bacteria) characteristic of stratified oligotrophic environments, predominated in oceanic waters (Blanco-Bercial et al. 2006, Cabal et al. 2008) forming a multivorous food-web (Isla et al. 2004). Under scarcity of food, the copepod community grazes unselectively, food is consumed in proportion to its abundance and quantity of food is more important than its quality (Halvorsen et al. 2001). Concomitant to the observed patterns of community assemblages, bulk properties such as total abundance and biomass and the average size of meso-zooplankton showed neat offshore decreasing trends (Isla et al. 2004, Nogueira et al. 2004, Blanco-Bercial et al. 2006).

The along-shelf dynamics in the upper layers (< 200 m) of the western Iberian shelf waters, defined by the surface equatorward Portuguese Current and the slope (sub-surface) poleward Iberian Poleward Current (IPC) (Peliz et al. 2003) (Figure 1A) was considered in several studies. Stöhr et al. (1996, 1997) pointed out the relevance of these opposing flows in the maintenance of a stable population of *Calanus finmarchicus* along the CanCUS, from NW Africa to the Galician sub-region, through a depth-stratified recirculation of different development stages. This research was among the first to recognize the relevance of the along-shelf circulation patterns on plankton dynamics in the NW Iberian region of the CanCUS (Sordo et al. 2001 suggested that for the HAB dinoflagellate species *Gymnodinium catenatum* and *Dinophysis acuminata*). Zooplankton community structure and species composition showed characteristic features in the zones influenced by the IPC, usually in mid-shelf and slope waters (Blanco-Bercial et al. 2006, Cabal et al. 2008). The phytoplankton community in the IPC was dominated by pico- and nano plankton size-fractions (autotrophic bacteria and small flagellates) (Álvarez-Salgado et al. 2003). Appendicularians, thaliaceans (*Salpa fusiformis*) and small copepods (e.g., *Oithona nana*), with preference for warm temperatures and specialized to feed on small-sized food items, as well as species related to high salinity and oceanic habitats, such as *Metridia lucens*, *Pleuromamma gracilis* and *Calocalanus styliremis*, showed higher relative abundances at the zone influenced by the IPC. Also, tropical/sub-tropical species, such as *Scolecithricella dentata* (Blanco-Bercial et al. 2006), *Clausocalanus farrani*, *Phaenna spinifera* and *Calocalanus contractus* (Cabal et al. 2008), were identified in that domains, as indicator species of the northward along-shelf transport by the IPC and displacement of isotherms (topicalization; Ibairbaiz et al. 2019).

2.2 *Long-Term Programs*

Information on the temporal variability of zooplankton in the Galician sub-region arises largely from the time-series program RADIALES (https://www.seriestemporales-ieo.net/) (Valdés et al. 2007, 2021). Framed in this program, a series of oceanographic stations distributed in across-shelf sections from ría to mid-shelf, are visited monthly since 1988 off 'Ría de A Coruña' (RC) ('Rías Altas') and 'Ría de Vigo' (RV) ('Rías Baixas') (Figure 1B) to assess temporal variability (and spatial differences across- and along-shelf) of hydrographic conditions and plankton components and processes (Bode et al. 2012a). Given the actual extent and periodicity of observations and the multidisciplinary of the program, the time series from RADIALES (which also include across-shelf sections in the Cantabrian Sea, southern Bay of Biscay, off Cudillero, Gijón and Santander) are suited to discern modes of variability at seasonal and long-term scales (Bode et al. 2012a, Buttay et al. 2016, 2017), including detection of ecosystem regime shifts (Bode et al. 2020) and their underlying abiotic (e.g., meteo-climatic) and biotic (e.g., community interactions) drivers. A comprehensive description of the modes of temporal variability of meteorological/climatic, hydrographic and plankton components in the Galician sub-region and the Cantabrian Sea is compiled in Bode et al. (2012a) and recently reviewed by Valdés et al. (2021). The running program provides data to assess the state of pelagic habitats (and other ecosystem components) within the frame of the MSFD, informing on base-line (or reference) states, climatic values and seasonal (phenology) and long-term variability of (zoo)plankton-based indicators of biodiversity and food-web functioning (MSFD descriptors D1 and D4). Data and results from the program also contribute to international ocean observing networks, such as GOOS or EMODnet, and expert groups on hydrography and plankton, especially for ICES and OSPAR (Valdés et al. 2021). Climatic values (general and seasonal statistics for ca. 1995–2010) of zooplankton bulk properties (total biomass and abundance), species composition (ranked species checklist and relative abundances) and abundance by taxonomic (functional) groups and copepod species can be consulted in the first assessment report of the environmental state of biodiversity in NW and N Iberian coastal-shelf waters (Velasco et al. 2012).

Seasonality was recognized as a prominent mode of temporal variability from the analysis of several years in the section of RC at the start of the program regarding bulk properties and species composition (Valdés et al. 1990, 1991, Bode et al. 1998), a succession of functional groups and trophic structure (Bode and Álvarez-Osorio 2004), and processes such as grazing impact of copepods on phytoplankton (herbivory) (Bode et al. 2003) and nitrogen regeneration through heterotrophy, which showed the relevance of the micro- versus meso-zooplankton community (microbial-loop versus food-chain) (Bode et al. 2004). Valdés et al. (2007) published the first comprehensive study of temporal variability of zooplankton in the Galician sub-region and the Cantabrian Sea, based on the analysis of 19 stations, covering neritic, shelf and oceanic domains, sampled monthly from ca. 1995 to 2000. The authors described seasonal and interannual patterns of bulk properties (meso-zooplankton biomass), assemblages and species (*Temora longicornis* and *T. stylifera*). They highlighted distinct temporal patterns across neritic-oceanic environments and along the NW and N Iberian arch, attributing them to differences in the intensity and timing of seasonal stratification-mixing regimes in the Galician and Cantabrian Sea (Valdés et al. 2007, Bode et al. 2012b). These two zones conform distinct sub-regions according to the differentiated occurrence of upwelling (e.g., extended seasonality and 1–2 weeks event-scale cycles versus reduced seasonality and shorter events), physiographic configuration (e.g., presence of rías versus narrow and steep shelf indented by canyons) and predominant type of temperate climate (e.g., warm versus oceanic). Later studies, with extended time horizons and covering both sub-regions (Velasco et al. 2012, Bode et al. 2013), both sections in Galicia (Bode et al. 2012b) or only the section of RV (Buttay et al. 2016, 2017), confirmed average seasonal patterns of bulk properties and taxonomic groups, their across- and along-shelf spatial differences, and highlighted long-term and interannual variability driven by local to basin-scale meteo-climatic factors (Bode et al. 2009, 2012b, Buttay et al. 2016, 2017).

Seasonal patterns of relative composition of taxonomic groups and copepod species were, in general terms, similar in the northern and southern areas of the Galician sub-region (Valdés et al. 2007, Velasco et al. 2012). The average rank order of recurrence of taxonomic groups (percentage of presence in the time-series samples) was: copepods (100% recurrence), appendicularians, siphonophores and eufausiaceans (> 75%), chaetognaths, cnidarians, foraminiferans and cladocerans (ca. 50%) (Velasco et al. 2012). The contribution of copepods to the total holo-zooplankton abundance was higher in the northern (RC) section (range between

80% and 90% during summer and winter, respectively) than in the southern RV (range from 60% to 85%). Despite these seasonal differences, the ranked list of recurrent species was similar along the sub-region: *Acartia clausi, Paracalanus parvus, Oithona plumifera, Calanus helgolandicus, Pseudocalanus elongatus* and *Oncaea media* (> 75%), *Centropages* spp., *Paraeuchaeta hebes, Temora longicornis, Oithona similis* and *Centropages chiercihae* (> 50%). Some species were, however, recurrent in the northern (> 50%) but not in the southern shelf (ca. 25%), such as *Euterpina acutifrons, Ctenocalanus vanus, Ditrichocorycaeaus anglicus* and *Calocalanus styliremis*. It is worth mentioning that *Temora stylifera* integrated the list of highly recurrent species in the Cantabrian Sea but not in the Galician sub-region (15% and 30% of recurrence in RV and RC, respectively). Also, *Centropages typicus, Eucalunus elongatus* and *Corycaeus anglicus* were recurrent in the Cantabrian Sea but not in the Galician sub-region (> 50% and < 25%, respectively) (Velasco et al. 2012).

The main characteristics of zooplankton seasonality, summarized from the studies based on the RADIALES program, were:

1) The amplitude of seasonal signals of biomass and total and relative abundances of the main groups decreased across-shelf from neritic to oceanic domains (except for doliolids, salps and eufausiaceans), and along-shelf from southern (RV section) to northern (RC section) areas of the Galician sub-region and, more conspicuously, between the Galician sub-region and the Cantabrian Sea sections. The across-shelf decreasing trend is more acute for meroplankton groups (larval stages of cirripeds, decapods, gastropods, bivalves and echinoderms).

2) The average shape of the annual cycles of bulk variables and abundance of main groups tended to be unimodal in the southern area (RV), peaking around summer but bimodal in the northern (RC), with peaks around spring and autumn.

3) The period of higher total biomass and abundance (considering the median of each time series as the reference value) extended from ca. March to October in the Galician sub-region, matching the average seasonality of upwelling. This period shortened in the Cantabria Sea sub-region, where the higher biomass and abundance were usually recorded from March to June.

4) The growth season is extended during the whole upwelling season for some groups, such as copepods, appendicularians and eufausiaceans (in the mid-shelf), while it is reduced, usually from summer to early autumn, for others, such as cladocerans, cnidarians, doliolids, chaetognaths and salps. Considering copepod species, the growth season extended during the upwelling season for *A. clausi, C. helgolandicus, O. similis* and *P. parvus*, while it was more restricted for other species, which showed maximum abundances around spring (*C. vanus, Clausocalanus* spp.), summer (*T. longicornis, P. elongatus*) or autumn (*T. stylifera, O. media*).

5) The general characteristics of seasonality (amplitude, shape and timing) showed, however, conspicuous year-to-year changes linked to the variability of the main meteorological and hydrographic drivers, with a preponderant role

of upwelling dynamics and runoff pulses, interacting at seasonal and shorter sub-seasonal scales (Buttay et al. 2017). This year-to-year variability increased as the aggregation level of the considered variable decreased from bulk properties and taxonomic (functional) groups to species.

Long-term changes in zooplankton and its relationship with meteo-climatic and hydrographic variability have been the subject of several studies. Bode et al. (2009) combined data from sections RV and RC of RADIALES (1994–2006 and 1989–2006, respectively) and from the CPR standard area F4 (CPR-F4: 1958–2006) to estimate annual trends and cross-correlations for a set of abiotic and biotic variables: upwelling index (UI, sub-regional scale) and North-Atlantic Oscillation (NAO, basin-scale) and several attributes of the plankton community (phytoplankton color index, depth-integrated chlorophyll and abundance of diatoms, dinoflagellates, meso-zooplankton, copepods and the recurrent/characteristic species *A. clausi*, *C. helgolandicus* and *T. stylifera*). Posterior investigations considered an enlarged list of sub-regional (sea surface temperature from satellites; SST) and climatic indices (Atlantic Multidecadal Oscillation; AMO) and extended the analysis to diversity indices (species richness and Shannon for the copepod community; S and H, respectively) (Bode et al. 2012b) and abundance of gelatinous zooplankton (tunicates and medusae) (Bode et al. 2013). UI (NAO, AMO and SST) showed decreasing (increasing) linear trends between 1985 and 2006, concomitant with a weakening of northerly upwelling favorable winds and enhanced surface warming under the predominance of positive NAO phases in that period (Álvarez-Salgado et al. 2008). No significant or feeble trends were detected for most of the plankton time series. Only diatoms and *T. stylifera* seemed to follow the long-term pace of meteorological and hydrographic variables: the former decreased, penalized by weakening of upwelling, while the latter increased, favored by warming and strengthened stratification (Bode et al. 2009). Copepods and gelatinous time series showed periods of alternating dominance, noticeable in all the series, with extensions that were about 4–7 and < 3 years, respectively; copepods increased in coastal sites since the 90's, while the gelatinous displayed an extending period of dominance during the 80' (Bode et al. 2013). Copepod diversity (S) decreased in oceanic waters (CPR-F4) from 1958 to 2006 (and also in the contiguous Bay of Biscay, CPR-E4) (Nogueira et al. 2012), and in the mid-shelf off A Coruña (RC) from 1994 to 2006 (Bode et al. 2012b). These long-term linear trends represented ca. 4% of the variance of their respective time series. The seasonal signals amounted to ca. 28 and 45% in, respectively, the oceanic area (F4) and the coastal-shelf site (RC), and showed a bimodal pattern, more marked in RC, characterized by a primary peak in spring and a secondary one in autumn. Most of the plankton variables analyzed showed lagged (1–5 years) response to interannual meteorological and hydrographic variability, contrasting to results for higher latitudes that showed more direct, linear relationships. Delayed responses may reflect non-linear dynamics and multi-scale effects, which involved interacting temporal scales, e.g., between seasonal and short-term upwelling cycles (Buttay et al. 2017), as well as spatial scales, e.g., connections between processes at local, sub-regional and basin-scales (Bode et al. 2012b). The observed resilience of the plankton community to long-term variability of upwelling and climatic indices (acting at sub-regional and basin-scales, respectively) was attributed to the high dynamism characteristic of EBUS (Bode et al. 2009) in the Galician sub-region is imposed by the event-scale dynamics of upwelling (cycles), buffered by the strong seasonality (main mode of temporal variability) of meteorological and hydrographic drivers.

Recent analyses of the time series of the RV section (1995–2011) (Buttay et al. 2016, 2017) confirmed previously described general temporal and spatial patterns of zooplankton at different organization levels and highlighted the predominance of abiotic, weather-induced factors operating at multiple, interacting scales to drive zooplankton dynamics and community assembly and quantifying the intensity and synchronicity level of the seasonal signals of the functional groups and copepod species that shape the community. The authors suggested that observed interannual changes in the amplitude and phase of the seasonal cycles of community components, which showed marked shifts around 2000 and 2005 (increased amplitude and synchronization), were modulated by the combination of interannual changes of the seasonal cycles of upwelling (UI) and outflow from river Miño (a proxy for runoff and precipitation regime in the southern area of the sub-region) and the event-scale dynamics of upwelling cycles. Between 2000 and 2005, reinforced seasonality of UI and Miño outflow in combination with reduced duration of upwelling cycles (from an average periodicity of 11 to 7 days) resulted in:

1) Reduced losses of zooplankton in winter due to enhanced coastal retention promoted by intense downwelling and the formation of a neat shelf frontal zone associated with the development of the western Iberian buoyant plume (WIBP) (Peliz et al. 2002);

2) Favorable growth conditions during the growth season due to enhanced food supply (primary production) associated with higher (average) upwelling intensity but reduced offshore exportation linked to shorter periodicity of upwelling cycles.

The interaction of drivers at multiple scales resulted in higher amplitude and synchronization of zooplankton seasonality at all organization levels. Synchrony (versus compensatory) dynamics accentuate (relax) competition at the level of functional groups and species and may affect the stability and resilience of the community, causing its progress to another state or regime.

Regime shifts, defined as fast but perdurable changes in ecosystem structure and function (Möllmann et al. 2015), were also investigated for the northern part of the sub-region (RC, 1990–2018) (Bode et al. 2020). In this study, the authors identified two major shifts in the plankton community. The first had a turning point around 1997, separating cold-dry and warm-wet periods associated with distinct climatic conditions. Total phytoplankton biomass and meso-zooplankton abundance did not show significant differences between periods, but phytoplankton mean size (related to community structure) and trait diversity (functionality), and mero- to holo-zooplankton ratio were lower in the cold-dry phase. The other detected turning point took place around 2001 and was attributed to local hydrographic factors: continental inputs and remineralization and production rates were lower before than after this year. Community changes were measurable only for the zooplankton community, which showed a lower micro- to meso-zooplankton ratio and meso-zooplankton trait diversity, and higher meso-zooplankton mean size in the period of low (local) continental inputs.

These recent analyses of RV and RC time series confirmed the predominance of bottom-up, meteorological, climatic and hydrographic variability on zooplankton dynamics in the Galician sub-region but stressed the relevance of the combined effect of driven factors and of the interactions between these factors in a wide range of interacting temporal (Buttay et al. 2016, 2017) and spatial (Bode et al. 2020) scales. The response of the community to this multi-factorial, multiscale forcing was noticeable regarding dynamical (e.g., level of synchrony) and structural (e.g., composition of functional groups and species) aspects of the zooplankton community.

3. Update of Zooplankton Time Series From the Coastal Time-Series Observing System 'Ría de Vigo'

The 'Ría de Vigo' (RV), the southernmost of the Galician 'Rías Baixas' (Figure 1B, Table 1), is located in a highly populated area (ca. 2,700 inhabitants·km^{-2}) and houses important industrial (e.g., shipyards), aquaculture (e.g., mussel production), fisheries (e.g., harvesting precious mollusks and crustacea) and leisure activities (e.g., tourism). These activities involve multiple anthropogenic impacts and pressures on the ecosystem.

3.1 *Zooplankton Sampling in the RV Section*

The zooplankton community started to be sampled systematically in 1995 in the RV section. Samples were collected monthly within the framework of the ongoing time series monitoring project RADIALES. Three oceanographic stations (Figure 1C) are sampled by means of double-oblique hauls from the surface down to 5 m above the sea floor, using a double 40 cm diameter Bongo net with 200 μm mesh size, equipped with flow meters for the calculation of the volume of water filtered and TD (temperature-depth) sensor to record the temperature profile and depth reached in the haul. The sample from one cod-end of the net was preserved in 4% sodium tetraborate-buffered formaldehyde for taxonomic identification and counting (abundance, ind.m^{-3}); the other cod-end was retained in a 200 μm mesh for

Table 1. The Galician 'Rías Baixas'.

Ria	Length (km)	Surface (km^2)	Volume (m^3)	Max. Depth (m)	River	Mussel Platforms ('bateas')	
						N	%[1]
Vigo	33	175	3100	42	Oitavén	498	15
Pontevedra	23	145	3240	40	Lérez	343	10
Arousa	26	230	4300	65	Ulla-Umia	2319	68
Muros-Noia	12	120	2700	46	Tambre	122	3

[1] The 'Rías Altas' of Ares and Betanzos are not included; they represent ca. 4% of the total number of 'bateas' in Galicia.

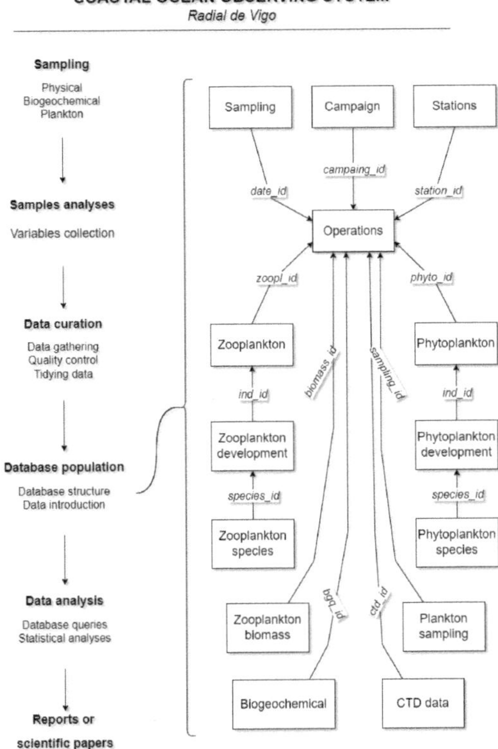

Figure 2. Sequence of operations carried out in the RV section, part of the time-series program RADIALES (left). Structure of the database VGOHAdB (right), representing the flow diagram between data tables to manage the diverse suite of metadata and physical, biogeochemical and plankton variables from sampling in the RV section. Information contained in the data tables is given in Annexe 2.

dry-weight biomass estimation (BZ_T, mg·m^{-3}). Sub-samples for taxonomic identification were taken until at least 1,000 zooplankton organisms per sample were identified to the lowest possible taxonomic level under a stereomicroscope. Counts in the sub-samples were converted to full-sample numbers per cubic meter for both total abundance (NZ_T) and abundance per taxa (NZ_i). Sampling and sample processing methods were consistent throughout the time series, and zooplankton identifications were always carried out by the same expert taxonomists.

3.2 *Database (VGOHAdB)*

Physical, biogeochemical and biological (plankton) data are organized in the database VGOHAdB (Figure 2). Data are prepared as tidy data (Wickham 2003) before being incorporated into the database to fit the database structure. Accordingly, every row is transformed into a unique and identified tuple with one or many identifiers. In this curation process, the raw data are split into different tables mirroring the database structure (Annex 2). This data curation is made mainly through different R and bash scripts that read pre-formatted Excel files and generate the SCSV (semicolon-separated values) files, which will be imported into the database. Data processing is done in a GIT version control system that warrants the control of changes and data recovery.

3.3 Data Analysis

Data analysis is carried out, as far as possible, in R using the *Tidyverse* tools (Wickham et al. 2019, 2022). Regular data analysis workflow starts with queries to the VGOHAdB database to extract target information. In order to ease the workflow and considering that data analyses are done in R, the package *dbplyr* is used to carry out those queries and to store the information in the IDE, RStudio Desktop (RStudio Team, 2020). Data extractions are made with *dbplyr*, queries with the target data of interest are run, and data are imported into RStudio as 'Tibbles'. Considering the atomized structure of the data, multiple database tables are queried simultaneously, joining them together by the identifiers of the different database tables and operations identifiers (Annex 2). Once the data is downloaded into the IDE, the connection with the database is cut, and data are ready to be analyzed.

3.4 Database and Data Analysis

We present several examples to illustrate the convenience of combining database management and time-series analysis tools to characterize the dynamics of zooplankton community attributes; (1) long-term and year-to-year variability of the seasonality of total zooplankton abundance (Figure 3), (2) temporal variability of taxonomic groups (functional or life-forms): copepods and cladocerans (Figure 4), (3) long-term and seasonal variability of indicator species *Temora longicornis* and *T. stylifera* (Figure 5).

1) Long-term and year-to-year variability of the seasonal mode of variation: Figure 3 illustrates a possible sequence of analysis of the time series downloaded from the database VGOHAdB through a combination of different statistical procedures contained in several R packages. In the example, the analysis was applied to assess the interannual variability of seasonality of total zooplankton abundance (NZ_T, ind/m³) in the outer reaches of Ría de Vigo (oceanographic station St-01, Figure 1C). The sequence of procedures for the analysis of NZ_T was as follows (A to G plots in Figure 5):

A) Screening of raw time-series data, checking for gaps (coded as NA), outliers and anomalous data. The circled value in the plot corresponded to an anomalous high peak of echinoderm eggs and larvae; this observation was ignored in subsequent analyses. The decision about data removal and substitution (time-series purge) should be made according to the objectives of the study and constraints imposed by the type of statistical/numerical analysis to be carried out;

B) Statistical distributions of monthly data (e.g., skewness, kurtosis, departure from normality, etc.) (violin plots) and basic monthly statistics (median,

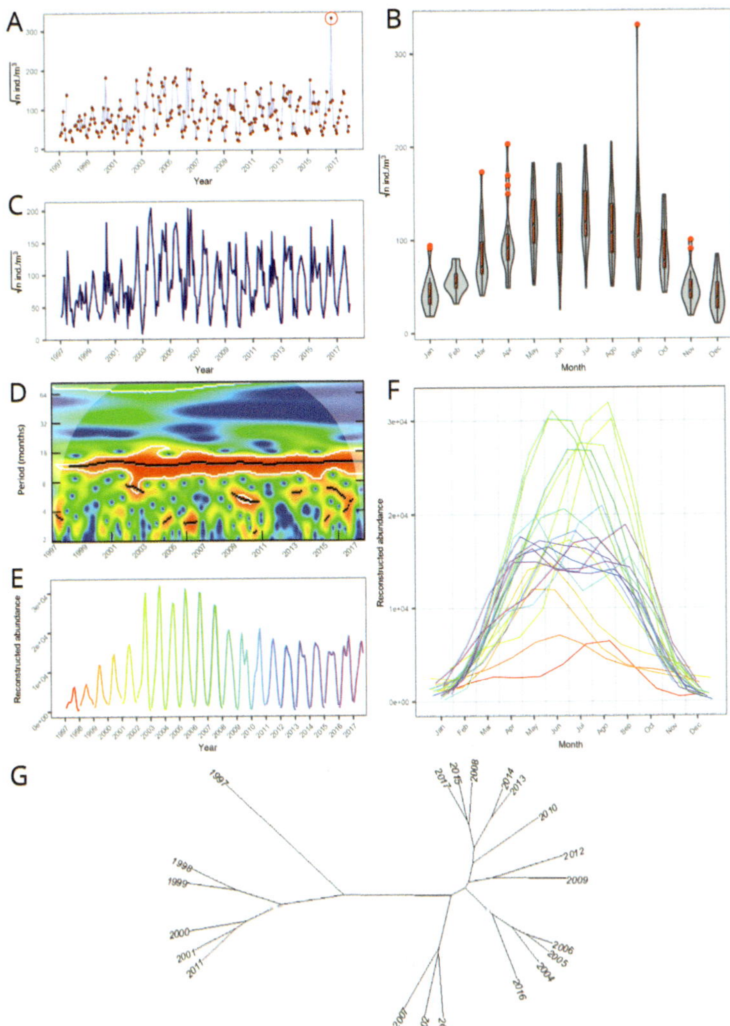

Figure 3. Example of a step-by-step regular workflow for analyzing zooplankton time-series: total zooplankton abundance (ind·m⁻³) in Ría de Vigo (St. 01, Figure 1) from 1997 to 2019. (A) Raw time-series: inspection of the series (general patterns, gaps) and detection of 'anomalous' data and outliers (circled value in September 2017); (B) Violin-boxplots (monthly): statistical distribution of (monthly) values (box-plot inserted marking the median, 1st and 3rd quartiles and 1st and 9th deciles); (C) Purged (anomalous and/or outliers removed), interpolated (linear) and regularised (30 days intervals) time-series; (D) Wavelet scalogram of the purged, interpolated and regularized time-series (C) showing the significant periodic modes (T around 12, 8, 6 and 4 months); (E) Time-series of the reconstructed seasonal mode of variation (annual and semi-annual components from wavelet analysis); (F) Superimposed annual cycles extracted from the time-series of the seasonal mode; line colours match the ones of the time-series of the reconstructed seasonal mode (E); (G) Phyilogram of the similarity among the reconstructed annual seasonal cycles (F).

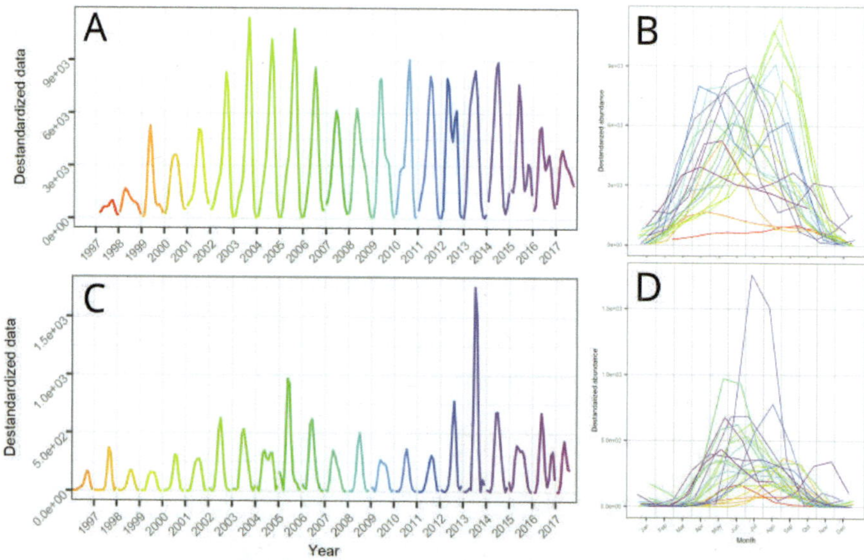

Figure 4. Time-series of abundance (square root transformed, ind·m⁻³) of the reconstructed seasonal mode (left) and the ensemble of seasonal cycles (right), (A–B) for copepods and (C–D) for cladocerans.

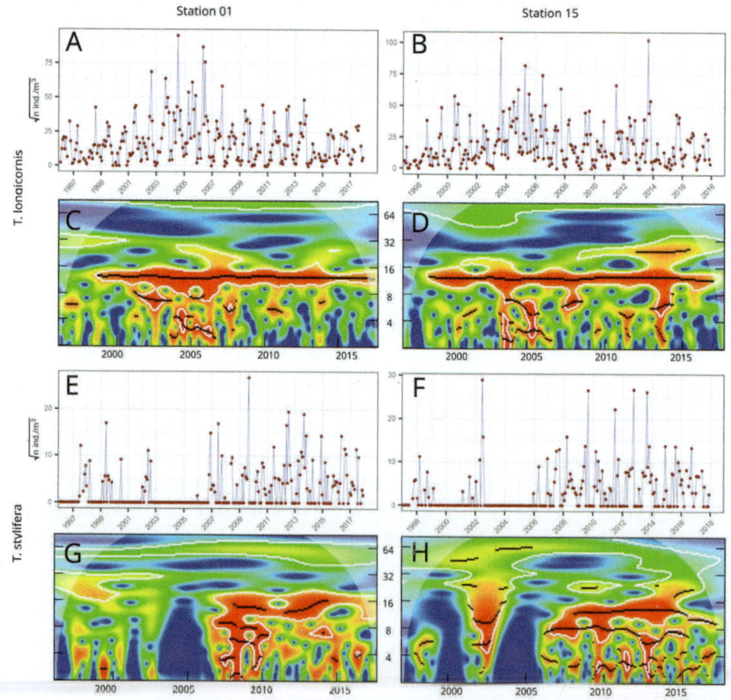

Figure 5. Raw time-series of abundance (square root transformed, ind·m⁻³) of *Temora longicornis* (A–B for stations St 01 and 15 respectively) and *T. stylifera* (E–F). Scalograms of the regularised time series of *T. longicornis* (C–D) and *T. stylifera* (G–H).

1st and 3rd quartiles and 1st and 9th deciles) (inserted box-plots in Figure 3B) help to decide data transformations (for NZ_T, normalization by square-root);

C) Time-series interval regularization is required for the application of most of time-series analysis techniques. Diagnosis of time-series characteristics (e.g., time units, number of gaps, minimum, maximum and average intervals, etc.), time-interval regularization and interpolation were accomplished with package *pastecs* (Grosjean and Ibanez 2018). In the example, the NZ_T time series was regularized to a 30-day interval (close to the average 31 days), and NA was filled up by time-weighted linear interpolation, maximizing the number of actual values maintained in the regularized time series and setting the start and end of the time series in January 1997 and December 2017 (244 time-records);

D) Wavelet analysis of the regularized series. The scalogram of the wavelet power spectrum points out the main cyclical modes of temporal variability (y-axis: periodicity in months, T) and their variability through time (x-axis: time in months, t). Wavelet analysis was carried out with the package *WaveletComp* (Roesch and Schmidbauer 2018);

E) Time series of the seasonal mode of variation were reconstructed from the annual and semi-annual components extracted from wavelet analysis, showing the interannual variability of seasonality (wavelets for T = 12 and 6 months);

F) Stacked annual cycles extracted from the time series of the seasonal mode of variation, showing the year-to-year varying amplitude, timing and shape of the annual cycles;

G) Phylogram from hierarchical clustering of the seasonal cycles according to the similarity (Canberra distance) of the reconstructed annual cycles (packages *hclust* and *ape*).

2) Long-term and seasonal variability of life-form pairs: copepods and cladocerans. Figure 4 shows the time series of the seasonal mode of variation (Figures 4A and 4C) and stacked annual cycles (Figures 4B and 4D) for copepods and cladocerans (upper and lower rows, respectively), obtained according to the sequential time-series analysis scheme described above (Figure 3A–G). The series showed contrasting patterns. In terms of interannual variation of the seasonal mode, both series showed an increment in the amplitudes of their respective seasonal signals between 2002 and 2006 (Figure 3E for NZ_T) (Buttay et al. 2016, 2017). However, the last years of the series showed a distinct pattern, particularly from 2012 onwards, when the amplitude of the seasonal mode for cladocerans (Figure 4C) increased relative to that for copepods (Figure 4A). This reduction of the copepods to cladocerans ratio may be expected under the scenario of decreasing upwelling and increasing surface warming and stratification. Regarding seasonality patterns, the ensemble of annual cycles evidenced the larger extent of the growth season for copepods (Figure 4B), from spring to autumn along the upwelling season, and the uncoupling between the

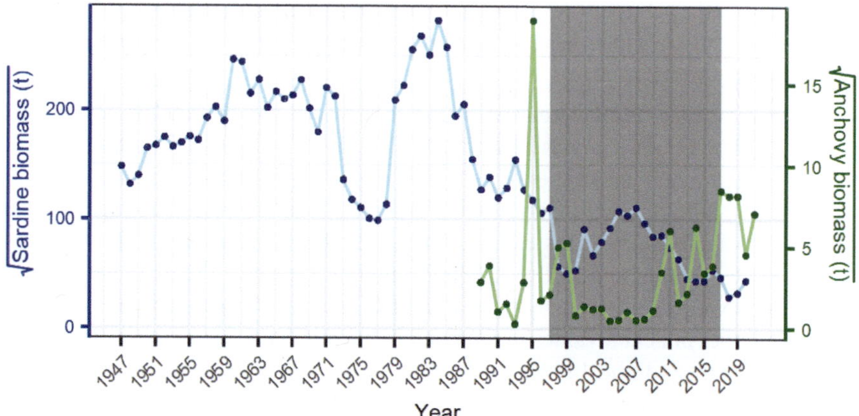

Figure 6. Annual landings (tonnes) of sardine (left y-axis, blue line) and anchovy (right, green) from 1947 to 2020 in the IX-a North ICES area.

seasonal cycles of both groups, with annual peak values centered in summer for cladocerans (Figure 4D). Seasonal differences reflected the higher diversity of the copepod community, which resulted in a larger suite of adaptive traits to cope with the high dynamism characterizing coastal-shelf habitats in EBUS.

3) Temporal dynamics of indicator species *Temora longicornis* and *T. stylifera*. The raw time series of these congeneric species and their scalograms (for the regularized series) are shown in Figure 5 for stations St 01 and 15. The time series of the two species showed contrasting dynamics. The time course of *T. longicornis* raw time series (Figure 5A and 5B for St 01 and 15, respectively) followed in general terms that of copepods (Figure 4A) and total zooplankton abundance, Figure 3A–B, E), exhibiting four distinct long-term periods (local trends) (Buttay et al. 2016), those identified by Buttay et al. (2016, 2017), with turning points around 2002 and 2007 separating periods of low, high and medium seasonal amplitude respectively, and a new one around 2014 marking the start of a period in which amplitude values of the annual signals were similar to those recorded at the beginning of the series. The raw series of *T. stylifera* (Figure 5E–F for St 01 and 15, respectively) showed two evident local trends, with a turning point around 2007 separating long-term periods of low versus high recurrence and abundance, respectively before and after the population dynamics shift. The scalograms for *T. longicornis* (Figure 5C–D for St 01 and 15) emphasized the significance of seasonality (T around 12 and 6 months) along the time series and of sub-seasonal modes during the years of greater seasonal amplitude in the long-term (local trend) period between ca. 2002 and 2007. For *T. stylifera*, the scalograms (Figure 5G–H for St 01 and 15) pointed out its very low and sporadic recurrence before the turning in point in 2007, and the increased recurrence and consolidation of a seasonal (T = 12 and 6 months), although fluctuating (significant at T < 4 months), mode of temporal variation afterwards. It is worth mentioning that the increased recurrence and consolidation of seasonality observed for *T. longicornis* since 2007 are coincident with, respectively, the decrease of the biomass of the stocks of sardine (*Sardina*

pilchardus) and the increase of that for anchovy (*Engraulis encrasicolus*) in the northern part of the ICES area IXa (Figure 6).

4. Concluding Remarks

Zooplankton community assembly and dynamics (considering temporal fluctuations but also spatial distributions) showed a significant relationship with the variability of several meteorological and climatic factors. In the Galician sub-region of the CanCUS, the influences of a temperate climate, upwelling and downwelling regimes and continental freshwater inputs play a prominent role. These factors operate in combination, on a wide range of interacting temporal/spatial scales, from short-term/local to long-term/basin-scales, determining in turn the hydrodynamic and hydrographic conditions of the pelagic habitat inhabited by zooplankton, such as the across- and along-shelf exchanges and temperature distributions (abiotic determinants of the habitat), as well as the fate of phytoplankton production and species composition (biotic) that constitute the main food provision for most holo- and mero-zooplankton species.

The spatial distribution of zooplankton community attributes traced, in general terms, the hydrographic structures resulting from the action of the physical oceanographic processes occurring in the sub-region. At local and meso-scales, the changing aspects of the across-shelf upwelling cell and the two-layered positive circulation characteristic of the rías shape contrast coastal and shelf habitats separated by neat but movable frontal zones. At regional scales, the along-shelf physical processes associated with upwelling dynamics in EBUS, namely the IPC along the Iberian section of the CanCUS, as well as the influence of the plume from river Miño (WIBP), also define contrasting spatial domains along the Galician sub-region and marked differences between this and the contiguous Cantabrian Sea (southern Bay of Biscay).

Seasonality is the principal mode of temporal variability of zooplankton, noticeable at the level of bulk (aggregated) properties, functional groups and species. This seasonality resulted from the temperate character of the ecosystem and the seasonal and sub-seasonal, event-scale dynamics of coastal upwelling (upwelling cycles) and, to a lesser extent, freshwater inputs from river discharges at the head of the rías and from river Miño. This mode of variation showed, however, significant year-to-year variability, displayed as changes in the characteristics of the seasonal mode, amplitude, timing and shape, resulting from the combination of the annual and semi-annual (and other sub-seasonal) cyclical components.

Long-term variability accounted for a minor but significant contribution to total temporal variance. A series of tipping points marking shifts in the time series were detected for several variables, reflecting trends and interannual variability of climatic variables (indices) operating at basin-scale as well as of meteorological and hydrographic processes occurring at local and regional scales. Worth mention is the apparent long-term increment of recurrence and consolidation of seasonality of the *T. stylifera* (in regards to the congeneric *T. longicornis*) since ca. 2007 and of the relative proportion of cladocerans relative to copepods since ca. 2012. This relative increment of thermophilic species and groups is concomitant with a reported decrease in upwelling intensity and increased SST in the Galician sub-region.

Time series of plankton components are essential to diagnose the state of pelagic habitats, as it is conceived in recent European policies (MSFD). They also offer the possibility of using them as prognosis tools when coupled with ocean-coastal ecosystem models. The strength of time series depends on the maintenance of the existing time-series programs framed on coastal-ocean observing systems, such as the COOS 'Ría de Vigo', which integrate variables (from meteorological and hydrographic drivers to ecosystem), approaches (observational, experimental and modeling) and purposes (research components and environmental or resources assessment).

5. Acknowledgment

We are indebted to all scientists and technicians who contributed to the ongoing time-series coastal-ocean observing system 'Radial de Vigo' and its development since 1987, some of them now retired. We extend our gratitude to those who participated in the development and maintenance of the other oceanographic sections distributed along the NW and N Iberian coastal and shelf waters that also form part of the time-series program RADIALES. We also owe a debt of gratitude to the crews of the research vessel RV 'J. M. Navaz' (out of operation since March 2018), RV 'R. Margalef' and RV 'A. Alvariño' who made possible the observations at sea. We acknowledge Dra. Isabel Riveiro and Dr. Fernando Ramos for kindly providing sardine and anchovy landings in ICES area IXa-North.

Annex 1. Beginnings and development of zooplankton research in the Galician sub-region of the CanCUS

Beginnings: 1880–1950

In 1880, the collapse of the sardine fisheries on the French Atlantic coasts led to a deep economic crisis in the French canning sector derived from the alarming drop in catches. To alleviate that situation, the proposed solution was the exploitation of fishing grounds and the installation of new canning factories (mixed capital companies) in Galicia and Portugal, whose fishing grounds seemed to be unaffected. At that time, a series of cruises were carried out by Prince Albert I of Monaco on board *L'Hirondelle* (1885–1886), *Princess Alice* (1894 and 1896) and *Princess Alice II* (1908) on the Galician coast and shelf waters (Figure 7). As a result, Pouchet and Guerne published 1887 '*Sur la nourriture de la sardine*', where the authors analyzed gut contents of sardines fished off A Coruña and which can be considered the first work dealing with zooplankton in Galicia. Roque Carús Falcón published in 1903

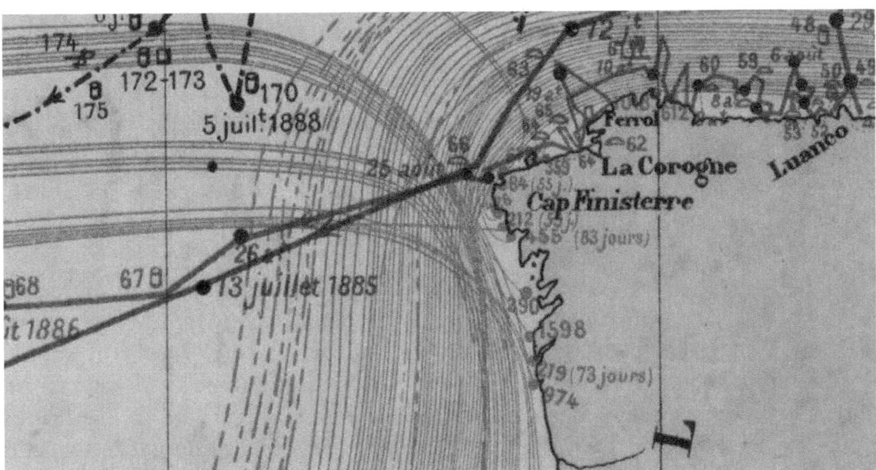

Figure 7. Navigation routes followed by the 'L'Hirondelle' yacht close Galicia (from *Itinéraires du yacht "L'Hirondelle" dans l'océan Atlantique nord en 1885, 1886, 1887, 1888* (Université de Burdeux; https://1886.u-bordeaux-montaigne.fr/s/1886/item/245653#?c=&m=&s=&cv=&xywh=-425%2C-77%2C14441%2C10697).

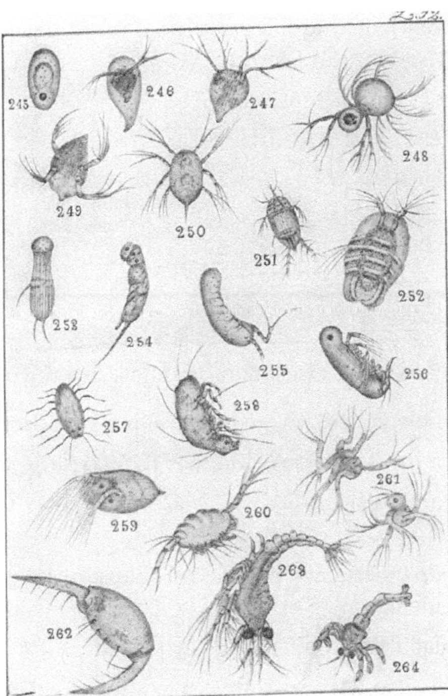

Figure 8. Example of zooplankton drawings by Roque Carús Falcón (from "Los misterios de la Naturaleza. Investigaciones sobre el micro-plakton de la Ría de Arosa". Biblioteca Digital Hispánica (Biblioteca Nacional de España).
http://bdh.bne.es/bnesearch/CompleteSearch.do;jsessionid=076C2C86297E0EDE8D8FF93F-79089D06?showYearItems=&field=autor&advanced=&exact=&textH=&completeText=&text=%-22Car%c3%bas+Falc%c3%b3n%2c+Roque+%22&pageSize=1&pageSizeAbrv=30&pageNumber=1).

the first illustrated guide including zooplankton: *"Los misterios de la naturaleza. Investigaciones sobre el microplancton de la Ría de Arosa"* (RA in Figure 1B). Despite the limited means at his disposal (a small boat, a bucket to collect seawater and a microscope), the 240 pages publication included a series of high-quality drawings on his observations (Figure 8).

Development From 1950

The impulse of zooplankton research in Galicia was driven by the activity of a series of scientists, some of them of international reputation, the foundation of several institutions devoted to marine investigation and the accomplishment of several multidisciplinary research projects, some in collaboration with renowned international institutions. Among the pioneers of zooplankton research, it should be highlighted Ángeles Alvariño (e.g., Alvariño 1955), a recognized expert on chaetognaths (e.g., Alvariño 1960) who started at IEO but developed most of his career in the USA (Scripps Institution of Oceanography and National Oceanic and Atmospheric Administration) (González-Garcés 2016). Ramón Margalef (Prat et al. 2015), who made outstanding contributions to the ecology of phytoplankton (Margalef's mandala) (Margalef 1978, Margalef et al. 1979, Litchman and Klausmeier 2008, Glibert 2016), also contributed to the knowledge of zooplankton species (Massuti and Margalef 1950, Margalef and Durán 1953).

Foundation of Research Facilities

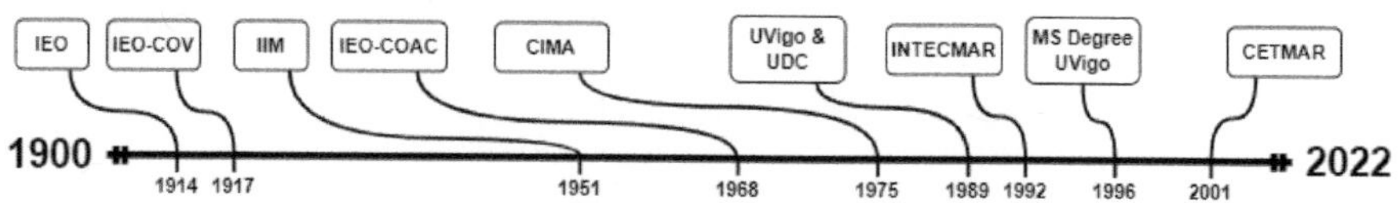

1914 IEO. Instituto Español de Oceanografía (https://www.ieo.es/en/home)

1917 COV. Centro Oceanográfico de Vigo (IEO) (https://www.ieo.es/es/web/vigo/home)

1952 IIM. Instituto de Investigaciones Marinas (CSIC) (https://www.iim.csic.es/)

1968 COAC. Centro Oceanográfico de A Coruña (IEO) (http://www.ieo.es/es/web/coruna/)

1975 CIMA. Centro de Investigacións Mariñas (https://mar.xunta.gal/gl/formacion/centros-de-investigacion/cima)

1989 UDC. Universidade de A Coruña (https://www.udc.es/)

1989 UVIGO. Universidade de Vigo (https://www.uvigo.gal/en). Degree in Marine Sciences in 1996 (https://mar.uvigo.es/)

1992 INTECMAR. Instituto Tecnolóxico para o Control do Medio Mariño de Galicia (http://www.intecmar.gal/)

2001 CETMAR. Centro Tecnolóxico do mar (https://cetmar.org/?lang=en)

Annex 2. Metadata and data tables in the VGOHAdB database (Figure 2)

Sampling: Timestamp of each sampling, as well as the full date extended in columns (year, month, day, hour and minute) and a unique identifier for each date.

Campaign: Includes all relevant information of each campaign/survey/project from which there is data and a unique identifier for each campaign.

Stations: Incorporate geographical information about the sampling stations, such as latitude and longitude or nearest distance to the coast and a unique identifier for each date.

Plankton Sampling: Contains metadata information related to sampling and analytical methods of the different variables (e.g., Niskin bottles, plankton nets, etc.).

Operations: Relates all information from sampling. This central table coalesces information from the three predictor tables (stations, campaign and sampling) into tuples and establishes a unique identifier for each.

Response variables: Provides the actual values of the different variables that have been obtained after sample processing and analysis:

- **Zooplankton:** Abundance of each zooplankton species (or taxa in some cases) identified and counted (number of individuals per cubic meter).

- **Development:** Assigns for each zooplankton species (or taxa) a tuple of different life-form traits, such as size, development stage, habitat, etc.

- **Taxonomy:** Assigns for each different zooplankton species (or taxa) a tuple with all the complete taxonomic affiliation, the aphiaID and the WoRMS link (that ensures that taxonomic names are updated).

- **Zooplankton biomass:** Total zooplankton abundance and the biomass of the samples (within the ca. 0.2 to 2 mm size range), as well as supplementary information necessary to abundance and biomass estimation (e.g., filtered volume).

- **Biogeochemical:** Information on biogeochemical variables (e.g., concentration of inorganic nutrients) registered for each sampling date, location and water column depth.

- **CTD:** Includes data (generally as water column profiles) recorded by automatic probes, such as CTD (conductivity-temperature-depth) or fluorescence.

References

Álvarez, I., Prego R., deCastro M. and Varela M. 2012. Galicia upwelling revisited: out-of-season events in the rias (1967–2009). Ciencias Marinas 38(1B): 143–159.

Álvarez-Salgado, X.A., Figueiras, F.G., Pérez, F.F., Groom, S., Nogueira, E., Borges, A.V. et al. 2003. The portugal coastal counter current off NW spain: new insights on its biogeochemical variability. Progress in Oceanography 56: 281–321.

Álvarez-Salgado, X.A., Labarta, U., Fernández-Reiriz, M.J., Figueiras, F.G., Rosón, G., Piedracoba, S. et al. 2008. Renewal time and the impact of harmful algal blooms on the extensive mussel raft culture of the Iberian coastal upwelling system (SW Europe). Harmful Algae 7: 849–855. doi: 10.1016/j.hal.2008.04.007.

Alvariño, A. 1955. Zooplankton investigations. Report of the Council, Journal of the Marine Biological Association of the United Kingdom, Cambridge 34(3): 665–666.

Alvariño, A. 1960. Chaetognats and the California current. CalCOFI. Lake Arrowhead: 34–35.

Ambar, I. and Fiuza, A. 1994. Some features of the portugal current system: a poleward slope undercurrent, an upwelling related southward flow and an autumn-winter poleward coastal surface current, 2nd international conference on air-sea interaction and on meteorology and oceanography of the coastal zone, preprints, September 22–27. American Meteorological Society, pp. 286–287.

Arístegui, J., Barton, E.D., Álvarez-Salgado, X.A., Miguel, A., Santos, P., Figueiras, F. et al. 2009. Sub-regional ecosystem variability in the canary current upwelling. Prog. Oceanogr. 83(1–4): 33–48.

Bagøien, E., Miranda, A., Reguera, B. and Franco, J.M. 1996. Effects of two paralytic shellfish toxin producing dinoflagellates on the pelagic harpacticoid copepod Euterpina acutifrons. Marine Biology 126: 361–369. doi: 10.1007/BF00354618.

Barquero, S., Cabal, J.A., Anadón, R., Fernández, E., Varela, M. and Bode, A. 1998. Ingestion rates of phytoplankton by copepod size fractions on a bloom associated with an off-shelf front off NW spain. Journal of Plankton Research 20(5): 957–972. doi: 10.1093/plankt/20.5.957.

Barreiro, A., Guisande, C., Maneiro, I., Vergara, A.R., Riveiro I. and Iglesias, P. 2007. Zooplankton interactions with toxic phytoplankton: some implications for food web studies and algal defence strategies of feeding selectivity behaviour, toxin dilution and phytoplankton population diversity. Acta Oecologica 32(3): 279–290. doi: 10.1016/j.actao.2007.05.009.

Batten, S.D., Fileman, E.S. and Halvorsen, E. 2001. The contribution of microzooplankton to the diet of mesozooplankton in an upwelling filament off the north west coast of spain. Progress in Oceanography 51: 385–398. doi: 10.1016/S0079-6611(01)00076-3.

Beck, H.E., Zimmermann, N.E., McVicar, T.R., Vergopolan, N., Berg, A. and Wood, E.F. 2018. Data descriptor: present and future köppen-geiger climate classification maps at 1-km resolution. SCIENTIFIC DATA | 5: 180214 | DOI: 10.1038/sdata.2018.214.

Bedford, J., Johns, D., Greenstreet, S. and McQuatters-Gollop, A. 2018. Plankton as prevailing conditions: a surveillance role for plankton indicators within the marine strategy framework directive. Marine Policy 89: 109–115. doi: 10.1016/j.marpol.2017.12.021.

Blanco-Bercial, L., Álvarez-Marqués, F. and Cabal, J.A. 2006. Changes in the mesozooplankton community associated with the hydrography off the north-western Iberian Peninsula. ICES Journal of Marine Science 63(5): 799–810. doi: 10.1016/j.icesjms.2006.03.007.

Bode, A., Varela, M., Fernández, E., Arbones, B., González, N., Carballo, R. et al. 1994. Biological characteristics of the plankton associated to a shelf-break front off the galician coast. Gaia 8: 9–18.

Bode, A., Álvarez-Ossorio, M.T. and González, N. 1998. Estimations of mesozooplankton biomass in a coastal upwelling area off NW Spain 20(5): 1005–1014.

Bode, A., Álvarez-Ossorio, M.T., Barquero, S., Lorenzo, J., Louro, A. and Varela, M. 2003. Seasonal variations in upwelling and in the impact of copepods on phytoplankton off a coruña (Galicia, NW Spain). Journal of Experimental Marine Biology and Ecology 297: 85–105. doi: 10.1016/S0022-0981(03)00370-8.

Bode, A., Carrera, P. and Lens, S. 2003. The pelagic food-web in the upwelling ecosystem of Galicia (NW Spain) during spring: natural abundance of stable carbon and nitrogen isotopes. ICES Journal of Marine Science 60: 11–22. doi: 10.1006/jmsc. 2002. 1326.

Bode, A., Álvarez-Ossorio, M.T., Carrera, P. and Lorenzo, J. 2004. Reconstruction of trophic pathways between plankton and the North Iberian sardine (*Sardina pilchardus*) using stable isotopes. Scientia Marina 68: 165–178. doi: 10.3989/scimar.2004.68n1165.

Bode, A. and Álvarez-Ossorio, T.A. 2004. Taxonomic versus trophic structure of mesozooplankton: a seasonal study of species succession and stable carbon and nitrogen isotopes in a coastal upwelling ecosystem. ICES Journal of Marine Science 61: 563–571. doi: 10.1016/j.icesjms.2004.03.004.

Bode, A., Barquero, S., González, N., Álvarez-Ossorio, M.T. and Varela, M. 2004. Contribution of heterotrophic plankton to nitrogen regeneration in the upwelling ecosystem of a coruña (NW Spain). Journal of Plankton Research 26(1): 11–28. doi: 10.1093/plankt/fbh003.

Bode, A., Álvarez-Ossorio, M.T., Cunha, M.E., Garrido, S., Peleteiro, J.B., Porteiro, C. et al. 2007. Stable nitrogen isotope studies of the pelagic food web on the Atlantic shelf of the Iberian Peninsula. Progress in Oceanography 74: 114–131. doi: 10.1016/j.pocean.2007.04.005.

Bode, A., Álvarez-Ossorio, M.T., Cabanas, J.M., Miranda, A. and Varela, M. 2009. Recent trends in plankton and upwelling intensity off Galicia (NW Spain). Progress in Oceanography 83: 342–350. doi: 10.1016/j.pocean.2009.07.025.

Bode, A., Álvarez-Ossorio, M.T., Miranda, A., López-Urrutia, A. and Valdés, L. 2011. Comparing copepod time-series in the north spain: spatial autocorrelation of community composition. Progress in Oceanography 97–100: 108–119. doi: 10.1016/j.pocean.2011.11.013.

Bode, A., Lavín, A. and Valdés, L. 2012a. Cambio climático y oceanográfico en el atlántico del norte de España. Temas de Oceanografía Nº5. Instituto Español de Oceanografía, Ministerio de Ciencia e Innovación. ISBN: 978-84-95877-08-6.

Bode, A., Álvarez-Ossorio, M.T., Miranda, A. and Ruíz-Villareal, M. 2013. Shifts between gelatinous and crustacean plankton in a coastal upwelling region. ICES Journal of Marine Science 70(5): 934–942. doi: 10.1093/icesjms/fss193.

Bode, A., Álvarez, M., García-García, M.L., Louro, M.A., Nieto-Cid, M., Ruíz-Villareal, M. et al. 2020. Climate and local hydrography underlie recent regime shifts in plankton communbities off Galicia (NW Spain). Oceans 1: 181–197. doi: 10.3390/oceans1040014.

Bode, A., Lamas, A.F. and Monpeán, C. 2020. Effects of upwelling intensity on nitrogen and carbon fluxes through the planktonic food web off a coruña (Galicia, NW Spain) assessed with stable isotopes. Diversity 12: 121. doi:10.3390/d12040121.

Bucklin, A., Peijnenburg, K., Kosobokova, K.N., O'Brien, T.D., Blanco-Bercial, L., Cornils, A. et al. 2021. Toward a global reference database of COI barcodes for marine zooplankton. Marine Biology 168(78): 1–26 doi: 10.1007/s00227-021-03887-y.

Buttay, L., Cazelles, B., Miranda, A., Casas, G., Nogueira, E. and González-Quirós, R. 2017. Environmental multi-scale effects on zooplankton inter-specific synchrony. Limnology and Oceanography 62: 1355–1365.

Buttay, L., Miranda, A., Casas-Rodríguez, G., González-Quirós, R. and Nogueira, E. 2016. Long-term and seasonal zooplankton dynamics in the northwest Iberian shelf and its relationship with meteo-climatic and hydrographic variability. Journal Plankton Research 38(1): 106–121.

Buttay, L., Vasseur, D.A., González-Quirós, R. and Nogueira, E. 2022. Nutrient limitation can explain a rapid transition to synchrony in an upwelling-driven diatom community. Limnol. Oceanogr. 67: S298–S311. doi: 10.1002/lno.12033.

Cabal, J., González-Nuevo, G. and Nogueira, E. 2008. Mesozooplankton species distribution in the NW and N Iberian shelf during spring 2004: relationship with frontal structures. Journal of Marine Systems 72: 282–297. doi: 10.1016/j.jmarsys.2007.05.013.

Cáceres-Martínez, J. and Figueras, A. 1998. Distribution and abundance of mussel (Mytilus galloprovincialis Lmk) larvae and post-larvae in the ria de vigo (NW Spain). Journal of Experimental Marine Biology and Ecology 229: 277–287. doi: 10.1016/S0022-0981(98)00059-8.

Ceballos, S., Cabal, J.A. and Álvarez-Marqués, F. 2004. Reproductive strategy of calanoides carinatus and calanus helgolandicus during a summer upwelling event off NW spain. Marine Biology 145(4): 739–750. doi: 10.1007/s00227-004-1357-z.

Chiba, S., Tadokoro, K., Sugisaki, H. and Saino, T. 2006. Effects of decadal climate change on zooplankton over the last 50 years in the western subarctic north Pacific. Global Change Biology 12(5): 907–920. doi: 10.1111/j.1365-2486.2006.01136.x.

Fernández de Puelles, M.L., Valdés, L., Varela, M., Alvarez-Ossorio, M.T. and Halliday, N. 1996. Diel variations in the vertical distribution of copepods off the north coast of Spain. ICES Journal of Marine Science 53(1): 97–106. doi: 10.1006/jmsc.1996.0009.

Fileman, E. and Burkill, P. 2001. The herbivorous impact of microzooplankton during two short-term lagrangian experiments off the NW coast of Galicia in summer 1998. Progress in Oceanography 51: 361–383. doi: 10.1016/S0079-6611(01)00075-1.

Fraga, F. 1981. Upwelling off the galician coast, northwest spain. pp. 176–182. In Coastal Upwelling. Coastal and Estuarine Science, Richards F.A. (Ed.), Vol. 1 (Washington, DC: AGU).

Fraga, F. and Margalef, R. 1979. Las rias gallegas. Estudio y Explotación del mar en Galicia. Santiago de Compostela, 101–121.

Fusté, X. and Gili, J.M. 1991. Distribution pattern of decapod larvae off the north-western Iberian Peninsula coast (NE Atlantic). Journal of Plankton Research 13(1): 217–228. doi: 10.1093/plankt/13.1.217.

García-Soto, C., Pingree, R.D. and Valdés, L. 2002. Navidad development in the southern bay of biscay: climate change and swoddy structure from remote sensing and in situ measurements. Journal of Geophysical Research 107: 1–29.

Gilcoto, M., Alvarez-Salgado, X.A. and Pérez, F.F. 2001. Computing optimum estuarine residual fluxes with a multiparameter inverse method (OERFIM): application to the ria de vigo (NW Spain). Journal of Geophysical Research-Oceans 106: 31303–31318.

Gomez-Gesteira, M., Beiras, R., Presa, P. and Vilas, F. 2011. Coastal processes in northwestern Iberia, Spain. Estuarine, Coastal and Shelf Science 64(4): 721–737.

González, A.F., Otero, J., Guerra, A., Prego, R., Rocha, F. and Dale, A.W. 2005. Distribution of common octopus and common squid paralarvae in a wind-driven upwelling area (ria of vigo, northwestern spain). Journal of Plankton Research 27(3): 271–277. doi: 10.1093/plankt/fbi001.

González-Garcés, A. 2016. Ángeles alvariño gonzález, investigadora marina de relevancia mundial. Temas de Oceanografía. Instituto Español de Oceanografía. 118 pp. ISBN: 978-84-95877-54-3.

González-Nuevo, G. and Nogueira, E. 2005. Intrusions of warm and salty waters onto the NW and N Iberian shelf in early spring and its relationship to climate variability. Journal of Atmospheric and Ocean Science 10(4): 361–375.

Grosjean, P. and Ibanez, F. 2018. Pastecs: package for analysis of space-time ecological series.

Guisande, C., Frangópulos, M., Carotenuto, Y., Maneiro, I., Riveiro, I. and Vergara, A.R. 2002a. Fate of paralytic shellfish poisoning toxins ingested by the copepod Acartia clausi. Marine Ecology Progress Series 240: 105–115. doi: 10.3354/meps240105.

Guisande, C., Maneiro, I., Riveiro, I., Barreiro, A. and Pazos, Y. 2002b. Estimation of copepod trophic niche in the field using amino acids and marker pigments. Marine Ecology Progress Series 239: 147–156. doi: 10.3354/meps239147.

Halvorsen, E., Hirst, A.G., Batten, S.D., Tande, K.S. and Lampitt, R.S. 2001. Diet and community grazing by copepods in an upwelled filament off the NW coast of spain. Progress in Oceanography 51: 399–421. doi: 10.1016/S0079-6611(01)00077-5.

Harris, R., Wiebe, P., Lenz, J., Skjoldal, H.R. and Huntley, M. (eds.). 2000. ICES Zooplankton Methodology Manual. Academic Press. ISBN 0-12-327645-4. 684 pp.

Hays, G.C., Richardson, A.J. and Robinson, C. 2005. Climate change and marine plankton. Trends in Ecolohgy and Evolution 20(6): 337–44. doi: 10.1016/j.tree.2005.03.004.

Ibarbalz, F.M., He Henry, N., Brandao, M.C., Lombard, F., Bowler, C. and Zingere, L. 2019. Global trends in marine plankton diversity across kingdoms of life. Cell 179: 1084–1097. doi: 10.1016/j.cell.2019.10.008.

Isla, J.A. and Anadón, R. 2004. Mesozooplankton size-fractionated metabolism and feeding off NW spain during autumn: effects of a poleward current. ICES Journal of Marine Science 61: 526–534. doi: 10.1016/j.ecss.2004.04.011.

Kämpf, J. and Chapman, P. (eds.). 2016. Upwelling system of the world. A Scientific Journey to the most Productive Marine Ecosystems. Springer, ISBN 978-3-319-42522-1. 434 pp.

Kehayias, G. (ed.). 2014. Zooplankton: Species Diversity, Distribution and Seasonal Dynamics. Nova Science Publishers. ISBN 1629486809. 137 pp.

Litchman, E. and Klausmeier, C.A. 2008. Trait-based community ecology of phytoplankton. Annual Review of Ecology, Evolution and Systematics 39: 615–639. doi: 10.1146/annurev.ecolsys.39.110707.173549.

López-López, L., Miranda, A., Casas, G., Preciado, M.I. and Tel, E. 2014. Comparison of mesozooplankton assemblages across quasi-synoptic oceanographic features on the north-western Iberian shelf break. Hydrobiologia 741: 193–203. doi: 10.1007/s10750-014-1924-3.

Macho, G., Molares, J. and Vázquez, E. 2005. Timing of larval release by three barnacles from the NW Iberian Peninsula. Marine Ecology Progress Series 298: 251–260. doi: 10.3354/meps298251.

Margalef, R. 1978. Life-forms of phytoplankton as survival alternatives in an unstable environment. Oceanologica Acta 1(4): 493–509.

Margalef, R. and Durán, M. 1953. Microplancton de vigo, de octubre de 1951 a septiembre de 1952. Publicaciones del Instituto de Biología Aplicada 13: 5–78.

Massuti, M. and Margalef, R. 1950. Introducción al estudio del plancton marino. Patronato Juan de la Cierva de Investigación Técnica (CSIC). Barcelona, 184 pp.

McQuatters-Gollop, A., Atkinson, A., Aubert, A., Bedford, J., Best, M., Bresnan, E. et al. 2019. Plankton lifeforms as a biodiversity indicator for regional-scale assessment of pelagic habitats for policy. Ecological Indicators 101: 913–925. doi: 10.1016/j.ecolind.2019.02.010.

Möllman, C., Folke, C., Edwards, M. and Conversi, A. 2015. Marine regime shifts around the globe: Theory, drivers and impacts. Philosophical Transactions of the Royal Society B 370.

Neshyba, S.J., Mooers, C.N.K., Smith, R.L. and Barber, R.T. 1989. Poleward flows along eastern ocean boundaries. Coastal and Estuarine Studies, vol. 34. Springer–Verlag. 374 pp.

Nogueira, E., González-Nuevo, G., Bode, A., Varela, M., Morán, X.A.G. and Valdés, L. 2004. Comparison of biomass and size spectra derived from optical plankton counter data and net samples: application to the assessment of mesoplankton distribution along the Northwest and North Iberian Shelf. ICES Journal of Marine Science 61: 508–517. doi: 10.1016/j.icesjms.2004.03.018.

Nogueira, E., González-Nuevo, G. and Valdés, L. 2012. The influence of phytoplankton productivity, temperature and environmental stability on the control of copepod diversity in the north east Atlantic. Progress in Oceanography 97–100: 92–107. doi: 10.1016/j.pocean.2011.11.00.

Peliz, A., Dubert, J., Haidvogel, D.B. and LeCann, B. 2003. Generation and unstable evolution of a density-driven eastern poleward current: the Iberian poleward current. Journal of Geophysical Research 108: (C8, 3268).

Peliz, A., Rosa, T.L., Santos, M.P. and Pissarra, J.L. 2002. Fronts, jets, and counter-flows in the western Iberian upwelling system. Journal of Marine Systems 35(1–2): 61–77. doi: 10.1016/S0924-7963(02)00076-3.

Perry, R., Batchelder, H., Mackas, D., Chiba, S., Durbin, E., Greve, W. et al. 2004. Identifying global synchronies in marine zooplankton populations: issues and opportunities. ICES Journal of Marine Sciences 61(4): 445–456. doi: 10.1016/j.icesjms.2004.03.022.

Picado, A., Lorenzo, M.N., Alvarez, I., deCastro, M., Vaza, N. and Diasa, J.M. 2016. Upwelling and chl-a spatiotemporal variability along the Galician coast: dependence on circulation weather types. Int. J. Climatol. 36: 3280–3296. DOI: 10.1002/joc.4555.

Poulet, S.A., Laabir, M. and Chaudron, Y. 1996. Characteristic features of zooplankton in the bay of biscay. Scientia Marina 60(Supl. 2): 79–95.

Prat, N., Joandomènec, J. and Francesc, P. 2015. Ramón Margalef, ecólogo de la biosfera. Una biografía científica. Universidad de Barcelona. 188 pp. ISBN: 978-84-475-3747-1.

Ríos, A.F., Pérez, F.F. and Fraga, F. 1992, Water masses in the upper and middle north Atlantic ocean east of the Azores. Deep-Sea Research, 39: 645–658.

Riser, C.W., Wassmann, P., Olli, K. and Arashkevich, E. 2001. Production, retention and export of zooplankton faecal pellets on and off the Iberian shelf, north-west spain. Progress in Oceanography 51: 423–441. doi: 10.1016/S0079-6611(01)00078-7.

Robinson, K.L., Ruzicka, J.J., Decker, M.B., Brodeur, R.D., Hernandez, F.J., Quiñones, J. et al. 2014. Jellyfish, forage fish, and the world's major fisheries. Oceanography 27(4): 104–115. doi: 10.5670/oceanog.2014.90.

Rocha, F., Guerra, A., Prego, R. and Piatkowski, U. 1999. Cephalopod paralarvae and upwelling conditions off Galician waters (NW Spain). Journal of Plankton Research 21(1): 21–33. doi: 10.1093/plankt/21.1.21.

Rodríguez, J.M., Cabrero, A., Gago, J., Guevara-Fletcher, C., Herrero, M., Hernández de Rojas, A. et al. 2015. Vertical distribution abd migration of fish larvae in the NW Iberian upwelling system during the winter mixing period: implications for cross-shelf distribution. Fisheries Oceanography 24(3): 274–290. doi:10.1111/fog.12107.

Roesch, A. and Schmidbauer, H. 2018. Wavelet comp: computational wavelet analysis.

Roura, A., Álvarez-Salgado, X.A., González, A.F., Gregori, M., Rosón, G. and Guerra, A. 2013. Short-term meso-scale variability of mesozooplankton communities in a coastal upwelling system (NW spain). Progress in Oceanography 109: 18–32. doi: 10.1016/j.pocean.2012.09.003.

RStudio Team. 2020. RStudio: integrated development environment for R. RStudio, PBC, Boston, MA.

Smayda, T.J. and Trainer, V.L. 2011. Dinoflagellate blooms in upwelling systems: seeding, variability, and contrasts with diatom bloom behavior. Progress in Oceanography 85(2010): 92–107. doi: 10.1016/j.pocean.2010.02.006.

Sordo, I., Barton, E.D., Cotod, J.M. and Pazos, Y. 2001. An inshore poleward current in the NW of the Iberian Peninsula detected from satellite images, and its relationship with G. catenatum and D. acuminate blooms in the Galician Rías. Estuarine, Coastal and Shelf Science 56(6): 787–799. doi:10.1006/ecss.2000.0788.

Steinberg, D.K. and Landry, M.R. 2017. Zooplankton and the ocean carbon cycle. Annual Review of Marine Science 9: 413–44. doi: 10.1146/annurev-marine-010814-015924.

Stöhr, S., Hagen, E., John, H.-Ch., Mittelstaedt, E., Schulz, K., Vanicek, M. et al. 1997. Poleward plankton transport along the moroccan and Iberian continental slope. Berichte der Biologischen Anstalt Helgoland, 12.

Stöhr, S., Schulz, K. and John, H. Ch. 1996. Population structure and reproduction of calanus helgolandicus (copepoda, calanoida) along the Iiberian and moroccan slope. Helgoländer Meeresuntersuchungen 50: 457–475. doi: https://doi.org/10.1007/BF02367161.

Taylor, A.H., Allen, J.I. and Clark, P.A. 2002. Extraction of a weak climatic signal by an ecosystem. Nature 416: 629–632.

Tenore, K.R., Alonso-Noval, M., Alvarez-Ossorio, M.T., Atkinson, L.P., Cabanas, J.M., Cal, R.M. et al. 1995. Fisheries and oceanography off Galicia, NW Spain: mesoscale spatial and temporal changes in physical processes and resultant patterns of biological productivity. Journal of Geophysical Research 100 C6: 10943–10966.

Uberos, S.R., Vergara, A.R., Dominguez-Petit, R. and Saborido-Rey, F. 2021. Larval fish community in the north-westen Iberian upwelling system during the summer period. Oceans 2: 700–722. doi: 10.3390/oceans2040040.

Valdés, L., Álvarez-Ossorio, M.T., Lavín, A., Varela, M. and Carballo, R. 1991. Ciclo anual de parámetros hidrográficos, nutrientes y plancton en la plataforma continental de la couña (NO, España). Boletín del Instituto Español de Oceanografía 71(1): 91–138.

Valdés, L., Bode, A., Latasa, M., Nogueira, E., Somavilla, R., Varela, M.M. et al. 2021. Three decades of continuous ocean observations in north Atlantic spanish waters: The RADIALES time series project, context, achievements and challenges. Progress in Oceanography 198: 102671. doi: 10.1016/j. pocean.2021.102671.

Valdés, L., López-Urrutia, A., Cabal, J., Álvarez-Ossorio, M.T., Bode, A., Miranda, A. et al. 2007. A decade of sampling in the bay of biscay: what are the zooplankton time series telling us? Progress in Oceanography 74: 98–114. doi: 10.1016/j.pocean.2007.04.016.

Valdés, L., Román, M.R., Álvarez-Ossorio, MT., Gauzens, A.L. and Miranda, A. 1990. Zooplankton composition and distribution off the coast of Galicia, spain. Journal of Plankton Research 12(3): 629–643. doi: 10.1093/plankt/12.3.629.

Vandromme, P., Nogueira, E., Huret, M., López-Urrutia, A., González-Nuevo González, G., Sourisseau, M. et al. 2014. Springtime zooplankton size structure over the continental shelf of the bay of biscay. Ocean Science 10: 821–835.

Varela, M., Álvarez-Ossorio, M.T., Bode, A., Prego, R., Bernárdez, P. and García-Soto, C. 2010. The effects of a winter upwelling on biogeochemical and planktonic components in an area close to the galician upwelling core: The Sound of Corcubión (NW Spain). Journal of Sea Research 64(3): 260–272. doi: 10.1016/j.seares.2010.03.004.

Velasco et al. 2012. Estrategia marina, demarcación marina noratlántica. Parte IV Descriptores de Buen Estao Ambiental. Descriptor 1: Biodiversidad. Evaluación Inicial y Buen Estado Ambiental (331 pp + 157 pp. Annexes) (https://www.miteco.gob.es/es/costas/temas/proteccion-medio-marino/estrategias-marinas/demarcacion-noratlantica/).

von Richthofen, F. 1886. Fuhrer für Forrschungsreisende. Oppenheim, Berlin.

Wickham, H. 2003. Tidy Data. Journal of Statistical Software 59.

Wickham, H., Averick, M., Bryan, J., Chang, W., McGowan, L.D., François, R. et al. 2019. Welcome to the tidyverse. Journal of Open Source Software 4: 1686.

Wickham, H., Girlich, M. and Ruiz, E. 2022. Dbplyr: A 'dplyr' back end for databases.

Wooster, W.S., Bakun, A. and Mclain, D.R. 1976. The seasonal upwelling cycle along the eastern boundary of the north atlantic. Journal of Marine Research 34: 131–141.

2.2

Ecological Application of Biomarkers to Mesozooplankton Communities in the Mediterranean Sea

Maria Protopapa,[1,*] *Lidia Yebra,*[2] *Rolf Koppelmann*[3] and *Soultana Zervoudaki*[1]

1. Introduction

Plankton are useful indicators of marine ecosystem health due to their significant role in the functioning of marine ecosystems and biogeochemical cycles, owing to their key position at the food web base and rapid response to environmental change (Hussain et al. 2020). According to Reid and Edwards (2001), plankton has two roles with respect to climate: first, as an indicator of climate change in present-day populations and the fossil record and, second, as a factor contributing to climate change through, e.g., its role in the CO_2 cycle. Temperature is one of the main factors of climate change as well as a key parameter playing a critical role in physiological zooplankton rates (e.g., enzymatic reactions, respiration and growth rates), which in turn impact body size, generation time and production rates (Packard et al. 1974, Peters 1983, Mauchline 1998). Another factor affected by climate change is the wind. The wind has an effect on water turbulence, which in turn influences the behavior of individuals, encounter rates between prey and predator (Mackenzie and Legget 1991) and feeding behavior (Margalef 1997). Wind also promotes the upwelling of nutrient-enriched waters to surface layers, e.g., allowing for the highest plankton production in the SW Mediterranean Sea (Mercado et al. 2007, Yebra et al. 2017a). These microscale effects of hydroclimatic forcing on plankton occur permanently, with each species integrating the effects over its generation time and transferring the integrated effect to the next generation. Therefore, the effects of hydroclimatic

[1] Institute of Oceanography, Hellenic Centre for Marine Research (HCMR), 46.7 Km Athens-Sounio av., 19013 Anavyssos, Attiki, Greece.

[2] Centro Oceanográfico de Málaga (IEO, CSIC), Puerto Pesquero s/n, 29640 Fuengirola, Spain.

[3] University of Hamburg, Institute of Marine Ecosystem and Fishery Science, Große Elbstraße 133, 22767 Hamburg, Germany.

* Corresponding author: mariaprot@hcmr.gr

variability on individual organisms may ultimately have a strong impact on the whole ecosystem (Beaugrand 2005).

Planktonic copepods represent the dominant taxa in the Mediterranean Sea zooplankton assemblages, and they generally constitute ~ 55–95% of the total community (Wells 1984, Longhurst 1989, Siokou-Frangou et al. 2010, Yebra et al. 2022). As key components in aquatic food webs, they form the main link in the transfer of energy through the food web to the top carnivores (Rupper et al. 2003). They act as the main grazers of small autotrophic, heterotrophic nanoplankton and microplankton species as well as food sources for higher trophic levels (Cushing 1990, Huys and Boxshall 1991, Mauchline 1998, Calbet 2001, Calbet and Saiz 2005, Neffati et al. 2013). In addition, they play an essential role in the control of fish recruitment since copepod eggs, nauplii and copepodite stages are prey for larval and adult fish (Cushing 1990, García and Palomera 1996, Yebra et al. 2019). According to several studies, copepods can be excellent bioindicators for climate change and ecosystem shifts (Beaugrand 2005) as well as for pollution in eutrophic coastal waters (Daly Yahia et al. 2004, Ben Lamine et al. 2015, Serranito et al. 2016, Drira et al. 2018). They can provide very useful indicators to highlight ecotoxicological alterations at relatively low levels of the food chain. This can thus prevent biomagnification phenomena, which can amplify community damage to the higher trophic levels until an entire ecosystem is damaged (Minutoli et al. 2009, Rumengan and Ohji 2012, Rodriguez et al. 2018).

To assess the ecological role of mesozooplankton, measurements of abundance and diversity, as well as physiological rates and trophic positions, are necessary. Since direct measurements of essential vital rates of zooplankton, such as respiration, production, and feeding, are not always possible; the use of biomarkers can gain knowledge into these parameters. The term "biological marker" was first introduced in the 1950s' and since then, many different definitions have been given (Porter 1957, Basu et al. 1960). What is certain, however, is that they can be applied to organisms of all zoological phyla, including wild-collected zooplankton, with less laboratory manipulation.

Here, we review the use of biomarkers such as the aminoacyl-tRNA synthetases (AARS) activity, electron transport system (ETS) activity, stable isotopes (SI) and fatty acids (FA) to study zooplankton communities of the Mediterranean Sea (MS). Alongside, we provide an insight into the studies using the above biomarkers to highlight the ecological role of zooplankton in the MS so far. Also, we underline the importance of using these biomarkers in monitoring programs not only to track good environmental status but also to study any environmental changes as a result of climate change. Combining these approaches promises to add scientific value to zooplankton ecology.

2. Zooplankton Biomarkers in Use

Knowledge of zooplankton respiration rates is essential for quantifying and modeling energy and carbon fluxes within marine food webs (Gómez et al. 1996). Zooplankton respiration has been widely studied by the scientific community (e.g., Conover 1960, Ikeda 1970, King and Packard 1975, Gómez et al. 1996,

Hernández-León and Gómez 1996, Hernández-León and Ikeda 2005, Del Giorgio and Williams 2005, Minutoli and Guglielmo 2009, 2012, Yebra et al. 2018, Herrera et al. 2019, Protopapa et al. 2019a). Electron transport system (ETS) activity is a specific and highly-sensitive method to estimate zooplankton respiration rates and, thus, carbon requirements (Packard 1971, King and Packard 1975, Bamstedt 1980, Bidigare et al. 1982). According to Minutoli et al. (2009), high or low ETS values in zooplankton may indicate growing or declining populations, respectively, or the beginning or end of a phytoplankton bloom the zooplankton is feeding on. In fact, the relationships between respiratory activity measured by the ETS, growth, reproduction, crowding and starvation have long been demonstrated (Ikeda and Motoda 1978, Schalk 1988).

The zooplankton production rate is defined as the increase in biomass of a population over a specific period (Downing and Rigler 1984) and can be determined as the product of total biomass by the weight-specific growth rate of the population (Kimmerer 1987). However, the growth of a population is not easily measured in the field, as the direct incubation of organisms on board oceanographic vessels is time-consuming and may lead to artifacts and errors in the assessment of the growth rates. Therefore, several direct and indirect methods, along with allometric models, have been developed to estimate the growth and production of zooplankton, with different success (see reviews in Yebra et al. 2017b, Kobari et al. 2019). Among the available biochemical methods to assess zooplankton growth rates, the AARS method (Yebra and Hernández-León 2004) stands out, as it is a simple, quick and non-radioactive assay that measures the activity of the enzymes aminoacyl-tRNA synthetases (AARS). Since these enzymes catalyze the first step of protein synthesis, their activity correlates with the somatic growth rate of the organisms. Further, these enzymes are present in all cells, from bacteria to fish, allowing the application of the AARS method to mixed populations within the wide planktonic spectrum (Yebra et al. 2017b).

Fatty acids (FA) are lipid components that can be used as trophic biomarkers since they are, in many circumstances, incorporated into consumers in a conservative manner, thereby providing information on predator-prey relations (Alfaro et al. 2006). Furthermore, as opposed to the more traditional gut content analyses, which provide information only on recent feeding, FA provides information on the dietary intake and the food constituents, leading to the sequestering of lipid reserves over a longer period (Hakanson 1984, Kirsch et al. 1998, Auel et al. 2002). The concept of FA being transferred conservatively through aquatic food webs was first suggested in 1935 by Lovern (1935). However, their use as a reliable method for tracing the food source through multiple food web linkages has been noted in the early 70s by Lee and his team (Lee et al. 1971). They found that the dietary FA are incorporated and unmodified into marine copepods' storage lipids, whereas the total lipid content of copepods is correlated with phytoplankton concentrations. According to Dalsgaard et al. (2003), specific FA may help interpret trophic relations in aquatic systems, as the group-specific FA composition of primary producers varies significantly (Volkman et al. 1989, Ahlgren et al. 1992). Consequently, it is important to comprehend how much the FA composition of zooplankton is determined by taxonomic affiliation, changed by diet and modified by starvation or temperature. It is also essential to

Table 1. Natural abundance of carbon and nitrogen stable isotopes.

Atomic number	Symbol	Mass number	Abundance %
6	C	12	98.89
		13	1.11
7	N	14	99.63
		15	0.37

know whether zooplankton maintain a semi-constant FA profile relative to their diets or, alternatively, bio-convert some FA into other FA molecules (Brett et al. 2009).

Stable isotopes are non-radioactive forms of an atom. The position of an atom in the elementary system depends on the number of protons in the nucleus. The number of neutrons in the nucleus affects the weight of an atom, and several isotopes of an atom exist. Processes in aquatic food webs can be studied by analyzing the stable isotope composition of prey and predators. In nature, lighter isotopes are more abundant than the heavier isotopes (Table 1), but heavier isotopes are slower than lighter isotopes during physical fractionation, and bindings with heavier isotopes are stronger than bindings with lighter isotopes. Hence, heavier isotopes are enriched within the food web by the organism relative to its diet (Michener and Schell 1994, Fry 2006), and the isotopic signature of an organism provides integrated information about its feeding habits over longer times. For ecological research, mainly carbon and nitrogen isotopes are used as food web markers. Stable nitrogen isotopes generally determine the trophic level (TL) of organisms (Minagawa and Wada 1984, Hobson and Welch 1992), while stable carbon isotopes are used to track the diet of organisms since there is less fractionation between ^{13}C and ^{12}C from prey to predator (France and Peters 1997, Post 2002).

Stable isotopes are measured by mass spectrometry, and the ratio of the heavier element to the lighter element is calculated against a standard. For carbon, Pee Dee Belemnite (PDB) collected from the upper cretaceous Peedee formation in South Carolina is generally used as a standard. The ^{13}C/^{12}C ratio in the standard is quite high due to its higher content of ^{13}C compared to most natural substances, which often results in negative values against the standard. For nitrogen, atmospheric nitrogen is used as a standard.

Stable isotope values are expressed in δ-notations as parts per thousand (‰), where R is the ratio of ^{13}C/^{12}C and ^{15}N/^{14}N for the sample and the standard:

$$\delta^{13}C \text{ or } \delta^{15}N\ (‰) = (R_{sample}/R_{standard} - 1) * 1{,}000$$

Although the increase in $\delta^{15}N$ between TLs can be variable (Vander Zanden and Rasmussen 2001, McCutchan et al. 2003), a mean trophic fractionation of 3.4‰ is widely applicable (Post 2002), and $\delta^{15}N$ values can be used to detect the trophic level of an organism against a baseline value. Different factors like C/N ratios (Schwamborn et al. 2015) or oxygen content (Czudaj et al. 2020) may influence the increase in $\delta^{15}N$ between trophic levels. For $\delta^{13}C$, the increase between trophic levels is 0.4–1.0‰ much smaller. Also, $\delta^{13}C$ in the atmosphere is around –7‰, and fractionation in C3 plants is 21‰ quite high, resulting in a $\delta^{13}C$ of –28‰ for most terrestrial plants (see Peterson and Fry 1987). For C4 plants, e.g.. common for tropical grasslands, the fractionation is 6‰ much smaller, resulting in a $\delta^{13}C$ of –13‰. In the ocean, values of dissolved and particulate organic matter are around –23‰ and –22‰ in $\delta^{13}C$, respectively. Since $\delta^{15}N$ is measured against atmospheric nitrogen as a standard, the value for $\delta^{15}N$ in the atmosphere is zero (see Peterson and Fry 1987). The $\delta^{15}N$ values for terrestrial plants indicate a wide spectrum from –8 to 3‰. Dissolved nitrogen in the ocean is around 1‰, whereas deep-water nitrogen ranges between 4–6‰. Released ammonia by predators is approximately reduced by 3‰ against the prey (Checkley and Miller 1989). Depending on the source of nitrogen, particulate organic matter can vary from –2 to 11‰ in the oceans (Peterson and Fry 1987). This indicates a strong need to identify baseline values when stable isotope analyses from different regions will be compared.

3. The Application of Biomarkers in the Mediterranean Sea

In order to understand the ecology and distribution of zooplankton in the Mediterranean Sea, its unique features have to be considered. The size, location and morphology of the MS and external forcing cause complex physical dynamics, as has been thoroughly described in Siokou-Frangou et al. (2010). Overall, nutrients and chlorophyll *a* pools classify the basin as oligotrophic to ultraoligotrophic (Krom et al. 1991). Some of the most productive areas of the MS are located in the Alboran Sea, its westernmost basin (e.g., Mercado et al. 2007, Yebra et al. 2017a, 2020), whereas the eastern MS is thought to be one of the most oligotrophic areas of the world (Zohary and Robarts 1992, Siokou-Frangou et al. 2010).

The MS is characterized by strong longitudinal environmental gradients (Danovaro et al. 1999) with increasing nutrient depletion and strong productivity gradients decreasing from west to east (Christaki et al. 2001), leading to a west-to-east decrease of zooplankton standing stocks (e.g., Mazzocchi et al. 1997, Kovalev et al. 1999, Dolan et al. 2002, Siokou-Frangou 2004, Koppelmann and Weikert 2007). Considering its relatively small surface area and volume (Bianchi and Morri 2000), high diversity (7% of the world's marine biodiversity, Coll et al. 2012) and high rate of endemism (average of total endemics: 20.2%, Coll et al. 2010) has been noticed. Hence, all these characteristics should likely be reflected in the structure and dynamics of plankton communities (Siokou-Frangou et al. 2010).

Though the aforementioned biomarkers have been used widely in the past decade in other areas of the world (e.g., for the Pacific Ocean: Packard et al. 1971, for the Atlantic Ocean: Packard et al. 1974, Koppelmann and Weikert 1999, for the Indian Ocean: Koppelmann et al. 2000, for the Antarctic: Hernández-León et al. 2000), it seems that they are not widely used in the Mediterranean Sea, and this is something that we are investigating in this work.

We employed a systematic qualitative approach through narrative analysis and synthesis to describe the current knowledge on the development and application of zooplankton biomarkers in environmental assessments and monitoring in the Mediterranean Sea. We conducted wide-ranging searches of multiple peer-reviewed sources and databases related to the knowledge of zooplankton biomarkers, such as ETS, AARS, stable isotopes and fatty acids in the Mediterranean Sea. The primary search tools were Google Scholar, Science Direct and an intuitive graph-based tool known as 'connected papers' (https://www.connectedpapers.com/). The latter tool allowed us to quickly detect the most relevant articles, access their titles and abstracts and screen these to determine their relevance to this study's overall objectives.

3.1 Electron Transport System (ETS) Activity

More than 50 years ago, Packard et al. (1971) successfully introduced ETS methodology in the study of plankton communities. When compared to the other biomarkers reported in this work, ETS seems to be the most used in the MS, with five references in the western MS, two in the eastern MS and two across the entire MS.

Alcaraz and Packard (1989) were the first ones to introduce the ETS method in the MS. This pioneer work compared ETS activity and direct measurements of zooplankton respiration of the Catalan Sea, indicating the existence of variability in both measurements and highlighting the necessity for more research so that ETS can someday be used as an estimator of metabolic activity in pelagic systems.

Later, Quiñones et al. (1994) applied ETS to study the inverse relationship between respiration and body size from bacteria to zooplankton at a frontal station in the Alboran Sea.

The next work came after 30 years, again in the western MS, by Herrera et al. (2014). In this work, the ETS technique was applied to estimate potential respiration and carbon demand from the zooplankton community (different size fractions: 53–200, 200–500, > 500 μm) in the upper 200 m of the water column in two areas with different oceanographic conditions: the Balearic and Algerian subbasins. They found that the zooplankton in both areas are not well-fed and that they are living under oligotrophic stress.

A few years later, Yebra et al. (2017) applied ETS and AARS as biomarkers to study the influence of hydrochemistry and trophic conditions on the coastal zooplankton community metabolic rates along the N Alboran Sea. They found that mesozooplankton production variability in these coastal waters during the summer was driven by the trophic conditions rather than by hydrology. Another important finding was that biomass-specific metabolic rates were driven by sea temperature.

Further offshore, Yebra et al. (2018) used metabolic rates to assess diel migrant zooplankton respiration and production at mesopelagic depths, proving the importance of anticyclonic eddies in the total carbon export to mesopelagic depths and the capacity of the biological pump for trapping atmospheric CO_2 in the Alboran Sea.

In more recent work, Yebra et al. (2020) studied the influence of hydrochemistry and trophic conditions on the coastal zooplankton community's biomass and metabolic activities along the Spanish Mediterranean coastal waters. This pioneering study combined the assessment of zooplankton biomass and metabolic rates together with their potential prey and predators' abundance in the field, adding further evidence of the strong link between coastal zooplankton and small pelagic fisheries species and improving our knowledge of the potential mechanisms driving the decline of small pelagic fisheries in the W MS.

Minutoli and Guglielmo (2009) and Minutoli et al. (2017) were the two works conducted across the entire MS. The first one used the ETS method to study spatial and diel variability in zooplankton respiratory ETS activity from the Strait of

Gibraltar to the easternmost station near the Isle of Rhodes. The carbon requirements per unit of zooplankton biomass indicated an increasing gradient from west to east, possibly amplifying the features of an impoverished region. The second work used the ETS method to evaluate zooplankton carbon requirements from the sinking flux from the Strait of Gibraltar to the island of Crete. Sea temperature was also a critical factor for the ETS activity and, consequently, for the mean specific carbon demands. Additionally, zooplankton from the Eastern and Western Mediterranean have contributed similarly to community carbon demands and the carbon losses from the sinking flux in the water column.

In the eastern basin, Koppelmann et al. (2004) were the first to use the ETS approach to study deep-sea mesozooplankton community respiration and its relation to particle flux in the eastern MS. In this work, once again, sea temperature played an important role in affecting standardized carbon consumption rates. Additionally, absolute rates reflected the oligotrophic character of the basin. In 2019, Protopapa et al. (2019a) published a second work in the Eastern MS, where ETS was used to estimate mesozooplankton carbon requirements and proved that hydrographical features locally affect the mesozooplankton communities under oligotrophic conditions. Additionally, their results were similar to the ones reported by Herrera et al. (2014) for the Western Mediterranean but very low compared to the values reported by Minutoli and Gugliemo (2009).

3.2 Aminoacyl-tRNA Synthetases (AARS) Activity

The AARS method has been applied widely in the Atlantic, Pacific and Indic Oceans, and in Antarctic waters, to assess both growth and production rates of pelagic zooplankton species and guilds (see Yebra et al. 2017b, Kobari et al. 2022). However, its application in the Mediterranean Sea has been limited to a few studies in the Western Mediterranean and a single study in the Eastern Mediterranean basin.

Studies in the western basin include: (i) the first assessment of zooplankton production rates in the Alboran Sea, showing that this basin is among the most productive regions in the MS (Yebra et al. 2017a); (ii) a pioneer application of the AARS method to assess diel vertical migrant zooplankton production down to 800 m, which allowed inferring the carbon flux to the mesopelagic zone due to both respiration and mortality at depth (Yebra et al. 2018); (iii) the incipient incorporation of the AARS method as a tool within the 2010–2012 Spanish Marine Strategy Framework Directive (MSFD) monitoring program, which highlighted the potential cues behind the differentiated trophic status of the NW and SW Mediterranean basins (Yebra et al. 2020).

In the eastern basin, a first attempt was made to investigate the carbon requirements and zooplankton production of mesozooplankton using enzymatic activity indices (ETS and AARS). According to the results, very low zooplankton production rates were detected but close to results from other studies, such as by Zervoudaki et al. (2007) for the northern Aegean Sea (Protopapa et al. 2019a).

The works mentioned here have been further described in the ETS chapter above. It seems that researchers tend to use AARS and ETS together since their combined application was first demonstrated by Yebra et al. (2004), as they can obtain simultaneously and from the same samples complementary information on the production and carbon requirements of the target populations.

3.3 Fatty Acids (FA)

The FA method has been applied widely from the Atlantic to the Pacific Ocean and from the Arctic to the Antarctic. The studies have been focused mainly on the lipids of larger calanoid copepods, which dominate the zooplankton biomass and are particularly important in northern temperate and polar latitude pelagic food webs (Sargent and Henderson 1986, Dalsgaard et al. 2003). On the contrary, knowledge on zooplankton FA is still scarce in the Mediterranean Sea; four studies have been conducted in the western and one in the Eastern Mediterranean Sea.

In the mid-90s, Najdec et al. (1994) conducted a pioneer work in the northern Adriatic Sea, taking samples from sediment traps in parallel with laboratory experiments in order, among other things, to evaluate the importance of zooplankton lipids in mass flux. Indeed, a high percentage of zooplankton lipids remained entrapped within the fecal pellet membrane in the form of droplets and appeared as a component of mass flux, indicating that the seafloor can be seeded by easily available organic matter. Furthermore, entrapping of fecal pellets by marine snow and their subsequent remineralization by heterotrophs can also be an additional source of DOM for the upper layers of the water column.

In the same year, another pioneer work was published by Serrazanetti et al. (1994), which was more focused on the study of lipid constituents present in zooplankton in the waters of Trieste, in addition to providing information concerning the distribution of these compounds in the trophic chain.

In the late 90s, Mayzaud et al. (1999) worked on the lipid and fatty acid composition of the northern krill *Meganyctiphanes norvegica* from the Ligurian Sea with respect to sex and season. Both sexes showed similar lipid structures but seasonal differences were observed. Additionally, it was underlined the significance of fatty acids as potential markers of feeding behavior. A few years later, Rossi et al. (2006) used FA in phytoplankton, zooplankton and anchovy larvae to study the relationship among the different trophic levels on the Catalan coast. It seems that anchovy larvae have an additional food resource that is complementary to a zooplankton diet. More than ten years after this study, Protopapa et al. (2019b) used combined analyses of FA and stable isotopes (SI) to characterize food preferences among copepod species/taxa and to trace their food sources in the eastern MS. Overall, a good agreement between FA and SI in some species of copepods was noticed with omnivory prevailing as a feeding mode, demonstrating a high degree of opportunistic feeding of copepods in these ultra-oligotrophic waters.

3.4 Stable Isotopes (SI)

Stable isotopes are used to identify trophic relationships in terrestrial, limnic and oceanic ecosystems worldwide (e.g., Boecklen et al. 2011, Fry 2006, Gladyshev 2009, Rundel et al. 1989). In this paper, we present stable isotope analyses of mesozooplankton in the Western and Eastern Mediterranean Sea with a focus on nitrogen isotopes as tracers for food web and trophic level investigations. As pointed out before, baseline or reference values are needed to determine the trophic position of an organism. This can either be phytoplankton as the primary food source for epipelagic species or particulate organic matter as a food source for deep-living organisms or the stable isotope values of a species is used for which the trophic position is well known. Generally, baseline values in the Mediterranean Sea are lower than in the open ocean. Kerherve et al. (2001) measured very low values of $\delta^{15}N$ (near 0‰) in settling particles in the Western Mediterranean and concluded that fixation of atmospheric nitrogen may play a significant role. Also, Pantoja et al. (2002) and Sachs and Repeta (1999) assumed that the low $\delta^{15}N$ values in the eastern Mediterranean were caused by nitrogen fixation. Koppelmann et al. (2003, 2009) measured a $\delta^{15}N$ of 0.43–1.74‰ in suspended particles in the epipelagic ocean and 0.68–2.17‰ in sinking particulate organic matter in deep-water sediment traps at a deep-water site in the Levantine Basin, Eastern Mediterranean. The authors also discussed that nitrogen fixation plays a role in the Mediterranean Sea. However, Krom et al. (2004) proposed an alternative explanation for the low $\delta^{15}N$ in the P-limited eastern Mediterranean. When phosphate is entirely consumed, light PON of 3.5–4.0‰, but heavy nitrate of 17–20‰ remains in the system, which can cause an export of light PON. Overall, the $\delta^{15}N$ baseline in the Mediterranean Sea is lower than in most other open ocean regions.

This low $\delta^{15}N$ baseline is also reflected throughout the food web in mesozooplankton. Banaru et al. (2013) analyzed size-fractionated zooplankton from 80 to > 2,000 µm in the upper 50 m of the Western Mediterranean Sea (Bay of Marseille) and detected that $\delta^{15}N$ generally increased with increasing size with some exceptions in the > 2,000 µm size class, where high numbers of gelatinous organisms including salps were found. The measured values ranged between 1.08 and 4.83‰ across all size ranges and seasons. Similar low $\delta^{15}N$ values and an increase in size were detected by Koppelmann et al. (2003) at a deep site in the Levantine Basin (eastern Mediterranean Sea) with values between 2.0 and 3.1‰ in the upper 250 m in April 1999. With increasing depth, the $\delta^{15}N$ of zooplankton increased up to 11.5‰ at more than 3,000 m depth. Since $\delta^{15}N$ of particulate organic matter increased only from 1.2‰ in 700 m depth to 1.7–2.5‰ in 2,700 m depth, there is a strong indication that the food web in the deep Levantine Sea is more complex or that a food source additional to particulate organic matter exists. These results were later confirmed by food web analyses in October 2001 in the same region (Koppelmann et al. 2009). Also, Polunin et al. (2001) found an increase in $\delta^{15}N$ with increasing depth for a bathyal benthic community in the Western Mediterranean Sea. Recently, Quintanilla et al. (2020) studied stable isotopes of both sardine larvae ($\delta^{15}N$ 5.78–6.94‰) and zooplankton ($\delta^{15}N$ 200–500 µm 2.50–5.30‰, 500–1000 µm 2.73–4.85‰) as their prey, in the Alboran Sea, and found that the isotopic signature of larvae was highly influenced by the mesozooplankton community structure, indicating dietary changes throughout their ontogenic development.

Protopapa et al. (2019), as thoroughly described in the FA chapter, combined fatty acid trophic markers and stable isotopes to analyze the trophic position of several copepods in the epi- and mesopelagic zones of the eastern Mediterranean.

The isotopic structure was analyzed using the Stable Isotope Analysis package SIAR (Jackson et al. 2011), and isotopic niche width and overlaps were calculated.

4. Biomarkers as Indicators of the Environmental Status

To safeguard the environmental status, the European Union has implemented the Water Framework Directive (WFD; 2000/60/EC) and the Marine Strategy Framework Directive (MSFD; 2008/56/EC) legislations, which promote the use of biological tools to detect the quality of aquatic systems. These legislations are aimed at the protection, maintenance and restoration of marine environments.

Zooplankton is not identified as a biological quality element (BQE) of WFD and is seldom considered in monitoring initiatives (Perry et al. 2004, Tett et al. 2008, Rombouts et al. 2013). However, several authors have stated that this group may be used for the assessment of aquatic systems due to its sensitivity to stressors and can be a prospective bioindicator of water quality and eutrophication status (e.g., Webber and Webber 1998, Buchanan 1993, Olson et al. 2005). The Marine Strategy Framework Directive undertakes that zooplankton monitoring can be useful in identifying environmental changes and anthropogenic invasion in nature (Serranito et al. 2016). Since zooplankton is very sensitive to changes, the response to the disturbance will appear in a short time with respect to the higher trophic levels. The short response and their ubiquity make them potential bioindicators for different water bodies (Serranito et al. 2016).

Recently, Ndah et al. (2022) highlighted the immense potential for using zooplankton in ecological assessments in the context of the MSFD and classified the existing indicators and related indices into two broad categories: (i) the holistic biomass-based indices (including the mean size and total stock and the plankton lifeform index) captured essential changes in ecosystem structure and functioning and (ii) the stressor-response indices, targeted specific issues linked to the local effects of climate and anthropogenic pressures. The latter category of indicators is vital for understanding the current state of marine ecosystems and for predicting future ecological changes, but the available information is currently uncoordinated and scattered across the literature (Ndah et al. 2022).

Discussions about bioindicators and biomarkers are often unclear. Even though these concepts are related, the fundamental difference is that biomarkers concentrate on measurement attributes, while bioindicators require validation in addition to measurement (McCarty et al. 2003). In addition, a biomarker may be considered a good tool for environmental assessment if it reflects the following criteria: (i) it should be reliable, with a short life span and easy to sample; (ii) relatively cheap and easy to perform; (iii) should be sensitive to pollutant exposure and/or effects in order to serve as an early warning parameter (Fossi and Marsili 1997).

Zooplankton is found in all marine environments, making it possible to select zooplankton species/groups for biomarkers analysis. Furthermore, biomarker analyses are based mainly on intermediate metabolites that provide information about the real effects of environmental pressures. The above review underlined the significance of the aforementioned biomarkers in estimating the carbon requirements of zooplankton and the trophic relationships in the food web. In addition, the majority of studies identified that sea temperature played an essential role in affecting carbon consumption rates and, consequently, the effectiveness of the biological pump under the pressure of climate change. The biomarkers showed in this study follow the abovementioned criteria and, therefore, could be proposed as indices of trophic status and ecological integrity of the marine environment. We recognized that the ecological indicator development and application is relatively young and growing; this review, therefore, serves as an information source for scientists, environmental managers and policymakers interested in developing and using new zooplankton indices to implement the MSFD in European waters and globally.

5. Conclusions

- According to our research, biomarkers such as ETS, AARS, FA and SI are underutilized in the Mediterranean Sea. Further research is needed in order to gain a better knowledge of how the different trophic and hydrological regimes in the Mediterranean Sea impact the above biomarkers.

- The aforementioned biomarkers can be used in monitoring programs to investigate the impact of climate change on zooplankton and the efficiency of the biological pump.

- The above biomarkers can also be included in studies for their further use as indicators of the good environmental status of the pelagic habitat, according to the MSFD.

References

Ahlgren, G., Gustafsson, I.-B. and Boberg, M. 1992. Fatty acid content and chemical composition of freshwater microalgae. J. Phycol. 28: 37–50.

Alcaraz, M. and Packard, T.T. 1989. Zooplankton ETS activity and respiration in the Catalan Sea (western Mediterranean). Sci. Mar. 53(2–3): 247–250.

Alfaro, A.C., Thomas, F., Sergent, L. and Duxbury, M. 2006. Identification of trophic interactions within an estuarine food web (northern new Zealand) using fatty acid biomarkers and stable isotopes. Estuar. Coast Shelf Sci. 70: 271–286.

Auel, H., Harjes, M., da Rocha, R., Sttibing, D. and Hagen, W. 2002. Lipid biomarkers indicate different ecological niches and trophic relationships of the arctic hyperiid amphipods themisto abyssorum and *T. libellula*. Polar Biology 25: 374383.

Bamstedt, U. 1980. ETS activity as an estimator of respiratory rate of zooplankton populations. The significance of variations in an environmental factor. J. Exp. Mar. Biol. Ecol. 42: 267–283.

Banaru, D., Carlotti, F., Barani, A., Grégori, G., Neffati, N. and Harmelin-Vivien, M. 2014, Seasonal variation of stable isotope ratios of size-fractionated zooplankton in the Bay of Marseille (NW Mediterranean Sea). J. Plankton Res. 36: 145–156.

Basu, P.K., Miller, I. and Ormsby, H.L. 1960. Sex chromatin as a biologic cell marker in the study of the fate of corneal transplants. American Journal of Ophthalmology 49(3): 513–515. doi:10.1016/0002-9394(60)91653-6. PMID 13797463.

Beaugrand, G. 2005. Monitoring pelagic ecosystems using plankton indicators. ICES J. Mar. Sci. 62(3): 333–338. https://doi.org/10.1016/j.icesjms.2005.01.002.

Ben Lamine, Y., Pringault, O., Aissi, M., Cherif, E., Mahmoudi, E., Kefi-Daly Yahia, O. et al. 2015. Environmental controlling factors of copepod communities in the gulf of tunis (south western Mediterranean Sea). Cah. Biol. Mar. 56: 213–229.

Bianchi, C.N. and Morri, C. 2000. Marine biodiversity of the Mediterranean Sea: situation, problems and prospects for future research. Mar. Pollut. Bull. 40: 367–376.

Bidigare, R.R., King, F.D. and Biggs, D.C. 1982. Glutamate dehydrogenase and respiratory electron transport system activities in the Gulf of Mexico zooplankton. J. Plankton Res. 4: 895–912.

Boecklen, W.J., Yarnes, C.T., Cook, B.A. and James, A.C. 2011, On the use of stable isotopes in trophic ecology. Annual Review of Ecology, Evolution, and Systematics 42: 411–440.

Brett, M., T., Müller-Navarra D., C. and Persson J. 2009. Crustacean zooplankton fatty acid composition, in arts M., T., Brett M., T., and Kainz, M., Lipids in Aquatic Ecosystem Book. Springer pp. 115–147.

Buchanan, C. (ed.). 1993. Development of zooplankton community environmental indicators for chesapeake bay. Report for the USEPA, Chesapeake Bay Program and Maryland Department of the Environment. ICPRB Report 93–2. Retrieved in Nov 2020, from https://www.potomacriver.org/wp-content/uploads/2014/12/ICP93-2_Buchanan.pdf.

Calbet, A. and Saiz, E. 2005. The ciliate–copepod link in marine ecosystems. Aquatic Microbial Ecology 38: 157–167.

Calbet, A. 2001. Mesozooplankton grazing effect on primary production: a global comparative analysis in marine ecosystems. Limnology and Oceanography 46: 1824–1830.

Checkley, D.M. and Miller, C.A. 1989, Nitrogen isotope fractionation by oceanic zooplankton. Deep-Sea Res. I 36: 1449–1456.

Coll, M., Piroddi, C., Steenbeek, J. et al. 2010. The biodiversity of the Mediterranean Sea: estimates, patterns and threats. PLoS ONE 5: e11842.

Coll, M., Piroddi, C., Albouy, C., Lasram, F.B.R., Cheung, W.W.L., Christensen, V. et al. 2012. The Mediterranean Sea under siege: spatial overlap between marine biodiversity, cumulative threats and marine reserves. Glob. Ecol. Biogeogr. 21: 465–480.

Conover, R. 1960. The feeding behavior and respiration of some marine planktonic crustacea. Biol. Bull. 119: 399–415.

Christaki, U., Giannakourou, A., Van Wambeke, F. and Gregori, G. 2001. Nanoflagellate predation on auto- and heterotrophic picoplankton in the oligotrophic Mediterranean Sea. J. Plankton Res. 23: 1297–1310.

Cushing, D.H. 1990. Plankton production and year-class strength in fish populations: an update of the match/mismatch hypothesis. Adv. Mar. Biol. 26: 249–293. https://doi. org/10.1016/S0065-2881(08)60202-3.

Czudaj, S., Giesemann, A., Hoving, H.J., Koppelmann, R., Lüskow, F., Möllmann, C. et al. 2020. Spatial variation in the trophic structure of micronekton assemblages from the eastern tropical north Atlantic in two regions of differing productivity and oxygen environments. Deep Sea Res. I 163: 103275.

Daly Yahia, M.N., Souissi, S. and Kefi-Daly Yahia, O. 2004. Spatial and temporal structure of planktonic copepods in the Bay of Tunis (southwestern Mediterranean Sea). Zool. Stud. 43: 366–375.

Dalsgaard, J., St. John, M., Kattner, G. et al. 2003. Fatty acid trophic markers in the pelagic marine environment. Adv. Mar. Biol. 46: 225–340.

Danovaro, R., Dinet, A., Duineveld, G. and Tselepides, A. 1999. Benthic response to particulate fluxes in different trophic environments: a comparison between the Gulf of Lions–Catalan Sea (western-Mediterranean) and the cretan Sea (eastern- Mediterranean). Prog. Oceanogr. 44: 287–312.

Del Giorgio, P. and Williams, P. 2005. Respiration in Aquatic Ecosystems. Oxford University Press, New York.

Dolan, J.R., Claustre, H., Carlotti, F., Plouvenez, S. and Moutin, T. 2002. Microzooplankton diversity: relationships of tintinnid ciliates with resources, competitors and predators from the Atlantic coast of Morocco to the Eastern Mediterranean, Deep-Sea Res. Pt. I, 49: 1217–1232.

Downing, J.A. and Rigler, H. 1984. A Manual on Methods for the Assessment of Secondary Productivity in Fresh Waters. Blackwell Scientific Publications, Oxford, UK.

Drira, Z., Kmiha-Megdiche, S., Sahnoun, H., Tedetti, M., Pagano, M. and Ayadi, H. 2018. Copepod assemblages as a bioindicator of environmental quality in three coastal areas under contrasted anthropogenic inputs (Gulf of Gabes, Tunisia). J. Mar. Biol. Assoc. U.K. 98(8): 1889–1905. https://doi.org/10.1017/S0025315417001515.

European Commission (EC). 2000. Directive of the European parliament and of the council 2000/60/EC establishing a framework for community action in the field of Water Policy. PE-CONS 3639/1/00

European Commission. 2020. Background document for the marine strategy framework directive on the determination of good environmental status and its links to assessments and the setting of environmental targets. Accompanying the Report from the Commission to the European Parliament and the Council on the implementation of the Marine Strategy Framework Directive (Directive 2008/56/EC), Commission Staff Working Document, SWD(2020)62), Brussels.

Fossi, M.C. and Marsili, L. 2017. The use of non destructive biomarkers in the study of marine mammals. Biomarkers 2(4): 205–16. doi: 10.1080/135475097231571.

France, R.L. and Peters, R.H. 1997. Ecosystem differences in the trophic enrichment of ^{13}C in aquatic food webs. Canadian Journal of Fisheries and Aquatic Sciences 54(6): 1255–1258.

Fry, B. 2006. Stable Isotope Ecology. Springer, New York.

García, A. and Palomera, I. 1996. Anchovy early life history and its relation to its surrounding environment in the western Mediterranean basin. Sci. Mar. 60(2): 155–166.

Gladyshev, M.I. 2009. Stable isotope analyses in aquatic ecology (a review). Journal of Siberian Federal University 2(4): 381–402.

Gómez, M., Torres, S. and Hernández-León, S. 1996. Modification of the electron transport system (ETS) method for routine measurements of respiratory rates of zooplankton. S. Afr. J. Mar. Sci. 16: 15–20.

Hakanson, J.L. 1984. The long and short term feeding condition in field-caught *Calanus pacificus*, as determined from the lipid content. Limnology and Oceanography 29: 794804.

Hernández-León, S. and Gómez M. 1996. Factors affecting the respiration/ETS ratio in marine zooplankton. J. Plankton Res. 18: 239–255.

Hernández-León, S., Almeida, C., Portillo-Hahnefeld, A., Gómez, M. and Montero, I. 2000. Biomass and potential feeding, respiration and growth of zooplankton in the bransweld strait (Antarctic Peninsula) during austral summer. Polar Biol. 23: 679–690.

Hernández-León, S. and Ikeda, T. 2005. A global assessment of mesozooplankton respiration in the ocean. J. Plankton Res. 27: 153–158.

Herrera, A., Gómez, M., Packard, T.T., Fernández and de Puelles, M.L. 2014. Zooplankton biomass and electron transport system activity around the Balearic Islands (western Mediterranean). J. Mar. Syst. 138: 95–103. doi: 10.1016/j.jmarsys.2014.06.006.

Herrera, I., Yebra, L., Antezana, T., Giraldo, A., Farber-Lorda, J. and Hernández-León, S. 2019. Vertical variability of *Euphausia distinguenda* metabolic rates during diel migration into the oxygen minimum zone of the eastern Tropical Pacific off Mexico. J. Plankton Res. 00(00): 1–12. doi:10.1093/plankt/fbz004.

Hobson, K.A. and Welch, H.E. 1992. Determination of trophic relationships within a high artic marine food web using delta ^{13}C and delta ^{15}N analysis. Mar. Ecol. Prog. Ser. 84: 9–18.

Hussain, M.B., Laabir, M. and Daly-Yahia, M.N. 2020. A novel index based on planktonic copepod reproductive traits as a tool for marine ecotoxicology studies. Science of the Total Environment 727: 138621. https://doi.org/10.1016/j.scitotenv.2020.138621.

Huys, R. and Boxshall, G.A. 1991. Copepod Evolution. Ray Society, London.

Ikeda, T. 1970. Relationship between respiration rate and body size in marine plankton animals as a function of the temperature of habitat. Bulletin of the Faculty of Fisheries Hokkaido University 21: 91–112.

Ikeda, T. and Motoda, S. 1978. Estimated zooplankton production and their ammonia excretion in the Kuroshio and adjacent Seas. Fish. Bull. 76: 357–367.

Jackson, A.L., Inger, R., Parnell, A.C. and Bearhop, S. 2011. Comparing isotopic niche widths among and within communities: SIBER - Stable Isotope Bayesian Ellipses in R. Journal of Animal Ecology 80: 595–602.

Kerhervé, P., Minagawa, M., Heussner, S. and Monaco, A. 2001. Stable isotopes (^{13}C/^{12}C and ^{15}N/^{14}N) in settling organic matter of the northwestern Mediterranean Sea: biogeochemical implications. Oceanologica Acta 24 Suppl. S77–S85.

Kimmerer, W.J. 1987. The theory of secondary production calculations for continuously reproducing populations. Limnol. Oceanogr. 32: 1–13.

King, F. and Packard, T. 1975. Respiration and the activity of respiratory electron transport system in marine zooplankton. Limnol. Oceanogr. 20: 849–859.

Kirsch, P.E., Iverson, S.J., Bowen, W.D., Kerr, S.R. and Ackman, R.G. 1998. Dietary effects on the fatty acid signature of whole atlantic cod (*Gadus morhua*). Canadian Journal of Fisheries and Aquatic Sciences 55: 1378–1386.

Kobari, T., Sastri, A.R., Yebra, L., Liu, H. and Hopcroft, R.R. 2019. Evaluation of trade-offs in traditional methodologies for measuring metazooplankton growth rates: assumptions, advantages and disadvantages for field applications. Progress in Oceanography 178: 102137. doi: 10.1016/j.pocean.2019.102137.

Kobari, T., Sastri, A. and Yebra, L. (eds.). 2022. Report of working group 37 on zooplankton production methodologies, applications and measurements in PICES regions. PICES Sci. Rep. No. 63: North Pacific Marine Science Organization, Sidney, BC, Canada, xx pp. Available at https://meetings.pices.int/publications/scientific-reports.

Koppelmann, R. and Weikert, H. 1999. Temporal changes of deep-sea mesozooplankton abundance in the temperate NE atlantic and estimates of the carbon budget. Mar. Ecol. Prog. Ser. 179: 27–20.

Koppelmann, R., Schäfer, P. and Schiebel, R. 2000. Organic carbon losses measured by heterotrophic activity of mesozooplankton and $CaCO_3$ flux in the bathypelagic zone of the Arabian Sea. Deep-Sea Res. II 47: 169–187.

Koppelmann, R., Weikert, H. and Lahajnar, N. 2003. Vertical distribution of mesozooplakton and its $\delta^{15}N$ signature at a deep-sea site in the Levantine Sea (eastern Mediterranean) in April 1999. J. Geophys. Res. 108. doi:10.1029/2002JC001351.

Koppelmann, R., Weikert, H., Halsband-Lenk, C. and Jennerjahn, T. 2004. Mesozooplankton community respiration and its relation to particle flux in the oligotrophic eastern Mediterranean. Global Biogeochemical Cycles 18: GB1039. doi.org/10.1029/2003GB002121.

Koppelmann, R. and Weikert, H. 2007. Spatial and temporal distribution patterns of deep-sea mesozooplankton in the eastern Mediterranean–indications of a climatically induced shift? Marine Ecology 28: 259–275.

Koppelmann, R., Böttger-Schnack, R., Möbius, J. and Weikert, H. 2009. Trophic relationships of zooplankton in the eastern Mediterranean based on stable isotope measurements. Journal of Plankton Research 31(6): 669–686.

Kovalev, A.V., Kideys, A.E., Pavlova, E.V., Shmeleva, A.A., Skryabin, V.A., Ostrovskaya, N.A. et al. 1999. Composition and abundance of zooplankton of the eastern Mediterranean Sea. *In*: Malanotte-Rizolli, P. and Eremeev, V.N. (eds.). The Eastern Mediterranean as a Laboratory Basin for the Assessment of Contrasting Ecosystems. Springer Science Business Media, Dordrecht.

Krom, M.D., Kress, N., Brenner, S. and Gordon, L.I. 1991. Phosphorus limitation of primary productivity in the eastern Mediterranean Sea. Limnol. Oceanogr. 36: 424–432.

Krom, M.D., Herut, B. and Mantoura, R.F.C. 2004, Nutrient budget for the eastern Mediterranean: implications for phosphorus limitation. Limnol. Oceanogr. 49: 1582–1592.

Lee, R.F., Nevenzel, J.C. and Paffenhofer, G-A. 1971. Importance of wax esters and other lipids in the marine food chain: phytoplankton and copepods. Mar. Biol. 9: 99–108.

Longhurst, A. 1985. The structure and evolution of plankton communities. Prog. Oceanogr. 15: 1–35.

Longhurst, A.R. and Harrison, W.G. 1989. The biological pump: profiles of plankton production and consumption in the upper ocean. Prog. Oceanogr. 22: 47–123.

Lovern, J.A. 1935. C. Fat metabolism in fishes. VI. The fats of some plankton crustacea. Biochemical Journal 29: 847–849.

Mackenzie, B.R. and Legget, W.C. 1991. Quantifying the contribution of small-scale turbulence to the encounter rates between larval fish and their zooplankton prey: effects of wind and tide. Marine Ecology Progress Series 73: 149–160.

Margalef, R. 1997. Turbulence and marine life. Scientia Marina 61(Suppl 1): 109–123.

Mauchline, J. 1998. The Biology of Calanoid Copepods. Academic Press, San Diego. Adv. Mar. Biol. 33: 1–170.

Mayzaud, P., Virtue, P. and Albessard, E. 1999. Seasonal variations in the lipid and fatty acid composition of the euphasiid *Meganyctiphanes norvegica* from the Ligurian Sea. Mar. Ecol. Prog. Ser. 186: 199–210.

Mazzocchi, M.G., Christou, E.D., Fragopoulu, N. and Siokou-Frangou, I. 1997. Mesozooplankton distribution from sicily to cyprus (eastern Mediterranean): I. general aspects, Oceanol. Acta 20: 521–535.

McCutchan, J.H.J., Lewis, W.M.J., Kendall, C. and McGrath, C.C. 2003. Variation in trophic shift for stable isotope ratios of carbon, nitrogen, and sulfur. Oikos 102: 378–390.

Mercado, J.M., Cortés, D., García, A. and Ramírez, T. 2007. Seasonal and interannual changes in the planktonic communities of the northwest Alboran Sea (Mediterranean sea). Prog. Oceanogr. 74: 273–293. doi: 10.1016/j.pocean.2007.04.013.

Michener, R.H. and Schell, D.M. 1994, Stable isotopes ratios as tracers in marine aquatic food webs. pp. 138–157. *In*: Lajtha, K. and Michener, R.H. (eds.). Stable Isotopes in Ecology and Environmental Research. Blackwell Scientific Publications, Oxford.

Minagawa, M. and Wada, E. 1984. Stepwise enrichment of ^{15}N along food chains: further evidence and the relation between δ^{15}N and animal age. Geochim. Cosmochim. Acta 48: 1135–1140.

Minutoli, R. and Guglielmo, L. 2009. Zooplankton respiratory electron transport system (ETS) activity in the Mediterranean Sea: spatial and diel variability. Mar. Ecol. Prog. Ser. 38: 199–211.

Minutoli, R., Fossi, M.C., Zagami, G., Granata, A. and Guglielmo, L. 2008. First application of biomarkers approach in the zooplanktonic copepod *Acartia latisetosa* for the early management and conservation of transitional waters ecosystems. Transitional Water Bulletin 1: 45–52.

Najdek, M., PuSkaric, S. and Bochdansky, A.B. 1994. Contribution of zooplankton lipids to the flux of organic matter in the northern Adriatic Sea. Mar. Ecol. Prog. Ser. 111: 241–249.

Ndah, A.B., Meunier, C.L., Kirstein I.V., Gobel, J., Ronn, L. and Boersma M. 2022. A systematic study of zooplankton-based indices of marine ecological change and water quality: application to the European marine strategy framework directive (MSFD). Ecological Indicators 135: 108587. https://doi.org/10.1016/j.ecolind.2022.108587.

Neffati, N., Daly Yahia-Kefi, O., Bonnet, D., Carlotti, F. and Daly Yahia, M.N. 2013. Reproductive traits of two calanoid copepods: *Centropages ponticus* and *Temora stylifera*, in autumn in Bizerte Channel. J. Plankton Res. 35(1): 80–96. https://doi.org/10.1093/plankt/fbs071.

Olson, M.M., Wood, R. and Sellner, K.G. 2005. Zooplankton/food web monitoring for adaptive multi-species management near-term. Recommendations: second report of a CRC-sponsored Workshop, January 12–13, Chesapeake Research Consortium -CRC Publication 05–159c.

Packard, T.T. 1971. The measurement of electron transport system activity in marine phytoplankton. J. Mar. Res. 29: 235–244.

Packard, T., Harmon, D. and Boucher, J. 1974. Respiratory electron transport activity in plankton from upwelled waters. Tethys 6: 213–222.

Pantoja, S., Repeta, D.J., Sachs, J.P. and Sigman, D.M. 2002. Stable isotope constraints on the nitrogen cycle of the Mediterranean Sea water column. Deep-Sea Res. I 49: 1609 1621.

Perry, R., Batchelder, H., Mackas, D., Chiba, S., Durbin, E., Greve, W. et al. 2004. Identifying global synchronies in marine zooplankton populations: issues and opportunities. ICES J. Mar. Sci. 61(4): 445–456. https://doi.org/10.1016/j.icesjms.2004.03.022.

Peters, R.H. 1983. The Ecological Implications of Body Size. Cambridge University Press, Cambridge.

Peterson, B.J. and Fry, B. 1987. Stable isotopes in ecosystem studies. Annual Review of Ecological Systems 18: 293–320.

Polunin, N.V.C., Morales-Nin, B., Pawsey, W.E., Cartes, J.E., Pinnegar, J.K. and Moranta, J. 2001. Feeding relationships in Mediterranean bathyal assemblages elucidated by stable nitrogen and carbon isotope data. Mar. Ecol. Prog. Ser. 220: 13–23.

Porter, K.A. 1957. Effect of homologous bone marrow injections in x-irradiated rabbits. British Journal of Experimental Pathology 38(4): 401–412. PMC 2082598. PMID 13460185.

Post, D.M. 2002. Using stable isotopes to estimate trophic position: models, methods, and assumptions. Ecology 83: 703–718.

Protopapa, M., Zervoudaki, S., Tsangaris, C., Velaoras, D., Koppelmann, R., Psarra, S. et al. 2019a. Zooplankton distribution and electron transport system activity in the cretan passage, eastern Mediterranean. Deep. Res. Part II. https://doi.org/10. 1016/j.dsr2.2019.03.001.

Protopapa, M., Koppelmann, R., Zervoudaki, S., Wunsch, C., Peters, J., Parinos, C. et al. 2019b. Trophic positioning of prominent copepods in the epi- and mesopelagic zone of the ultra-oligotrophic eastern Mediterranean Sea. Deep. Res. Part II 164: 144–155.

Quintanilla, J.M., Laiz-Carrión, R., García, A., Quintanilla, L.F., Cortés, D., Gómez-Jakobsen, F. et al. 2020. Early life trophodynamic influence on daily growth patterns of the Alboran Sea sardine (*Sardina pilchardus*) from two distinct nursery habitats (Bays of Málaga and Almería) in the western Mediterranean Sea. Marine Environmental Research 162: 105195. 10.1016/j. marenvres.2020.105195.

Quiñones, R.A., Blanco, J.M., Echevarría. F., Fernández-Puelles, M.L., Gilabert, J., Rodríguez, V. et al. 1994. Metabolic size spectra at a frontal station in the Alboran Sea. Working Group 4 Report. Scientia Marina 58(1–2): 53–58.

Reid, P.C. and Edwards, M. 2001. Plankton and climate. pp. 2194e2200. *In*: Encyclopaedia of Sciences. Ed. by J. Steele. Academic Press, Oxford.

Rodríguez, L.P., Caliani, I., Brugnano, C., Granata, A., Guglielmo, R., Guglielmo, L. et al 2018. Biomarkers employment in planktonic copepods for early management and conservation of aquatic ecosystems: The case of the 'Capo Peloro' lakes (southern Italy). 18: 161–169. DOI: 10.1016/j. rsma.2017.10.002.

Rombouts, I., Beaugrand, G., Artigas, L.F., Dauvin, J.C., Gevaert, F., Goberville, E. et al. 2013. Evaluating marine ecosystem health: case studies of indicators using direct observations and modelling methods. Ecol. Indic. 24: 353–365.

Rossi, S., Sabatıs, A., Latasa, M. and Reyes, E. 2006. Lipid biomarkers and trophic linkages between phytoplankton, microzooplankton and the anchovy (*Engraulis encrasicolus*) larvae in the NW Mediterranean. J. Plank. Res. 28: 551–562.

Rumengan, I.F.M. and Ohji, M. 2012. Ecotoxicological risk of organotin compounds on zooplankton community. Coastal Marine Sciences 35: 129–135.

Rundel, P.W., Ehleringer, J.R. and Nagy, K.A. 1989. Stable Isotopes in Ecological Research. Springer-Verlag, New York.

Rupper, E.E., Fox, R.S. and Barnes, R.D. 2003. Invertebrate Zoology. A Functional Evolutionary Approach, 7th edition. Brooks Cole-Thomson Learning, Belmont CA, p. 963.

Sachs, J.P. and Repeta, D.J. 1999, Oligotrophy and nitrogen fixation during eastern Mediterranean sapropel events. Science 286: 2485–2488.

Sargent, J.R. and Henderson, R.J. 1986. Lipids. pp. 59–108. *In*: Corner, E.D.S. and O'Hara, S.C.M. (eds.). The Biological Chemistry of Marine Copepods. Vol. 1. Clarendon Press, Oxford.

Schalk, P.H. 1988. Respiratory electron transport system (ETS) activities in zooplankton and micronekton of the Indo-Pacific region. Mar. Ecol. Prog. Ser. 44: 25–35.

Schwamborn, R. and Giarrizzo, T. 2015. Stable isotope discrimination by consumers in a tropical mangrove food web: how important are variations in C/N ratio? Estuaries and Coasts 38(3): 813–825.

Serranito, B., Aubert, A., Stemmann, L., Rossi, N. and Jamet, J.L. 2016. Proposition of indicators of anthropogenic pressure in the bay of toulon (Mediterranean Sea) based on zooplankton time-series. Cont. Shelf Res. 121: 3–12. https://doi.org/10.1016/j. csr.2016.01.016.

Serrazanetti, G.P., Pagnucco, C., Conte, L.S., Artusi, R., Fonda-Umani S. and Bergami C. 1994. Sterols and fatty acids in zooplankton of the Gulf of Trieste. Comp. Biochem. Physiol. 107B(3): 443–446.

Siokou-Frangou, I. 2004. Epipelagic mesozooplankton and copepod grazing along an east-west transect in the Mediterranean Sea, Rapports de la Commission Internationale pour l'Exploration Scientifique de la Mer Mediterranee 37: 439–2004.

Siokou-Frangou, I., Christaki, U., Mazzocchi, M.G., Montresor, M., d' Alcala, M.R. et al. 2010. Plankton in the open Mediterranean Sea: a review. Biogeosciences 7: 1543–1586.

Tett, P., Carreira, C., Mills, D.K., van Leeuwen, S., Foden, J., Bresnan, E. et al. 2008. Use of a phytoplankton community index to assess the health of coastal waters. ICES J. Mar. Sci. 65(8): 1475–1482. https://doi.org/10.1093/icesjms/fsn161.

Vander Zanden, M.J. and Rasmussen, J.B. 2001. Variation in d^{15}N and d^{13}C trophic fractionation: implications for aquatic food web studies. Limnol. Oceanogr. 46: 2061–2066.

Volkman, J.K., Jeffrey, S.W., Nichols, P.D., Rogers, G.I. and Garland, C.D. 1989. Fatty-acid and lipid composition of 10 species of microalgae used in mariculture. J. Exp. Mar. Biol. Ecol. 128 : 219–240.

Webber, D.F. and Webber, M.K. 1998. The water quality of kingston harbour: evaluating the use of the planktonic community and traditional water quality indices. Chem. Ecol. 14(3–4): 357–374. https://doi.org/10.1080/02757549808037614.

Wells, P.G. 1984. Marine ecotoxicological test with zooplankton. pp. 215–256. *In*: Persoone, G., Jaspers, E. and Claus, C. (eds.). Ecotoxicological Testing for the Marine Environment, Vol. 1, State University of Ghent, Laboratory for Biological Research in Aquatic Pollution, Bredene, Belgium.

Yebra, L. and Hernández-León, S. 2004. Aminoacyl-tRNA synthetases activity as a growth index in zooplankton. J. Plankton Res. 26: 351–356.

Yebra, L., Hernández-León, S., Almeida, C., Bécognée, P. and Rodríguez, J.M. 2004. The effect of upwelling filaments and island-induced eddies on indices of feeding and metabolism in copepods. Progress in Oceanography 62: 151–169. 10.1016/j.pocean.2004.07.008.

Yebra, L, Kobari, T., Sastri, A.R., Gusmao, F. and Hernández-León, S. 2017b. Advances in biochemical indices of zooplankton production. Advances in Marine Biology 76(4): 157–240. doi: 10.1016/bs.amb.2016.09.001.

Yebra, L, Putzeys, S., Cortés, D., Mercado, J.M., Gómez-Jakobsen, F., León, P. et al. 2017a. Trophic conditions govern summer zooplankton production variability along the SE Spanish coast (SW Mediterranean). Estuarine, Coastal and Shelf Science 187: 134–145. doi: 10.1016/j.ecss.2016.12.024.

Yebra, L., Herrera, I., Mercado, J.M., Cortés, D., Gómez-Jakobsen, F., Alonso, A. et al. 2018. Zooplankton production and carbon export flux in the western Alboran Sea gyre (SW Mediterranean). Progress in Oceanography 167: 64–77. doi: 10.1016/j.pocean.2018.07.009.

Yebra, L., Hernández de Rojas, A., Valcárcel-Pérez, N., Castro, M.C., García-Gómez, C., Cortés, D. et al. 2019. Molecular identification of the diet of sardina pilchardus larvae in the SW Mediterranean Sea. Marine Ecology Progress Series 617–618: 41–52. 10.3354/meps12833.

Yebra, L., Espejo, E., Putzeys, S., Giráldez, A., Gómez-Jakobsen, F., León, P. et al. 2020. Zooplankton biomass depletion event reveals the importance of small pelagic fish top-down control in the Western Mediterranean coastal waters. Frontiers in Marine Science 7: 608690. doi: 10.3389/fmars.2020.608690.

Yebra, L., Puerto, M., Valcárcel-Pérez, N., Putzeys, P., Gómez-Jakobsen, F., García-Gómez, C. et al. 2022. Spatio-temporal variability of the zooplankton community in the SW Mediterranean 1992–2020: linkages with environmental drivers. Progress in Oceanography 203: 102782. https://doi.org/10.1016/j.pocean.2022.102782.

Zervoudaki, S., Christou, E.D., Nielsen, T.G., Siokou-Frangou, I., Assimakopoulou, G. et al. 2007. The importance of small-sized copepods in a frontal area of the Aegean Sea. J. Plankton Res. 29: 317–338.

Zohary, T. and Robarts, R. 1992. Bacterial numbers, bacterial production, and heterotrophic nanoplankton abundance in a warm core eddy in the eastern Mediterranean, Mar. Ecol.- Prog. Ser. 84: 133–137.

2.3

Metagenomic-Based Analysis Providing a Complementary Data Method for Plankton Ecological Studies

Marco Simões,[1,2,*] *Sónia Cotrim Marques,*[1]
Paulo Maranhão[1] and *Maria Jorge Campos*[1]

1. Introduction

The world's ocean fauna is dominated in terms of abundance and biomass by the drifting organisms collectively referred to as plankton (Bucklin et al. 2004). The term 'plankton' is commonly defined as a group of organisms that float in the water column and have limited swimming capability, being insufficient to withstand currents (Anon 2000). Together with the phytoplankton and bacterioplankton, zooplankton (planktonic animals) constitute the plankton community. Zooplankton are exceptional in that they occur in all marine waters, throughout all depths and, for many species, across widespread biogeographical distributions (e.g., Barnard et al. 2004, Razouls et al. 2022). Ranging in size from microns (protozooplankton) to centimetres and metres (mesozooplankton, including chains of Thaliacea), the animals making up the zooplankton are taxonomically and structurally diverse and thus perform a variety of ecosystem functions, such as consumers, producers and preys (Bathmann et al. 2001, Lima-Mendez et al. 2015). Undoubtedly, the most important role of zooplankton is as the major grazers in aquatic food webs, being critical intermediaries in the flow of energy and matter through marine food chains, from primary producers to consumers at higher trophic levels, such as fish, marine mammals and turtles (Richardson 2008, Sterner 2009). It includes holoplanktonic organisms, the permanent members of the zooplankton, species spending their whole life in the pelagic realm. The holoplankton contains representatives of nearly every taxonomical group, such as euphausiids (krill), copepods (crustacean, over 10,000 species alone are known),

[1] MARE/ARNET, School of Tourism and Maritime Technology, Polytechnic of Leiria, 2520-630 Peniche, Portugal.

[2] CIIMAR – Interdisciplinary Marine and Environmental Research Centre 4450–208 Matosinhos.

* Corresponding author: marco.a.simoes@ipleiria.pt

various pelagic (free-swimming) sea snails and slugs, salps, jellyfish and a small number of the marine worms. To most people, jellyfish are probably the most visible and best-known of this group. The most famous is the Bluebottle or Portuguese Man-of-War *Physalia physalis*. However, copepods comprise the bulk of the numbers in the holoplankton, occurring in every aquatic environment, even outnumbering insects by possibly three orders of magnitude (Uttieri 2018, Walter and Boxshall 2022). On the other way, the meroplankton are temporary residents of the plankton community. The meroplankton is much more diverse than the holoplankton because several benthic or pelagic species spend their larval and juvenile stages in the plankton (Anon 2000). In temperate areas, most species, including fish, as well as many benthic invertebrates, e.g., polychaetes, molluscs, echinoderms, bryozoans, and barnacles, and decapods among the crustaceans reproduce with pelagic larvae (Grantham et al. 2003). Especially for sessile organisms this ensures dispersal over great distances and a good ability for exploitation of new territories (Scheltema 1988). Besides, currents transport the offspring to new areas, which is especially important for survival of immobile benthic forms since the larvae do not compete with the parents for scarce resources such as food or space (Fetzer and Arntz 2008).

Compared with the terrestrial environment, the pelagic realm has few physical barriers obstructing the mixing of planktonic species. Nevertheless, there are some hydrographic barriers between different water masses, which have distinct physico-chemical conditions and ecological properties (Beaugrand and Ibañez 2002). Many of the problems associated with studying zooplankton have long been recognised. These animals are heterogeneously distributed due to physical and biological processes. In coastal areas, the environmental conditions fluctuate at different time and space scales, subjecting organisms that inhabit the pelagic realm to tidal, diurnal and seasonal environmental changes. Only a few species can develop in coastal ecosystems, and their dynamics and distribution vary substantially between years and decades and are affected both directly and indirectly by the short- and long-term hydrographic and physical conditions of their environment (Fromentin and Ibanez 1994, Licandro and Ibanez 2000). While some zooplanktonic species have a wide distribution encompassing a broad spectrum of environmental variables, others are restricted to such narrow limits determined by their specific tolerance to temperature, salinity, food availability (Cervetto et al. 1999, Gaudy et al. 2000, Leandro et al. 2006) and other factors that can be used as biological indicators of the particular water-mass types they inhabit (Beaugrand et al. 2002, Bonnet and Frid 2004). Temperature, for example, can greatly influence the community structure and production of zooplankton (Leandro et al. 2006), which, in turn, can lead to large seasonal, annual, and decadal changes in population size and geographic distribution (Ibarbalz et al. 2019). Moreover, biological mechanisms such as active vertical migrations (DVM) may also account for a significant part of the temporal variation in zooplankton community structure. Common among all freshwater and marine zooplankton taxa, DVM probably represents the biggest synchronised animal migration in terms of biomass on the planet and is of paramount importance for ecosystem functioning and carbon cycling (Hays 2003). As a result of DVM, organisms aggregate periodically at certain depths. Zooplankton usually aggregates near the surface by night and at great depth by day, but patterns differ by taxa and stage (e.g., Dauvin et al. 1998, David et al. 2005, Queiroga et al. 1997, Rawlinson et al. 2004). Much current research focuses on distinguishing the physical or biological signals (e.g., predator cues) that alter DVM behaviours and DVM-driven patterns of distribution (Hays 2003). However, the degree of migration can be modified by zooplankton behavioural responses to variations in predators, food, temperature, oxygen and their rhythms (Armengol and Miracle 2000, Haupt et al. 2009, Queiroga et al. 1997). Nonetheless, the study of these rather complex hydrodynamic processes is often complicated because of their multiple confounding interactions.

At times, abrupt and dramatic changes in ecosystems occur in response to subtle climate or physical forcing (Malhi et al. 2020, Perry et al. 2004). Such abrupt reorganisation, known as regime shift, can transform systems from one stable state to another (Hare and Mantua 2000). The assemblages of species in ecological communities reflect interactions among organisms as well as between organisms and physical forcing. We might expect, therefore, that rapid climatic change or extreme climatic events (such as drought and flood) can alter community composition (Walther et al. 2002). The effects of climatic variation over decadal to multidecadal time scales on marine ecosystems worldwide have been reported in recent decades (Chiba et al. 2006, Molinero et al. 2008, Perry et al. 2004, Walter and Boxshall 2022). Indeed, although the concept of regime shifts has been discussed for several decades, interest has increased during the 1990s and into this century. One reason is that ecosystems contain species of commercial importance whose abundance, distribution and productivity differ among them (Batten and Welch 2004). In addition, although time series of physical variables are generally available, corresponding time series of biological variables to examine the ecosystem response to regime shifts are much rarer and only now reach sufficient length for analysis. Climate change also will have consequences on the pelagic community (Hays et al. 2005). Because there are many processes separating climate and higher trophic levels, a retrospective study of lower trophic levels should elucidate the mechanisms linking variation in climate, hydrographic environment and ecosystems. As Hays et al. (2005) stated, zooplankton may be very useful, being particularly good indicators of climate change in the marine environment for several reasons. If we consider their pivot role in linking primary producers with fish, their response could be expanded to the food web, thereby affecting the functioning of the marine ecosystem. At a global scale, many of the standard climate indices now used by ocean researchers were defined over the past ten years and compared to variations on local zooplankton populations, e.g., North Atlantic Oscillation (NAO) (Hurrell 1995, Planque and Taylor 1998), northern hemisphere temperature (NHT) (Beaugrand et al. 2002, Heyen et al. 1999), Pacific Decadal Oscillation and the Northern Oscillation Index (PDO) (McGowan et al. 2003). Based on regional information, regime shifts have been identified all over the world, and attempts have been made to construct a global picture of the climate-ecosystem link. The impacts of global warming on zooplankton are manifested by a biogeographical re-distribution of certain crustacean plankton assemblages and were reflected in a northward extension of warm-water species and a decrease in the numbers of colder water species as temperature warms (Beaugrand et al. 2002). In addition, clearer evidence of the consequence of global warming comes from an earlier timing of important life cycle events or phenology (Chiba et al. 2006, Edwards and Richardson 2004, Mackas et al. 1998). Phenological change in one trophic level can result in a trophic mismatch that could resonate to higher trophic levels, an important mechanism of variation in ecosystem functions (Edwards and Richardson 2004). Winter temperatures are projected to continue to increase more than summer temperatures, narrowing the annual water temperature range (European Environmental Agency – EEA 2014, IPCC Working Group 1 et al. 2013). This is likely to cause plankton species' ranges to shift and increase the vulnerability of some marine ecosystems to non-native invasive species (e.g., Marques et al. 2008, Uttieri et al. 2020). Further, increases or decreases in precipitation and runoff may create extreme events, floods or droughts respectively, increasing in frequency worldwide (Gleick 2003, Mirza 2003). Historical evidence shows that zooplankton does not recover easily from catastrophes. When a population crash occurred across the oceans 65 million years ago, it took approximately 3 million years for the zooplankton to recover (Dybas 2006). How the zooplankton around the world will be affected in the long-term by abrupt climate change is difficult to predict. We do know that with less zooplankton, ecosystems from the poles to the tropics and from freshwater to the salty seas will be negatively impacted. Because of their pivot role in aquatic ecosystems any change in the zooplankton diversity and abundance will have consequences on the marine food web and on other trophic levels. Despite a lot of effort have been conducted all over the world our knowledge concerning the climate impact change in zooplankton has increased only modestly. Yet, it is very important to note that many of these potential alterations can only be detected by continuous monitoring, as they occur abruptly and without warning.

In contrast to other marine species, such as commercial fish species and many intertidal benthic organisms, for which much data exist, zooplankton has short life spans, and assessments of their abundance are unaffected by fishing effort (Batten and Welch 2004). Hence, most zooplankton population changes can be attributed to environmental causes. Moreover, most species are short-lived (from a few months to one year), so population size is less influenced by the persistence of individuals from previous years. Besides, recruitment and mortality rates are slow enough that major population fluctuations are not missed by sampling at monthly intervals. This leads to a tight coupling between environmental change and zooplankton dynamics. Third, unlike most fish and marine mammals, changes in zooplankton population size are rapid enough to track seasonal-to-interannual changes in environmental conditions (e.g., David et al. 2005, Kimmel and Roman 2004, Marques et al. 2018). Indeed, zooplankton can show dramatic changes in distribution because they are free-floating and can respond easily to changes in temperature and oceanic current systems by expanding and contracting their ranges (Beaugrand and Reid 2003, Chavez et al. 2003, Hurrell 1995). Recent evidence suggests that zooplankters are more sensitive indicators of change than are even environmental variables themselves because the nonlinear responses of biological communities can amplify subtle environmental perturbations (Murphy et al. 2020). Finally, because many fish are dependent on a zooplankton food source during their pre-recruit life history stages, zooplankton anomalies may be a useful leading indicator of what will happen to commercial fish stocks several years later (Beaugrand and Reid 2003, Lomartire et al. 2021). Therefore, besides zooplanktonic organisms being an important food resource of pelagic ecosystems, they also serve as an integrator of hydroclimatic forcing, providing an accurate diagnosis of ecosystem state (Beaugrand 2005). For these reasons, the use of zooplanktonic species as indicators of specific water masses has long been stressed at a global (Beaugrand 2005, Russell 1939) and local scale (Bonnet and Frid 2004, Marques et al. 2018, Whitman et al. 2004). Furthermore, zooplankton is very abundant and can be quantified by relatively simple and intercomparable sampling methods. Because of their rich species diversity, the vast volume of water they inhabit, their trophodynamic role in aquatic food webs, and because they include the meroplanktonic stages of vast benthic organisms and fish, the zooplankton should be included in any monitoring and biodiversity assessment programme.

Although marine ecosystems are now intensively studied from the perspective of management, long-term series are not so numerous, especially for zooplankton. Yet, organisational support for zooplankton studies is needed. The implementation of monitoring programmes over much longer periods than traditional compliance monitoring is recognised as an important issue for understanding certain ecological phenomena, such as extreme events (e.g., drought and flood) in which a long time is required to detect potential regime shifts or trends (Dolbeth et al. 2007, Dybas 2006). The importance and relevance of long-term zooplankton data are not understood by many decision-makers and funding agencies. In contrast to fisheries, there is no mandated requirement to sample zooplankton. For example, Directive 2000/60/EC of the European Parliament established a legal framework for community action in the field of water policy. This framework required sampling of phytoplankton, phytobenthos, benthic invertebrates and fish but not zooplankton (Borja 2005). Since zooplankton observations are not required, it suggests that they probably will not be made in the context of this policy. This will reduce organisational support for zooplankton sampling, with the result that insights into climate and anthropogenic forcing of zooplankton variations and early warning of substantial temporal shifts of aquatic ecosystems will be harder to achieve.

On-going plankton monitoring programmes around the world can act as sentinels to identify future changes in marine ecosystems. Crucial to identifying these future changes are the maintenance of the plankton time series and the funding of projects that continue to enhance the unique data sets these time series provide. The analysis of change in response to long-term variation in climatic variables requires datasets over much longer periods than traditional compliance monitoring. If climate change is having an effect on the marine environment (and there is a consensus that this is real), then long-term data and time series can also be used to try to understand and predict responses. Understanding the characteristics and drivers of long-term fluctuation of zooplankton populations may provide opportunities for adaptive management that will maintain healthy marine ecosystems. If large changes in the productivity of the ecosystem occur, it is important to recognise them early in order to provide warnings to fishery and resource managers and potentially to adopt measures to mitigate the changes. Moreover, Taylor et al. (2002) suggest that subtle ecosystem effects of climate change may be amplified by complex biological interactions of the system. Thus, changes in zooplankton or other biological constituents may be better early indicators of regime shifts than physical changes.

As mentioned, the study of plankton communities is of extreme importance for the diverse reasons specified. However, their monitoring through classic methods is limited due to characteristics like the fragile nature, small body size, a large number of taxa (Bucklin et al. 2010) and the existence of cryptic species that may result in difficult identification and an underestimation of species biodiversity (Knowlton 1993). Therefore, its characterisation becomes difficult, expertise-dependent and time-consuming (Carvalho et al. 2010).

Molecular techniques have been applied as an alternative approach to overcome the difficulties presented by classic methods. With molecular tools and with the recent advances in massive sequencing technologies, metagenomics presents an advantage in obtaining new and complementary information. Through the large amount of data that this method offers, metagenomics allows us to unveil in a more complete way the key importance of plankton on the aquatic ecosystems and fight scarcity of data, showing potential to help in the monitoring of aquatic environment on the characterisation of interspecies relationships and the biogeochemical cycles of the aquatic environment.

2. What is Metagenomics?

Metagenomics is defined as the direct genetic analysis of genomes contained in an environmental sample (Thomas et al. 2012). The term 'metagenome' was first introduced to describe the combined genomes of natural microbes in the soil environment (Handelsman et al. 1998). Since then, microbial communities from diverse niches (including oceans, lakes, soils, thermal vents, hot springs, the mouth and gastrointestinal tract) have been involved in metagenome-based studies over a very short period (Riesenfeld 2004, Xu 2006). Now, the term 'metagenome' has been extended to refer to all the genetic material recovered directly from environmental samples, which can be analysed in a way analogous to the study of a single genome (Handelsman 2004, Xu 2006). The genomic study of single organisms has extended our understanding from single genes to the collective genes of an organism, whereas metagenomics goes beyond the genome of single organisms and enables us to study genomes of the community (Sharma et al. 2008). More importantly, metagenomics transcends the limitations of classical genomics and microbiology and provides valuable strategies to study the relationships among genes, genomes, organisms and communities present in the natural world (Handelsman 2004).

Metagenomics is the study of genomic material collected directly from natural environmental samples. It is also termed environmental genomics, ecogenomics and community genomics. However, metagenomics is most used in describing this emerging field of study (Yan and Yu 2011). Although the term 'metagenomics' was coined in 1998 (Handelsman et al. 1998), the concept that organisms could be identified without cultivation by retrieving and sequencing their DNA directly from natural samples dates much earlier (e.g., Pace et al. 1985, 1986, Thomas et al. 2012, Yan and Yu 2011). The most commonly used target gene for bacterial identification is 16S rRNA (or 16S rDNA), which is the gold standard in microbial typing (Baker et al. 2003, Pel et al. 2018).

All metagenomic studies begin with community DNA extraction from diverse members present in a particular environment (Yan and Yu 2011). Metagenome information is then analysed directly or cloned before investigations of the structure, function and dynamics of natural microbial communities (Thomas et al. 2012). Over the past decades, metagenomic studies have provided valuable information in several fields, including microbiology, medicine, energy, biotechnology, agriculture, environmental remediation and gut microbial ecology (e.g., Bharti and Grimm 2021, Loza et al. 2022, Wang et al. 2017). Metagenomics has bridged the gap between genetics and ecology, demonstrating that the genes of single organisms are related to the genes of other species or even to the entire community (Handelsman 2004).

Because of the ability to characterise these gene networks and the respective communities as a whole, the application of metagenomics to phytoplankton, zooplankton and higher trophic levels is so important and useful for the aquatic environments' future.

2.1 *Metagenomics Applied to Phytoplankton*

The most important group of organisms responsible for marine primary production is phytoplankton; however, it accounts for less than 1% of the total Earth's photosynthetic biomass (Simon et al. 2009). Besides supporting the fisheries stocks and contributing to the marine ecosystems, through the biogeological cycles, these groups of organisms also have an important role in the food web (first level of the food pyramid in the trophic chain) and in the climate (Delmont et al. 2022, Pierella Karlusich et al. 2021). Phytoplankton comprise a very large number of autotrophic organisms, obtaining energy through photosynthesis and are phylogenetically diverse. Such organisms live in the water column down to 200 m and have dimensions that can range between 0.4 µm to 200 µm (Simon et al. 2009). There are three main clades of phytoplankton in the ocean: dinoflagellates, coccolithophores, and diatoms (Pierella Karlusich et al. 2022, Rynearson and Palenik 2011, Sjöqvist 2022). These include organisms such as cyanobacteria, *Prochorococcus*, diatoms, pranisophytes and prymnesiophytes (Cuvelier et al. 2010).

Metagenomics methods, in opposition to PCR-dependent methods, such as metabarcoding, were primarily used to characterise ocean biodiversity; however, metagenomics was mainly used to assess the biodiversity of prokaryotic organisms. Since eukaryotic DNA appears in low copy numbers, with sequences several times larger than prokaryotic sequences and with a large portion of noncoding DNA, the match for eukaryotic sequences in DNA reference databases is not very promising unless there is a very close relation between an organism or the genome of the organism is totally sequenced. For these reasons, it is expected that metagenome results will yield fewer results when matched against the databases (Delmont et al. 2021, Singer et al. 2020). When studies from metabarcoding and metagenomics data were compared for marine biomonitoring, the number of metagenomic sequences, compared to metabarcoding ones, originate less level of identification. Most metabarcoding reads can be identified, at least to the kingdom level, and the level of confidence is also higher for the metabarcoding sequences (Singer et al. 2020). Metabarcoding planktonic studies, based on PCR amplification of hypervariable regions of the 16S or 18S rRNA genes, have several drawbacks associated with these techniques, mainly due to the mismatches that can occur during PCR, which can account for differences in the relative and genuine abundance of organisms (Pierella Karlusich et al. 2022). Also, the study of rRNA genes does not allow complete intraspecific diversity patterns, and organisms with the same rRNA genes may present different genomes and phenotypes (Sjöqvist 2022). Current knowledge indicates that phytoplankton is characterised by high levels of intraspecific variation, and the lack of relevant gene annotation in the available databases might hamper the understanding of its diversity (Sjöqvist 2022).

With metagenomic approaches, it is possible to sequence DNA belonging to organisms present in the environments (eDNA). The Tara Oceans programme performed a large sampling of plankton in the upper layers of the oceans to study marine plankton globally. In this project, large numbers of marine plankton metagenomes and metatranscriptomes were produced that can be used to perform several further studies (Delmont et al. 2021, Pierella Karlusich et al. 2022, Tully et al. 2017). More than 35,000 samples from 210 ecosystems were generated, and more than 7.2 Tbp of metagenomic data were obtained from the Tara Oceans programme (Tully et al. 2017). Delmont et al. 2021 were able to assemble more than 700 eukaryotic genomes from data acquired through this project. So far, more than 40,000 taxa of phytoplankton have been revealed together with several other data, such as abundance and diversity patterns and other environmental parameters (Vincent et al. 2022), and it is estimated, from the TARA data, that more than 90% of the most abundant protists in the ocean belong to unknown species with no cultivated representatives (Sjöqvist 2022).

Metagenomic methods applied to the study of phytoplankton have been used to (i) identify and determine abundance and diversity of phytoplankton, (ii) to determine the biochemistry, metabolism, and new genes (Monier et al. 2012, Pierella Karlusich et al. 2021), (iii) to do ancient DNA studies (Armbrecht et al. 2021), (iv) to study blooms (Francis et al. 2019, Nowinski et al. 2019) and (v) to find bacterial-viral/phytoplankton association (Fu et al. 2020, de Oliveira Santos et al. 2021). In this chapter, we will approach the most recent studies regarding the identification, determination of abundance and diversity of phytoplankton by metagenomic approaches.

The study of environmental biodiversity, using metagenomic, is independent of previous knowledge of genomic sequences of the organisms; therefore, it is possible to use this technique to obtain knowledge regarding new phylogenetic groups of organisms (Johnson and Martiny 2015). However, environmental genomics that uses direct assemblies of metagenomic reads must originate reads encompassing several genes. These can be difficult to achieve for eukaryotic organisms such as phytoplankton since these organisms possess genomes that are several times larger than the genomes of bacteria and archaea. Therefore, the level of sequence coverage needed to study these organisms is higher, but the number of sequences generated in these studies is larger and presents higher genomic heterogeneity (Vorobev et al. 2020). Nevertheless, Delmont et al. (2022) were able to produce eukaryotic metagenome assembly genomes (MAGs) from environmental communities. This proved that the MAGs can be used to predict marine eukaryotic diversity (Massana and López-Escardó 2022). In the study by Delmont et al. (2022), it was possible to detect several taxonomic groups, such as copepods, metazoans, stramenopiles, haptophytes and chlorophytes, despite not being possible to identify dinoflagellates, marina alveolates and radiolaria. Also, due to the binning methodology used in the process, the 18S rRNA was not present in MAGs (Massana and López-Escardó 2022).

In a study carried out by Duncan et al. (2022), which investigated the differences in phytoplankton in microbiomes from the Artic and Atlantic oceans, the findings showed that adjacent sampling sites in each ocean had between 51–88% of MAGs in common; however, MAGs from both oceans were very different. Eukaryotic MAGs were more diverse in the Arctic, while in the Atlantic, prokaryotic MAGs were more diverse. However, protein families from prokaryotes and eukaryotes were shared between 70% of Artic and Atlantic MAGs, while Artic-only protein families were mostly found in eukaryotic and Atlantic-only protein families were mainly found in prokaryotes. These data suggest that phytoplankton in these two regions are genetically defined by the environment.

In Monterey Bay, California, a coastal ocean embayment with distinct hydrographic seasons, a metagenomic and amplicon-dependent study showed that phototrophic organisms, including cyanobacteria, decrease abundance with depth, in alignment to a relative decrease in aerobic anoxygenic phototrophy-related genes. During the three oceanographic seasons found in the region, cyanobacteria were abundant members of the community during the oceanic and winter seasons; 98% of the community was classified at the order level as Synechococcales and the genus level as *Synechococcus* CC9902, present throughout the year but not during upwelling in surface waters, and *Prochlorococcus marinus* (strain MIT 9313) during late summer and fall. This study was conducted from 2014 to 2016 (Reji et al. 2020). During the study, the relative abundances of *Bathycoccus* and *Ostreococcus* were comparable and 2- to 3-fold higher than that of *Micromonas*.

Studies of eukaryotic organisms involving metagenomic methodologies are not very frequent since samples obtained for metagenomic studies are dominated by prokaryotic representatives; however, these representatives of the marine ecosystem can also be studied through metagenomics sequencing, namely picoeukaryotes, organisms with ≤ 3 μm in its maximal dimension with representatives in all major algal groups as green algae, haptophytes, stramenopiles and dinoflagellates (Reyes-Prieto et al. 2019). Rashid et al. (2018) investigated the seasonal changes in the community in Ofunato Bay, located in the north-eastern Iwate Prefecture of Japan, known for having the highest diversity of marine organisms in the world. The study found that, independently of depth and sampling stations, the relative abundance of *Bathycoccus* and *Ostreococcus* were identical but 2 to 3-fold higher than *Micromonas*. These organisms belonged to the order Mamiellales in the class Mamiellophyceae under the phylum of Chlorophyta and had associated changes in relative abundancy according to season, being more abundant during both winter and summer; however, their abundancy was not dependent on sampling location and depth. In the same area, the Ofunato Bay, a study comprising both 16S rRNA genes and metagenomes found that *Bathycoccus, Ostreococcus* and *Micromonas* accounted for 8.8–15.2%, 3.6–4.9% and 2.1–3.1% of total read counts, while cyanobacteria contributed to 0.8–1.2% of the total WGS reads, with the most abundant organism being *Synechococcus,* followed by *Prochlorococcus* and *Nostoc,* followed by *Prochlorococcus* and *Nostoc.* The results regarding cyanobacteria results were not in accordance with the 16s reads also done in this study (Reza et al. 2018).

The communities of *Synechococcus* were studied by the global metagenome dataset from the Ocean Sampling Day that occurred on 21 June 2014. A total of 150 metagenomes, totalling 41 Gbp, was used to assess the relative abundance of the *Synechococcus/Cyanobium* and *Prochlorococcus* clades. Since *Prochlorococcus* was only abundant in a small number of sampling sites, it was not considered in the analysis. Eight strains belonging to clades I to IV of *Synechococcus* lineages, which dominate *Synechococcus* community in cold and temperate coastal areas, were studied. The study concluded that one or two taxa dominated the *Synechococcus/Cyanobium* community and below latitude 35°N clade I and IV were dominant, apart from the Mediterranean Sea, where clade III was mainly present, and clade II dominated latitudes below 35°N. Other results were obtained in this study regarding clade dominance according to the thermophysiology of the community, salinity and nutrient availability (Doré et al. 2022).

2.2 Metagenomics Applied to Zooplankton

Zooplanktonic organisms, single-celled microeukaryotes and small multicellular zooplankton are very important in the food webs of marine- and freshwater ecosystems, representing the biggest planktonic biomass of the world's ocean (Dortch and Packard 1989, Gasol et al. 1997). It supports fisheries and shapes the biogeochemical cycles of the planet, from primary production, organic matter recycling, sequestration of carbon and the transfer of organic material to higher food webs trophic levels (Barton et al. 2013, Stefanni et al. 2018). However, the large number of species, morphological characteristics and highly variable genome size have been obstacles for deep and complex studies.

In metagenomic analysis, DNA samples are analysed as a whole, providing more information than metabarcoding and, consequently, a better characterisation. Through this molecular tool, it is possible to identify the species present in samples and obtain insights into the metabolic activities and functional roles of the environment. In this way, it allows us to study the impacts of environmental changes (including climate change) or blooms on zooplankton populations, as well as their responses and adaptations to the different geographical conditions of the oceans. Therefore, understudied or not-studied species-specific interactions and species responses/ adaptations might be unveiled. Blooms of any organism always have impacts on the zooplankton populations. The metagenomic analysis allows characterising those impacts on zooplankton species, establishing a negative or positive correlation between bloom, like cyanobacteria, phytoplankton or tunicate blooms, with the rest of zooplankton species, like molluscs, fish, copepods, jellyfish, amphipods and worms (Lehman et al. 2021, Sordino et al. 2020).

Lehman et al. (2021) found that cyanobacteria bloom might have affected zooplankton communities' survival through the production of toxins or metabolites and diatoms abundance, interfering with the quality of zooplankton feeding due to the lack of omega-3 (Lehman et al. 2021).

In the case study of Sordino et al. (2020), while phytoplankton bloom was evaluated, tunicates abundance and metatranscriptomics changed. As soon as phytoplankton bloom started, the number of tunicates-assigned unigenes increased, occurring an evolution in time course. In the beginning, feeding and development-related genes were the most representative. Then, during the bloom peak and downstream, growth and reproductive-related gene expression increased. At the end of the bloom, abiotic stress-associated gene expression increased because of low nutrient availability (Sordino et al. 2020).

Thus, metagenomic analysis is an important tool for (i) monitor and predict zooplankton communities' response to biotic and abiotic changes, climate change and anthropogenic activities and (ii) establish positive/negative correlations or symbiotic relationships between the different organisms that compose communities (bacterioplankton, phytoplankton and zooplankton) (Lehman et al. 2021, Sordino et al. 2020). Therefore, it provides useful information that can be used to predict organisms' response and respective influence in the trophic cascade and food web. Generally, blooms are related to imbalances in the biogeochemical cycles, such as the increased amounts of iron. Zooplankton is involved in several processes that shape those cycles, but unlike microbes and viruses, knowledge about it is scarce due to their variable genome sizes, the predominance of noncoding sequences and the very large number of taxa, not existing available comprehensive gene catalogues (Carradec et al. 2018).

To overcome this data scarcity, transcriptome data sets from cultured marine eukaryotes and zooplankton have been developed, allowing to identify not only taxonomically zooplankton species but also genes associated with light-based energetic processes, transport of nutrients, carbohydrate metabolism, flagellar movement, multicellularity, cell-cell contact, chitin metabolism and muscular movement on environmental samples (Carradec et al. 2018). However, the majority (51.2%) of unigenes were new, presenting no matches in public sequence databases, which limits the insights that the gene catalogue may offer (Carradec et al. 2018).

Still about biogeochemical cycles, several studies have focused on identifying independently the key species involved in the biological carbon pump (Guidi et al. 2016). However, the results have demonstrated that carbon export depends on multiple biotic interactions, being necessary for holistic analysis of the entire planktonic ecosystem for a better understanding of the involved mechanisms (Guidi et al. 2016).

Advanced sequencing technologies are useful for surveying whole planktonic communities and associated molecular functions in a detailed way, monitoring and predicting ecosystem functions related to ocean biogeochemistry (Guidi et al. 2016, Karsenti et al. 2011, Strom 2008, Worden et al. 2015). However, the description of sub-communities driven by abiotic interactions has not been robustly taken into account by the applied models. To overcome this issue, the application of new models has been tested (Aylward et al. 2015, Guidi et al. 2016, Langfelder and Horvath 2008).

Guidi et al. (2016) applied a system biological approach known as Weighted Gene Correlation Network Analysis (WGCNA) to the set of these data. Through the application of this approach, Guidi et al. (2016) observed that prokaryotes and viruses are important for carbon export and highlighted the central role of copepods and diatoms. In addition, they were able to predict 89% of the variability in carbon export through gene information, largely unknown, that belonged to prokaryotes and viruses (Guidi et al. 2016). This demonstrates not only the potential of metagenomic data for monitoring the ocean but also the research that still needs to be done to clarify these complex interactions between plankton and the biogeochemical cycles of the ocean.

Metagenomic analysis has also been used to study the population structure of some zooplankton species. To confirm or clarify the results obtained by allozymes, mitochondrial DNA (mtDNA) or microsatellites approaches, metagenomics has been applied to gene expression variations at the population scale (Laso-Jadart et al. 2020) or sequencing of the fragments obtained from Restriction site-associated DNA sequencing (Rad-Seq) method (Deagle et al. 2015).

Variation of gene expression is an adaptation mechanism of organisms (including zooplankton), which is controlled by selective genetic factors or by environmental changes (Sato et al. 2016, Whitehead 2012). To study population structure, population-scale allelic differential expression (psADE) of genes can be measured (Laso-Jadart et al. 2020). It consists of the difference between the allelic abundance at the genomic and transcriptomic levels (Laso-Jadart et al. 2020). To measure psADE would require the sequencing of several individuals at genomic and transcriptomic levels individually, but the alternative metagenomic and metatranscriptomic sequencing approaches allow direct population insight like genetic structure and loci under psADE, natural selection or both (Laso-Jadart et al. 2020).

In the case study of Laso-Jadart et al. 2020, functional transcriptomic analysis showed a weak to moderate differentiation between the seven populations from the Artic Sea of the copepod *Oithona similis*, which is widespread around the world and presents strong adaptive capacities to environmental variations (Laso-Jadart et al. 2020).

Metagenomic analysis applied to the Rad-Seq method allows to identify allelic variations between and within individuals, such as the Single Nucleotide Polymorphisms (SNPs) from markers distributed throughout the genome (Davey et al. 2011, Narum et al. 2013). Therefore, it is possible to assemble a library of marker sequences to identify allelic variants, even in species without a reference genome (Deagle et al. 2015).

Deagle et al. (2015) used this technique to study the genetic structure of Antarctic krill. They observed that it was possible to discriminate individuals, but it was not possible to define a population structure.

Until recently, zooplankton and bacteria have been treated as separate elements of the marine food web, linked only by nutrient cycling and trophic cascades

(Tang et al. 2010). However, they establish a complex and sophisticated symbiosis, and metagenomic analysis has helped to unveil it.

Metagenomic analysis has allowed clarifying differences between zooplankton-associated bacteria and ambient water bacteria communities. Zooplankton-associated bacteria present high abundance but low richness compared to ambient water, composed of only a few dominant taxa that vary according to zooplankton species (De Corte et al. 2018, Xia et al. 2020). This suggests that zooplankton-associated bacteria consist of a specialised community that depends on the acidic environment provided, which requires bacteria's capacity to tolerate low pH (De Corte et al. 2018, Xia et al. 2020) and the ability to metabolise the compounds released by zooplankton (De Corte et al. 2018, Schroeder et al. 2020, Xia et al. 2020). Besides this, this molecular approach has demonstrated that zooplankton may serve as a microbial seed bank of marine rare species and due to the zooplankton widespread, it might serve as a vector to disperse these bacteria (including pathogens) across a broad geographical range (Xia et al. 2020).

The expressed genes confirm the bacterial community specialisation. Several bacteria-associated genes related to surface attachment, pili and fimbriae formation, chitin-recognition proteins, genes related to pH regulation and polysaccharides and amino-sugars metabolisation, genes related to pathways that occur in anaerobic conditions and genes encoding enzymes that participate in iron and phosphorus recycling pathways are expressed by these bacteria communities (De Corte et al. 2018, Xia et al. 2020).

Therefore, metagenomics has allowed the expansion of the role of these communities, understanding (i) the complex and specialised relationship between zooplankton and their associated bacteria and (ii) the importance of the zooplankton-associated bacteria on biogeochemical cycles that generally are underrepresented in water's bacteria (like anaerobic processes) (De Corte et al. 2018, Xia et al. 2020).

Although all the results obtained from metagenomic and metatranscriptomics analysis are promising, little is known about species-specific interactions and species' role in biogeochemical cycles. This is due to (i) the lack of studies and (ii) the limitations of the traditional genome assembly approach associated with big zooplanktons' genome complexity and large amount of sequence data (Vorobev et al. 2020). In addition, genome reference-dependence of these approaches has also been a limitation because, although several collections of reference marine eukaryote organisms' sequences have been created, the majority are derived from cultured organisms, resulting in incomplete databases (Sibbald and Archibald 2017, De Vargas et al. 2015).

A novel method has been developed, based on reference-independent that bins co-abundant gene groups (CAGs) with similar variations of abundance profile across several metagenomic samples, producing genomes from metagenomic data (Nielsen et al. 2014, Vorobev et al. 2020). CAGs might be composed of metagenomic species (MGS) or metagenomics-based transcriptomes (MGT).

Through the metagenomic transcriptomes (MGT) approach, Vorobev et al. (2020) were able to study taxonomic diversity, identifying eukaryotes, bacteria, archaea and viruses. However, the resolution of taxonomic identification varied between organisms. This could be due to the lack of data about specific zooplankton groups in reference databases (Vorobev et al. 2020). In addition, taxonomic diversity differed significantly compared to the collections of reference transcriptomes derived from culture organisms. This might be due to the coastal origin of cultured strains and the absence of zooplankton in the collection-selected organisms (Vorobev et al. 2020).

This approach also allows the comparison and identification of new ecotypes of two representative species through differences in the average nucleotide identity (ANI) (Vorobev et al. 2020). It is also possible to study functional insights about plankton communities and interspecies relationships through the MGTs expressing genes coding for key enzymes, which participate in processes of production or degradation (Vorobev et al. 2020). Therefore, it is possible to establish plankton interspecies relationships involved in biogeochemical cycles of the ocean elements, e.g., nitrogen.

This method presents some problems, such as the number of small gene clusters that cannot reliably cover a full eukaryotic transcriptome, leading to the fragmentation of the MGTs (Vorobev et al. 2020). This might be due to the presence/absence of some genes in some subpopulations and insufficient sequencing depth for some samples, resulting in partial genomic coverage (Vorobev et al. 2020). A depth sequencing level below or near the limit of detection results in the lack of corresponding reads for some genes, which can lead to incomplete coverage of the transcriptome by metagenomic reads (Vorobev et al. 2020). However, the MGT approach presents advantages compared to metagenome and metatranscriptome assembly methods, including the access to genomic data of organisms not available in culture because of the culture-independent assembly and clustering of sequence data and *de novo* definition of gene clusters that allows for the reconstruction of transcriptomes with no need for references (Vorobev et al. 2020).

Metagenomics has also been used to study the most abundant biological entities on Earth, viruses. Aquatic viruses might be classified into three groups: single-stranded DNA (ssDNA), double-stranded DNA (dsDNA) and single-stranded RNA (ssRNA). Several studies have demonstrated the important roles of viruses in biogeochemistry, ecosystem functions and zooplankton (Ian Hewson et al. 2013, 2018).

There are several reports about viruses that can infect marine metazoan and non-metazoan (Dunlap et al. 2013, Gudenkauf et al. 2014, Hewson et al. 2014, Hewson et al. 2012), causing mortality (Gudenkauf and Hewson 2016, Munn 2006, Suttle 2007). In some cases, zooplankton populations are almost decimated, which might affect the dynamics of marine ecosystems, habitat ecology and biogeochemistry, specifically the conversion of particulate into dissolved matter (Carpenter 1990, Middelboe and Lyck 2002, Riemann and Middelboe 2002). Therefore, although studies about viruses associated with marine organisms are scarce due to difficulties associated with traditional methods, it is important to study viral diversity and how it affects zooplankton communities (Gudenkauf and Hewson 2016, Ian Hewson et al. 2013), and metagenomics allows overcoming those difficulties.

Sequencing technology advances have allowed the study of viruses in marine invertebrates. Sequencing environmental samples and whole tissue of organisms, new viral families have been identified in marine zooplankton (e.g., Dunlap et al. 2013, Ng 2010, Rosario and Breitbart 2011). Metagenomics analysis can also be applied to detect free-living viruses in open ocean seawater and freshwater (López-Bueno et al. 2009, Rosario et al. 2009). Viral community composition, biogeographic and seasonal occurrence across aquatic environments, and how viruses interact with and influence host dynamics can also be studied (Ian Hewson et al. 2013, 2018).

The composition of viral communities might be relevant to the study. Viral communities of aquatic environments depend on the production of new virus particles through the infection of native organisms from an environment (autochthonous infection) and import from the surrounding habitats (allochthonous sources), and their decrease (IanHewson et al. 2012, Hewson et al. 2018, Köstner et al. 2017). The allochthonous sources can lead to the appearance of viruses that were not known as zooplankton pathogens, highlighting the importance of metagenomics in virus screening.

Viruses may interact in diverse ways, such as latent infection (causing little change in host abundance), chronic infection, imparting host fitness cost, or dramatic epidemics that result in rapid host population decline (Ian Hewson et al. 2013).

Besides the influence of viruses in zooplankton dynamics, metagenomics also allows the study of the dynamics of the virus in the host population. There are reports in which higher viral load was associated with lower population abundance, suggesting that lower host density may induce higher viral load to allow successful transmission of viral particles upon host lysis (Ian Hewson et al. 2013).

However, the lack of information about viral genomes in databases may decrease confidence in metavirome works (Labonté and Suttle 2013, Weynberg et al. 2014). Gudenkauf and Hewson (2016) reported a wide range of virus percentage identification, from 0.06% to 88.03%, in viromes prepared from different zooplankton organisms like echinoderms, arthropods, cnidarians and urochordates. This large range of identification might be a consequence of the lack of representative viral genomes in databases (Gudenkauf and Hewson 2016, Weynberg et al. 2014).

In addition to the lack of reference sequences in databases, viral genomes do not have conserved genes (ribosomal ribonucleic acids, rRNA) like cellular microorganisms, which makes it difficult to study through the PCR approach (Rohwer and Edwards 2002). Through viral metagenomics, it is possible to study the composition of viral assemblage without prior knowledge of gene sequence (Rosario and Breitbart 2011), allowing to identify the presence of viruses (Kapoor et al. 2008, Li et al. 2010, Terry Fei Fan Ng et al. 2009) and to investigate microbe-microbe, microbe-host interactions, pathogenesis and disease diagnosis (Gudenkauf and Hewson 2016). Data about targets for molecular quantification of viruses (Thurber et al. 2009), their prevalence in populations (Terry Fei Fan Ng et al. 2009) and distribution in the environment are also provided by metagenomics (Ian Hewson et al. 2012, Rosario et al. 2009).

Several viruses are already known to infect zooplankton; metagenomics made the task of knowing them easier to perform despite the small number of data associated with databases. Bidnaviruses and asfiviruses in Echinodermata, circoviruses and densoviruses in Chordata, iridoviruses in Echinodermata and Chordata and arteriviruses and nucleopolyhedroviruses in Arthropoda were identified to infect these zooplankton phyla (Gudenkauf and Hewson 2016).

Until recently, metagenomics studies only allowed the identification of viruses associated with animals other than marine zooplankton. Single-stranded RNA viruses, like arteriviruses that were only associated with mammals, were found in shrimp and fish (Gudenkauf and Hewson 2016). Single-stranded DNA viruses like bacilladnaviruses, bidnaviruses, circoviruses, densoviruses were associated for the first time with marine invertebrates, showing a higher widespread capacity for infection than previously thought (Gudenkauf and Hewson 2016, Ian Hewson et al. 2013). Double-stranded DNA viruses like asfiviruses, nucleopolyhedroviruses and iridoviruses were also found recently in marine invertebrates (Gudenkauf and Hewson 2016).

Special attention needs to be taken when viruses are identified. Metagenomics identifies all the viruses present in samples, not distinguishing between the virus of zooplankton and the virus present in the digestive system of zooplankton organisms (Gudenkauf and Hewson 2016). Bacteriophage and phycodnaviruses, which infect bacteria and algae, respectively, are also identified (Gudenkauf and Hewson 2016). However, if the aim is to characterise viruses that infect zooplankton, then in that case, it might not make sense to include them in the analysis (Gudenkauf and Hewson 2016). It is important to differentiate it, although this kind of result can provide useful information about the environment (Gudenkauf and Hewson 2016). For example, the presence of phycodnaviruses in invertebrates may suggest exposure to infected algae and the presence of this virus in the surrounding water (Gudenkauf and Hewson 2016). The presence of mimiviruses might suggest the presence of amoebas in/on zooplankton hosts or in the environment (Gudenkauf and Hewson 2016). Retroviruses might indicate infection in the hosts (Gudenkauf and Hewson 2016).

It is also possible to study the transmission of viruses by metagenomic techniques. By analysing free-living particles in water or soil, it is possible to understand if viral transmission occurs vertically (host individual to host individual) or horizontally (via free particles in the water column) (Ian Hewson et al. 2013).

Metagenomic analysis can also help in the clarification of taxonomic identification of parasites that infect zooplankton. Morphological identification can classify this kind of organism as belonging to a certain group or species for years or centuries. Phylogenetic analysis based on conserved genes, such as 18S rRNA or ITS, can help identify species, supporting morphological identification or clarifying it. However, sometimes these genes do not offer sufficient resolution, remaining uncertain taxonomic position (Lohr et al. 2010, Lu et al. 2020).

The application of metagenomic analysis may help in these cases, sequencing the whole genome and providing more data about proteins. Therefore, by assembling genome and multi-protein analysis, metagenomics can help clarify the taxonomic identification of organisms.

3. Acknowledgements

This work was funded by national funds through FCT (Fundação para a Ciência e a Tecnologia) I.P., under the project MARE (UIDB/04292/2020 and UIDP/04292/2020), the project LA/P/0069/2020 granted to the Associate Laboratory ARNET, the project eFishing MAR20-01-77P3-FEAMP-000006 and the grant awarded by FCT to Marco Simões (2020.07688.BD).

References

Anon. 2000. ICES Zooplankton Methodology Manual.

Armbrecht, L., Hallegraeff, G., Bolch, C.J.S., Woodward, C. and Cooper, A. 2021. Hybridisation capture allows DNA damage analysis of ancient marine eukaryotes. Scientific Reports 11(1). doi: 10.1038/s41598-021-82578-6.

Armengol, X. and Miracle, M.R. 2000. Diel vertical movements of zooplankton in lake la cruz (Cuenca, Spain). Journal of Plankton Research 22(9). doi: 10.1093/plankt/22.9.1683.

Aylward, Frank O., John M. Eppley, Jason M. Smith, Francisco P. Chavez, Christopher A. Scholin and Edward F. DeLong. 2015. Microbial community transcriptional networks are conserved in three domains at ocean basin scales. Proceedings of the National Academy of Sciences of the United States of America 112(17). doi: 10.1073/pnas.1502883112.

Baker, G.C., Smith, J.J. and Cowan, D.A. 2003. Review and re-analysis of domain-specific 16S Primers. Journal of Microbiological Methods 55(3).

Barnard, R., Batten, S.D., Beaugrand, G., Buckland, C., Conway, D.V.P., Edwards, M. et al. 2004. Continuous plankton records: plankton atlas of the north Atlantic ocean (1958–1999). II. biogeographical charts. Marine Ecology Progress Series (SUPPL.).

Barton, Andrew D., Andrew J. Pershing, Elena Litchman, Nicholas R. Record, Kyle F. Edwards, Zoe V. Finkel et al. 2013. The biogeography of marine plankton traits. Ecology Letters 16(4).

Bathmann, U., Bundy, M.H., Clarke, M.E., Cowles, T.J., Daly, K., Dam, H.G. et al. 2001. Future marine zooplankton research—a perspective: marine zooplankton colloquium. Marine Ecology Progress Series 222.

Batten, Sonia D. and David W. Welch. 2004. Changes in oceanic zooplankton populations in the north-east pacific associated with the possible climatic regime shift of 1998/1999. in Deep-Sea Research Part II: Topical Studies in Oceanography. Vol. 51.

Beaugrand, Grégory. 2005. Monitoring pelagic ecosystems using plankton indicators. In ICES Journal of Marine Science. Vol. 62.

Beaugrand, Grégory and Frédéric Ibañez. 2002. Spatial dependence of calanoid copepod diversity in the north Atlantic ocean. Marine Ecology Progress Series 232. doi: 10.3354/meps232197.

Beaugrand, Grégory and Philip C. Reid. 2003. Long-term changes in phytoplankton, zooplankton and salmon related to climate. Global Change Biology 9(6). doi: 10.1046/j.1365-2486.2003.00632.x.

Beaugrand, Grégory, Philip C. Reid, Frédéric Ibañez, J. Alistair Lindley and Martin Edwards. 2002. Reorganization of north Atlantic marine copepod biodiversity and climate. Science 296(5573). doi: 10.1126/science.1071329.

Bharti, Richa and Dominik G. Grimm. 2021. Current challenges and best-practice protocols for microbiome analysis. Briefings in Bioinformatics 22(1).

Bonnet, D. and Frid, C. 2004. Seven copepod species considered as indicators of water-mass influence and changes: results from a northumberland coastal station. In ICES Journal of Marine Science. Vol. 61.

Borja, Ángel. 2005. The european water framework directive: a challenge for nearshore, coastal and continental shelf research. Continental Shelf Research 25(14). doi: 10.1016/j.csr.2005.05.004.

Bucklin, A., de Vargas, C., Hopcroft, R.R., Madin, L.P., Thuesen, E.V., Wiebe, P.H. et al. 2004. Census of marine zooplankton (CMarZ)-science plan. CMarZ Website (July).

Bucklin, Ann, Brian D. Ortman, Robert M. Jennings, Lisa M. Nigro, Christopher J. Sweetman, Nancy J. Copley et al. 2010. A 'rosetta stone' for metazoan zooplankton: DNA barcode analysis of species diversity of the Sargasso Sea (northwest Atlantic ocean). Deep-Sea Research Part II: Topical Studies in Oceanography 57(24–26). doi: 10.1016/j.dsr2.2010.09.025.

Carpenter, R.C. 1990. Mass mortality of diadema antillarum - i. Long-term effects on sea urchin population-dynamics and coral reef algal communities. Marine Biology 104(1). doi: 10.1007/BF01313159.

Carradec, Quentin, Eric Pelletier, Corinne Da Silva, Adriana Alberti, Yoann Seeleuthner, Romain Blanc-Mathieu et al. 2018. A global ocean atlas of eukaryotic genes. Nature Communications 9(1). doi: 10.1038/s41467-017-02342-1.

Carvalho, G.R., Creer, S., Michael J. Allen, Costa, F.O., Tsigenopoulos, C.S., Le Goff-Vitry, M., Magoulas, A. et al. 2010. Genomics in the discovery and monitoring of marine biodiversity. Introduction to Marine Genomics.

Cervetto, Guillermo, Raymond Gaudy and Marc Pagano. 1999. Influence of salinity on the distribution of Acartia Tonsa (copepoda, calanoida). Journal of Experimental Marine Biology and Ecology 239(1). doi: 10.1016/S0022-0981(99)00023-4.

Chavez, Francisco P., John Ryan, Salvador E. Lluch-Cota and Miguel Ñiquen, C. 2003. Climate: from anchovies to sardines and back: multidecadal change in the pacific ocean. Science 299(5604).

Chiba, Sanae, Kazuaki Tadokoro, Hiroya Sugisaki and Toshiro Saino. 2006. Effects of decadal climate change on zooplankton over the last 50 years in the western subarctic north pacific. Global Change Biology 12(5). doi: 10.1111/j.1365-2486.2006.01136.x.

De Corte, Daniele, Abhishek Srivastava, Marja Koski, Juan Antonio L. Garcia, Yoshihiro Takaki, Taichi Yokokawa et al. 2018. Metagenomic insights into zooplankton-associated bacterial communities. Environmental Microbiology 20(2). doi: 10.1111/1462-2920.13944.

Cuvelier, Marie L., Andrew E. Allen, Adam Monier, John P. McCrow, Monique Messié, Susannah G. Tringe et al. 2010. Targeted metagenomics and ecology of globally important uncultured eukaryotic phytoplankton. Proceedings of the National Academy of Sciences of the United States of America 107(33). doi: 10.1073/pnas.1001665107.

Dauvin, Jean Claude, Eric Thiébaut and Zixian Wang. 1998. Short-term changes in the mesozooplanktonic community in the seine ROFI (region of freshwater influence) (eastern english channel). Journal of Plankton Research 20(6). doi: 10.1093/plankt/20.6.1145.

Davey, John W., Paul A. Hohenlohe, Paul D. Etter, Jason Q. Boone, Julian M. Catchen and Mark L. Blaxter. 2011. Genome-wide genetic marker discovery and genotyping using next-generation sequencing. Nature Reviews Genetics 12(7).

David, Valérie, Benoît Sautour, Pierre Chardy and Michel Leconte. 2005. Long-term changes of the zooplankton variability in a turbid environment: the gironde estuary (France). Estuarine, Coastal and Shelf Science 64(2–3). doi: 10.1016/j.ecss.2005.01.014.

Deagle, Bruce E., Cassandra Faux, So Kawaguchi, Bettina Meyer and Simon N. Jarman. 2015. Antarctic krill population genomics: apparent panmixia, but genome complexity and large population size muddy the water. Molecular Ecology 24(19). doi: 10.1111/mec.13370.

Delmont, Tom O., Morgan Gaia, Damien D. Hinsinger, Paul Fremont, Chiara Vanni, Antonio Fernandez Guerra et al. 2021. Functional repertoire convergence of distantly related eukaryotic plankton lineages revealed by genome-resolved metagenomics. BioRxiv.

Delmont, Tom O., Morgan Gaia, Damien D. Hinsinger, Paul Fremont, Chiara Vanni, Antonio Fernandez Guerra et al. 2022. Functional repertoire convergence of distantly related eukaryotic plankton lineages abundant in the sunlit ocean. Cell Genomics 2(Issue 5).

Dolbeth, M., Cardoso, P.G., Ferreira, S.M., Verdelhos, T., Raffaelli, D. and Pardal, M.A. 2007. Anthropogenic and natural disturbance effects on a macrobenthic estuarine community over a 10-year period. Marine Pollution Bulletin 54(5). doi: 10.1016/j.marpolbul.2006.12.005.

Doré, Hugo, Jade Leconte, Ulysse Guyet, Solène Breton, Gregory K. Farrant, David Demory et al. 2022. Global phylogeography of marine synechococcus in coastal areas reveals strikingly different communities than in the open ocean. BioRxiv 2022.03.07.483242. doi: 10.1101/2022.03.07.483242.

Dortch, Quay and Theodore T. Packard. 1989. Differences in biomass structure between oligotrophic and eutrophic marine ecosystems. Deep Sea Research Part A, Oceanographic Research Papers 36(2). doi: 10.1016/0198-0149(89)90135-0.

Duncan, Anthony, Kerrie Barry, Chris Daum, Emiley Eloe-Fadrosh, Simon Roux, Katrin Schmidt et al. 2022. Metagenome-assembled genomes of phytoplankton microbiomes from the arctic and Atlantic oceans. Microbiome 10(1). doi: 10.1186/s40168-022-01254-7.

Dunlap, Darren S., Terry Fei Fan Ng, Karyna Rosario, Jorge G. Barbosa, Anthony M. Greco, Mya Breitbart et al. 2013. Molecular and microscopic evidence of viruses in marine copepods. Proceedings of the National Academy of Sciences of the United States of America 110(4). doi: 10.1073/pnas.1216595110.

Dybas, Cheryl Lyn. 2006. On a collision course: ocean plankton and climate change. BioScience 56(8).

Edwards, Martin and Anthony J. Richardson. 2004. Impact of climate change on marine pelagic phenology and trophic mismatch. Nature 430(7002). doi: 10.1038/nature02808.

European Environmental Agency–EEA. 2014. Projected changes in annual, summer and winter temperature.

Fetzer, Ingo and Wolf E. Arntz. 2008. Reproductive strategies of benthic invertebrates in the kara Sea (russian arctic): adaptation of reproduction modes to cold water. Marine Ecology Progress Series 356. doi: 10.3354/meps07271.

Francis, T. Ben, Karen Krüger, Bernhard M. Fuchs, Hanno Teeling and Rudolf I. Amann. 2019. Candidatus Prosiliicoccus vernus, a spring phytoplankton bloom associated member of the flavobacteriaceae. Systematic and Applied Microbiology 42(1). doi: 10.1016/j.syapm.2018.08.007.

Fromentin, J.M. and F. Ibanez. 1994. Year-to-year changes in meteorological features of the french coast area during the last half-century. Examples of two biological responses. Oceanologica Acta 17(3).

Fu, He, Christa B. Smith, Shalabh Sharma and Mary Ann Moran. 2020. Genome sequences and metagenome-assembled genome sequences of microbial communities enriched on phytoplankton exometabolites. Microbiology Resource Announcements 9(30). doi: 10.1128/mra.00724-20.

Gasol, Josep M., Paul A. Del Giorgio and Carlos M. Duarte. 1997. Biomass distribution in marine planktonic communities. Limnology and Oceanography 42(6). doi: 10.4319/lo.1997.42.6.1353.

Gaudy, Raymond, Guillermo Cervetto and Marc Pagano. 2000. Comparison of the metabolism of Acartia clausi and A. Tonsa: influence of temperature and salinity. Journal of Experimental Marine Biology and Ecology 247(1). doi: 10.1016/S0022-0981(00)00139-8.

Gleick, Peter H. 2003. Global freshwater resources: soft-path solutions for the 21st century. Science 302(5650). doi: 10.1126/science.1089967.

Grantham, Brian A., Ginny L. Eckert and Alan L. Shanks. 2003. Dispersal potential of marine invertebrates in diverse habitats. Ecological Applications 13(1 SUPPL.). doi: 10.1890/1051-0761(2003)013[0108:dpomii]2.0.co;2.

Gudenkauf, Brent M., James B. Eaglesham, William M. Aragundi and Ian Hewson. 2014. Discovery of urchin-associated densoviruses (family parvoviridae) in coastal waters of the big Island, hawaii. Journal of General Virology 95(PART3). doi: 10.1099/vir.0.060780-0.

Gudenkauf, Brent M. and Ian Hewson. 2016. Comparative metagenomics of viral assemblages inhabiting four phyla of marine invertebrates. Frontiers in Marine Science 3(MAR). doi: 10.3389/fmars.2016.00023.

Guidi, Lionel, Samuel Chaffron, Lucie Bittner, Damien Eveillard, Abdelhalim Larhlimi, Simon Roux et al. 2016. Plankton networks driving carbon export in the oligotrophic ocean. Nature 532(7600). doi: 10.1038/nature16942.

Handelsman, Jo. 2004. Metagenomics: application of genomics to uncultured microorganisms. Microbiology and Molecular Biology Reviews 68(4). doi: 10.1128/mmbr.68.4.669-685.2004.

Handelsman, Jo, Michelle R. Rondon, Sean F. Brady, Jon Clardy and Robert M. Goodman. 1998. Molecular biological access to the chemistry of unknown soil microbes: a new frontier for natural products. Chemistry and Biology 5(10). doi: 10.1016/S1074-5521(98)90108-9.

Hare, Steven R. and Nathan J. Mantua. 2000. Empirical evidence for north pacific regime shifts in 1977 and 1989. Progress in Oceanography 47(2–4).

Haupt, Florian, Maria Stockenreiter, Michaela Baumgartner, Maarten Boersma and Herwig Stibor. 2009. Daphnia diel vertical migration: implications beyond zooplankton. Journal of Plankton Research 31(5). doi: 10.1093/plankt/fbp003.

Hays, Graeme C. 2003. A review of the adaptive significance and ecosystem consequences of zooplankton diel vertical migrations. In Hydrobiologia. Vol. 503.

Hays, Graeme C., Anthony J. Richardson and Carol Robinson. 2005. Climate change and marine plankton. Trends in Ecology and Evolution 20(6 SPEC. ISS.).

Hewson, I., Brown, J.M., Burge, C.A., Couch, C.S., LaBarre, B.A., Mouchka, M.E. et al. 2012. Description of viral assemblages associated with the gorgonia ventalina holobiont. Coral Reefs 31(2). doi: 10.1007/s00338-011-0864-x.

Hewson, Ian, Jorge G. Barbosa, Julia M. Brown, Ryan P. Donelan, James B. Eaglesham, Erin M. Eggleston et al. 2012. Temporal dynamics and decay of putatively allochthonous and autochthonous viral genotypes in contrasting freshwater lakes. Applied and Environmental Microbiology 78(18). doi: 10.1128/AEM.01705-12.

Hewson, Ian, Kalia S.I. Bistolas, Jason B. Button and Elliot W. Jackson. 2018. Occurrence and seasonal dynamics of RNA viral genotypes in three contrasting temperate lakes. PLoS ONE 13(3). doi: 10.1371/journal.pone.0194419.

Hewson, Ian, Jason B. Button, Brent M. Gudenkauf, Benjamin Miner, Alisa L. Newton, Joseph K. Gaydos et al. 2014. Densovirus associated with sea-star wasting disease and mass mortality. Proceedings of the National Academy of Sciences of the United States of America 111(48). doi: 10.1073/pnas.1416625111.

Hewson, Ian, Gabriel Ng, Wen Fang Li, Brenna A. LaBarre, Isabel Aguirre, Jorge G. Barbosa et al. 2013. Metagenomic identification, seasonal dynamics, and potential transmission mechanisms of a daphnia-associated single-stranded DNA virus in two temperate lakes. Limnology and Oceanography 58(5). doi: 10.4319/lo.2013.58.5.1605.

Heyen, H., Fock, H. and Greve, W. 1999. Detecting Relationships between the interannual variability in ecological time series and climate using a multivariate statistical approach-a case study on helgoland roads zooplankton. Climate Research 10(3). doi: 10.3354/cr010179.

Hurrell, James W. 1995. Decadal trends in the north Atlantic oscillation: regional temperatures and precipitation. Science 269(5224). doi: 10.1126/science.269.5224.676.

Ibarbalz, Federico M., Nicolas Henry, Manoela C. Brandão, Séverine Martini, Greta Busseni, Hannah Byrne et al. 2019. Global trends in marine plankton diversity across kingdoms of life. Cell 179(5). doi: 10.1016/j.cell.2019.10.008.

IPCC Working Group 1, I., Stocker, T.F., Qin, D., Plattner, G.K., Tignor, M., Allen, S.K. et al. 2013. IPCC, 2013: Climate change 2013: The physical science basis. Contribution of Working Group I to the Fifth Assessment Report of the Intergovernmental Panel on Climate Change. IPCC AR5.

Johnson, Zackary I. and Adam C. Martiny. 2015. Techniques for quantifying phytoplankton biodiversity. Annual Review of Marine Science 7. doi: 10.1146/annurev-marine-010814-015902.

Kapoor, Amit, Joseph Victoria, Peter Simmonds, Elizabeth Slikas, Thaweesak Chieochansin, Asif Naeem et al. 2008. A highly prevalent and genetically diversified picornaviridae genus in south asian children. Proceedings of the National Academy of Sciences of the United States of America 105(51). doi: 10.1073/pnas.0807979105.

Karsenti, Eric, Silvia G. Acinas, Peer Bork, Chris Bowler, Colomban de Vargas, Jeroen Raes et al. 2011. A holistic approach to marine eco-systems biology. PLoS Biology 9(10). doi: 10.1371/journal.pbio.1001177.

Kimmel, David G. and Michael R. Roman. 2004. Long-term trends in mesozooplankton abundance in chesapeake bay, usa: influence of freshwater input. Marine Ecology Progress Series 267. doi: 10.3354/meps267071.

Knowlton, N. 1993. Sibling species in the Sea. Annual Review of Ecology and Systematics 24.

Köstner, Nicole, Lisa Scharnreitner, Klaus Jürgens, Matthias Labrenz, Gerhard J. Herndl and Christian Winter. 2017. High viral abundance as a consequence of low viral decay in the baltic Sea redoxcline. PLoS ONE 12(6). doi: 10.1371/journal.pone.0178467.

Labonté, Jessica M. and Curtis A. Suttle. 2013. Previously unknown and highly divergent SsDNA viruses populate the oceans. ISME Journal 7(11). doi: 10.1038/ismej.2013.110.

Langfelder, Peter and Steve Horvath. 2008. WGCNA: an r package for weighted correlation network analysis. BMC Bioinformatics 9. doi: 10.1186/1471-2105-9-559.

Laso-Jadart, Romuald, Kevin Sugier, Emmanuelle Petit, Karine Labadie, Pierre Peterlongo, Christophe Ambroise et al. 2020. Investigating population-scale allelic differential expression in wild populations of oithona similis (Cyclopoida, claus, 1866). Ecology and Evolution 10(16). doi: 10.1002/ece3.6588.

Leandro, Sérgio Miguel, Henrique Queiroga, Laura Rodríguez-Graña and Peter Tiselius. 2006. Temperature-dependent development and somatic growth in two allopatric populations of acartia clausi (Copepoda: Calanoida). Marine Ecology Progress Series 322. doi: 10.3354/meps322189.

Lehman, Peggy W., Tomofumi Kurobe, Khiet Huynh, Sarah Lesmeister and Swee J. Teh. 2021. Covariance of phytoplankton, bacteria, and zooplankton communities within microcystis blooms in san francisco estuary. Frontiers in Microbiology 12. doi: 10.3389/fmicb.2021.632264.

Li, Linlin, Amit Kapoor, Beth Slikas, Oderinde Soji Bamidele, Chunlin Wang, Shahzad Shaukat et al. 2010. Multiple diverse circoviruses infect farm animals and are commonly found in human and chimpanzee feces. Journal of Virology 84(4). doi: 10.1128/jvi.02109-09.

Licandro, Priscilla and Frédéric Ibanez. 2000. Changes of zooplankton communities in the gulf of tigullio (Ligurian Sea, western Mediterranean) from 1985 to 1995. Influence of Hydroclimatic Factors. Journal of Plankton Research 22(12). doi: 10.1093/plankt/22.12.2225.

Lima-Mendez, Gipsi, Karoline Faust, Nicolas Henry, Johan Decelle, Sébastien Colin, Fabrizio Carcillo et al. 2015. Determinants of community structure in the global plankton interactome. Science 348(6237). doi: 10.1126/science.1262073.

Lohr, Jennifer N., Christian Laforsch, Henrike Koerner and Justyna Wolinska. 2010. A daphnia parasite (caullerya mesnili) constitutes a new member of the ichthyosporea, a group of protists near the animal-fungi divergence. Journal of Eukaryotic Microbiology 57(4). doi: 10.1111/j.1550-7408.2010.00479.x.

Lomartire, Silvia, João C. Marques and Ana M.M. Gonçalves. 2021. The key role of zooplankton in ecosystem services: a perspective of interaction between zooplankton and fish recruitment. Ecological Indicators 129.

López-Bueno, Alberto, Javier Tamames, David Velázquez, Andrés Moya, Antonio Quesada and Antonio Alcamí. 2009. High diversity of the viral community from an antarctic lake. Science 326(5954). doi: 10.1126/science.1179287.

Loza, Antonio, Fernando García-Guevara, Lorenzo Segovia, Alejandra Escobar-Zepeda, Maria del Carmen Sanchez-Olmos, Enrique Merino et al. 2022. Definition of the metagenomic profile of ocean water samples from the gulf of mexico based on comparison with reference samples from sites worldwide. Frontiers in Microbiology 12. doi: 10.3389/fmicb.2021.781497.

Lu, Yameng, Eduard Ocaña-Pallarès, David López-Escardó, Stuart R. Dennis, Michael T. Monaghan, Iñaki Ruiz-Trillo et al. 2020. Revisiting the phylogenetic position of caullerya mesnili (ichthyosporea), a common daphnia parasite, based on 22 protein-coding genes. Molecular Phylogenetics and Evolution 151. doi: 10.1016/j.ympev.2020.106891.

Mackas, David L., Robert Goldblatt and Alan G. Lewis. 1998. Interdecadal variation in developmental timing of neocalanus plumchrus populations at ocean station P in the subarctic north pacific. Canadian Journal of Fisheries and Aquatic Sciences 55(8). doi: 10.1139/f98-080.

Malhi, Yadvinder, Janet Franklin, Nathalie Seddon, Martin Solan, Monica G. Turner, Christopher B. Field et al. 2020. Climate change and ecosystems: threats, opportunities and solutions. Philosophical Transactions of the Royal Society B: Biological Sciences 375(1794).

Marques, Sónia Cotrim, Ulisses Miranda Azeiteiro, Sérgio Miguel Leandro, Henrique Queiroga, Ana Lígia Primo, Filipe Martinho et al. 2008. Predicting zooplankton response to environmental changes in a temperate estuarine ecosystem. Marine Biology 155(5). doi: 10.1007/s00227-008-1052-6.

Marques, Sónia Cotrim, Miguel Ângelo Pardal, Ana Lígia Primo, Filipe Martinho, Joana Falcão, Ulisses Azeiteiro et al. 2018. Evidence for changes in estuarine zooplankton fostered by increased climate variance. Ecosystems 21(1). doi: 10.1007/s10021-017-0134-z.

Massana, Ramon and David López-Escardó. 2022. Metagenome assembled genomes are for eukaryotes too. Cell Genomics 2(5): 100130. doi: https://doi.org/10.1016/j.xgen.2022.100130.

McGowan, John A., Steven J. Bograd, Ronald J. Lynn and Arthur J. Miller. 2003. The biological response to the 1977 regime shift in the california current. Deep-Sea Research Part II: Topical Studies in Oceanography 50(14–16). doi: 10.1016/S0967-0645(03)00135-8.

Middelboe, Mathias and Peter G. Lyck. 2002. Regeneration of dissolved organic matter by viral lysis in marine microbial communities. Aquatic Microbial Ecology 27(2). doi: 10.3354/ame027187.

Mirza, M. Monirul Qader. 2003. Climate change and extreme weather events: can developing countries adapt? Climate Policy 3(3). doi: 10.3763/cpol.2003.0330.

Molinero, Juan Carlos, Frédéric Ibanez, Sami Souissi, Emmanueelle Buecher, Serge Dallot and Paul Nival. 2008. Climate control on the long-term anomalous changes of zooplankton communities in the northwestern Mediterranean. Global Change Biology 14(1). doi: 10.1111/j.1365-2486.2007.01469.x.

Monier, Adam, Rory M. Welsh, Chelle Gentemann, George Weinstock, Erica Sodergren, E. Virginia Armbrust et al. 2012. Phosphate transporters in marine phytoplankton and their viruses: cross-domain commonalities in viral-host gene exchanges. Environmental Microbiology 14(1). doi: 10.1111/j.1462-2920.2011.02576.x.

Munn, Colin B. 2006. Viruses as pathogens of marine organisms - from bacteria to whales. Journal of the Marine Biological Association of the United Kingdom 86(3).

Murphy, Grace E.P., Tamara N. Romanuk and Boris Worm. 2020. Cascading effects of climate change on plankton community structure. Ecology and Evolution. doi: 10.1002/ece3.6055.

Narum, Shawn R., Alex Buerkle, C., John W. Davey, Michael R. Miller and Paul A. Hohenlohe. 2013. Genotyping-by-sequencing in ecological and conservation genomics. Molecular Ecology 22(11).

Ng, T.F. 2010. Discovery of novel viruses from animals, plants, and insect vectors using viral metagenomics by. Discovery.

Ng, Terry Fei Fan, Charles Manire, Kelly Borrowman, Tammy Langer, Llewellyn Ehrhart and Mya Breitbart. 2009. Discovery of a novel single-stranded dna virus from a sea turtle fibropapilloma by using viral metagenomics. Journal of Virology 83(6). doi: 10.1128/jvi.01946-08.

Nielsen, H. Bjørn, Mathieu Almeida, Agnieszka S. Juncker, Simon Rasmussen, Junhua Li, Shinichi Sunagawa et al. 2014. Identification and assembly of genomes and genetic elements in complex metagenomic samples without using reference genomes. Nature Biotechnology 32(8). doi: 10.1038/nbt.2939.

Nowinski, Brent, Jessie Motard-Côté, Marine Landa, Christina M. Preston, Christopher A. Scholin, James M. Birch et al. 2019. Microdiversity and temporal dynamics of marine bacterial inconstitucionalissimamente genes. Environmental Microbiology 21(5). doi: 10.1111/1462-2920.14560.

de Oliveira Santos, Leandro, Iamê Alves Guedes, Sandra Maria Feliciano de Oliveira e. Azevedo, and Ana Beatriz Furlanetto Pacheco. 2021. Occurrence and diversity of viruses associated with cyanobacterial communities in a brazilian freshwater reservoir. Brazilian Journal of Microbiology 52(2). doi: 10.1007/s42770-021-00473-8.

Pace, Norman R., David A. Stahl, David J. Lane and Gary J. Olsen. 1986. The analysis of natural microbial populations by ribosomal RNA sequences.

Pace, N.R., Stahl, D.A., Lane, D.J. and Olsen, G.J. 1985. Analyzing natural microbial populations by RRNA sequences. ASM American Society for Microbiology News 51(1).

Pel, Joel, Amy Leung, Wendy W.Y. Choi, Milenko Despotovic, W. Lloyd Ung, Gosuke Shibahara et al. 2018. Rapid and highly-specific generation of targeted dna sequencing libraries enabled by linking capture probes with universal primers. PLoS ONE 13(12). doi: 10.1371/journal.pone.0208283.

Perry, R. Ian, Harold P. Batchelder, David L. Mackas, Sanae Chiba, Edward Durbin, Wulf Greve et al. 2004. Identifying global synchronies in marine zooplankton populations: issues and opportunities. in ICES Journal of Marine Science. Vol. 61.

Pierella Karlusich, Juan José, Eric Pelletier, Lucie Zinger, Fabien Lombard, Adriana Zingone et al. 2022. A robust approach to estimate relative phytoplankton cell abundances from metagenomes. Molecular Ecology Resources. doi: 10.1111/1755-0998.13592.

Pierella Karlusich, Juan José, Eric Pelletier, Lucie Zinger, Fabien Lombard, Adriana Zingone et al. 2021. A robust approach to estimate relative phytoplankton cell abundance from metagenomic data. BioRxiv 2021.05.28.446125. doi: 10.1101/2021.05.28.446125.

Planque, Benjamin and Arnold H. Taylor. 1998. Long-term changes in zooplankton and the climate of the north Atlantic. in ICES Journal of Marine Science. Vol. 55.

Queiroga, Henrique, John D. Costlow and Maria Helena Moreira. 1997. Vertical migration of the crab carcinus maenas first zoea in an estuary: implications for tidal stream transport. Marine Ecology Progress Series 149(1–3). doi: 10.3354/meps149121.

Rashid, Jonaira, Atsushi Kobiyama, Md Shaheed Reza, Yuichiro Yamada, Yuri Ikeda, Daisuke Ikeda et al. 2018. Seasonal changes in the communities of photosynthetic picoeukaryotes in ofunato bay as revealed by shotgun metagenomic sequencing. Gene 665. doi: 10.1016/j.gene.2018.04.071.

Rawlinson, K.A., Davenport, J. and Barnes, D.K.A. 2004. Vertical migration strategies with respect to advection and stratification in a semi-enclosed lough: a comparison of mero- and holozooplankton. Marine Biology 144(5). doi: 10.1007/s00227-003-1261-y.

Razouls, C., Desreumaux, N., Kouwenberg, J. and de Bovée, F. 2022. Biodiversity of marine planktonic copepods (morphology, geographical distribution and biological data). Retrieved June 1, 2022 (https://copepodes.obs-banyuls.fr/en/).

Reji, Linta, Bradley B. Tolar, Francisco P. Chavez and Christopher A. Francis. 2020. Depth-differentiation and seasonality of planktonic microbial assemblages in the monterey bay upwelling system. Frontiers in Microbiology 11. doi: 10.3389/fmicb.2020.01075.

Reyes-Prieto, A., Yoon, H.S. and Bhattacharya, D. 2019. Marine algal genomics and evolution. Encyclopedia of Ocean Sciences.

Reza, Md Shaheed, Atsushi Kobiyama, Yuichiro Yamada, Yuri Ikeda, Daisuke Ikeda, Nanami Mizusawa et al. 2018. Taxonomic profiles in metagenomic analyses of free-living microbial communities in the ofunato bay. Gene 665. doi: 10.1016/j.gene.2018.04.075.

Richardson, Anthony J. 2008. In hot water: zooplankton and climate change. in ICES Journal of Marine Science. Vol. 65.

Riemann, Lasse and Mathias Middelboe. 2002. Viral lysis of marine bacterioplankton: implications for organic matter cycling and bacterial clonal composition. Ophelia 56(2).

Riesenfeld, Christian S., Patrick D. Schloss and Jo Handelsman. 2004. Metagenomics: genomic analysis of microbial communities. Annual Review of Genetics 38.

Rohwer, Forest and Rob Edwards. 2002. The phage proteomic tree: a genome-based taxonomy for phage. Journal of Bacteriology 184(16). doi: 10.1128/JB.184.16.4529-4535.2002.

Rosario, Karyna and Mya Breitbart. 2011. Exploring the viral world through metagenomics. Current Opinion in Virology 1(4).

Rosario, Karyna, Siobain Duffy and Mya Breitbart. 2009. Diverse circovirus-like genome architectures revealed by environmental metagenomics. Journal of General Virology 90(10). doi: 10.1099/vir.0.012955-0.

Russell, F.S. 1939. Hydrographical and biological conditions in the north sea as indicated by plankton organisms. ICES Journal of Marine Science 14(2). doi: 10.1093/icesjms/14.2.171.

Rynearson, Tatiana A. and Brian Palenik. 2011. Learning to read the oceans. Genomics of marine phytoplankton. Advances in Marine Biology 60.

Sato, Mitsuhiko P., Takashi Makino and Masakado Kawata. 2016. Natural selection in a population of drosophila melanogaster explained by changes in gene expression caused by sequence variation in core promoter regions. BMC Evolutionary Biology 16(1). doi: 10.1186/s12862-016-0606-3.

SCHELTEMA, RUDOLF S. 1988. Initial evidence for the transport of teleplanic larvae of benthic invertebrates across the east pacific barrier. The Biological Bulletin 174(2). doi: 10.2307/1541781.

Schroeder, Anna, David Stanković, Alberto Pallavicini, Fabrizia Gionechetti, Marco Pansera and Elisa Camatti. 2020. DNA metabarcoding and morphological analysis- assessment of zooplankton biodiversity in transitional waters. Marine Environmental Research 160. doi: 10.1016/j.marenvres. 2020.104946.

Sharma, Pooja, Hansi Kumari, Mukesh Kumar, Mansi Verma, Kirti Kumari, Shweta Malhotra et al. 2008. From bacterial genomics to metagenomics: concept, tools and recent advances. Indian Journal of Microbiology 48(2). doi: 10.1007/s12088-008-0031-4.

Sibbald, Shannon J. and John M. Archibald. 2017. More protist genomes needed. Nature Ecology and Evolution 1(5).

Simon, Nathalie, Anne Lise Cras, Elodie Foulon and Rodolphe Lemée. 2009. Diversity and evolution of marine phytoplankton. Comptes Rendus - Biologies 332(2–3).

Singer, Gregory A.C., Shahrokh Shekarriz, Avery McCarthy, Nicole Fahner and Mehrdad Hajibabaei. 2020. The utility of a metagenomics approach for marine biomonitoring. BioRxiv.

Sjöqvist, Conny. 2022. Evolution of phytoplankton as estimated from genetic diversity. Journal of Marine Science and Engineering 10(4).

Sordino, Paolo, Salvatore D'Aniello, Eric Pelletier, Patrick Wincker, Valeria Nittoli, Lars Stemmann et al. 2020. Into the bloom: molecular response of pelagic tunicates to fluctuating food availability. Molecular Ecology 29(2). doi: 10.1111/mec.15321.

Stefanni, Sergio, David Stanković, Diego Borme, Alessandra de Olazabal, Tea Juretić, Alberto Pallavicini et al. 2018. Multi-marker metabarcoding approach to study mesozooplankton at basin scale. Scientific Reports 8(1). doi: 10.1038/s41598-018-30157-7.

Sterner, R.W. 2009. Role of zooplankton in aquatic ecosystems. Encyclopedia of Inland Waters.

Strom, Suzanne L. 2008. Microbial ecology of ocean biogeochemistry: a community perspective. Science 320(5879).

Suttle, Curtis A. 2007. Marine viruses- major players in the global ecosystem. Nature Reviews Microbiology 5(10). doi: 10.1038/nrmicro1750.

Taylor, Arnold H., J. Icarus Allen and Paul A. Clark. 2002. Extraction of a weak climatic signal by an ecosystem. Nature 416(6881). doi: 10.1038/416629a.

Thomas, Torsten, Jack Gilbert and Folker Meyer. 2012. Metagenomics- a guide from sampling to data analysis. Microbial Informatics and Experimentation 2(1). doi: 10.1186/2042-5783-2-3.

Thurber, Rebecca V., Matthew Haynes, Mya Breitbart, Linda Wegley and Forest Rohwer. 2009. Laboratory procedures to generate viral metagenomes. Nature Protocols 4(4). doi: 10.1038/nprot.2009.10.

Tully, Benjamin J., Rohan Sachdeva, Elaina D. Graham and John F. Heidelberg. 2017. 290 Metagenome-assembled genomes from the mediterranean sea: a resource for marine microbiology. PeerJ 2017(7). doi: 10.7717/peerj.3558.

Uttieri, M., Aguzzi, L., Aiese Cigliano, R., Amato, A., Bojanić, N., Brunetta, M. et al. 2020. WGEUROBUS– working group 'towards a european observatory of the non-indigenous calanoid copepod pseudodiaptomus marinUS.' Biological Invasions 22(3). doi: 10.1007/s10530-019-02174-8.

Uttieri, Marco. 2018. Trends in Copepod Studies - Distribution, Biology and Ecology.

De Vargas, Colomban, Stéphane Audic, Nicolas Henry, Johan Decelle, Frédéric Mahé, Ramiro Logares et al. 2015. Eukaryotic Plankton diversity in the sunlit ocean. Science 348(6237). doi: 10.1126/science.1261605.

Vincent, Flora, Federico M. Ibarbalz and Chris Bowler. 2022. Global marine phytoplankton revealed by the tara oceans expedition. Advances in Phytoplankton Ecology.

Vorobev, Alexey, Marion Dupouy, Quentin Carradec, Tom O. Delmont, Anita Annamalé, Patrick Wincker et al. 2020. Transcriptome reconstruction and functional analysis of eukaryotic marine plankton communities via high-throughput metagenomics and metatranscriptomics. Genome Research 30(4). doi: 10.1101/gr.253070.119.

Walter, T.C,. Boxshall, G. 2022. World of copepods database. Retrieved June 1, 2022 (https://www.marinespecies.org/copepoda).

Walther, Gian Reto, Eric Post, Peter Convey, Annette Menzel, Camille Parmesan, Trevor J.C. Beebee et al. 2002. Ecological responses to recent climate change. Nature 416(6879).

Wang, Baohong, Mingfei Yao, Longxian Lv, Zongxin Ling and Lanjuan Li. 2017. The human microbiota in health and disease. Engineering 3(1).

Weynberg, Karen D., Elisha M. Wood-Charlson, Curtis A. Suttle and Madeleine J. H. van Oppen. 2014. Generating viral metagenomes from the coral holobiont. Frontiers in Microbiology 5(MAY). doi: 10.3389/fmicb.2014.00206.

Whitehead, Andrew. 2012. Comparative genomics in ecological physiology: toward a more nuanced understanding of acclimation and adaptation. in Journal of Experimental Biology. Vol. 215.

Whitman, Richard L., Meredith B. Nevers, Maria L. Goodrich, Paul C. Murphy and Bruce M. Davis. 2004. Characterisation of lake michigan coastal lakes using zooplankton assemblages. Ecological Indicators 4(4). doi: 10.1016/j.ecolind.2004.08.001.

Worden, Alexandra Z., Michael J. Follows, Stephen J. Giovannoni, Susanne Wilken, Amy E. Zimmerman, and Patrick J. Keeling. 2015. Rethinking the marine carbon cycle: factoring in the multifarious lifestyles of microbes. Science 347(6223).

Xia, Xiaomin, Sze Ki Leung, Shunyan Cheung, Shuwen Zhang and Hongbin Liu. 2020. Rare bacteria in seawater are dominant in the bacterial assemblage associated with the bloom-forming dinoflagellate noctiluca scintillans. Science of the Total Environment 711. doi: 10.1016/j.scitotenv.2019.135107.

Xu, Jianping. 2006. Microbial ecology in the age of genomics and metagenomics: concepts, tools and recent advances. Molecular Ecology 15(7).

Yan, Qing Yun and Yu He YU. 2011. Metagenome-based analysis: a promising direction for plankton ecological studies. Science China Life Sciences 54(1).

CHAPTER 3
A Worldwide Perspective on Zooplankton Biodiversity

3.1

Zooplankton From the Portuguese Coast and Northern Estuaries (Western Iberian Peninsula)

Fernando Morgado[1],* and *Luis R. Vieira*[2]

1. Introduction

The Portuguese coast, located on the Western Iberian Peninsula, has an extensive shoreline, with an approximate extension of 900 km (between Caminha and Vila Real de Santo António) (Vasconcelos et al. 2011), that constitutes the largest Exclusive Economic Zone in Europe (1,727,408 Km²) (corresponds to 1.25% of the entire ocean area under the jurisdiction of countries), including Azores archipelago, with 953,633 km² and Madeira archipelago, with 446,108 km² (corresponding to a total for Portugal of 3,027,408 km²) (Dias 2005, Ferreira et al. 2016). The entire Portuguese National Maritime Space corresponds to about 4% of the Atlantic and 1% of the global ocean (Bessa Pacheco 2013) and includes nearly 50% of the volume of the EU's marine waters (MAM 2014). The Portuguese coast represents the northeastern Atlantic limit of the Iberian Coast ecoregion that covers the southwestern shelf seas and adjacent deeper Eastern Atlantic Ocean waters of the EU ecoregion, which includes parts of three Exclusive Economic Zones of European Union Member States (France, Spain and Portugal) and is part of the Iberia/Canary upwelling ecosystem, associated with the North Atlantic anticyclonic gyre (Wooster et al. 1976, Haynes et al. 1993), a small portion of high seas (ICES 2020). The Western Iberia coast region is characterised by several prominent topographic features, such as capes, promontories and submarine canyons (Relvas et al. 2007) that expose this region to the open-ocean dynamics and to the atmospheric climate events, including fronts and low-pressure systems, that enters the European continent mainly from the Atlantic Ocean (IPCC 2014, Rubal

[1] The Center for Environmental and Marine Studies (CESAM) and Department of Biology University of Aveiro, Campus Universitário de Santiago, Aveiro, Portugal.

[2] Interdisciplinary Center of Marine and Environmental Research (CIIMAR). University of Porto, Terminal de Cruzeiros do Porto de Leixões, Av. General Norton de Matos s/n, 2250–208 Matosinhos, Portugal.

* Corresponding author: fmorgado@ua.pt

et al. 2013, Campuzano et al. 2018). Recent studies on climate-driven changes have highlighted the high sensitivity of the South European marine coastal environment to climate variability (IPCC 2014, Rubal et al. 2013). The Western Iberian and the Portuguese Continental Shelf typically are narrow and show a width smaller than 30 km and, as a consequence, the oceanography in this ecoregion is characterised by marked seasonal mixing and stratification of water masses typical of temperate seas, and the ecological processes influenced by both terrestrial nutrient loads and ocean-margin exchanges (upwellings) (Peliz et al. 2002). This general pattern is modified over the shelf by wind-driven upwelling, river outflow and tidal-related processes, increasing the productivity of the system with large variations across the region. It is a complex area characterised by different mesoscale structures that are observed at a seasonal timescale, like fronts, jets, upwelling filaments and countercurrents (Haynes et al. 1993, Peliz et al. 2002, Relvas and Barton 2005). The coastal upwellings enhance the importance of exchanges across the shelf border and act as a significant source of nutrients to the coastal zone (Peliz et al. 2002), and the input of freshwater from rivers and estuaries from the Western Iberian Buoyant Plume is also an important shaping event under downwelling-favourable winds (ICES 2020). Habitats further offshore are shaped by the influence of Atlantic waters and Western Iberia (ICES 2020). Generally, these atmospheric systems have a north-south orientation and influence the observed water levels along the entire coast (Saunders 1982, Fiúza 1982, Frouin et al. 1990, Haynes and Barton 1990, Jorge da Silva 1992, Ambar and Fiúza 1994, Fiúza et al. 1998, Pérez et al. 2001, Campuzano et al. 2018). The large-scale features of the circulation of the West Iberian and Portuguese coast and the upwelling events have been elucidated during recent years (Fiúza et al. 1982, Frouin et al. 1990, Haynes and Barton 1990, Haynes et al. 1993, Castro et al. 2000, Hoinka and Castro 2003, Peliz et al. 2002, Peliz et al. 2005, Relvas et al. 2007). The coast of mainland Portugal is high-mesotidal and wave-dominated, with energy decreasing in sheltered west-east trending sections (Andrade and Freitas 2002). The open coast includes areas with mesotidal exposed Atlantic coast, moderated exposed Atlantic coast and mesotidal sheltered Atlantic coast characteristics (Ferreira et al. 2005). The coastline comprises a wide geomorphological diversity with ponds, rocky shores, coastal lagoons and estuaries that represent key ecological and socio-economic elements throughout its extension (Andrade and Freitas 2002, Ferreira and Matias 2013). The wind, tidal force, river discharge, the continental shelf and the coastal region morphology can influence the local ocean dynamics, in particular the patterns that are registered in the western Iberian Peninsula coast (Fiúza et al. 1982).

The Portuguese coast extension and geographical position at the Atlantic Ocean has played a crucial role throughout its history and have a vital interest in seas and oceans also in the present (Strategic Ocean Commission 2004, Governo Português 2006-Estratégia Nacional para o mar-2013–2020, Portuguese Government 2016, 2019, Forum Oceano 2020, Directorate-General for Maritime Policy 2017, 2020a, b). The maritime character of Portugal is evidenced not only by the possession of an extensive Exclusive Economic Zone (EEZ) but also by the extension of the maritime coast in relation to the surface that emerged (Portuguese Government 2016, 2019, National Strategy for the Sea 2013–2020, Resolução do Conselho de Ministros no.

163/2006; Forum Oceano 2020, Directorate-General for Maritime Policy (DGPM) 2017, 2020a, b). Portugal has one of the largest maritime territories in the world, representing 51% of the marine waters under the jurisdiction of the Member States of the European Union in spaces adjacent to the European continent (European Parliament 2018, Portuguese Government 2016, 2019, Forum Oceano 2020, Directorate-General for Maritime Policy (DGPM) 2017, 2020a, b). The Portuguese coast, bounded by the Atlantic Ocean along its entire western coast and by the Mediterranean Sea to the south, is the longest in the European Union (Portuguese Government 2016, 2019, Forum Oceano 2020, Directorate-General for Maritime Policy (DGPM) 2017, 2020a, b). The zooplankton of the Portuguese coast and estuarine areas were extensively studied in terms of their spatial and temporal structure (composition, distribution and diversity) and also related to environmental and biological processes interactions, resulting from exchanges between the estuarine and the neritic waters of the Platform, such as circulation regimes and the physicochemical characteristics of water, currents, populations dispersion, tidal cycles and the lunar periods. Studies on links between environmental multi-stressors and anthropogenic impacts on the land-ocean system and studies for modelling the reproductive potential of zooplankton by physicochemical factors and extrinsic biological factors involved in the life cycles of key species were also developed in order to monitor the reproductive cycle of species, biological factors (availability of food, predation and competition and growth and mortality rates).

2. History of Marine Biology Research in Portugal

Throughout history, the marine biology research approach in Portugal was marked by private research agendas and national strategic objectives that began from the final years of the Constitutional Monarchy period (1863–1908), guided by an advisory in nature vision and largely dependent on the private initiative of professors, researchers and naturalists and the transition to the First Portuguese Republic (1910–1926), directed towards a more scientific vision and applied research (Saldanha 1982, 1984, 1994, 1997, Saldanha and Ré 1997, Amorim 2005, Bernardo 2013, Lopes 2017, Pinto 2017, Amorim and Pinto 2019, Salgueiro 2021). The oceanographic and marine sciences in Portugal began in the 1890s as a result of oceanographic expeditions and hydrographic missions off the Portuguese coast, performed by Barbosa du Bocage (1823–1907) (Lisbon Polytechnic School and head of the zoology section of the National Natural History Museum), Augusto Nobre and Magalhães Ramalho (1890s), who contributed to relevant advances in the field of oceanography, oceanographic expeditions and hydrographic missions (Almaça 1966, 1997, Madruga 2013, Salgueiro 2021), and also other organised by King Dom Carlos I, between 1896 and 1907, focused on the study of Portuguese marine fauna (Saldanha 1982, 1984, 1997, DGEEC 2011). These oceanographic expeditions resulted in the publication of important international naturalistic and scientific works focused on fisheries, oyster farming and marine species description (Bocage 1864, 1868), the first microscopic photographs of organisms such as plankton (Bragança 1899, 1904) and the creation of museum collections and exhibitions (Costa 1918, Deacon 1997, Jardim et al. 2014, Salgueiro 2021) and with publications also in marine physics, oceanographic processes and marine biology, economy of the sea and ethnography (Silva 1882, 1891).

During the First Portuguese Republic, very important initiatives for the development of a more embracing and public marine research were the creation of centres for research in marine biology, the 'Foz Maritime Zoology Station' in 1913 in the University of Porto, the 'Navy Ministry's Vasco da Gama Aquarium' in 1898 in Lisbon, the 'Maritime Section of the University of Lisbon Bocage Museum', the precursor of the current 'Guia Maritime Laboratory' (Salgueiro 2021) and the creation in 1919 of the 'Estação de Biologia Marinha' in Lisbon, which gave place in 1975 to the Portuguese Institute for Marine and Fisheries Research (INIP) (Dos Santos e Garrido 2003). These centres for research and some pioneer universities (the University of Coimbra, the University of Lisbon and the University of Porto) began to develop science and original experimental work (Salgueiro 2017, 2021), which allowed the emergence of outstanding pioneer researchers in the history of marine and oceanography sciences, knowledge dissemination and scientific research, such as Artur Ricardo Jorge, Antero de Seabra, Celestino da Costa and Alfredo de Magalhães Ramalho (Almaça 2000, Pinto 2017, Salgueiro 2017, 2021), and the development of important and innovative works for marine and planktonic research, namely bacterioplankton, phytoplankton and zooplankton composition and dynamics of Portuguese shelf waters and estuaries (Carrisso 1911, Candeias 1926, Silva 1952, Pinto 1950, Alvarino 1957, Silva 1959, 1962a, b, Saldanha 1997, Almaça 2000, Pinto 2017, Salgueiro 2017, 2021, Paulino et al. 2019). Until the second quarter of the 20th century, Portugal has already theoretical and practical studies in zoology, botany, histology, and embryology and marine biology was associated with scientific knowledge dissemination, fisheries, fish farming, management of water resources, the creation of scientific collections, expeditions and study missions, support for economic activities and public education (Saldanha 1997, Almaça 2000, Pinto 2017, Salgueiro 2017, 2021, Paulino et al. 2019). At the international level, Portugal was already a strategic partner in international matters of the sea, namely in the International Council for the Exploration of the Sea (ICES), the Intergovernmental Oceanographic Commission of the United Nations, Education, Science and Culture Organization (UNESCO) and in the United Nations Convention on the Law of the Sea (1974–1982). At the national level, there was no structured scientific policy (Ruivo 1998, 2006a, b), only in the 70s did the planned and coordinated scientific and technological research begin to be supported by adequate policies (Gago 1990, Gonçalves 1996, Ruivo 1998, 2006a, b). Concerning marine scientific research, only the Hydrographic Institute (IH) and the Fisheries and Sea Research Institute (IPIMAR) brought together financial, structural and marine navigation infrastructure conditions to develop scientific research (Ruivo 1998, 2006a, b, Gonçalves 1996, Independent World Commission for the Oceans 1998, Amorim and Pinto 2019).

Portugal's integration into the European Community allowed, in addition to accelerated industrial and technological development, rational management of natural resources in order to guarantee a balanced development, avoiding their dilapidation and the exploitation of marine resources (living, not energy) supported by research in both fundamental and applied fields that constituted the essential scientific basis for the aforementioned balanced development policy (Council of Ministers 1998).

The European Commission defined, in 2005, the sea as a strategic objective, having drawn up among many other initiatives in 2007 the Integrated European Maritime Policy (EC 2006), followed in 2008 by the Marine Strategy Framework Directive and the European Strategy for Marine Research and Maritime (EC 2008). The Marine Strategy Framework Directive (DQEM – Directive 2008/56/EC of the European Parliament and the Council) is one of the most important legislative instruments to guarantee the integrity of ecosystems of the European Union, establishing itself as the environmental pillar of the Maritime Policy integrated. More recently, the sea is one of the defined priority areas in the European Strategy for the decade 2010–2020 – Horizon 2020 (EC 2011). In this context, since then, Portugal has developed a huge set of political initiatives, noteworthy of the creation, in 1998 of the Intersectoral Oceanographic Commission and the Dynamics Program for Marine Sciences and Technologies to investigate and update the state of the art of marine sciences in Portugal, carry out a political diagnosis and present a proposal for the delimitation of the continental shelf of Portugal (Council of Ministers 1998).

Within the scope of the European Science Foundation, Portugal joined the Marine Board, a prospective body in the field of marine sciences and EurOcean, the European Information Center on Marine Science and Technology, for the identification, collection and production of relevant scientific information (Carvalho 2010). The Strategic Oceans Commission was also created in 2003 (Council of Ministers 2003) in order to define the National Strategy for the Ocean and in 2004, was elaborated in the report 'The Ocean: A National Design for the 21st Century' (Council of Ministers 2006), which led to the creation in 2005 of the Mission Structure for the extension of the Continental Shelf, responsible for Portuguese representation at European level in discussions related to future European policy, to develop the National Strategy for the Sea and to prepare the extension of the boundaries of the Portuguese area of influence on the continental shelf (Council of Ministers 2006). Portugal defined, in 2007, the sea as a strategic vector for the country's development, having created an Interministerial Commission for Sea to coordinate national action in this area and drawing up and implementing a National Strategy for the Sea and acting in line with the most recent European policies (Council of Ministers 2007) and, in order to integrate the European vision and the application of a holistic policy for the management of oceans and seas, create the DG MARE, in 2006, for the management of fisheries and maritime affairs (Council of Ministers 2007). In order to efficiently protect the European marine environment and promote the sustainable use of the oceans, the Marine Strategy Framework Directive (DQEM – Directive 2008/56/EC of the European Parliament and of the Council) was transposed in 2008 into national legislation (Decree Law no. 108/2010, amended and republished by Decree-Law no. 136/2013, later amended by Decree-Law no. 143/2015). Its application is limited to marine waters under European jurisdiction, in accordance with the United Nations Convention on the Law of the Sea (UNCLOS) (Resolution of the Assembly of the Republic No. 60-B/97) and integrated into what was defined in the Water Framework Directive (WFD – Directive 2000/60/EC). Since 2014, The National Ocean Strategy (2013–2020) has been put in place for all the Portuguese maritime space (DR 2014a) that constitutes a base law for maritime spatial planning and management of the Portuguese National Maritime Space in order to Portugal's sustainable development

(DR 2014b). Recently, the implementation and operationalisation of the Maritime Spatial Planning Directive and UN Sustainable Development Goals, as required by the 2030 European Agenda for Sustainability (EU 2019), aim to achieve broader goals that, besides global decline of marine living resources, marine habitats ecosystem functioning and the impacts of biodiversity loss and climate change an integrated approach needs to be implemented to achieve global sustainability of economies, societies and communities and prosperity and to reduce poverty and social and economic inequalities (Belgrano and Villasante 2020).

Since then, strategic vectors have been defined that settle on three pillars: knowledge, spatial planning and ordering and promotion and defence of active role in national interests (Costa and Gonçalves 2010). For the integrated management of the ocean, including coastal zone protection, use and sustainable development of the ocean and its resources, scientific-technological training and research integrated systems for data collection, information management, knowledge and monitoring of the ocean, including coastal areas (Strategic Ocean Commission 2004), were employed. Portugal showed high specialisation in marine sciences from 2000–2010 (namely in areas such as Fisheries and Marine and Aquatic Biology, Oceanography and Ocean Engineering, in which the country reinforced its specialisation). On the other hand, Portugal also showed relevant scientific expertise in domains such as the Environment and Biology, which have high potential for relevant national clusters of a similar nature, technological or economic, such as those of the Sea and Biotechnology (FCT 2013). In Portugal, Marine Biology and Biological Oceanography have experienced considerable development in the last 15 to 20 years, reaching qualitative levels that bring us closer to other European countries. The ocean and coastal, estuarine and lagoon areas have been primarily studied in their various aspects (biological, physical, chemical, geological and conservationist). The works already carried out or being carried out in the field of marine sciences and technologies in Portuguese waters are diverse, having addressed with greater or lesser depth, practically in all the fields of investigation (Vieira 2019). In the field of R&D in Portugal, even if there are still some shortages of human and technological resources, there is already a good European framework on the part of the scientific community. Portugal is already a reference in research in marine science and has been a pioneer in the field of innovation to protect the maritime environment on the surface and nearby (IOC-UNESCO).

3. Plankton Research on the Portuguese Coast (Western Iberia Coast): A Brief Historical Synopsis

Plankton research in Portugal followed the intermittent pattern presented by the rest of the scientific research in the country until the 70s of the 20th century. Since the first quarter of the 20th century, marine biology research in Portugal has been negligible, and only the University of Coimbra and two higher technical schools, the Porto Polytechnic Academy and the Lisbon Polytechnic School, developed science and original experimental work (Salgueiro 2017, 2021). A constant growth in the number of works of plankton in Portuguese waters was verified since the first quarter of the 20th century, especially in the 1990's, more moderate for the 60' and 70' decades (Saldanha 1997, Dos Santos e Garrido 2003, DGEEC 2011, Salgueiro 2021). The subjects studied in the first publications were focused on the description of new occurrences, morphology and taxonomy, while in the last decades, the laboratory studies and interdisciplinary ecology studies dominated (Saldanha 1997, Dos Santos e Garrido 2003, DGEEC 2011, Salgueiro 2021). The first study on plankton dates back to 1880, carried out by Paul Langerhans in the coastal waters of Madeira and studied the Chaetognatha and Apendicularia (Langerhans 1880a, b). As the result of the oceanographic campaigns off the Portuguese coast organised by King Dom Carlos I between 1896 and 1907 to study the Portuguese marine fauna, some important scientific works were published (Bragança 1899, 1904), the first microscopic photographs of organisms such as plankton (Saldanha 1997, DGEEC 2011, Salgueiro 2021). Following these marine and oceanographic expeditions and research approaches and publications, the first official centres for research in marine biology were established, the 'Foz Maritime Zoology Station' in 1913, in the University of Porto, the 'Navy Ministry's Vasco da Gama Aquarium' in 1898, Ministry of the Navy in Lisbon and the 'Maritime Section of the University of Lisbon Bocage Museum', precursor of the current 'Guia Maritime Laboratory' (Salgueiro 2021). At this time, in 1911, a significative and innovative advance in plankton research was developed at the Faculty of Science of the University of Coimbra by Luís Wittnich Carrisso, who conducted a study of the relation of Dinoflagellata, diatoms, foraminifera and copepods with environmental variables to pursue his doctoral thesis (Carrisso 1911). Structured Plankton research in Portugal also began in Lisbon in 1919, with the creation of the 'Estação de Biologia Marinha' which gave place in 1975 to the Portuguese Institute for Marine and Fisheries Research (INIP) (Dos Santos e Garrido 2003) and later in the 1920's, in the Porto University together with various foreigners' institutions' plankton studies started with studies in the Sezimbra bay and a list of Copepoda from Portuguese coast (Candeias 1926, 1930).

In the second quarter of the 20th century , there were also important and pioneer contributions to plankton studies in Portugal, such as studies on the phytoplankton communities from the S. Martinho do Porto Bay (Pinto and Silva 1948) and the Portuguese Óbidos lagoon (Silva 1952), zooplankton from the Cascais Bay (Silva 1950) and in 1956, studies of the morphology, life cycle and environmental factors that promote toxin production in marine bloom-forming dinoflagellates (Silva 1959, 1962a, b). The Estela Sousa e Silva Algal Culture Collection (ESSACC) is currently the legacy of nearly 60 years of the National Institute of Health Dr Ricardo Jorge (INSA) scientific and public activity in marine and freshwater phytoplankton biology and toxicology (Paulino et al. 2009). However, until the second half of the 20th century, only INIP, IPIMAR (Fisheries Research Institute) and the University of Lisbon were responsible for most of the oceanic campaigns and publications for plankton studies, with publications mostly in national journals. Some studies have addressed seasonal bacterioplankton, phytoplankton and zooplankton composition and dynamics of Portuguese shelf waters and estuaries (Azores and Madeira included) (Pinto 1950, Alvarino 1957, Carmona 1972, Dias et al. 1976, Matos 1977, Alvarino 1980, Muzavor 1981, 1986, Vilela et al. 1990) and studies of the ichthyoplankton only started in Portugal in 1979 (Ré 1984a, b, 1987, 1989, Ré et al. 1990, Cunha and Figueiredo 1988, Figueiredo and Santos 1989, Meneses and Ré 1991, Andres et al. 1992, John and Ré 1993, John et al. 1996). Plankton studies on the Portuguese coast (Western Iberia coast) had a great development after the 1980s with works on ichthyoplankton and planktonic production cycles and on the relation between plankton and fisheries resources, such as the occurrence of eggs and planktonic stages of marine fishes from the coast of Algarve (southern Portuguese coast), occurrence of eggs and newly hatched larvae of *Abudefduf luridus* from the Azores, occurrence of postlarval stages of *Lebetus* spp., occurrence, mortality and dimensions of sardine eggs (*Sardina pilchardus*), the identification of the first planktonic larval stage of (*Engraulis encrasicolus*), of the first planktonic larval stages fish species, ecology of posture and diel spawning time of planktonic phase of *Sardina pilchardus* (Ré 1979, 1980, 1981a, b, Ré et al. 1982, Ré 1986a, b, 1989, Ré et al. 1990). Several studies have addressed the composition and dynamics of the zooplankton of the Western Iberia coast that includes the regions off Galiza and the Portuguese shelf waters (INIP 1986, Cunha 1993, John 1993, Santos et al. 2001, Queiroga et al. 2005, Sobrinho-Gonçalves et al. 2013) and also in the Azores and Madeira archipelagos.

The Azores archipelago location in the middle of the North Atlantic Ocean, a large marine territory where islands, with distinct geographic and geological features, recognised for its key role as a transitionary habitat for large open-ocean animals, such as cetaceans, sharks, pelagic fish or sea-turtles (Prieto et al. 2017) and to harbour valuable deep-sea resources and ecosystems, such as deep-sea fish (Menezes et al. 2006) or cold-water coral aggregations (Pham et al. 2015), this had led to several recent zooplankton biomass studies (Sobrinho-Gonçalves and Isidro 2001), mesozooplankton and copepod community composition from areas nearby but away from Azores region (Ottens et al. 1991, Head et al. 1999, 2002, Kahru et al. 1991, Schiebel et al. 2002, Huskin et al. 2001), studies on zooplankton biomass of the seamounts located within the Azorean Exclusive Economic Zone (EEZ) (Martin and Christiansen 2009), size distribution and vertical migration data and variability of zooplankton communities at Condor seamount and surrounding areas (Carmo et al. 2013). The Madeira archipelago is one of the four Macaronesian island systems in the Eastern Atlantic (north to south: Azores, Madeira, Canary Islands and Cape Verde, located at the North Atlantic Subtropical Gyre's eastern boundary) (Freitas et al. 2019). The island system is mainly influenced by the Azores Current, which joins the Canary Current north and around Madeira Island, with dominant northeastern trade winds and typical oligotrophic conditions (Longhurst et al. 1995). Marine plankton biodiversity research in Madeira has been mainly restricted to phytoplankton (Narciso et al. 2019), Ctenophora, the macro gelatinous zooplankton, represented mostly by *Eurhamphaea vexilliger*, *Ocyropsis crystallina*, *Cestum veneris* and *Beroe* sp. (Gueroun 2021). Recently, a study characterised microplastics and mesozooplankton abundances off the south coast of Madeira Island (Sambolino 2022).

A list of the key-research works on plankton in the Portuguese coast is presented in Table 1.

Table 1. History of research works on plankton in the Portuguese coast.

Authors	Research Group	Species	Seasonality	Forcing Factors	Coastal Region
Langerhans 1880a, b		Chaetognatha and Apendicularia		Marine plankton biodiversity	Off the Portuguese coast, the Madeira archipelago
Bragança 1899, 1904		Microscopic photographs of planktonic organisms	–	Marine plankton biodiversity	Off Portuguese coast
Carrisso 1911	Faculty of Science, University of Coimbra	Dinoflagellata, diatoms, foraminifera and copepods	–	relationship between plankton and environmental variables	Off Portuguese coast
Candeias 1926	Porto University	Copepoda taxonomic composition		Marine plankton biodiversity	Off Portuguese coast
Alvarino 1957		Zooplancton del Atlantico Ibérico		Campaña del 'Xauen' en el verano de 1954	Off Portuguese coast
Silva 1959, 1962a, b	Porto University	Toxin production in marine bloom-forming dinoflagellates	Morphology, life cycle, and environmental factors	plankton and environmental variables that promote toxin blooms	Off Portuguese coast
Dias et al. 1976		Zooplankton biovolumes	Planktonic production cycles	Relationship between plankton and fisheries resources	Off the Portuguese coast, the Azores Archipelago
Ré 1981a	Faculty of Science, University of Lisbon	Ichthyoplankton, occurrence of postlarval stages of *Lebetus* (Pisces: Gobiidae)	Planktonic production cycles	Relationship between plankton and fisheries resources	Off the Portuguese coast, the central region
Ré 1981b	Faculty of Science, University of Lisbon	Ichthyoplankton, seasonal occurrence, mortality and dimensions of sardine eggs (*Sardina pilchardus*)	Planktonic production cycles	Relationship between plankton and fisheries resources	Off the Portuguese coast, the central region

Table 1 contd. ...

...Table 1 contd.

Authors	Research Group	Species	Seasonality	Forcing Factors	Coastal Region
Muzavor 1981		Copepod composition	Planktonic production cycles	Relationship between plankton and fisheries resources	Off the Portuguese coast, the Azores Archipelago
Harris 1982		Feeding behaviour of *Calanus* and *Pseudocalanus* (experimental manipulated enclosed systems)	Planktonic production cycles	Relationship between plankton and fisheries resources	Off the Portuguese coast, the Azores Archipelago
Neto and Paiva, 1984	UCTRA, University of Algarve	Zooplankton from southern Portuguese coast			Off southern Portuguese coast
Sobral et al. 1985		Zooplankton biovolumes	Planktonic production cycles	Relationship between plankton and fisheries resources	Off the Portuguese coast, the Azores Archipelago
Ré 1986a	Faculty of Science, University of Lisbon	Planktonic production cycles, Ecology of posture and planktonic phase of *Sardina pilchardus*	Planktonic production cycles	Relationship between plankton and fisheries resources	Off the Portuguese coast, the central region
Ré 1986b	Faculty of Science, University of Lisbon	Ichthyoplankton (*Sardina pilchardus* and *Engraulis encrasicolus*), identification of the first planktonic larval stages	Planktonic production cycles	Relationship between plankton and fisheries resources	Off the Portuguese coast, the central region
INIP 1986	Oceanography and Plankton Group (INIP)	Zooplankton composition and distribution	Planktonic production cycles	Relationship between plankton and fisheries resources	Off Portuguese coast
Cunha and Figueiredo 1988	Oceanography and Plankton Group (INIP)	Reproductive cycle of *S. pilchardus*	Planktonic production cycles (1971/1987)	Relationship between plankton and fisheries resources	Off the central Portuguese coast

Ré 1989	Faculty of Science, University of Lisbon	Ichthyoplankton (*Sardina pilchardus*) Sardine diel spawning time	Planktonic production cycles	Relationship between plankton and fisheries resources	Off Portuguese coast
Figueiredo and Santos 1989	Oceanography and Plankton Group (INIP)	Reproductive biology of *Sardina pilchardus*	Planktonic production cycles seasonal maturity evolution (1986 to 1988)	Relationship between plankton and fisheries resources	Off Portuguese coast
Ré et al. 1990	Faculty of Science, University of Lisbon & Oceanography and Plankton Group (INIP)	Ichthyoplankton (*Sardina pilchardus*) Sardine spawning	Planktonic production cycles	Relationship between plankton and fisheries resources	Off Portuguese coast
Meneses and Ré, 1991	Faculty of Science, University of Lisbon	Ichthyoplankton (*Sardina pilchardus*), sardine eggs	Infection of sardine eggs by a parasitic dinoflagellate *Ichthyodinium chabelardi*	Relationship between plankton and fisheries resources	Off Portuguese coast
Ottens 1991		Planktonic foraminifera		North Atlantic water mass indicators	Off the Portuguese coast, the Azores Archipelago
Kahru et al. 1991		Particle (plankton) size structure		Azores front, North Atlantic Bloom Study	Off the Portuguese coast, the Azores Archipelago
Andres et al. 1992	Faculty of Science, University of Lisbon	Ichthyoplankton, Fish larvae and gammaridae plankton	Planktonic production cycles	Relationship between plankton and fisheries resources	Off the Northern Portuguese coast

Table 1 contd. ...

...Table 1 contd.

Authors	Research Group	Species	Seasonality	Forcing Factors	Coastal Region
Cunha 1993	Oceanography and Plankton Group (INIP)	Off the northwestern Portuguese coast (Espinho and Figueira da Foz), zooplankton biomass maintained high production levels during the spring, summer and autumn month Off the southwestern and south Portuguese coast (Sines and Lagos), the Zooplankton biomass pattern is much more irregular, with values decreasing significantly after the spring boom (May/June) Off the northwestern Portuguese coast, low production levels in January and February, and increased till June, followed by a small decrease in July. In the southern region, the pattern is more variable, with increases in the production cycle during spring (March/April), summer (July/August) and autumn (October)	Zooplankton biomass increases in March and maintains high levels till October, after which the biomass decreases 335 /μm sieve was higher, and the levels of production were more constant than those retained by the 505 / μm Contribution of the smaller species, which in general are herbivores, for the annual zooplankton production curve is higher, especially during the spring, summer and autumn months (April till October)	Upwelling bidimensional pattern Winter downwelling events Summer upwelling events (from March till October)	Off the northwestern Portuguese coast (Nazaré Canyon, Peniche and Figueira da Foz on the northern coast, Sines at the southwest and Lagos in the south)
John and Ré 1993	Faculty of Science, University of Lisbon	Ichthyoplankton, cross-shelf zonation, vertical distribution, and drift of fish larvae	Fish larvae	Across an upwelling front	Off the Northern Portuguese coast

Dam et al. 1993		Trophic role of mesozooplankton	Planktonic production cycles	North Atlantic blooms	Off the Portuguese coast, the Azores Archipelago
Cowles and Fessenden 1995		Copepod grazing and fine-scale distribution patterns	Planktonic production cycles	Relationship between plankton and fisheries resources	Off the Portuguese coast, the Azores Archipelago
John et al. 1996	Faculty of Science, University of Lisbon	Sardine larvae		Spring-upwelling event	Off Northern Portugal
Villa et al. 1997	UCTRA, University of Algarve	Phytoplankton biomass and zooplankton abundance	Planktonic production cycles, spawning of *Loligo vulgaris*	Relationship between plankton and fisheries resources	Off the south coast of Portugal (Sagres)
Chícharo 1998		Nutritional condition and starvation in *Sardina pilchardus* larvae	Environmental factors influence *Sardina pilchardus* larvae' nutritional condition	Relationship between plankton and fisheries resources	Off southern Portugal
Head et al. 1999		A comparative study of size-fractionated mesozooplankton biomass and grazing	Mesozooplankton biomass and grazing	Influence of mesoscale structures	Off the Portuguese coast, the Azores Archipelago
Santos et al. 2001	Oceanography and Plankton Group (IPIMAR)	Ichthyoplankton, Sardine and horse mackerel recruitment	Sardine and horse mackerel	Across an upwelling front	Off Portuguese coast
Sobrinho-Gonçalves and Isidro 2001		Ichthyoplankton composition	Planktonic production cycles	Relationship between plankton and fisheries resources	Off the Portuguese coast, the Azores Archipelago
Huskin et al. 2001		Mesozooplankton distribution and copepod grazing	Planktonic production cycles	Influence of mesoscale structures	Off the Portuguese coast, the Azores Archipelago
Head et al. 2002		Copepod composition	Planktonic production cycles	Relationship between plankton and fisheries resources	Off the Portuguese coast, the Azores Archipelago

Table 1 contd. ...

...Table 1 contd.

Authors	Research Group	Species	Seasonality	Forcing Factors	Coastal Region
Schiebel et al. 2002		Distribution of planktic foraminifera, shelled gastropods, and coccolithophorids	Planktonic production cycles	Impact of the Azores Front	Off the Portuguese coast, the Azores Archipelago
Borges et al. 2003	Oceanography and Plankton Group (IPIMAR)	Sardine regime shifts catches and wind conditions	Sardine regime shifts relation with catches and wind conditions (time series analysis)	Relationship between plankton and fisheries resources	Off Portuguese coast
Huskin et al. 2004		Copepod composition	Planktonic production cycles	Relationship between plankton and fisheries resources	Off the Portuguese coast, the Azores Archipelago
Santos et al. 2004	Oceanography and Plankton Group (IPIMAR)	Impact of a winter upwelling event on the distribution and transport of sardine (*Sardina pilchardus*) eggs and larvae off western Iberia: a retention mechanism	Winter upwelling event	Across an upwelling front	Off Portuguese coast
Queiroga et al. 2005	Department of Biology and CESAM, University of Aveiro	Inner upwelling front: *Acartia clausi* and *Temora longicornis*, first-zoeae *Carcinus maenas*, *Portumnus latipes* and *Atelecyclus undecimdentatus* Outer upwelling front: *Clausocalanus arcuicornis*, *Euchaeta hebes* and *Lensia subtilis*, zoeae II Polybinid crab. Widespread: *Muggiae atlantica*, *Centropages chierchiae*, *Calamus helgolandicus*, and *Sagitta friderici*	Summer	Across an upwelling front	Off Ria de Aveiro (northwestern Portuguese coast) (18° and 20° W and 37° and 59 N)

Santos et al. 2005		Sardine larvae (*Sardina pilchardus*)		Survival of sardine larvae	Off the Portuguese coast
Marta-Almeida et al. 2006	Department of Physics and Department Biology and CESAM, University of Aveiro	Crab larvae	Influence of vertical migration pattern on retention of crab larvae	Seasonal upwelling system	Off the Portuguese coast
Faria et al. 2006.	University of Algarve	Ichthyoplankton dynamics		Relationship between plankton and fisheries resources	Off the southern Portuguese coast
Chícharo et al. 2006		Interannual differences in ichthyofauna structure	Before and after the Alqueva Dam construction	Relationship between plankton and fisheries resources	Off the southern Portuguese coast
Santos et al. 2007	Oceanography and Plankton Group (IPIMAR)	Horizontal and vertical distribution of cirripede cyprid larvae	Cirripede cyprid larvae	In an upwelling system	Off the Portuguese coast
Peliz et al. 2007	Department of Physics and Department Biology and CESAM, University of Aveiro, ECOLA - Echanges Côte-Large, LEGOS - Laboratoire d'études en Géophysique et océanographie spatiales and IRD [France-Ouest] - Institut de Recherche pour le Développement	Crab larvae	Physical processes of crab larvae dispersal	Western Iberian Shelf	Off the Portuguese coast, Western Iberian Shelf

Table 1 contd. ...

...Table 1 contd.

Authors	Research Group	Species	Seasonality	Forcing Factors	Coastal Region
Queiroga et al. 2007		Oceanographic and behavioural processes controlling invertebrate larval dispersal and recruitment	Decapod larvae implications for offshore transport	Upwelling ecosystem	Western Iberia cost
Bode et al. 2007		Stable nitrogen isotope studies of the pelagic food web			Off the Iberian Peninsula
Santos et al. 2008	Oceanography and Plankton Group (IPIMAR)	Diel vertical migration of decapod larvae	Decapod larvae implications for offshore transport	Upwelling ecosystem	Portuguese coastal
Gaard et al. 2008		Copepod composition	Planktonic production cycles	Marine plankton biodiversity	Off the Portuguese coast, the Azores Archipelago
Garrido et al. 2009	Oceanography and Plankton Group (IPIMAR) and Faculty of Sciences Lisbon	Spatial distribution and vertical migrations of fish larvae communities		Relationship between plankton and fisheries resources	Off Northwestern Iberia
IPIMAR *website*, U-AMB Oceanography and Plankton Group (IPIMAR) in the framework of the International Year of Biodiversity (2010)	Oceanography and Plankton Group (IPIMAR)	*Acartia clausi, Acartia danae, Acartia grani, Acartia longiremis,* zoaee zoeae *Polybius henslowi,* zoeae *Liocarcinus* spp., *oikopleura* sp, *Meganyctiphanes norvegica, Sardina pilchardus* eggs and larvae, Cirripedia larvae, *Engraulius encrasicolus* eggs and larvae, *Trachurus trachurus* eggs and larvae, *Nyctiphanes couchii, Nematoscelis megalops, Euphausia krohnii, Obelia* spp.	Zooplankton are very abundant in Spring/Summer, Cirripedia larvae very abundant in Spring, *Nyctiphanes couchii, Nematoscelis megalops, Euphausia krohnii* are frequent in spring/summer, *Obelia* spp frequent in summer *Sardina Pilchardus* and *Engraulius encrasicolus* eggs and larvae are very abundant in autumn/spring, *Trachurus trachurus* eggs and larvae are very abundant in winter/spring	Seasonal variation	West Coast north of Lisbon

Miranda et al. 2011 (ICES Zooplankton Status Report 2010/2011)	Instituto Español de Oceanografía	Copepod species of warm waters (*Temora stylifera*) appeared in relatively large numbers during warmer periods (1997–1998, 2001–2002, and 2009) Other species decreased in the five years prior to 2010 (*Calanoides carinatus, Acartia clausi*)	Seasonal cycle of zooplankton biomass with high values from April to October, with a slight reduction in June and August and a clear reduction in winter. Zoopankton shows a single late summer biomass and abundance peak Interannual biomass anomalies reveal an increasing trend, following a period of low biomass observed in 1994–2001	Seasonal stratification of the water column masked upwelling events from April to September Upwelling events provide zooplankton populations favorable conditions for development in summer, which is the opposite of what occurs in other temperate seas in this season of the year Upwelling is highly variable in intensity and frequency, demonstrating substantial year-to-year variability	Coastal region of Galicia, Vigo (northwest Spain) (42.1417°N 8.9533°W)

Table 1 contd. ...

...Table 1 contd.

Authors	Research Group	Species	Seasonality	Forcing Factors	Coastal Region
Santos and Santos 2011 (ICES Zooplankton Status Report 2010/2011)	Oceanography and Plankton Group of the Instituto Português do Mar e da Atmosfera (IPMA)	Acartia spp. was the most abundant copepod present in the samples. Copepods are mainly represented by the genera *Acartia, Paracalanus, Oncaea* and *Oithona.* Other species (*Temora stylifera, T. longicornis,* and *Centropages* spp.) are also important but occur later in the season, which explains the high copepod abundance late in the year	The seasonal cycle of zooplankton biomass is characterized by a bimodal pattern, with peak biomass in April and August Copepod abundance remains high throughout the season, with the highest abundance from August through November Total copepod abundance has been decreasing, mainly caused by a reduction in the three most abundant genera: Acartia, Paracalanus, and Oithona. The trends in these species are positively correlated Following the decrease in the abundance of copepods, bivalve veligers have become more abundant in the area (Veliger abundance follows the increasing abundance of the invasive species *Ruditapes philippinarum* in the Tagus estuary since its establishment ten years ago	Bimodal pattern of seasonal upwelling Total copepod abundance and *in situ* temperature interannual anomalies oscillate together and result in a significant negative correlation	Off Cascais Bay outside the Tagus River estuary

Carmo et al. 2013		Zooplankton composition	Planktonic production cycles	Marine plankton biodiversity	Off the Portuguese coast, the Azores Archipelago
Sobrinho-Gonçalves et al. 2013	Oceanography and Plankton Group of the Instituto Português do Mar e da Atmosfera (IPMA)	Environmental forcing on the interactions of plankton communities		Upwelling system	Off Portuguese coast
Frias et al. 2014		Evidence of microplastics in samples of zooplankton			Off Portuguese coast
Kaufmann et al. 2015		Phytoplankton composition		Marine plankton biodiversity	Off the Portuguese coast, the Madeira Archipelago
Narciso et al. 2019		Phytoplankton composition		Marine plankton biodiversity	Off the Portuguese coast, the Madeira Archipelago
Gueroun 2021		Ctenophora (macro gelatinous zooplankton)	*Eurhamphaea vexilliger, Ocyropsis crystallina, Cestum veneris* and *Beroe* sp.	Marine plankton biodiversity	Off the Portuguese coast, the Madeira Archipelago
Sambolino et al. 2022		Microplastics and mesozooplankton abundances		Marine plankton biodiversity	Off the Portuguese coast, the Madeira Archipelago

4. Zooplankton Composition and Seasonal Distribution on the Portuguese Coast

The copepods *Acartia clausi* and *Paracalanus parvus* are the most abundant species of mesozooplankton in most areas of the Western Iberia coast, and the macrozooplankton species *Calanus helgolandicus* is also important in terms of biomass and the distribution of warm-water copepod species like *Temora stylifera* and *Calanoides carinatus* has moved northwards across the ecoregion (ICES 2020). The zooplankton composition on the Portuguese coast is represented mainly by Copepoda, Larvacea Cirripedia larvae, Cladocera, Echinodermata larvae, Siphonophorae, Cnidaria, Bivalvia larvae, Decapoda larvae, Gastropoda larvae, Chaetognatha, Euphausiacea, Polychaeta and Bryozoa larvae. The *taxa* group included typical species from neritic waters of the Palaeartic Atlantic and Mediterranean regions, and the species often dominate the zooplankton composition of European temperate estuaries (Christou and Verriopoulos 1993, Morgado et al. 2006a, Vieira et al. 2015). On the northern Portuguese coast, the species composition is dominated by the copepods *Acartia clausi*, *Temora longicornis*, *Oithona plumifera*, *Clausocalanus arcuicornis*, *Paracalanus parvus*, *Oithona similis* and *Centropages chierchiae*, the cladocera *Evadne nordmanni*, the siphonophore *Muggiaea atlantica* and the tunicates *Oikopleura dioica* and *Oikopleura longicauda*, showing higher densities in the summer and lower in winter (Morgado et al. 2006, Vieira et al. 2015).

The works developed off the cost in the Azores archipelago area showed that the zooplankton was dominated, during March and April by the *calanoid* copepoda, *Temora longicornis, Acartia clausi, Calanus helgolandicus, Corycaeus* sp., *Oithona* spp., *Centropages* spp., *Calanus* spp., *Oncaeidae, Lucicutia* spp., *Eucalanidae, Furcilia larva, Rhincalanus* spp. *nauplii, Acartia* spp., *Crustacea nauplii, Cladocera, Evadne* spp. and *Pseudevadne* spp., *tunicate,* Doliolida spp, and *Cirripedia nauplius, Bivalvia Veligera larva, Acantharia* and *Radiolaria* (Muzavor 1981, Sobral et al. 1985, Head et al. 2002, Huskin et al. 2004, Gaard et al. 2008, Carmo et al. 2013). In this coastal area the ichthyoplankton composition is dominated, February/March and May/June, by larvae of *Helicolenus dactylopterus, Macroramphosus scolopax, Phycis phycis, Lepidopus caudatus, Serranus cabrilla, Cerastoscopelus madeirinsis, Callionymus reticulatus, Cyclothone* spp., *Diogenichthys atlanticus, Lampanyctus pusillus* and *Myctophum punctatum* (Sobrinho-Gonçalves and Isidro 2001). Currently, long-term data is available on the zooplankton of the Azores archipelago from the Continuous Plankton Recorder survey that was towed in oceanic waters in the northern part of the ecoregion in the 1960s to the early 1980s (ICES 2019). After a decadal gap, the route went into operation again around the Azores ecoregion from 1997 onwards; long-term trends suggest that at the decadal scale, zooplankton populations are mainly influenced by large-scale natural climate variations, such as the Atlantic Multidecadal Oscillation and the North Atlantic Oscillation. Phytoplankton trends for this region show a general increase in smaller phytoplankton and a decrease in larger phytoplankton (e.g., large diatoms and dinoflagellates). For zooplankton, the abundance of Euphausids and Chaetognaths has generally declined over the decadal period, whereas the abundance of copepods has remained relatively stable. Of the main zooplankton community, the Appendicularians (larvaceans) have shown the largest increase in abundance over the last 50 years (ICES 2019).

4.1 The Northwest Portuguese Coast and Estuaries

Portugal has a vast number of important estuaries, noteworthy that those situated on the limits and parts of the rivers that flow into them from the northwestern (Minho estuary) and southeastern (Guadiana estuary) borders with Spain. The Portuguese estuaries are complex ecosystems, highly productive, very important in the transition of inland waters and the ocean, ensuring the biogeochemical recycling of solutes/compounds and play an important nursery role for several commercially important fish species, and their habitat use by juveniles (Vasconcelos et al. 2011). The estuaries of the Portuguese coast have very different physical and chemical characteristics in terms of shape, size, fluvial regime and hydrodynamics (Kjerfve et al. 1996, Bettencourt et al. 2004, Vasconcelos et al. 2011, Campuzano et al. 2018). They usually have a strong salinity range from salty to brackish waters, depending on the freshwater inputs and the level of water exchange with the sea where ocean and freshwater meet controlled by the tidal regime, estuary geomorphology and river discharge (Bettencourt et al. 2004, Campuzano et al. 2018, Neto et al. 2022). The most important estuaries on the Portuguese coast, from north to south, were the Minho, Douro, Ria Aveiro, Mondego, Tejo, Sado, Mira, Ria Formosa and Guadiana. These systems differ substantially in terms of their geomorphologic and hydrologic characteristics; Tejo and Sado are large systems with areas of 320 and 180 km^2, respectively, while Mira is the smallest with 5 km^2. River flow also differs markedly: Minho, Douro, and Tejo have mean flow values above 300 m^3 s^{-1}, contrasting with low freshwater flow in estuaries such as the Mira and Ria Formosa. Ria Aveiro and Ria Formosa are shallow coastal lagoon systems with large intertidal areas. Shallow areas are a common feature in all estuarine systems, with mean depths varying between 1 and 6 m. In general, they can be grouped as follows: Mesotidal stratified estuary – Minho estuary, Lima estuary and Douro estuary; mesotidal well-mixed estuary – Ria de Aveiro, Mondego estuary, Tagus estuary, Sado estuary, Mira estuary and Guadiana estuary; mesotidal shallow lagoon-Ria Formosa. Four essential typologies can be defined (Bettencourt et al. 2004, Neto et al. 2022) (Table 2).

The northwest Portuguese coast, at latitudes 40–42°N, is a highly energetic region, as may be inferred from the average wave power obtained with data from the wave analysis model archive of the European Centre for Medium-Range Weather Forecasts (Cruz 2008). On the Portuguese coast, waves usually arrive from the northwest, with an offshore mean significant wave height of 2–3 m and a mean wave period of 8–12 s. Storms generated in the North Atlantic are frequent in winter and can persist for up to 5 d, with significant wave heights as high as 8 m (Costa et al. 2001). The tides are semi-diurnal, ranging from 2 m to 4 m during spring tides. The mean sea level is +2 m chart datum (CD). The strong wave regime induces a southward alongshore transport of 1–2 × 10^6 m^3 year^{-1} (Oliveira 1997). The Portuguese northwest coast may be divided into two stretches based on the geomorphological characteristics. One, from Caminha to Espinho, consists of low rocky formations; the other, from Ovar to Marinha Grande, is mostly a low-lying open sandy shore, vulnerable to wave

Table 2. Four essential typologies can be defined in Portuguese estuaries (Sources: European Commission 2000, Bettencourt et al. 2004, Neto et al. 2022).

Estuarine Typology	Location	Characteristics	
Mesotidal Estuaries	Located in the central and northern part of the Portuguese coast.	With variable stratification of the water column but with less than 50% of the intertidal area and in the form of a channel.	Minho estuary, Lima estuary, Neiva estuary, Cávado estuary, Ave estuary, Leça, Douro estuary, Mondego estuary and Lis estuary.
Ria de Aveiro	Located in the central and northern part of the Portuguese coast.	With a more sprawling shape, well mixed and with an available intertidal area of approximately 50%.	Ria de Aveiro
Narrow Estuaries	Located in the southern part of the Portuguese coast.	Include the narrow southern mesotidal estuaries, channel-shaped, with a well-mixed water column and less than 50% intertidal area.	Mira estuary, Arade estuary and Guadiana estuary
More Sprawling Mesotidal Estuaries in the South	Located in the southern part of the Portuguese coast.	Well mixed but with an intertidal area of less than 50% of its total area.	Tejo estuary and Sado estuary

action and backed by dunes that have already been destroyed in some places. The Douro River estuary and the Aveiro lagoon are important morphological features and are strongly influenced by anthropogenic effects emanating from the cities of Porto and Aveiro (Silva et al. 2007). The northern Portuguese estuaries can be mesotidal stratified estuaries, such as the Minho estuary, Lima estuary and Douro estuary or mesotidal well-mixed estuaries, such as the Ria de Aveiro and Mondego estuary (European Commission 2000, Bettencourt et al. 2004, Neto et al. 2022).

4.2 Zooplankton Composition and Distribution in Northern Portuguese Estuaries

Following the first works carried out on the Portuguese coast, constant growth in the number of plankton in Portuguese estuarine, lagoon, river, lagoon and reservoir environments were verified since the first quarter of the 20th century, with a more moderate development until the 60' and 70' decades (Dos Santos e Garrido 2003, Salgueiro 2017, 2021) due to the works of IPIMAR (Fisheries Research Institute). Although a constant growth in the number of works of plankton in Portuguese waters was verified since the second quarter of the 20th century, especially in the 1990s, with many works on the Southern Portuguese estuaries developed by INIP and Lisbon Faculty of Science (Dos Santos e Garrido 2003, Salgueiro 2017, 2021). The other Portuguese universities began their works on plankton in a structured and multidisciplinary way only in the 80th decade of the 20th century following the '1st Meeting of Portuguese Planctologists', which took place at the University of Aveiro in 1989. This meeting formed a joint organisation of the Department of Biology of the University of Aveiro, the Faculty of Sciences of the University of Lisbon and INIP and constituted the first presentation of plankton research works in Portugal by Portuguese researchers, with some Spanish guests (Galicia), to share experiences and knowledge about the different works and projects being developed in the plankton domain. This meeting represented a very important milestone in Portuguese plankton research, allowed the development of partnerships that proved to be crucial for future research on plankton in Portugal and contributed to the promotion of national and international research consortia within Portuguese, Spanish and other international universities and research institutions to a more deep knowledge on plankton responses to changes in environmental conditions and useful to meet together researchers from other scientific areas within biology, oceanography, physics, chemistry and informatics. After this meeting, of note is the realisation of the '2nd Meeting of Portuguese Planctologists', which took place at the University of Algarve in 1992, with a wider organisation that included, in addition to the first organisers, the University of Algarve and the University of Coimbra. The '1st Iberian Plankton Congress' in 1995, which took place at the University of Coimbra, in a joint organisation within the Portuguese consortium (IMAR; University of Coimbra), the Faculty of Sciences of Lisbon, the Department of Biology of the University of Aveiro and INIP) and the Spanish Universities of Vigo and Corunha. The 'I International Plankton Symposium', was held in Espinho in 2001, the 'II Plankton Symposium' in 2003, which took place in Vigo, Spain; the 'III Plankton Symposium' in 2005, which took place in Figueira da Foz, Portugal, in a joint organisation of the Institute of Marine Research, IMAR (University of Coimbra), the Faculty of Sciences of Lisbon, the Department of Biology of the University of Aveiro and INIP. The 'Plankton Symposium IV – 1st Brazilian Plankton Congress' in 2006, organised by The University of Aveiro and a Brasilian Universities consortium, took place in João Pessoa, Pernambuco, Brazil. These national and international scientific meetings were fundamental for the great development observed in planktonic studies in Portugal to achieve a more comprehensive insight into the mechanisms governing the occurrence and abundance of dominant plankton taxa and which factors control plankton biodiversity and biocomplexity to predict major biological processes in the aquatic systems. As a corollary of these initiatives, a huge set of national and international networks was started, giving rise to numerous national and international scientific meetings, projects and studies that have been developed and publications reported to the Portuguese plankton researchers' community (Universities, Research Institutes and laboratories, NGOs). The results of studies carried out in the northern Portuguese estuaries are listed in Table 3.

Table 3. Dominant zooplanktonic groups and species and seasonality in the northern Portuguese estuaries.

	Zooplankton Groups	Zooplankton Species	Seasonality	Authors
Minho estuary	**Holoplankton** - Copepods, Cladocera, Isopoda, Cnidaria, Cnidaria, Cladocera, Mysidacea, Isopoda, Chaetognatha, Tunicata. **Meroplankton** - Cirripedia larvae, Mollusca larvae, Bryozoa larvae, Annelida larvae, Decapoda larvae, Echinodermata larvae and fish eggs and larvae	*A. clausi, A. tonsa, O. nana, O. plumifera, E. acutifrons, P. parvus, calanus helgolandicus, Centropages chierchiae, Centropages typicus, Oncae* spp.*, Corycaeus anglicus, Pseudocalanus elongatus, Temora longicornis, Podon polyphemoides, Podon leuckarti, Evadne nordmanni, Penillia avirostris, Oikopleura dioica, Doliolum* spp.*, Noctiluca scintilans, Diphyes díspar, Sarsia geminifera, Sarsia prolifera, Phialella quadrata, Phialidium hemisphaericum*	Late spring/summer and early autumn (June/ August)	Pinheiro and Fidalgo 2000, Pinto and Martins 2013, Fatela et al. 2014, Vieira et al. 2015, Baeta et al. 2017
Lima estuary	**Holoplankton** - Copepods, Cladocera, Isopoda, Cnidaria, Cnidaria, Cladocera, Mysidacea, Isopoda, Chaetognatha, Tunicata. **Meroplankton** - Cirripedia larvae, Mollusca larvae, Bryozoa larvae, Annelida larvae, Decapoda larvae, Echinodermata larvae and fish eggs and larvae	*A. clausi, A. tonsa, O. nana, O. plumifera, E. acutifrons, P. parvus, calanus helgolandicus, Centropages chierchiae, Centropages typicus, Oncae* spp.*, Corycaeus anglicus, Pseudocalanus elongatus, Temora longicornis, Podon polyphemoides, Podon leuckarti, Evadne nordmanni, Penillia avirostris, Oikopleura dioica, Doliolum* spp.*, Noctiluca scintilans, Diphyes díspar, Sarsia geminifera, Sarsia prolifera, Phialella quadrata, Phialidium hemisphaericum*	Late spring/summer (April/August) and early autumn (October)	Ramos et al. 2006a, 2006b, Ramos 2007, Ramos et al. 2009a, 2009b, 2010, Azevedo et al. 2013, Vieira et al. 2015, Baeta et al. 2017

Ria de Aveiro	**Holoplâncton** - dominated by adults, copepodits and nauplii of Copepoda, Siphonophora, Chaetognatha, Appendiculata and Hydromedusae. **Meroplâncton** - Mollusca larvae, cirripedia larvae, Isopoda (praniza and adults), Polychaeta larvae, Decapod larvae, Polychaeta larvae, Gastropoda larvae and Bivalvia larvae	*Paracalanus parvus, Pseudocalanus elongatus, Clausocalanus* spp., *Temora longicornis, Calanipeda aquaedulcis, Acartia clausi, Acartia bifilosa var, Acartia* spp., *Oithona nana, Oithona similis, Cyclopina gracilis, Oncaea media, Euterpina acutifrons* and *Tachidius discipes*	Spring (April) summer (July), and Autumn (October)	Morgado 1991, Morgado 1993, Azeiteiro and Morgado 1996, Morgado 1997, Pereira et al. 2000, Morgado et al. 2003a, 2003b, Morgado et al. 2006a,b, Queiroga et al. 1994, 1997, Queiroga 1996, Queiroga et al. 2005, Morgado et al. 2006, 2007, 2014, Leandro et al. 2006, 2007, 2014, Cardoso et al. 2013
Mondego estuary	**Holoplankton**- Copepoda adults, copepodits and nauplii, Cladocera, Mysidacea (Mesopodopsis slabberi), Dinoflagellate (Noctiluca scintillans) and Appendiculata (Oikopleura dioica) and Hydromedusae. **Meroplankton** - Mollusca larvae, cirripedia larvae, Isopoda (praniza and adults), Polychaeta larvae, and Decapod larvae, Polychaeta larvae, Gastropoda larvae, Bivalvia larvae	*Oithona nana, Oithona similis, Oithona plumifera, Acartia tonsa, Acartia clausi, Acartia bifilosa var. inermis, Euterpina acutifrons, Oithona similis, Temora longicornis, Clausocalanus arcuicornis, Paracalanus parvus, Calanus helgolandicus, harpacticoida n.id.*	Autumn and spring (October and May or October/November and April/May)	Ribeiro 1989, 1991, Ribeiro et al. 1996, Soares 1996, Gonçalves 1991, Gonçalves et al. 2003, Azeiteiro 1999, Azeiteiro et al. 1999, Azeiteiro et al. 1999, Azeiteiro et al. 2000, 2005, Vieira et al. 2002a,b, 2003a, b, Morgado et al. 2007, Pastorinho et al. 2003a, b, Primo et al. 2009, 2011, 2012, Gonçalves et al. 2010a, b, 2011, 2012a, b, c, Falcão et al. 2012, Gonçalves et al. 2015

5. Final Remarks

The region of the Iberian Peninsula is a transition zone between two distinct oceanographic regimes, where the western coast, which comprises the Galicia and Portugal coasts, is under the influence of the Canary-North African upwelling, characterised by a succession of nutrient inputs during most of spring and summer due to northern and northeastern winds (Arístegui et al. 2006). This region has shown during the last decades gradients of environmental changes that are reflected in plankton communities. The seasonal pattern of the mesozooplankton in off-shelf areas sees an annual maximum in spring and a secondary peak in late summer and early autumn. This pattern is modified in many coastal areas by the summer upwelling (Western Iberian shelf) and river plumes, which strongly influence shelf areas across the ecoregion. The seasonal upwelling and the plumes enhance zooplankton growth in summer and the observed bimodal pattern in this area (ICES 2020). The copepods *Acartia clausi* and *Paracalanus parvus* are the most abundant species of mesozooplankton in most areas of the Western Iberia coast, and the macrozooplankton species *Calanus helgolandicus* is also important in terms of biomass and the distribution of warm-water copepod species like *Temora stylifera* and *Calanoides carinatus* has moved northwards across the ecoregion (ICES 2020). The fact that throughout the year, there are high densities of nauplii and copepodites of several species of Copepoda indicates that the reproductive activity of this group occurs continuously. The density of most neritic species is significantly correlated with water salinity, while that of estuarine species is significantly correlated with water temperature.

From the point of view of geographic distribution, the dominant species in northern Portuguese coast and estuaries, mainly neritic and coastal, are mentioned for subtropical or temperate zones that occur with a wide distribution along the western basin of the Mediterranean and the Adriatic, the Atlantic coast of the United States of America and the North Atlantic. The abundance of neritic and coastal species points to the transport by the tidal current as one of the mechanisms responsible for the fluctuations in zooplankton density in estuarine areas. The zooplankton composition allows for the description of three types of oceanographic influences in the northern Portuguese estuaries: the Atlantic Contingent that comprises steno or euryhaline marine organisms, occurring in the outermost areas and only sporadically within the estuaries and whose presence is associated with oceanic water masses. The neritic contingent has a wide distribution, formed by euryhaline marine organisms, which penetrate estuaries to find favourable conditions for the development of their life cycles and where they can occur with high density, presenting different longitudinal distributions depending on their respective saline preference. The estuarine contingent comprises truly estuarine, steno or oligohaline organisms, which occur in the interior or upstream of the estuaries, seasonally or throughout the year.

References

Almaça, C. 1966. Publicações do prof. Dr. augusto nobre sobre oceanografia biológica. Porto: Imprensa Portuguesa.

Almaça, C. 1997. Augusto nobre and marine biology in Portugal. pp. 125–134. *In*: Saldanha, Luís and Ré, Pedro (ed.). One Hundred Years of Portuguese Oceanography: in the footsteps of King Carlos de Bragança. Lisboa: Museu Bocage.

Almaça, C. 2000. Museu bocage: ensino e exibição. Lisboa: Museu Bocage.

Alvarino, A. 1957. Zooplancton del atlantico ibérico campaña del "xauen" en el verano de 1954. Boletin Espanol de Oceanografia 82: 1–51.

Alvarino, A. 1980. Distribution of Zooplankton Predators and Anchovy Larvae. CalCOFI Report, Vol. XXI.

Ambar, I. and Fiúza, A.F.G. 1994. Some features of the Portugal current system: a poleward slope undercurrent, an upwelling-related summer southward flow and an autumn–winter poleward coastal surface current. pp. 286–287. *In*: Katsaros, K.B., Fiúza, A.F.G. and Ambar, I. (eds.). Proceedings of the Second International Conference on Air–Sea Interaction and on Meteorology and Oceanography of the Coastal Zone. American Meteorological Society.

Amorim, I. 2005. A pesca pacificada ou os primórdios da questão das pescarias no quadro de uma reflexão sobre os recursos naturais. Geol. Nova 11: 103–124.

Amorim, I. and Pinto, B. 2019. Portugal in the European network of marine science heritage and outreach (19th–20th centuries). Humanities 8(1): 2019. Available at: https://www.mdpi.com/2076-0787/8/1/14.

Andrade, C. and Freitas, M.C. 2002. Coastal zones. pp. 173–2019. *In*: Santos F.D., Forbes, K. and Moita, R. (eds.). Climate Change in Portugal. Scenarios, Impacts and Adaptation Measures - SIAM project, 5 Gradiva, Liboa, Portugal.

Andres, H.-G., John, H.-C. and Ré, P. 1992. Fish larvae and gammaridae plankton off northern Portugal during autumn 1987. Senckenbergiana Maritima 22(3/6): 179/201.

Aristegui, J., Alvarez-Salgado, X.A., Barton, E.D., Hernandez-Leon, S., Roy, C. and Santos, A.M. 2006. Chapter 23: oceanography and fisheries of the canary current/Iberian region of the eastern North Atlantic (18a,E).

Azeiteiro, U.M. and Morgado, F.M. 1996. A comparison of the macrozooplankton in two different channels of Ria de Aveiro (Northern Portugal). Ciênc Biol. Ecol. Syst. 16(1/2): 61–74.

Azeiteiro, U.M.M. 1999. Ecologia pelágica do braço sul do estuário do rio mondego. Tese de Doutoramento, FCT, Universidade de Coimbra. 220 p.

Azeiteiro, U.M., Marques, J.C. and Ré. 1999. Zooplankton annual cycle in the mondego river estuary (Portugal). Arquivos do Museu Bocage III 8: 239–264.

Azeiteiro, U.M., Marques, J.C. and Ré, P. 2000. Zooplankton assemblages in a shallow seasonally tidal estuary in temperate atlantic ocean (western Portugal: mondego estuary). Arquivos do Museu Bocage III 12: 57–376.

Azeiteiro, U.M., Marques, S.C., Vieira, L.R., Pastorinho, R., Ré, P., Pereira, M.J. et al. 2005. Dynamics of the *Acartia* genus (calanoida: copepoda) in a temperate shallow estuary (the mondego estuary: western coast of Portugal). Acta Adriat 46: 7–20.

Azevedo, I., Ramos, S., Mucha, A.P. and Bordalo, A.A. 2013. Applicability of ecological assessment tools for management decision-making: a case study from the Lima estuary (NW Portugal) Ocean Coast. Manag. 72: 54–63.

Baeta, A., Vieira, L.R., Lírio, A.V., Canhoto, C., Marques, J.C. and Guilhermino, L. 2017. Use of stable isotope ratios of fish larvae as indicators to assess diets and patterns of anthropogenic nitrogen pollution in estuarine ecosystems. Ecological Indicators 83: 112–121.

Bernardo, L. 2013. Cultura científica em Portugal: uma perspectiva histórica. Porto: Uporto.

Bessa Pacheco, M. 2013. Medidas da terra e do mar. Instituto Hidrográfico, Lisboa.

Bettencourt, A.M., Bricker, S.B., Ferreira, J.G., Franco, A., Marques, J.C., Melo, J.J. et al. 2004. Typology and reference conditions for portuguese transitional and coastal waters. Final report of project TICOR - development of guidelines for the application of the European Union Water Framework Directive. IMAR/INAG, Lisboa. ISBN 972-9412-67-7. 100 pp.

Bragança, Carlos de. Ichthyologia II: esqualos obtidos nas costas de Portugal durante as campanhas de 1896 a 1903. Lisboa: Imprensa Nacional, 1904.

Bragança, Carlos de. Pescas marítimas I: a pesca do atum no Algarve em 1898. Lisboa: Imprensa Nacional, 1899.

Bocage, B.du. 1868. Considerações acerca do melhor aproveitamento das ostreiras da margem esquerda do tejo e da cultura das nossas ostras. Lisboa: Tipografia da Academia.

Bocage, B.du. 1864. Note sur la découverte d'un zoophyte de la famille hyalochaetides sur la côte du Portugal. Proceedings of the Zoological Society of London [v.1864]: 265–269.

Bode, A., Alvarez-Ossorio, M.T., Cunha, M.E., Garrido, S., Peleteiro, J.B., Porteiro, C. et al. 2007. Stable nitrogen isotope studies of the pelagic food web on the atlantic shelf of the Iberian Peninsula. Prog. Oceanogr. 74: 115–131.

Borges, M., Santos, A., Crato, N., Mendes, H. and Mota, B. 2003. Sardine regime shifts off Portugal: a time series analysis of catches and wind conditions. Sci. Mar. 67: 235–244.

Bragança, C.de. 1904. Ichthyologia II: esqualos obtidos nas costas de Portugal durante as campanhas de 1896 a 1903, Lisboa: Imprensa Nacional.

Bragança, C.de. 1899. Pescas marítimas I: a pesca do atum no algarve em 1898. Lisboa: Imprensa Nacional.

Campuzano, F.J., Juliano, M., Sobrinho, J., de Pablo, H., Brito, D., Fernandes, R. et al. 2018. Coupling watersheds, estuaries and regional oceanography through numerical modelling in the western Iberia: thermohaline flux variability at the ocean-estuary Interface; Froneman, E.W. (ed.). IntechOpen: London, UK, pp. 1–17. https://doi.org/10.5772/intechopen.72162.

Candeias, A. 1926. Première liste des copépodes des côtes du Portugal. Bull. Soc. Port. Naturelle 10: 23–55.

Candeias, A. 1930. Estudos de plancton da baía de sezimbra. Travaux de la Station de Biol. Mar. de Lisbonne, p. 1–73, est. I-VI Lisboa.

Cardoso, P.G., Marques, S.C., D'Ambrosio, M., Pereira, E., Duarte, A.C., Azeiteiro, U.M. et al. 2013. Changes in zooplankton communities along a mercury contamination gradient in a coastal lagoon (Ria de Aveiro, Portugal). Mar. Pollut. Bull. 76: 170–177.

Carmo, V., Santos, M., Menezes, G.M., Loureiro, C.M., Lambardi, P. and Martins, A. 2013. Variability of zooplankton communities at condor seamount and surrounding areas, Azores (NE Atlantic). Deep-Sea Research II 98: 63–74.

Carrisso, L. 1911. Materiais para o estudo do plâncton da costa portuguesa I. Boletim da Sociedade Broteriana 26: 5–42.

Castro, C.G., Pérez, F.F., Álvarez-Salgado, X.A. and Fraga, F. 2000. Coupling between the thermohaline, chemical and biological fields during two contrasting upwelling events off the NW Iberian Peninsula. Continental Shelf Research 20(2): 189–210.

Chícharo, M. 1998. Nutritional condition and starvation in *Sardina pilchardus* (L.) larvae off southern Portugal compared with some environmental factors. Journal of Experimental Marine Biology and Ecology 225(1): 123–137.

Chícharo, M.A., Chícharo, L. and Morais, P. 2006. Inter-annual differences of ichthyofauna structure of the guadiana estuary and adjacent coastal area (SE Portugal/SW Spain): Before and after Alqueva dam construction. Estuarine Coastal and Shelf Science 70: 39–56.

Comissão Estratégica dos Oceanos. 2004. Relatório da comissão estratégica dos oceanos, 1.ª ed., Junho 2004.

Comissão Europeia. 2000. Directiva 2000/60/CE do Parlamento Europeu e do Concelho de 23 de Outubro de 2000, que estabelece um Quadro de Acção Comunitária no Domínio da Politica da Água. J. Ofic. Com. Europeias L 327: 1–72.

Comissão Mundial Independente para os Oceanos. 1998. O oceano nosso futuro – relatório da comissão mundial independente para os oceanos (1998), Expo 98/Fundação Mário Soares, Lisboa: 247 pp.

Costa, M., Silva, R. and Vitorino, J. 2001. Contribution to the knowledge of the wave climate in the portuguese coast, sines, Portugal. Actas das 2as Jornadas Portuguesas de Engenharia Costeira e Portuária. Associação Internacional de Navegação.

Cowles, T.J. and Fessenden, L.M. 1995. Copepod grazing and fine-scale distribution patterns during the marine light-mixed layers experiment. Journal of Geophysical Research 100: 667–6786.

Christou, E.D. and Verriopoulos, G.C. 1993. Analysis of the biological cycle of acartia clausi (Copepoda) in a meso-oligotrophic coastal area of the eastern Mediterranean Sea using time-series analysis. Mar. Biol. 115: 643e651.

Cruz, J. 2008. Ocean wave energy. Current Status and Future Perspectives. Green Energy and Technology, Springer, Berlin, pg. 431.

Cunha, M.E. and Figueiredo, I. 1988. Reproductive cycle of S. pilchardus in the central region off the Portuguese coast (1971/1987). ICES, C.M. 1988/H:61: 1–30.

Cunha, M.E. 1993. Variabilidade estacional do zooplancton na plataforma continental portuguesa. Boletim Uca 1: 229–241.

Dam, H.G., Miller, C.A. and Jonasdottir, S.H. 1993. The trophic role of mesozooplankton at 47°N, 20°W during the north Atlantic bloom experiment. Deep-Sea Research Part II 40: 197–212.

Deacon, M. 1997. British marine scientists in portuguese seas, 1868–1870. pp. 65–110. *In*: Saldanha, Luís and Ré, Pedro (eds.). One hundred years of Portuguese oceanography: in the footsteps of King Carlos de Bragança. Lisboa: Museu Bocage.

DGEEC. 2011. Sumários estatísticos IPCTN10 (inquérito ao potencial científico e tecnológico nacional), lisboa, DGEEC.

Dias, J.A. 2005. Evolução da zona costeira portuguesa: forçamentos antrópicos e naturais, encontros científicos. Turismo, Gestão, Fiscalidade 1: 7–27.

Dias, M.L., Olsen, K. and Østedt, O.J. 1976. Report on a cruise by the R.V. "G. O.SarsH to the acores and the coast of Portugal November/December 1975. Coun. Meet. Int. Coun. Explor. Sea (J:12): 1–17.

Direção-Geral de Política do Mar (DGPM). 2017. Conhecimento do mar, mapa da ciência e tecnologias do mar em Portugal. Lisboa.

Direção-Geral de Política do Mar (DGPM). 2020a. Agenda 2030/ODS14. Disponível em https://www.dgpm.mm.gov.pt/agenda-2030 [Consultado em: 16 abril 2020].

Direção-Geral de Política do Mar (DGPM). 2020b. Observatório da economia azul. Disponível em https://www.dgpm.mm.gov.pt/observatorio [Consultado em: 16 abril 2020].

Dos Santos, A. and Garrido, S. 2003. A bibliometric study of Portuguese plankton literature: a preliminary analysis. Relat. Cient. Téc. IPIMAR, Série digital (http://ipimar-iniap.ipimar.pt), nº 2: 9 pp.

dos Santos, A., Santos, A.M. and Conway, D.V.P. 2007. Horizontal and vertical distribution of cirripede cyprid larvae in an upwelling system off the Portuguese coast, Mar. Ecol. Prog. Ser. 329: 145–155.

dos Santos, A., Santos, A.M.P., Conway, D.V.P., Bartilotti C., Lourenço P. and Queiroga H. 2008. Diel vertical migration of decapod larvae in the portuguese coastal upwelling ecosystem: implications for offshore transport, Mar. Ecol. Prog. Ser. 359: 171–183.

DR. 2014a. Resolution of the council of ministers 12, 2014. Resolução do Conselho de Ministros no. 12/2014 de 12 de Fevereiro. Diário da República, I série 30: 1310–1336.

DR. 2014b. Law 17, 2014. Lei no. 17/2014 de 10 de abril. Diário da república, I série, no. 71: 2358–2362.

Falcão, J., Marques, S.C., Pardal, M.A., Marques, J.C., Primo, A.L. and Azeiteiro, U.M. 2012. Mesozooplankton structural responses in a shallow temperate estuary following restoration measures. Estuar. Coast. Shelf Sci. 112: 23–30.

Faria, A., Morais, P. and Chícharo, M.A. 2006. Icthyoplankton dynamics in the guadiana estuary and adjacent coastal area, south-east Portugal. Estuarine Coastal and Shelf Science 70: 85–97.

Fatela, F., Moreno, J., Leorri, E. and Corbett, R. 2014. High marsh foraminiferal assemblages' response to intra-decadal and multi-decadal precipitation variability, between 1934 and 2010 (Minho, NW Portugal). J. Sea Res. 93: 118–132.

Ferreira, J.G., Nobre, A.M., Simas, T.C., Silva, M.C., Newton, A., Bricker, S.B. et al. 2005. A methodology for defining homogeneous water bodies in estuaries–application to the transitional systems of the EU Water framework directive. Estuarine, Coastal and Shelf Science 66(3/4): 468–482.

Ferreira, O. and Matias, A. 2013. Portugal. *In*: Williams, A. and Pranzini, E. (eds.). Coastal Erosion and Protection in Europe, Routledge, 457pp, doi:10.4324/9780203128558.

Ferreira, M.A. 2016. Evaluating Performance of Portuguese Marine Spatial Planning. Ph.D. thesis, FCSH/UNL, Lisbon.

Figueiredo, I.M. and Santos, A.M.P. 1989. Reproductive biology of Sardina pilchardus (Walb.): seasonal maturity evolution (1986 to 1988). ICES C.M. 1989/H: 40: 4.

Fiúza, A.F.G. 1982. The Portuguese coastal upwelling system. pp. 45–71. *In*: Actual Problems of Ocenography in Portugal, JNICT/NATO Marine Sciences Panel, Junta Nacional de Investigação Científica e Tecnológica, Lisboa.

Fiúza, A.FD., de Macedo, M.E.M. and Guerreiro, R. 1982. Climatological space and time variation of the portuguese coastal upwelling. Oceanologica Acta 5(1): 31–40.

Fiúza, A.F.G. 1984. Hidrologia e dinâmica das águas costeiras de Portugal. Ph.D. thesis, University of Lisbon.

Fiúza, A.F.G., Hamann, M., Ambar, I., Del Rio, G.D., González, N. and Cabanas, J.M. 1998. Water masses and their circulation off western Iberia during May 1993. Deep-Sea Research Part I-Oceanographic Research Papers 45: 1127–1160.

Fórum Oceano. 2020. Desafios do mar 2030. Fórum Oceano – Associação da Economia do Mar. janeiro de 2020.

Fraga, F. 1992. Water Masses in the upper and middle north Atlantic ocean east of the azores (3).

Freitas, R., Romeiras, M., Silva, L., Cordeiro, R., Madeira, P., González, J.A. et al. 2019. Restructuring of the 'Macaronesia' biogeographic unit: a marine multi-taxon biogeographical approach. Sci. Rep. 9: 15792. doi: 10.1038/s41598-019-51786-6.

Frias, J.P.G.L., Otero, V. and Sobral, P. 2014. Evidence of microplastics in samples of zooplankton from portuguese coastal waters. Mar. Environ. Res. 95: 89–95.

Frouin, R., Fiúza, A.F.G., Ambar, I. and Boyd, T.J. 1990. Observations of a poleward surface current off the coasts of Portugal and spain during winter. Journal of Geophysical Research 95(C1): 679.

Gaard, E., Gislason, A., Falkenhaug, T., Søiland, H., Musaeva, E., Vereshchaka, A. et al. 2008. Horizontal and vertical copepod distribution and abundance on the mid-atlantic ridge in June 2004. Deep-Sea Res. II 55: 59–71.

Gago, J.M. 1990. Manifesto para a ciência em Portugal: ensaio, Gradiva, 1990.

Garrido, S., Santos A.M.P., dos Santos A. and Ré, P. 2009. Spatial distribution and vertical migrations of fish larvae communities off northwestern Iberia sampled with LHPR and Bongo nets. Estuar. Coast. Shelf Sci. 84(4): 463–475.

Garrido, S., Ben-Hamadou R., Santos A.M.P., Ferreira S., Teodósio M.A., Cotano U. et al. 2015. Born small, die young: intrinsic, size-selective mortality in marine larval fish. Sci. Rep. 5: 17065.

Gonçalves, F.J.M. 1991. Zooplâncton e ecologia larvar de crustáceos decápodes no estuário do rio Mondego. PhD thesis, University of Coimbra, Coimbra, Portugal. p.330.

Gonçalves, M.E. 1996. Mitos e realidades da política científica portuguesa, revista crítica de ciências sociais, nº 46, Centro de Estudos Sociais, pp. 47–67.

Gonçalves, F., Ribeiro, R. and Soares, A.M.V.M. 2003. Comparison between two lunar situations on emission and larval transport of decapod larvae in the Mondego estuary (Portugal). Acta Oecologica 24S: S183–S190.

Gonçalves, A.M.M., De Troch, M., Marques, S.C., Pardal, M.A. and Azeiteiro, U.M. 2010a. Spatial and temporal distribution of harpacticoid copepods in Mondego estuary. Journal of Marine Biological Association of the United Kingdom 90(7): 1279–1290.

Gonçalves, A.M.M., Pardal, M.A., Marques, S.C., De Troch, M. and Azeiteiro, U.M. 2010b. Distribution and composition of small-size zooplankton fraction in a temperate shallow estuary (western Portugal). Fresenius Environmental Bulletin 19(12b): 3160–3176.

Gonçalves, A.M.S.M. 2011. Small-sized Zooplankton in the Mondego Estuary Distribution and Trophic Ecology. Tese de Doutoramento, Universidade de Coimbra, 186 pp.

Gonçalves, A.M.M., Pardal, M.A., Marques, S.C., Mendes, S., Fernández-Gómez, M.J., Galindo-Villardón, M.P. et al. 2012a. Response to Climatic variability of Copepoda life history stages in a southern European temperate estuary. Zoological Studies 51(3): 321–335.

Gonçalves, A.M.M., Pardal, M.A., Marques, S.C., Mendes, S., Fernández-Gómez, M.J., Galindo-Villardón, M.P. et al. 2012b. Diel vertical behavior of copepoda community (naupliar, copepodites and adults) at the boundary of a temperate estuary and coastal waters. Estuarine, Coastal and Shelf Science 98: 16–30.

Goncalves, A.M.M., Azeiteiro, U.M., Pardal, M.A. and De Troch, M. 2012c. Fatty acid profiling reveals seasonal and spatial shifts in zooplankton diet in a temperate estuary. Estuarine, Coastal and Shelf Science. https://publons.com/publon/2330802/.10.1016/J.ECSS.2012.05.020.

Gonçalves, D.A., Marques, S.C., Primo, A.L., Martinho, F., Bordalo, M.D-B. and Pardal, M.A. 2015. Mesozooplankton biomass and copepod estimated production in a temperate estuary (Mondego estuary): effects of processes operating at different timescales. Zool. Stud. 54: 57.

Governo português. 2006. Estratégia nacional para o mar 2013–2020. 73 pp.

Governo Português. 2016. Programa operacional mar 2020 (resolução do conselho de ministros n.º 13/2016 de 16 de março), Lisboa: Diário da República.

Governo Português. 2019. Plano de situação de ordenamento do espaço marítimo nacional para as subdivisões continente, madeira e plataforma continental estendida (resolução do conselho de ministros n.º 203-A/2019 de 30 de dezembro), Lisboa: Diário da República. 105

Gueroun, Sonia K., Susanne Schäfer, Francesca Gizzi, Soledad Álvarez, João G. Monteiro, Carlos Andrade et al. 2021. Planktonic ctenophora of the madeira archipelago (northeastern Atlantic). Zootaxa 5081(3): 433–443.

Harris, R.P. 1982. Comparison of the feeding behaviour of calanus and pseudocalanus in two experimentally manipulated enclosed systems. Journal of the Marine Biological Association of the United Kingdom 62: 71–91.

Haynes, R. and Barton, E.D. 1990. A poleward flow along the atlantic coast of the Iberian Peninsula. Journal of Geophysical Research 95: 425–41.

Haynes, R., Barton, E.D. and Pilling, I. 1993. Development, persistence, and variability of upwelling filaments off the atlantic coast of the Iberian Peninsula. Journal of Geophysical Research 98(C12): 22681.

Head, R.N., Harris, R.P., Bonnet, D. and Irigoien, X. 1999. A comparative study of size fractionated mesozooplankton biomass and grazing in the North East Atlantic. Journal of Plankton Research 21: 2285–2308.

Head, R.N., Medina, G., Huskin, I., Anadon, R. and Harris, R.P. 2002. Phytoplankton and mesozooplankton distribution and composition during transects of the Azores Subtropical Front. Deep Sea Research Part II: Topical Studies in Oceanography 49(19): 4023–4034.

Hoinka, B.K.P. and Castro, M.D.E. 2003. The Iberian Peninsula Thermal Low. 1491–1511.

Huskin, I., Anadon, R., Medina, G., Head, R.N. and Harris, R.P. 2001. Mesozooplankton distribution and copepod grazing in the subtropical Atlantic near the Azores; influence of mesoscale structures. Journal of Plankton Research 23: 671–691.

Huskin, I., Viesca, L. and Anadón, R. 2004. Particle flux in the subtropical Atlantic near the Azores: influence of mesozooplankton. J. Plankton Res. 26(4): 403–415.

ICES. 2019. Azores ecoregion – Ecosystem overview. In Report of the ICES Advisory Committee, 2019. ICES Advice 2019, Section 3.1. https://doi.org/10.17895/ices.advice.5753.

ICES. 2020. ICES 2020 Bay of biscay and the Iberian coast ecoregion fisheries overview - data output file. Data Outputs. https://doi.org/10.17895/ices.data.7609.

INIP. 1986. Relatorio de dados de plâncton, 1986. Projecto Ciclos de produção planctonica na Costa Portuguesa e sua relação com os recursos pesqueiros. Cruzeiro 02040386-Março/Abril 1986.

International Year of Biodiversity. 2010. Biodiversity is Life, Biodiversity is Our Life. Biodiversity and UNESCO: Human Well-being through Science, Culture, Education and Communication. Unesco editions.

IOC-UNESCO. 2020. Global Ocean Science Report 2020–Charting Capacity for Ocean Sustainability. K. Isensee (ed.). Paris, UNESCO Publishing.

IPCC. 2014. Climate change 2014: Impacts, Adaptation, and Vulnerability. Part b: Regional Aspects. Contribution of Working Group II to the fifth Assessment Report of the Intergovernmental Panel on Climate Change, Barros, V.R. et al. (eds.). Cambridge University Press, Cambridge, United Kingdom and New York, NY, USA, 688 pp.

Jardim, M.E., Peres, I., Ré, P. and Costa, F. 2014. A prática oceanográfica e a coleção iconográfica do rei dom Carlos I. História, Ciências, Saúde – Manguinhos 21(3): 883–909. https://doi.org/10.1590/S0104-59702014000300006.

John, H.C. and Ré, P. 1993. Cross-shelf zonation, vertical distribution and drift of fish larvae off northern Portugal during weak upwelling. ICES, C.M. 1993/L:33: Session O: 18pp.

John, H.-C., Ré, P. and Zuelicke, C. 1996. Sardine larvae in a spring-upwelling event off northern Portugal. cienc. Biol. Ecol. Syst. 16: 193–198.

Jorge da Silva, A. 1992. Dependence of upwelling related circulation on wind forcing and stratification over the Portuguese northern shelf. ICES C. M., 1992/C: 17.

Kahru, M., Nomman, S. and Zeitzschel, B. 1991. Particle (plankton) size structure across the Azores front (Joint Global Ocean Flux Study, North Atlantic Bloom Study). Journal of Geophysical Research 96: 7083–7088.

Kaufmann, M., Santos, F. and Maranhão, M. 2015. Checklist of nanno- and microphytoplankton off Madeira Island (northeast Atlantic) with some historical notes. Nova Hedwigia 101: 205–232.

Langerhans, P. 1880a. Die wurmfauna von Madeira. III. Z. Wiss. Zool. 34: 87–136.

Langerhans, P. 1880b. Uber Madeira's appendicularien. Z. Wiss. Zool. 34: 144–146.

Leandro, S.M., Tiselius, P. and Queiroga, H. 2006. Growth and development of nauplii and copepodites of the estuarine copepod Acartia tonsa from southern Europe (Ria de Aveiro, Portugal) under saturating food conditions. Marine Biology 150: 121–129.

Leandro, S.M., Morgado, F., Pereira, F. and Queiroga, H. 2007. Temporal changes of abundance, biomass and production of copepod community in a shallow temperate estuary (Ria de Aveiro, Portugal). Estuarine, Coastal and Shelf Science 74: 215–222.

Leandro, S.M., Tiselius, P., Marques, S.C., Avelelas, F., Correia, C., Sa, P. et al. 2014. Copepod production estimated by combining in situ data and specific temperature-dependent somatic growth models. Hydrobiologia.

Longhurst, A. 1995. Seasonal cycles of pelagic production and consumption. Progress in Oceanography 36(2): 77–167.

Lopes, Q. 2017. A europeização de Portugal entre guerras: a Junta de Educação Nacional e a investigação científica. Casal de Cambra: Caleidoscópio.

Madruga, C. 2013. José vicente barbosa du bocage,1823–1907: a construção de uma persona científica. Dissertação (Mestrado em História e Filosofia das Ciências) – Universidade de Lisboa, Lisboa.

MAM. 2014. Programa de monitorização e programa de medidas da directiva-quadro estratégia Marinha: subdivisões continente, Açores, Madeira e Plataforma Continental Estendida. Governo de Portugal, Lisboa.

Marques, J.C., Graça, M.A. and Pardal, M.A. 2002. Introducing the mondego river basin. pp. 7–12. *In*: Pardal, M.A., Marques, J.C. and Graça, M.A. (eds.). Aquatic Ecology of the Mondego River Basin. Global Importance of Local Experience. Imprensa da Universidade, Coimbra.

Marques, S.C., Azeiteiro, U.M., Marques, J.C., Neto, J.M. and Pardal, M.A. 2006. Zooplankton and ichthyoplankton communities in a temperate estuary: spatial and temporal patterns. Journal of Plankton Research 28: 297–312.

Marques, S.C., Azeiteiro, U.M., Martinho, F. and Pardal, M.A. 2007a. Climate variability and planktonic communities: The effects of an extreme event (severe drought) in a southern European estuary. Estuarine, Coastal and Shelf Science 73: 725–734.

Marques, S.C., Pardal, M.A., Pereira, M.J., Gonçalves, F., Marques, J.C. and Azeiteiro U.M. 2007b. Zooplankton distribution and dynamics in a temperate shallow estuary. Hydrobiologia 587: 213–223.

Marques, S.C., Azeiteiro, U.M., Leandro, S.M., Queiroga, H., Primom A.L., Viegas, I. et al. 2008. Predicting zooplankton response to environmental changes in a temperate estuarine ecosystem. Marine Biology 155: 531–541.

Marques, S.C., Azeiteiro, U.M., Martinho, F., Viegas, I. and Pardal, M.A. 2009. Evaluation of estuarine mesozooplankton dynamics at a fine temporal scale: the role of seasonal, lunar, and diel cycles. Journal of Plankton Research 31: 1249–1263.

Marques, S.C.F.C. 2009. Structure of Estuarine Zooplankton Assemblages – Contribution of Environmental Factors and Climate Variability. PhD thesis, University of Coimbra, Coimbra, Portugal. p. 174.

Marques, S.C., Primo, A.L., Martinho, F., Azeiteiro, U.M. and Pardal, M.A. 2014. Shifts in estuarine zooplankton variability following extreme climate events: a comparison between drought and regular years. Mar. Ecol. Prog. Ser. 499: 65–76.

Marques, S.C., Pardal, M.Â., Primo, A.L., Martinho, F., Falcão, J., Azeiteiro, U.M. et al. 2018. Evidence for changes in estuarine zooplankton fostered by increased climate variance. Ecosystems 21: 56–67.

Marques, S.C., D'Ambrosio, M., Primo, A.L., Molinero, J. and Pardal. M.Â. 2019. Drivers of interannual abundance changes in gelatinous carnivore zooplankton in the Iberian Peninsula, (Portugal). Front. Mar. Sci. Conference Abstract: XX Iberian Symposium on Marine Biology Studies (SIEBM XX). doi: 10.3389/conf.fmars.2019.08.00187.

Marta-Almeida, M., Dubert, J., Peliz, A. and Queiroga, H. 2006. Influence of vertical migration pattern on retention of crab larvae in the shelf in a seasonal upwelling system. Mar. Ecol. Prog. Ser. 307: 1–19.

Martin, B. and Christiansen, B. 2009. Distribution of zooplankton biomass at three seamounts in the NE Atlantic. Deep Sea Research Part II: Topical Studies in Oceanography 56: 2671–2682.

Mendes, S., Fernández-Gómez, M.J., Resende, P., Pereira, M.J., Galindo-Villardón, M.P. and Azeiteiro, U.M. 2009. Spatio-temporal structure of diatom assemblages in a temperate estuary. A STATICO analysis. Estuarine, Coastal and Shelf Science 84: 637–644.

Meneses, I. and Ré, P. 1991. Infection of sardine eggs by a parasitic dinoflagellate Ichthyodinium chabelardi off Portugal. Boletim do Instituto Nacional de Investigação das Pescas 16: 63–72.

Menezes, G.M., Sigler, M.F., Silva, H.M. and Pinho, M.R. 2006. Structure and zonation of demersal fish assemblages off the Azores archipelago (mid-Atlantic). Mar. Ecol. Prog. Ser. 324: 241–260. doi: 10.3354/meps324241.

Miranda, A., Casa, G. and Bode, A. 2011. ICES zooplankton status report 2010/2011. ICES Cooperative Research Report. https://doi.org/10.17895/ices.pub.5487.

Moreno, J., Fatela, F., Cascalho, J., Moreno, F. and Drago, T. 2005. Living foraminiferal assemblages from the minho and coura estuaries (northern Portugal): a stressfull environment. Thalassas, an International Journal of Marine Sciences 21(1): 17–28.

Morgado, F. 1991. Zooplâncton da Ria de Aveiro. Composição e distribuição das comunidades do Canal de Mira num ciclo anual. Ver. Biol. U. Aveiro 4: 157–172.

Morgado, F. 1997. Ecologia do zooplâncton da Ria de Aveiro. Caracterização espacio-temporal, transporte longitudinal e dinâmica tidal, nictemeral e lunar. PhD thesis, Universidade de Aveiro, 385: pp.

Morgado, F., Melo, R., Queiroga, H. and Sorbe, J.C. 2003a. Zooplankton abundance in a coastal station off the Ria de Aveiro inlet (north-western Portugal): relation with tidal and day/night cycles. Acta Oecol. 24: 175–181.

Morgado, F., Antunes, C. and Pastorinho, R. 2003b. Distribution and patterns of emergence of suprabenthic and pelagic crustaceans from a shallow temperate estuary (Ria de Aveiro, Portugal). Acta Oecol. 24: 205–217.

Morgado, F., Pastorinho, R., Quintaneiro, C. and Ré, P. 2006a. Vertical distribution and trophic structure of the macrozooplankton in shallow temperate estuary (Ria de Aveiro, Portugal). Sci. Mar 70: 177–188.

Morgado, F., Ré, P., Silva, N. and Azeiteiro, U.M. 2006b. Comparison of the zooplankton from two different temperate tidal systems in western Portugal: the Mondego Estuary and Ria de Aveiro Lagoon. Int. J. Lakes. Rivers. 1: 65–74.

Morgado, F., Antunes, C., Rodrigues, E., Pastorinho, R., Vieira, L.R. and Azeiteiro, U.M. 2007. Composition and trophic structure of zooplankton in a shallow temperate estuary (Mondego Estuary, Western Portugal). Zool. Stud. 46: 57–68.

Morgado, F., Terdalkar, S., Gadelha, J.R. and Pereira, M.L. 2013. Histology and histochemistry of the reproductive potential of *Acartia clausi* (copepoda: calanoida). Microsc. Microanal. 19(Suppl 4): 91–92. ISSN 1431–9276.

Morgado, F., Vieira, L.R., Ré, P. and Soares, A.M.V.M. 2014. Atlas do zooplâncton estuarino e marinho da costa Atlântica. Colecção Biologicando, Editora Afrontamento, Porto, 167: pp.

Morgado, F., Posada, N.G., Chavez, M.G., Soares, A.M.V.M. and Lopez, M.A.G. 2015. Pattern recognition techniques for biological tissues differentiation in planktonic organisms. Microsc. Microanal. 21(S6): 72–73. ISSN 1435–8115.

Muzavor, S.N.X. 1981. Contribuição para o estudo do zooplâncton nas águas dos Açores. Arquipélago Série Ciências da Natureza 2: 153–163.

Narciso, Á., Caldeira, R., Reis, J., Hoppenrath, M., Cachão, M. and Kaufmann, M. 2019. The effect of a transient frontal zone on the spatial distribution of extant coccolithophores around the Madeira archipelago (northeast Atlantic). Estuarine, Coastal and Shelf Science 223: 25–38. https://doi.org/10.1016/j.ecss.2019.04.014.

Neto, T. and Paiva, I. 1984. Algumas considerações sobre o zooplâncton da costa Algarvia. 3° Congresso do Algarve 1(57): 443–441.

Neto, J.M., Caçador, I., Caetano, M., Chaínho, P., Costa, L., Gonçalves, A.M.M. et al. 2022. Capítulo 16, Estuários. pp. 381–421. *In*: Feio, M.J. and Ferreira, V. (eds.). Rios de Portugal comunidades, processos e alterações. Imprensa da Universidade de Coimbra.

Oliveira, I.B.M. 1997. Proteger ou não proteger ou sobre a viabilidade de diferentes opções face à erosão da costa oeste portuguesa. Compilation of Ideas on the Portuguese Coastal Zones, Portugal. Associação Eurocoast-Portugal, pp. 205–227.

Ottens, J.J. 1991. Planktonic foraminifera as north Atlantic water mass indicators. Oceanologica Acta 14: 123–140.

Pacheco, M.B. 2014. A geografia marítima de Portugal. pp. 25–35. *In*: O Mar no Futuro de Portugal: Ciência e Visão Estratégica, Centro de Estudos Estratégicos do Atlântico.

Parlamento Europeu. 2018. Governação internacional dos oceanos: uma agenda para o futuro dos nossos oceanos no contexto dos objetivos do desenvolvimento sustentável para 2030, Resolução 2017/2055(INI) de 16.1.2018, Jornal Oficial da União Europeia, 19.12.2018, Bruxelas.

Pastorinho, M.R., Antunes, C.P., Marques, J.C., Pereira, M.L., Azeiteiro, U.M.M. and Morgado, F.M. 2003a. Histochemistry and histology in planktonic ecophysiological processes determination in a temperate estuary (Mondego River Estuary, Portugal). Acta Oecologica 24: 235–243.

Pastorinho, R., Vieira, L., Ré, P., Pereira, M., Bacelar-Nicolau, P., Morgado, F. et al. 2003b. Distribution, production, histology and histochemistry in Acartia tonsa (copepoda: calanoida) as means for life history determination in a temperate estuary (Mondego estuary, Portugal). Acta Oecologica 24: S259–S273.

Paulino, S., Sam-Bento, F., Churro, C., Alverca, E., Dias, E., Valério, E. et al. 2009. The estela sousa e silva algal culture collection: a resource of biological and toxicological interest. Hydrobiologia 636: 489–492. https://doi.org/10.1007/s10750-009-9977-4.

Pham, C.K., Vandeperre, F., Menezes, G., Porteiro, F., Isidro, E. and Morato, T. 2015. The importance of deep-sea vulnerable marine ecosystems for demersal fish in the Azores. Deep Sea Research Part I: Oceanographic Research Papers 96: 80–88.

Peliz, Á., Rosa, T.L., Santos, A.M.P. and Pissarra, J.L. 2002. Fronts, jets, and counter-flows in the western Iberian uupwelling system. Journal of Marine Systems 35(1–2): 61–77.

Peliz, A., Dubert, J., Santos, A.M.P., Oliveira, P.B. and Cann, B.L. 2005. Winter upper ocean circulation in the western Iberian basin - fronts, eddies and poleward flows: an overview. Deep-Sea Research Part I: Oceanographic Research Papers 52(4): 621–46.

Peliz, Á., Marchesiello, P., Dubert, J., Marta-Almeida, M., Roy, C. and Queiroga, H. 2007. A study of crab larvae dispersal on the Western Iberian Shelf: Physical processes. Journal of Marine Systems 68: 215–236.

Pereira, F., Pereira, R. and Queiroga, H. 2000. Flux of decapod larvae and juveniles at a station in the lower canal de mira (Ria de Aveiro, Portugal) during one lunar month. Invertebrate Reproduction and Development 38: 183–20.

Pérez, F.F., Castro, C.G., Álvarez-Salgado, X.A. and Rios, A.F. 2001. Coupling between the Iberian basin-scale circulation and the Portugal boundary current system: a chemical study. Deep-Sea Research Part I-Oceanographic Research Papers 48: 1519–1533.

Pinheiro, A. and Fidalgo L. 2000. Contribution for the study of bacterioplankton of the river minho estuary (Portugal). Verh. Internat. Verein. Limnol. 27: 1890–1893.

Pinto, J. DosS. and Silva, E.DeS. 1948. O plancton da baía de S. martinho do porto- I. diatomáceas e dinoflagelados. Separata do Bol. da Soc. Port, de Cien. Nat., pp. 1–41, est. I-VI. Lisboa.

Pinto, J. DosS. 1950. Estudos de plancton, seu interesse científico e econômico. Junta de Investigações Coloniais, pp. 1–27, pr. I-V. Lisboa.

Pinto, R. and Martins, F.C. 2013. The Portuguese national strategy for integrated coastal zone management as a spatial planning instrument to climate change adaptation in the minho river estuary (Portugal NW-coastal zone). Environ. Sci. Policy 33: 76–96.

Pinto, B. 2017. Historical connections between early marine science research and dissemination: the case-study of aquarium vasco da gama (Portugal) from the late 19th century to mid-20th century. ICES: Journal of Marine Science 74(6): 1522–1530.

Prieto, R., Tobeña, M. and Silva, M.A. 2017. Habitat preferences of baleen whales in a mid-latitude habitat. Deep Sea Research Part II: Topical Studies in Oceanography 141C: 155–167.

Primo, A.L., Azeiteiro, U.M., Marques S.C., Martinho, F. and Pardal, M.A. 2009. Changes in zooplankton diversity and distribution pattern under varying precipitation regimes in a southern temperate estuary. Estuarine Coastal and Shelf Science 82: 341–347.

Primo, A.L., Azeiteiro, U.M., Marques, S.C. and Pardal, M.A. 2011. Impact of climate variability on ichthyoplankton communities: an example of a small temperate estuary. Estuar Coast Shelf Sci. 91: 484–91.

Primo, A.L., Azeiteiro, U.M., Marques, S.C., Ré, P. and Pardal, M.A. 2012. Vertical patterns of ichthyoplankton at the interface between a temperate estuary and adjacent coastal waters: seasonal relation to diel and tidal cycles. J. Mar. Syst. 95: 16–23.

Queiroga, H., Costlow, Jr. J.D. and Moreira, M.H. 1994. Larval abundance patterns of carcinus maenas (decapoda, brachyura) in canal de mira (ria de aveiro, portugal). Marine Ecology Progress Series 111: 63–72.

Queiroga, H. 1996. Distribution and drift of the crab carcinus maenas (L.) (decapoda, portunidae) larvae over the continental shelf off northern Portugal in April of 1991. Journal Plankton Research 18: 1981–2000.

Queiroga, H., Costlow, Jr. J.D. and Moreira, M.H. 1997. Vertical migration of the crab carcinus maenas first zoeae: implications for tidal stream transport. Marine Ecology Progress Series 149: 121–132.

Queiroga, H., Silva, C., Sorbe, J.C. and Morgado, F. 2005. Zooplankton distribution across an upwelling front on the northern Portuguese coast. Hydrobiologia 545: 195–207.

Queiroga, H., Cruz, T., Dos Santos, A., Dubert, J., González-Gordillo, I., Paula, J. et al. 2007. Oceanographic and behavioural processes controlling invertebrate larval dispersal and recruitment in the western Iberia upwelling ecosystem. Progress in Oceanography 74(2–3): 174–191.

Ramos, S., Cowen, R.K., Paris, C., Ré, P. and Bordalo, A.A. 2006a. Environmental forcing and larval fish assemblage dynamics in the Lima river estuary (northwest Portugal). Journal of Plankton Research 28(3): 275–286.

Ramos, S., Cowen, R.K., Ré, P. and Bordalo, A.A. 2006b. Temporal and spatial distributions of larval fish assemblages in the lima estuary (Portugal). Estuarine, Coastal and Shelf Science 66: 30–314.

Ramos, S. 2007. Ichthyoplankton of the Lima estuary (NW Portugal): ecology of the early life stages of pleuronectiformes. Tese de Doutoramento, Universidade do Porto, 190: pp.

Ramos, S., Ré, P. and Bordalo, A.A. 2009a. Environmental control on early life stages of flatfishes in the Lima estuary (NW Portugal). Estuar. Coast. Shelf Sci. 83: 252–264.

Ramos, S., Ré, P. and Bordalo, A.A. 2009b. New insights into the early life ecology of sardina pilchardus (Walbaum, 1792) in the northern Iberian Atlantic. Scientia Marina 73(3): 449–459.

Ramos, S., Ré, P. and Bordalo, A.A. 2010. Recruitment of flatfish species to an estuarine nursery habitat (Lima estuary, NW Iberian Peninsula). Journal of Sea Research 64(4): 473–486.

Ramos, S., Paris, C.B. and Angélico, M.M. 2017. Larval fish dispersal along an estuarine–ocean gradient. Canadian Journal of Fisheries and Aquatic Sciences 74: 9.

Ré, P., Cabral e Silva, R., Cunha, E., Farinha, A., Meneses, I. and Moita, T. 1990. Sardine spawning off Portugal. Boletim do Instituto Nacional de Investigação das Pescas 15: 31–44.

Ré, P. 1979. The eggs and planktonic stages of portuguese marine fishes. I- Ichthyoplankton from the coast of Algarve (May, 1977). Arquivos do Museu Bocage (2ª série) 7(3): 23–51.

Ré, P. 1980. The eggs and newly hatched larvae of Abudefduf luridus (Cuvier, 1830) (Pisces; Pomacentridae) from the Azores. Arquivos do Museu Bocage (2ª ser) 7(8): 109–116.

Ré, P. 1981a. On the off Portugal. Boletim da Sociedade Portuguesa de Ciências Naturais 20: 67–69.

Ré, P. 1981b. Seasonal occurrence, mortality and dimensions of sardine eggs, Sardina pilchardus (Walb.), off Portugal. Cybium, (3ªser) 5(4): 41–48.

Ré, P., Farinha, A. and Meneses, I. 1982. Ichthyoplankton from the coast of Peniche (Portugal) (1979/80). Arquivos do Museu Bocage, (série A) 1(16): 369–402.

Ré, P., Farinha, A. and Meneses, I. 1983. Anchovy spawning in Portuguese estuaries (Engraulis encrasicolus, Pisces: Engraulidae). Cybium 7: 29–38.

Ré, P. 1984a. Evidence of daily and hourly growth in pilchard larvae based on otolith growth increments, Sardina pilchardus (Walbaum, 1792). Cybium 8(1): 33–38.

Ré, P. 1984b. Ictioplâncton do estuário do tejo. Resultados de 4 anos de estudo (1979/1981). Arquivos do Museu Bocage (série A) 2(9): 145–174.

Ré, P. 1986. Sobre a identificação dos primeiros estados larvares planctónicos de Sardina pilchardus (Walbaum, 1792) e de Engraulis encrasicolus (Linnaeus, 1758). Ciência Biológica- Ecologia e Sistemática (Portugal) 6 (1/2): 135–140.

Ré, P. 1986a. Ecologia da postura e da fase plantónica de sardina pilchardus (Walbaum, 1792) na região central da costa portuguesa. Boletim da Sociedade Portuguesa de Ciências Naturais 23: 5–81.

Ré, P. 1986b. Ecologia da postura e fase planctónica de Engraulis encrasicolus (Linnaeus, 1758) no estuário do Tejo. Publicações do Instituto de Zoologia "Dr. Augusto Nobre", 196: 45pp.

Ré, P. 1986c. Otolith microstructure and the detection of life history events in sardine and anchovy larvae. Ciência Biológica- Ecologia e Sistemática 6(1/2): 9–17.

Re, P. 1987. Ecology of the planktonic phase of the anchovy, Engraulis encrasicolus (L.), within Mira estuary (Portugal). Invesigaciones Pesqueras 51: 581–598.

Ré, P. 1989. Sardine diel spawning time off Portugal (Abstract). Rapport et Procés-Verbaux des Réunions du Conseil Permanent International pour l'Exploration de la Mer 191: 444–445.

Re, P., Cabral e Silva, R., Cunha, E., Farinha, A., Meneses, I. and Moita, T. 1990. Sardine spawning off Portugal. Boletim do Instituto Nacional de Investigação das Pescas 15: 31–44.

Ré, P. 1999. Ictioplâncton estuarino da Península Ibérica. Guia de identificação dos ovos e estados larvares planctónicos. Câmara Municipal de Cascais, Cascais, 163 pp.

Relvas, P. and Barton, E.D. 2002. Mesoscale patterns in the cape são vicente (Iberian Peninsula) upwelling region. Journal of Geophysical Research 107: 3164.

Relvas, P. and Barton, E. 2005. A separated jet and coastal counterflow during upwelling relaxation off cape são vicente (Iberian Peninsula). Continental Shelf Research 25: 29–49.

Relvas, P., Barton, E.D., Dubert, J., Oliveira, P.B., Peliz, A., Silva, J.C.B. et al. 2007. Physical oceanography of the western Iberian ecosystem: latest views and 213 challenges. Progress in Oceanography 74: 174–171.

Resolução do Conselho de Ministros nº 163/2006, de 12 de Dezembro (Estratégia Nacional para o Mar).

Ribeiro, R. 1989. Ictioplâncton do estuário do Mondego. Resultados preliminares (Fevereiro de 1988 a Janeiro de 1989). Revista de Biologia da Universidade de Aveiro 4: 233–243.

Ribeiro, R. 1991. Ictioplâncton do estuário do Mondego e Resultados. Revista de Biologia: Actas do IºEncontro de Planctologistas Portugueses 4: 233–244.

Ribeiro, R., Reis, J., Santos, C., Fernando Gonçalves, F. and Soares, A.M.V.M. 1996. Spawning of anchovy Engraulis encrasicolus in the Mondego estuary, Portugal. Estuarine, Coastal and Shelf Science 42: 467–482.

Rubal, M., Veiga, P., Cacabelos, E., Moreira, J. and Sousa-Pinto, I. 2013. Increasing Sea surface temperature and range shifts of intertidal gastropods along the Iberian Peninsula. J. Sea Res. 77: 1–10.

Ruivo, M. 1998. Uma nova política do mar para Portugal no novo regime dos oceanos. pp. 168–169. In: Janus – Anuário de Relações Exteriores, Lisboa: Público e Universidade Autónoma de Lisboa.

Ruivo, M. 2006a. Política marítima europeia 2 – ciência e governação do oceano. pp. 53–60. In: Europa – Novas Fronteiras, Revista do Centro de Informação Europeia Jacques Delors, nº 20, Julho/ Dezembro, Lisboa: Principia.

Ruivo, M. 2006b. Geopolítica dos recursos haliêuticos e cooperação internacional. pp. 19–31. In: Álvaro Garrido (coord.), A Economia Marítima Existe, Lisboa: Âncora Editora.

Saldanha, L. 1982. O rei D. Carlos pioneiro da Oceanografia Europeia. Revista da Marinha 110: 3–8.

Saldanha, L. 1984. Portugal - L'Appel de l'Océan. Océans 131: 32–37.

Saldanha, L. 1994. Os Mares de Sesimbra e as Primeiras Explorações Oceanográficas Europeias. Sesimbra Cultural, 4 -Dez/94. pp. 52–55.

Saldanha, L. 1997. King carlos de bragança, the father of portuguese oceanography. pp. 19–38. In: Saldanha, L. and Ré, P. (eds.). One Hundred Years of Portuguese Oceanography. In the footsteps of King Carlos de Bragança. Publicações Avulsas do Museu Bocage, 2ª Série, nº 2, Lisboa.

Saldanha, L. and Ré, P. (eds.). 1997. One hundred years of portuguese oceanography. In the footsteps of King Carlos de Bragança. Publicações avulsas do Museu Bocage (2ª série), 2: 453 pp.

Salgueiro, Â. 2017. Ciência e universidade na Primeira república. Casal de Cambra: Caleidoscópio.

Salgueiro, Â. 2021. Oceans, science, and universities: scientific study of the sea during the first Portuguese Republic História, Ciências, Saúde-Manguinhos 28(2): 473–489.

Sambolino, A., Herrera, I., Alvarez, S., Rosa, A., Alves, F., Canning-Clode, J. et al. 2022. Seasonal variation in microplastics and zooplankton abundances and characteristics: The ecological vulnerability of an oceanic island system. Marine Pollution Bulletin 181: 11390.

Santos, A., Borges, M. and Groom, S. 2001. Sardine and horse mackerel recruitment and upwelling off Portugal. ICES J. Mar. Sci. 58: 589–596.

Santos, A., Peliz, A., Dubert, J., Oliveira, P., Angelico, M. and Ré, P. 2004. Impact of a winter upwelling event on the distribution and transport of sardine (Sardina pilchardus) eggs and larvae off western Iberia: a retention mechanism. Cont. Shelf Res. 24: 149–165.

Santos, A.J.P., Nogueira, J. and Martins, H. 2005. Survival of sardine larvae off the Atlantic Portuguese coast: a preliminary numerical study. ICES Journal of Marine Science 62: 634–644.

Santos, A.M.P., Chicharo, A., Santos, A., Moita, T., Oliveira, P.B., Peliz, A. et al. 2007. Physical-biological interactions in the life history of small pelagic fish in the western Iberia upwelling ecosystem. Progress in Oceanography 74(2–3): 192–209.

Santos, A. and Santos, A. 2011. ICES zooplankton status report 2010/2011. ICES Cooperative Research Report. https://doi.org/10.17895/ices.pub.5487.

Saunders, P.M. 1982. Circulation in the eastern north Atlantic. Journal of Marine Research 40: 641–657.

Schiebel, R., Waniek, J. and Zeltner, A. 2002. Impact of the azores front on the distribution of planktic foraminifera, shelled gastropods, and coccolithophorids. Deep-Sea Research II 49(19): 4035–4050.

Silva, A.B.da, 1891. Estado actual das pescas em Portugal. Lisboa: Imprensa Nacional.

Silva, E.DeS. 1950. Les tintinnides de la baie de cascais. Bull, de l'Inst. Oceanographique 979: 1–28. (numeração da separata). Monaco.

Silva, E.S. 1952. Estudos de pâncton na lagoa de Óbidos. I. diatomáceas e dinoflagelados. Revista da Faculdade de Ciências de Lisboa 2C(2): 5–44.

Silva, E.S. 1959. Some observations on marine dinoflagellate cultures. I. *Prorocentrum micans* Ehr. and Gyrodinium sp. Notas e Estudos do Instituto de Biologia Marinha 21: 1–20.

Silva, E.S. 1962a. Some observations on marine dinoflagellates cultures. II. *Glenodinium foliaceum* Stein and *Goniaulax diacantha* (Meun.) Schiller. Botanica Marina 2: 75–100.

Silva, E.S. 1962b. Some observations on marine dinoflagellates cultures. III *Goniaulax spinifera* (Clap. And Lach.) Dies., *Goniaulax tamarensis* Leb. and *Peridinium trochoideum* (Stein) Lemm. Notas e Estudos do Instituto de Biologia Marinha 26: 1–21.

Silva, E.S. 1967. *Cochlodinium heterolobatum* n. sp.: structure and some cytophysiological aspects. Journal of Protozoology 14(4): 745–754.

Silva, E.S. 1978. Endonuclear bacteria in two species of dinoflagellates. Protistologica 14(2): 113–119.

Silva, R., Coelho, C., Veloso-Gomes, F. and Taveira-Pinto, F. 2007. Dynamic numerical simulation of medium-term coastal evolution of the west coast of Portugal, Journal of Coastal Research 50: 263–267.

Soares, A. 1996. Spawning of anchovy *Engraulis encrasicolus* in the Mondego estuary, Portugal. Estuarine Coastal and Shelf Science 42(4): 467–482.

Sobral, M., Cabeçadas, G., Ferreira, A.M., Sampaio, M.A., Lima, F. and Raminhos, A. 1985. Programa de apoio às pescas nos açores: cruzeiro 020100979. Instituto Nacional de Investigação das Pescas p. 91.

Sobrinho-Gonçalves, L., Moita, M.T., Garrido, S. and Cunha, M.E. 2013. Environmental forcing on the interactions of plankton communities across a continental shelf in the eastern Atlantic upwelling system. Hydrobiologia 713: 167–182.

Sobrinho-Gonçalves, L. and Isidro, E. 2001. Fish larvae and zooplankton biomass around faial Island (Azores archipelago). A Preliminary Study of Species Occurrence and Relative Abundance. Arquipélago.

Vasconcelos, R.P., Reis-Santos, P., Costa, M.J. and Cabral, H.N. 2011. Connectivity between estuaries and marine environment: Integrating metrics to assess estuarine nursery function. Ecol. Indicat. 11: 1123–1133.

Vieira, L., Azeiteiro, U.M., Pastorinho, R., Morgado, F., Bacelar-Nicolau, P., Marques, J.C. et al. 2002a. Condições físico-químicas, nutrientes, clorofila a e fitoplâncton no estuário do Mondego. pp. 113–132. *In*: Prego, R., Da Costa Duarte, A., Panteleitchouk, A. and Santos, T.R. (eds.). Estudos sobre Contaminação Ambiental na Península Ibérica, Instituto Piaget, Lisboa.

Vieira, L.R., Ré, P., Morgado, F., Pereira, M., Marques, J.C. and Azeiteiro, U.M. 2002b. Distribution and production of Acartia bifilosa var. inermis from a temperate estuary (Mondego Estuary, Portugal). Arquivos do Museu Bocage 17: 421–440.

Vieira, L.R., Azeiteiro, U.M., Ré, P., Pastorinho, R., Marques, J.C. and Morgado, F. 2003a. Zooplankton distribution in a temperate estuary (Mondego estuary southern arm: western Portugal). Acta Oecol. 24: 163–173.

Vieira, L.R., Morgado, F., Ré, P., Nogueira, A., Pastorinho, R., Pereira, M. et al. 2003b. Population dynamics of *Acartia clausi* from a temperate estuary (Mondego Estuary, western Portugal). Invertebr. Reprod. Dev. 44: 9–15.

Vieira, L.R., Guilhermino, L. and Morgado, F. 2015. Zooplankton structure and dynamics in two estuaries from the Atlantic coast in relation to multi-stressor exposure. Estuar. Coast Shelf Sci. 167: 347–367.

Vieira, E. 2019. A evolução da ciência em Portugal (1987–2016) Fundação Francisco Manuel dos Santos, 382, Estudos da Fundação. ISBN 978-989-9004-17-7, 400pp.

Villa, H., Quintela, J., Coelho, M.L., Icely, J.D. and Andrade, J.P. 1997. Phytoplankton biomass and zooplankton abundance on the south coast of Portugal (Sagres), with special reference to spawning of *Loligo vulgaris*. Scientia Marina 61(2): 123–129.

Wooster, W.S., Bakun, A. and McLain, D.L. 1976. The seasonal upwelling cycle along the eastem boundary of the North Atlantic. J. Mar. Res. 34: 131–141.

3.2

An Overview of Zooplankton Biodiversity in the Southwest Atlantic Ocean During the Last Two Decades With an Emphasis on Brazilian Estuaries, Coastal and Oceanic Regions

Sérgio Luiz Costa Bonecker, * *Cristina de Oliveira Dias,*
Márcia Salustiano de Castro,
Pedro Freitas de Carvalho and *Ana Cristina Teixeira Bonecker*

1. Introduction

Zooplankton is the link between primary producers and higher trophic levels, as it is responsible for transferring part of the energy and biomass obtained through feeding to larger organisms. Thus, understanding the factors influencing their abundance, biomass, composition, and dynamics is fundamental to a general understanding of ecosystem functioning. In addition to the biological interactions that affect zooplanktonic organisms, environmental changes can interfere with the dynamics of these communities. Zooplanktonic organisms have a very short life cycle and a high growth rate and, therefore, react very quickly to short, medium, and long-term environmental changes in various aquatic systems. As zooplankton have a low tolerance for environmental changes, they quickly reflect any change, no matter how subtle (Valentin 2008). As they are organisms that cannot overcome currents, they can easily expand or decrease their geographic distribution due to these changes.

This chapter aims to describe how the knowledge of the zooplankton community has advanced in the last 20 years in the Southwest Atlantic Ocean (SWAO)

Departamento de Zoologia, Instituto de Biologia, Universidade Federal do Rio de Janeiro, Rio de Janeiro, Rio de Janeiro, Brazil.
Emails: crcldias@hotmail.com, mscastro@biologia.ufrj.br, pedropfcilha@yahoo.com.br, ana@biologia.ufrj.br
* Corresponding author: bonecker@biologia.ufrj.br

in estuarine, coastal, and oceanic regions and in relation to the ecology of these organisms and the influence of anthropic impacts on them.

2. Zooplankton Diversity in the Southwest Atlantic Ocean

The number of zooplanktonic organisms in the oceans is estimated at 7,000, with 40% to 60% of these species being in the SWAO (Boltovskoy et al. 2003, 2005). A historical trend was observed in the rates of species descriptions for 17 taxonomic groups, as they increased from the late nineteenth to the early twentieth centuries. In 1890, 40 articles on zooplankton were published, which increased to 70 in 1915 (Boltovskoy et al. 2003). According to the Thomson ISI Web of Science database (http://thomsonreuters.com), in 1990, 92 articles were counted; in 2001, there were 110 published articles, but in 2016 it decreased to only 91 articles (Souza et al. 2018).

The first atlas that brought together the zooplankton species of the SWAO was published in 1981 by Demetrio Boltovskoy (Boltovskoy 1981), who described methodological aspects and the taxonomy and distribution of 16 zooplanktonic groups. In 1999, Boltovskoy published a new edition of the Atlas of Marine Zooplankton, expanding the coverage area to the coast of Africa by gathering 28 zooplanktonic groups (Boltovskoy 1999). However, the Southwest Atlantic still remains one of the least known marine environments, especially for many holoplanktonic groups (Boltovskoy et al. 2005, Neumann-Leitão et al. 2008).

Zooplankton studies on the east coast of South America began with the data acquired during the European expeditions of the nineteenth and early twentieth centuries (Lopes et al. 2006, Boltovskoy and Valentin 2018). Since the 1950s, there has been an increase in the number of studies on plankton in Brazil (Brandini et al. 1997, Lopes et al. 2006). For logistical reasons, most of these studies were conducted in coastal regions, such as bays and estuarine systems (Yoneda 2000).

In 1994, Brazil began the program REVIZEE (Assessment of the Sustainable Potential of Living Resources in the Exclusive Economic Zone), which was a multidisciplinary study that aimed to improve the knowledge of the composition and biomass of the plankton and, consequently, the biological stocks of commercial interest along the ~ 8,000 km long Brazilian coast and continental shelf up to 200 nautical miles. The results of this program were published in 35 books, six on plankton collected throughout the Brazilian territory from 6°N to 35°S (REVIZEE 2022). According to the limits of the Exclusive Economic Zone (EEZ) defined by the program REVIZEE (MMA 2006), based on oceanographic and biological features, the Brazilian coast and its continental shelf can be divided into four large regions: North, Northeast, Central and Southeast-South. These studies helped to integrate this information with the hydrology (water masses, currents, circulation and seasonal patterns) of each region of the Brazilian coast.

Despite the large extension of the Brazilian coast, which represents approximately 57% of the entire SWAO, there are still very few long-term ecological plankton studies that would allow monitoring changes in zooplankton communities due to human activities. A government funding agency created the Long-Term Ecological Research Program (PELD) 20 years ago to address this deficiency. This ongoing program is used to allocate resources for long-term studies of both marine and coastal ecosystems in the Patos Lagoon estuary (30°–32°S) and in its adjacent coast, Guanabara Bay (22°50′S–43°10′W), oceanic islands in the upwelling area of Cabo Frio (23°S), in dune habitats and coastal lagoons of Northern Rio de Janeiro (22°23'S–41°45'W), and in the Abrolhos Archipelago (18°S) (Tabarelli et al. 2013). There are ongoing efforts to discover new species (e.g., Sinev et al. 2010, Nogueira Jr et al. 2013, Sorrentino et al. 2016) and new records (e.g., Carvalho and Bonecker 2008, Arruda et al. 2010, Bonecker et al. 2014, Figueiredo et al. 2018, Nogueira Jr. et al. 2013, Dias et al. 2019, 2020). More recently, molecular techniques have become an important tool to improve the taxonomy and expand the knowledge of new species (da Costa et al. 2011, da Cruz et al. 2021, Vieira-Menezes et al. 2021).

Among the holoplanktonic groups, copepods are the most abundant and diverse; cnidarians, ctenophores, appendicularians, salps, doliolids, chaetognaths, mollusks, branchiopods, euphausiids, and holoplanktonic decapods (*Belzebub faxoni* and *Lucifer typus*) are identified at specific levels. However, many zooplankton groups do not have specialists in Brazil for specific identification (e.g., ostracods, amphipods, isopods, mysids, and stomatopods).

These results show that there is still a limitation of knowledge on zooplankton biodiversity in the SWAO and the need to have more taxonomists to understand the ecology and biodiversity of these organisms in a better way.

3. Zooplankton Ecology in Estuarine, Coastal, and Oceanic Environments

The Brazilian coast extends from the Oiapoque river (5°52′N) to Chuí (33°45′S) and is composed of a great diversity of ecosystems, including estuaries and bays. Based on the REVIZEE Program, the coast is divided into four segments. On the northern coast, the neritic domain has high productivity owing to large contributions of freshwater, sediments, and nutrients from Delta do Parnaíba (02°37′–03°05′S and 42°29′–41°09′W), Golfão Maranhense (02°24′10″–02°46′37″S and 44°22′39″–44°22′39″W), and Amazon River. The northeast coast is approximately 2,000 km long and has a regular geographic profile along with the north and south portions of estuaries of big rivers like Parnaíba (Piauí State) and São Francisco (Sergipe State). Some important features of the northeast coast are the presence of reef barriers, oceanic islands and seamounts, including Rocas Atoll (03°52′S–33°49′W), Fernando de Noronha (03°51′S–32°25′W), and São Pedro-São Paulo (00°56'N and 29°22'W) archipelagos (Souza et al. 2018). The Brazilian central coast extends from 12°S to 22°S, and the continental shelf is irregular, varying from 35 km to 190 km. This region has peculiar topographic features like submarine banks forming the Abrolhos Archipelago and Vitória-Trindade chain (20°–21°S), oceanic islands, and the presence of Doce River delta (19°30′–19°50′S and 39°35′–40°05′W) (Bonecker 2006). Moving south, the Southeast-South region is limited by Cabo de São Tomé (22°S) and Arroio Chuí (34°30′S), it is 2,000 km long, and the continental shelf varies from 50 km to 250 km (Gasalla et al. 2007). This region harbors different ecosystems like Guanabara, Ilha Grande (22°50′–23°20′S and 44°00′–44°45′W),

Sepetiba bays (22°50′–23°05′S and 43°30′–44°10′W), and the largest lagoon complex in South America, consisting of the Patos, Mirim, and Mangueira lagoons. Moreover, the Cabo Frio upwelling system (23°S and 42°W) is one of the most important hydrographic processes on the Brazilian coast, enriching the waters of the continental shelf in this region. The southern end of the Southeast-South region is influenced by the waters of the Rio de la Plata estuary (35°–36°S), which is a high productive area (Mianzan et al. 2001, Acha et al. 2008).

These differences in topographic and hydrological characteristics along the Brazilian coast influence zooplanktonic communities, such as variations in densities observed in the north/northern (200 to 600 ind.m^{-3}), central (5,600 to 8,000 ind.m^{-3}), and southern (between 1,000 and 3,000 ind.m^{-3}) (MMA 2006, Dias et al. 2020) zones. The highest densities of zooplankton are observed mainly in the inner continental shelf and in areas with the greatest influence of river inputs, upwelling, and around banks and seamounts (MMA 2006). The areas with the highest concentrations of fish eggs indicate that the occurrence of spawning is neritic, and they are found in front of estuaries and oceanic regions near submarine banks (MMA 2006).

Zooplankton diversity also varies along the Brazilian coast, with low values in the northern region (2.0 bits/ind., Jablonski et al. 2006) and high in the northeast (3.5 to 4.2 bits/ind.) (MMA 2006). There is a trend toward lower diversity values in the coastal region (e.g., 2.6 bits/ind. in the northeast region) in relation to transitional or deeper waters, where there is a mixture of coastal and oceanic species (Lopes et al. 2006).

Copepods dominate the entire SWAO, representing more than 70% of the total species in the northern area, 60% in the northeast, and 85% in the central region (MMA 2006). Other abundant groups along the Brazilian coast are Appendicularia, Chaetognatha, Ostracoda, Molluska, Decapoda, Branchiopoda, and Thaliacea (MMA 2006).

Plankton density, biomass, and richness vary between estuarine, coastal, and oceanic zones. Particularly in the SWAO, the planktonic community is influenced by the different characteristics of these regions that are facing modifications over the years. The planktonic community that inhabits estuarine and coastal environments, areas highly enriched by continental waters, shows physical-chemical changes directly influenced by the intense human activity in these regions. By contrast, the hydrography of the oceanic regions of the SWAO is extremely complex. The oceanographic structure is characterized by low productivity (oligotrophic area) and is altered by the presence of islands (São Pedro-São Paulo Archipelago, Rocas Atoll, and Abrolhos Archipelago) and ocean banks (Vitória-Trindade chain), meanders, vortices, upwellings (Cabo Frio and Santa Grande Cape, 28°S), and large ocean currents that cause changes in local hydrodynamics (Arruda et al. 2013, Calil et al. 2021), contributing to the increase in planktonic biomass (MMA 2006). A significant portion of zooplankton production is related to the existence of several marine fronts characterized by different forcings and temporal and spatial scales. A study that corroborates the information mentioned above was conducted in the tropical islands in northeastern Brazil (Fernando de Noronha, São Pedro-São Paulo archipelagos, and Rocas Atoll), where high zooplanktonic biomasses were observed in stations under the influence of the islands, especially at night (Campelo et al. 2019).

Estuaries and coastal areas are among the most productive ecosystems in the world especially because of the input of sediments and organic matter from continental waters in these areas. This daily entry of freshwater and organic matter into estuaries makes these environments highly dynamic, and their chemical and physical characteristics can vary within a short period. All these factors directly influence the biota of these environments, including the planktonic community. The estuarine environments occupy approximately 70% of the tropical and subtropical coastal zones, covering the SWAO. The zooplankton community in the estuary and bay systems comprises eurythermic and euryhaline species, mostly with estuarine-coastal species with an oceanic influence, typical to the tropical and subtropical SWAO. However, typical estuarine species (e.g., *Pleopsis polyphemoides, Parvocalanus crassirostris, Pseudodiaptomus acutus, Dioithona oculata, Euterpina acutifrons,* and *Hemicyclops thalassius*), species originating from subtropical and tropical oceanic waters (e.g., *Nannocalanus minor, Paracalanus aculeatus, Corycaeus speciosus, Krohnitta pacifica, Krohnitta subtilis, Serratosagitta serratodentata,* and *Doliolum denticulatum*), and indicators of cold water or upwelling at Cabo Frio (e.g., *Calanoides carinatus, Ctenocalanus citer,* and *Oikopleura cornutogastra*) are often associated with the waters of this ecosystem (Bradford-Grieve et al. 1999, Dias and Araujo 2006).

In a recent study (Dias et al. 2018) in Guanabara Bay, the copepods *Temora turbinata, Acartia tonsa, Paracalanus quasimodo,* and *Temora stylifera* presented the highest densities throughout the study period, with the dominance of the two first species. The exotic species *T. turbinata* and *A. tonsa* until two decades ago were not even recorded in Guanabara Bay (Nogueira et al. 1989, Schutze and Ramos 1999, Valentin et al. 1999, Bonecker et al. 2012). Recently, *T. turbinata* and *A. tonsa* have been reported as the dominant species (Gomes et al. 2004, Gomes 2007, Guenther et al. 2012). This restructuring of the local community may be related to the constant degradation of the environment, particularly in its more internal areas that are being reflected along the entire ecosystem (Gomes 2007).

Acartia tonsa was reported in Guanabara Bay in studies from the 1990s, while *T. turbinata* was reported only in 2000 (Guenther et al. 2012, Bonecker et al. 2012). *Acartia lilljeborgii* was reported as representative of the Acartiidae family, previously including being the most abundant species (Wandeness et al. 1997, Valentin et al. 1999, Bonecker et al. 2012). However, recent studies show *A. tonsa* as a species is well distributed throughout the Guanabara Bay (Guenther et al. 2012), and this fact can be explained by its success in eutrophic environments due to the wide availability of food (Marques et al. 2006).

Seasonal variations do not have much influence on estuarine and planktonic dynamics in large parts of the tropical portion of the SWAO, owing to little variation in atmospheric temperatures. In the southern region of the SWAO, owing to the more marked seasonality, this factor can directly interfere with the planktonic community. For example, Patos Lagoon is the largest strangled lagoon in the world. Approximately 80% of this lagoon is freshwater, and at its southern limit, it connects to the ocean through a narrow inlet (800 m) and a deep channel (15 m). Tidal energy is scarce due to amphidromic conditions (Odebrecht and Castello 2001). The main force controlling water dynamics and salinity distribution in this lagoon is regional precipitation. The abundance and zooplankton composition of the Patos Lagoon estuary reflects the local hydrological regime with alternating periods of flooding, freshwater runoff, and water mixing (Teixeira-Amaral et al. 2017). The calanoid copepod *A. tonsa*, from a brackish/marine environment, is abundant in autumn-winter when rainfall is low and salinity is high. During spring-summer, when prolonged land runoff combined with high temperatures allow the establishment of a community, the estuary is composed mainly of the freshwater calanoid copepod *Notodiaptomus incompositus*, thus, changing the entire zooplanktonic community (Montú et al. 1997, Teixeira-Amaral et al. 2017).

As estuaries are environments where a mixture of salt and fresh water occurs and are influenced by tides, the biotic communities that inhabit the estuarine regions are also directly affected by these factors. As a general pattern, areas with higher salinity, in the outermost portion of estuaries, tend to have greater planktonic diversity. This is owing to the mixture of enriched coastal water close to the coast and the oceanic oligotrophic tropical water, which results in an ideal system for most species, increasing diversity and positively interfering with zooplankton distribution. Examples have already been recorded in southeastern (Sterza and Fernandes 2006, Dias et al. 2018) and northeastern (Eskinazi-Sant'anna 2000, Neumann-Leitão et al. 2019) Brazil.

The high richness of zooplanktonic species in the estuaries of the SWAO can be explained by the hypothesis of habitat heterogeneity. That is, environments with high dynamism with a constant variation of physical-chemical characteristics can offer more niches, and different species can adapt better to each area, increasing zooplankton diversity in these environments (Neumann-Leitão et al. 2018). As in other locations in the world, copepods are the most abundant and most diverse zooplanktonic organisms in the estuarine environments of the SWAO; this result has already been observed in several studies in the south (Salvador and Bersano 2017), central (Dias et al. 2018), and northeast (Neumann-Leitão et al. 2019) Brazilian coast. Other zooplanktonic groups and their species are commonly found in estuaries of the SWAO, such as the larval stages of decapod crustaceans, mollusks, and fish. In addition to these groups, despite being restricted to more saline regions, there is an increase in the diversity of gelatinous planktonic organisms, such as appendicularians, thaliaceans, and hydromedusae (Nogueira Jr et al. 2018).

Recent studies on zooplankton in estuaries along the Brazilian coast have focused on trophic relationships in estuarine environments. In these studies, organisms were grouped considering the functional traits of the species based on their ability to predict or explain the structure of communities and their responses to environmental conditions. From the traits, it is possible to define the functional groups and understand the trophic importance of the dominant planktonic group in the pelagic environment. Trophic relationships based on the functional groups of copepods were studied by Neumann-Leitão et al. (2018) along the Brazilian North coast (Amapá, Pará, and Maranhão states), Neumann-Leitão et al. (2019) in the Tropical Western Atlantic (Rio Grande do Norte state), and Araujo et al. (2020) in estuaries of the Rio de Janeiro State.

Oceanic regions in tropical environments, including the SWAO, are considered oligotrophic and consequently, planktonic organisms are smaller than the ones that occur in nutrient-rich ecosystems (Boltovskoy 1999). Lower concentrations of nutrients provide more stable environments, with less variation in dissolved oxygen, and that are less favorable to the predominance of a single species. These characteristics result in greater richness and equitability and, consequently, greater zooplankton diversity in these environments (Margalef 1961). Owing to the high cost and logistical difficulties, studies on the characterization, distribution, and ecology of the zooplankton community in the oceanic environment in Brazil are less frequent than in the coastal environment. These studies are usually associated with multidisciplinary projects funded by government agencies or in partnerships with the private sector that need to monitor the possible effects of their economic activities on the environment.

Among the studies conducted in the oceanic region in Brazil for analyzing the zooplankton community, it is important to highlight the pioneering studies of the REVIZEE Project mentioned earlier. Zooplankton collected from São Tomé Cape to the south of Bahia State generated two guides for the identification and distribution of 23 groups and 114 species of zooplankton (Bonecker 2006). This project also included studies from other oceanic regions of the Brazilian coast, such as the northeast region that covers the areas from Salvador to the mouth of the Parnaíba River. In this region, the macrozooplankton community (300 µm) was composed of 201 species from seven phyla (Larrazábal et al. 2009).

Another study that provided important knowledge for the Brazilian oceanic region was HABITATS (Campos Basin Characterization Project). Bonecker et al. (2014) identified 128 zooplanktonic species in two sampling periods and determined their frequency of occurrence in neritic and oceanic regions and different water masses. In this study, Copepoda was the group with the highest number of species, followed by Siphonophora. Bonecker et al. (2018) analyzed the day-night variation of zooplanktonic organisms and showed that the myctophid larvae *Lepidophanes guentheri* and euphausiaceans (*Euphausia americana, Euphausia similis,* and *Nematoscelis atlantica*) performed vertical migration. Other studies analyzed the productivity of some copepod species (Dias et al. 2015) and the distribution of Appendicularia species in different surface water masses and their function as bioindicators (Carvalho and Bonecker 2016).

Troina et al. (2020) developed a study in the continental shelf and oceanic regions from Uruguay to Cabo Frio and analyzed the spatial and temporal patterns of stable isotope compositions of carbon and nitrogen in zooplankton. They observed that the highest values occurred in the southernmost area. More recently, a study in the Brazilian Equatorial Margin showed differences among zooplanktonic communities of three sedimentary basins (Ceará Basin, Pará-Maranhão Basin, and Foz do Amazonas Basin) because of the influence of the amounts of suspended solids from the plume of Amazon River (Dias et al. 2020).

Although some studies have been conducted in the Brazilian oceanic region in the last few years, there is still a great knowledge gap about this area and its ecological processes involving the zooplankton community. This is owing to the large length of the Brazilian coast, logistical difficulties, and the high cost of obtaining plankton samples from deep waters.

4. Effects of Human Activities on Zooplankton in the Southwest Atlantic Ocean: Pollution, Microplastics, and Climate Changes

4.1 Human Activities as a Source of Pollution

Studies that address the decline in planktonic species richness due to global impacts on diversity are more abundant in the Northern Hemisphere than in the Southern Hemisphere (Chaudhary et al. 2021). Estuarine environments are the main ecosystems subject to urban or industrial impacts (Robb 2014).

Estuaries have been affected by the increase in emissions of pollutants of human origin that are released indiscriminately throughout the coastal ecosystems of the SWAO (Gilbert et al. 2011). The continuous discharge of sewage into estuaries, lagoons, and coastal bays leads to an increase in eutrophication (McGlathery et al. 2007, Fonseca et al. 2021) that consequently affects the entire zooplanktonic community.

All these directly affect the biota that resides in these environments, resulting in a considerable loss of diversity, which also includes planktonic organisms (Araujo et al. 2020). Because they are present in all estuaries and are sensitive to environmental variations, planktonic organisms are commonly used as biomonitoring tools. In a series of studies with zooplanktonic organisms, conducted since 2012 in four estuaries with different degrees of pollution in southeastern Brazil, it was possible to observe the effects of the eutrophication process on the zooplanktonic organisms. The structure of copepod assemblages also showed the main separation trend between the most polluted estuaries and those less polluted (Araujo et al. 2016). It was observed that copepod biomass and productivity are negatively influenced by pollution indicator parameters, which makes these organisms a useful tool for monitoring estuarine environments (Araujo et al. 2017). Another study conducted in the same estuaries showed that appendicularian densities, such as that of *Oikopleura longicauda*, are associated with estuarine environments with low concentrations of nutrients and suspended materials (Carvalho et al. 2016). In this region, Santos et al. (2017) used an anthropogenic pollution pressure index (PI) to assess the degree of degradation of estuaries and observed that fish larvae present in the planktonic community have a high sensitivity to impacts of different origins.

Dias et al. (2018) studied the main zooplanktonic groups in Guanabara Bay and observed that the rainfall regime leads to variations in water quality and significantly decreases zooplankton diversity in this environment. Furthermore, it has been observed that the dominance of copepods has changed considerably in recent years. The cause of this change is still not clear, but it may be related to environmental changes due to anthropic activities, mainly because of the introduction of ballast water (MMA 2009).

Studies conducted along the northeast region of Brazil have shown impacts on estuarine environments, causing changes in the planktonic community. In the study by Araujo et al. (2008), human impacts in the Sergipe River estuary (11°S–37°W) resulted in changes in the dynamics and trophic structure of zooplanktonic organisms. In another study conducted in an estuary in Rio Grande do Norte, in the Guaraíras lagoon complex (6°09′–06°14′S to 35°05′–35°10′W), the concentrations of phosphorus and chlorophyll were used to determine the degrees of eutrophication, and these variables were observed to interfere directly with zooplankton biodiversity (Almeida et al. 2012).

More recently, in the coastal region of Salvador in Bahia State, it was recorded that copepod diversity was significantly low in an area rich in nutrients of anthropic origin that was supplied by a submarine pipeline. *Calocalanus pavo, Undinula vulgaris,* and *T. stylifera* were the best indicators of the impacted area (Conceição et al. 2021). In the Suape estuary (8°23'45"S–34°58'04"W) in Pernambuco State, morphological anomalies were observed in the copepod *A. lilljeborgii* probably because of the high concentrations of toxic substances of industrial origin released into the aquatic environment without adequate treatment (Melo et al. 2021). Additionally, in 2021, oil droplets were recorded in the Tamandaré region (8°45'35"S–35°06'17"W) on the coast of Pernambuco State, and these substances directly affected important zooplanktonic organisms such as crab larvae and reef-building corals (Campelo et al. 2021). All these studies of different scales of pollution in estuarine regions along the Brazilian coast showed how these affected specific communities of planktonic groups and the local biodiversity.

Recent studies on zooplankton communities have focused on the relationship between the planktonic community and the effects of human activities. Araujo et al. (2017) analyzed the effects of pollution in four estuaries on the southeastern Brazilian coast of the copepod community using some water quality indicators. In this study, changes in copepod communities were observed for the different levels of environmental degradation in each estuary. Rocha et al. (2022) used ecological indexes for the zooplankton community to assess the degree of impact in the coastal region five years after a dam collapsed in southeastern Brazil. The authors found that the zooplankton community was an effective tool to assess the ecosystem health in a coastal area.

There is a need to search for indexes that can explain the ecological state of an ecosystem based on the analysis of organisms that can be used as indicators of environmental quality.

Some effects of anthropic activities can be observed in ecosystems distantly located from the source of pollution, such as impacts caused by mining. In November 2015, the failure of a tailing dam released approximately 50 million cubic meters of mud contaminated with metals into the environment, reaching the Doce River. This river flows through the Brazilian states of Minas Gerais and Espírito Santo and reaches the Atlantic Ocean in the southeastern region of Espírito Santo coast. After 16 days of the disaster and traveling 650 km, the contaminated mud reached the sea. Owing to the marine currents, winds, and rains in the region, the mud reached marine conservation units such as the Comboios Reserve, Costa das Algas Marine Protected Area, Santa Cruz Wildlife Refuge, and Abrolhos Marine National Park.

Studies conducted immediately after the accident showed the acute and chronic effects of mud contaminated with metals on plankton. Preliminary results from samples collected in the coastal region recorded contamination of zooplankton by heavy metals (Bianchini et al. 2016). Another study conducted five days after the mud reached the marine environment showed an acute impact on the zooplankton community, with a peak in abundance and a decrease in diversity near the mouth of the Doce River. *Parvocalanus* sp. and *Oithona nana*, which are opportunistic, represented 61% of this abundance and were correlated with the concentrations of Fe, Pb, Cu, and particulate Zn. A change in the composition of the community was also observed after the passage of a cold front in the region due to the resuspension of the sediment and consequent increase in turbidity (Fernandes et al. 2020).

Approximately 10% of the larval fish collected at the mouth of the Doce River after the dam collapse had orange sediment attached. In addition, some specimens had damaged the digestive tract. The analysis of metals in these organisms showed a significantly higher concentration, mainly of manganese and iron, with those observed in larvae collected in the same area before the accident (Bonecker et al. 2019).

High concentrations of metals in the coastal region during the greatest inflow of water from the Doce River were associated with lower densities of zooplankton and ichthyoplankton. However, during the low discharge of the river, a lower concentration of metals was observed in the coastal marine region and higher densities of zooplankton and ichthyoplankton. Meteoceanographic events, such as the passage of a cold front and the incidence of winds in the region, lead to the resuspension of sediments contaminated with metals, thus affecting the planktonic community (Bonecker et al. 2021). The continuous monitoring of planktonic organisms in the coastal region adjacent to the mouth of the Doce River is essential for the assessment of the chronic effects of metal pollution in the marine region caused by the dam failure.

Oceanic regions are less subject to impacts caused by human activities. However, with the increasing demand for fossil fuels and the consequent increase in oil mining activity in the environment, there is an increase in concern about the possible negative effects that can be generated (Cordes et al. 2016). The monitoring of the zooplankton community in these areas that are still poorly studied, especially in deep-sea ecosystems, is important for obtaining a better understanding of the region and the species that are uncommon in planktonic samples. These studies are essential to monitor any changes that occur in the community due to oil mining activities and serve as a baseline for decision-making in the event of an accident. Although little studied, data on the possible impacts of these activities in oceanic regions have already been described for the Brazilian coast in some scientific conferences (e.g., Cotta 2004, Santos 2012, Nascimento et al. 2021). In addition, the data generated by monitoring programs are used to develop scientific studies on zooplankton ecology by contributing to the improvement of knowledge of these organisms along the Brazilian coast (Bonecker et al. 2014, Bonecker and Castro 2018, Dias et al. 2020).

4.2 Microplastics (MPs)

Plastic is a durable, versatile, and low-cost product for production and is therefore widely used in various industry sectors, including cosmetics, fisheries, mariculture, tourism, and domestic use for various applications (Arienzo et al. 2021). Plastic production has rapidly increased since 1920, in 2014, approximately 311 million tons of this product were produced worldwide, reaching 330 million metric tons in 2016 and 360 million metric tons in 2018 (Amelia et al. 2021, Arienzo et al. 2021).

Pollution by plastic is one of the main environmental problems, and if production continues to increase, the forecast states that by 2050, there will be more plastics than fish in the oceans (MacArthur et al. 2016, Arienzo et al. 2021). Plastic products made from polyethylene (PE), polypropylene (PP), or polyvinyl chloride (PVC) can take between 100 and over 1000 years to degrade. Owing to the action of solar radiation, plastic generates fragments that can be categorized according to their size: macro- ($\geq$ 25 mm), meso- (< 25 mm to 5 mm), micro- (< 5 mm to 1 µm), and nano-sized (< 1 µm) plastics. These fragments can also be classified according to their morphology, density, and material used for their production (Amelia et al. 2021, Arienzo et al. 2021).

Microplastics are commonly present in aquatic environments, owing to their transport over long distances due to their low density (Arienzo et al. 2021). They can be found in the environment in the form of biofilms, microbeads, foams, and mostly fibers, and their removal is very difficult. In the marine environment, MPs can be transported by marine currents, waves, and tides and can accumulate in the coastal region, remote oceanic areas, or large-scale subtropical gyres. In addition to causing pollution by their presence in the environment, they adsorb pollutants such as heavy metals, persistent organic pollutants, hydrocarbons, and pathogens that can be transferred to marine organisms via ingestion, affecting the food chain (Amelia et al. 2021, Arienzo et al. 2021). The capacity to absorb pollutants depends on the physical and chemical properties of MPs, such as color, density, and age and also on the environmental conditions (Amelia et al. 2021).

Microplastics can be ingested by various marine animals such as seabirds, whales, turtles, fishes, bivalves, benthic invertebrates, small crustaceans, protozoa, zooplankton, and larval fish. As zooplankton is a crucial food source for many secondary consumers, they represent a route by which MPs can enter the food chain and be transferred to higher trophic levels. Factors that affect the bioavailability of MPs to zooplankton organisms are abundance, size, shape, aggregation, age, density, color, and selection of species depending on the life stage. Zooplankton organisms can be affected by MPs in many ways: reduce copepod fecundity and decrease herbivory; negative effects on growth, development, and lifespan; delay development in the copepod *Tigriopus japonicus*; and blocking the digestive systems and filtering structures of gelatinous organisms (Botterell et al. 2019, Arienzo et al. 2021). The presence of algae seems to increase the uptake of MPs by copepods (Castro et al. 2018).

Studies to better understand how MPs behave in the environment and their direct and indirect effects on the biota and the ecosystem as a whole have been conducted worldwide. More than 80 countries have already conducted studies focusing on MPs and their ability to adsorb pollutants, including toxicity tests with marine organisms and ingestion of these components by mero- and holoplankton under laboratory and field conditions (Botterell et al. 2019, Amelia et al. 2021). This great diversity of studies is important to assist in the search for solutions to mitigate the impacts caused and to guide global decisions to combat one of the biggest sources of pollution in the aquatic environment.

The first reports of the presence of plastic fragments in birds and on the sea surface date from the 1960s and 1970s, respectively (Arienzo et al. 2021). Since then, studies on environmental pollution by MPs and their effects on organisms have increased, highlighting this growing concern.

The same occurred in Brazil when, in 1970, the first work on MPs was conducted, and from 2009 onward, publications were made. Castro et al. (2018) surveyed studies on MPs developed in the last decade in Brazil and observed that between 2009 and 2017, 35 studies were published, mainly in the southeast and northeast Brazilian coasts. Among these, only six (Ivar do Sul et al. 2013, 2014a, 2014b, Lima et al. 2014, 2015a, 2015b, Castro et al. 2016) quantified the presence of MPs with respect to planktonic organisms. Since 2017, more than 65 studies on MPs in Brazil have been conducted in aquatic environments, among which only one (Figueiredo and Vianna 2018) evaluated the influence of MPs on marine zooplanktonic organisms. Figueiredo and Vianna (2018) observed that in Guanabara Bay, the number of MPs was greater than that in other marine ecosystems, and the sizes recorded were large for ingestion by copepods but were too diluted to be ingested by fish larvae and Chaetognatha. Another study conducted in Guanabara Bay showed that pollution by MPs in this environment is among the highest reported in the literature and that the presence of MPs with sizes smaller than 1 mm could serve as food for micro- and mesoplankton (Olivatto et al. 2019).

Despite the increasing volume of work on MPs in Brazil, little is known about the effects of MPs on Brazilian marine zooplanktonic organisms. This shows the gap in studies on pollution in coastal environments for the continental shelf of the Brazilian coast.

4.3 Climate Changes

Understanding how the marine environment is affected by environmental changes is critical, as the oceans are responsible for absorbing more than 80% of the heat added to the climate. Climate change has been attracting more and more interest due to the consequences that the increase in the temperature of the Earth's surface can cause. There is already evidence of an increase in the global average temperature of the atmosphere and oceans.

Oceans play a key role in the climate system by storing most of the solar energy that reaches Earth, in the heat distribution, evaporation, and by participating in the carbon cycle. They are responsible for absorbing more than 80% of the heat added to the climate in the last 55 years (Levitus et al. 2005). The impacts of climate change can have serious consequences for the marine environment and its biota. According to Franco et al. (2020), climate change could also affect the life cycle, migration patterns, recruitment, variability, seasonality, phenology, and distribution of marine species, which could strain already vulnerable coastal communities (Poloczanska et al. 2016, Pecl et al. 2017).

Some impacts on the marine environment include the heating of the seawater surface, causing the stratification of the water column and preventing the cycling of nutrients; the increase in atmospheric temperature, favoring higher levels of precipitation; the melting of glaciers and the rise in sea level and the acidification and deoxygenation of seawater (Brierley and Kingsford 2009, Gruber 2011). Global warming causes the displacement of planktonic communities, leading species from cold waters to disappear while those from warm areas occupy new regions. This changes all trophic relationships and the phenomenon that has already been observed in the Northern Hemisphere (Castro 2008). In the SWAO, one of the most biologically productive areas of the world ocean (Lutz et al. 2010), there is strong scientific evidence suggesting that the Brazil current has been intensifying and shifting southwards during the past decades in response to changes in near-surface wind patterns, leading to intense ocean warming along the path of the Brazil current, the South Brazil Bight, and in the Río de la Plata (Franco et al. 2020). The SWAO presents one of the largest marine warming hotspots worldwide (Hobday and Pecl 2014), and high-resolution models project sea surface temperature will rise by at least 3°C by 2099 (Popova et al. 2016).

The accumulation of polluting gases generates an increase in the average temperature of the Earth's atmosphere, causing the air to absorb more humidity, thereby increasing the rate of precipitation. In the tropics, the increase in precipitation is also correlated with the increase in seawater surface temperature (Xie et al. 2009). A greater amount of rain causes an intensification of the inland water drain, reducing local salinity, increasing turbidity, and eutrophication of coastal regions. There is also a change in the pattern of atmospheric circulation and an increase in the strength of the winds, causing a greater incidence of coastal upwellings. Elevated atmospheric temperatures can still cause glaciers to melt, thus causing sea levels to rise (Brierley and Kingsford 2009).

Ocean acidification is directly linked to the increase in atmospheric CO_2 emissions caused by burning fossil fuels, which is aggravated by deforestation, thus releasing more carbon dioxide into the atmosphere. About 30% to 40% of the carbon dioxide present in the atmosphere is dissolved in the oceans. In the first instance, this absorption by the oceans can help to improve climate effects, but it has a negative effect on various oceanic organisms, impacting them to varying degrees. This is due to the chemical reactions of the dissolution of large amounts of CO_2 in water, which affects the formation of calcium carbonate (Steffen et al. 2015, Sodré et al. 2016). It causes the oceans to become increasingly acidic due to the chemical reaction that occurs between CO_2 and seawater, forming carbonic acid (H_2CO_3) (Orr et al. 2009). Studies have shown that states of lower environmental calcium carbonate saturation can have a dramatic effect on some calcifying species, including oysters, clams, sea urchins, shallow-water corals, deep-water corals, deep-sea corals, and calcareous plankton (Barbosa 2020). The fact that there are more free hydrogen ions directly influences the pH of the water by decreasing it and thus making the seawater more acidic (Steffen et al. 2015, Sodré et al. 2016). The increase in acidity alters the distribution and increases the density of gelatinous organisms in the world's oceans (Condon et al. 2012).

The increase in acidity added to the effect of low saturation of calcium carbonate on living beings at different levels. These biological effects on marine organisms include a decrease in metabolic rate and immune response, which makes them highly susceptible to pathogens. Calcium carbonate is also essential for the skeleton of corals, mollusks, and crabs, in addition to its functionality at the plankton level as a precursor for external structure formation. Thus, consequences occur in a cascade effect, affecting every trophic level (Steffen et al. 2015, Sodré et al. 2016).

It also affects the tolerance to changes in ocean water temperature, interfering with the distribution of more sensitive species (PBMC 2021). For example, it was also noticeable that CO_2 increases ocean noise, which affects the acoustic properties of seawater, in turn affecting animals that use echolocation (Steffen et al. 2015, Sodré et al. 2016).

The decrease In the concentration of dissolved oxygen (DO) in the oceans is the result of the increase in temperature and the consequent stratification of the oceans. As the temperature increases, the DO solubility is reduced, and it becomes less available to the biota. In this way, climate change can cause the expansion of low-oxygen zones, leading to the reduction of local biodiversity and putting areas of high fisheries productivity at risk, causing enormous economic and ecological damage. The intensities of acidification or deoxygenation vary according to the region of the impacted site, with high-latitude regions being more subject to acidification and low-latitude regions being more vulnerable to deoxygenation (Diaz and Rosenberg 2008).

The impact of climate change is not limited to its direct effects on the oceans. Coastal regions, where shallower waters and eutrophication processes (natural or artificial) predominate, are more subject to temperature variations than oligotrophic oceanic regions, where populations are controlled by the availability of nutrients (Moisan et al. 2002). The climate of continental areas also has indirect effects on the fertility of coastal areas through increased rainfall and continental discharges, resulting in an increase in nutrient rates and planktonic productivity in these areas of influence. As an example of how phenomena caused by climate change may also affect planktonic communities, we can mention the influence of the Rio de la Plata estuary on the adjacent oceanic region. According to Castro (2008), the warming of water causes an increase in the discharge of continental water, altering the planktonic composition, which in turn has several effects on the oceanic food chain, leading to the appearance of exotic and toxic species.

There is strong evidence of change in the abundance and structure of planktonic communities in recent decades. Zooplankton responses to climate change include the decline in krill by more than an order of magnitude in the last 25 years and the alteration of the synchronism between prey and predators and between the occurrence of fish larvae (meroplankton) and their food peak abundance, causing increased mortality of these larvae and a decrease in the production of fishery resources. The ocean warming along the path of the Brazil current is presumably responsible for the poleward shift of commercially important pelagic species in the region and the long-term shift from cold water to warm water species in the industrial fisheries of Uruguay (Franco et al. 2020).

In Brazil, evidence indicates that the strong decline in pelagic fisheries production (e.g., sardines) would be caused in part by the weakening of the Cabo Frio upwelling (Valentin 2008). Discussions are ongoing on whether the surface heating of Antarctic waters would be responsible for the decrease in deep currents and as a consequence of the upwelling process on both sides of the SWAO. Regarding fisheries production in the northeast region of Brazil, partial results showed a sensitivity to global climate events, with greater production in La Niña years and lower production in El Niño years. However, each species responds differently to these deviations in weather

(Barbalho et al. 2013). In the southern region of Brazil, some researchers have studied climate change impacts on the catch volume of shrimp in the region, associated with events of high freshwater discharges from rivers in the Patos Lagoon due to El Niño. These events cause losses in the shrimp catch volume and, consequently, a negative impact on the local fishing economy (Soares 2016).

Marine zooplankton research in Brazil has been primarily descriptive, with most studies focusing on community structure analysis and related issues (Lopes 2007). The biggest challenge encountered in studies that address the consequences of these changes on the zooplankton community is probably to establish a cause-and-effect relationship. Considering the importance of this biota, the relative scarcity of long-time study series of plankton in the oceans is surprising. The key problem in documenting and understanding the plankton response to climate change is the difficulty in maintaining the continuity of sampling time series. The evolutionary trend of biomass and plankton composition in the waters of the SWAO, which can be attributed to climate change, can only be detected by applying long-term programs.

Examples of studies of mesozooplankton based on long-time series are found in the different climatic regions of the globe, whether cold or temperate-cold (Valentin et al. 2020). Because it does not systematically monitor plankton, as is done in the North Atlantic, where it has been periodically monitored for about 50 years, Brazil cannot assess the impact of climate change on the marine food chain and fisheries resources (Valentin 2008, Castro 2008). Studies based on these data have shown that global climate change is affecting the distribution of planktonic communities, with consequences on the entire marine food chain and, therefore, on fisheries (Castro 2008). Regarding long-time series of studies in Brazil, we can mention one conducted in Guanabara Bay (Central region) and Patos Lagoon (South region; Valentin et al. 2020). The numerous studies on the zooplankton of Guanabara Bay are old, and only in the last 20 years have more specific studies been conducted on certain groups (chaetognaths, cladocerans, Decapoda larvae, ichthyoplankton, and bacterioplankton), but these are not long-time series studies (Valentin et al. 2020). The main mesozooplankton components in Guanabara Bay were analyzed over a decade (2003–2013, Valentin et al. 2020). These authors found high interannual variability in the density of the mesozooplankton and plankton trophic networks of the bay', which reflected the seasonal factor manifested by the hydrological changes that reveal the global climatic conditions.

The Patos Lagoon estuary and adjacent marine coast in southern Brazil were studied, and the results gathered over almost 20 years indicate the importance of the large-scale climatic phenomenon El Niño Southern Oscillation (ENSO) on the functioning of the estuarine ecosystem (Odebrecht et al. 2017). The authors present the state of the art on long-term changes induced by the natural and human-related factors in the adjacent marine coast while also contributing to the prediction of possible responses to future climate changes and the conservation of coastal ecosystems.

In Brazil, there are national or regional, multi-institutional programs involving studies on plankton, all being limited in time (e.g., Brazilian program REVIZEE, AMBES, and HABITATS). International cooperation programs together with

Argentina, Uruguay, and the USA were conducted to monitor the surface temperature of the South Atlantic Ocean. These projects were supported by the Program GOOS (Global Oceanic Observation System) of the COI (Intergovernmental Oceanographic Commission). The SACC Consortium (South Atlantic Climate Change) also participated with support from the IAI (Inter-American Institute for Global Change Research), and the NSF (National Science Foundation). These programs aimed to determine the physical mechanisms that control biological processes in highly productive regions of the western South Atlantic and their variability on intra-annual timescales. However, none of these projects included plankton samplings (Castro 2008, Valentin 2008).

In the last 20 years, the studies conducted in the SWAO show that there are still some gaps, mainly concerning the knowledge of the biodiversity of several estuaries, coastal, and oceanic regions. There is a need to seek important tools such as indexes that allow the assessment of the health of an ecosystem and the strengthening of long-term environmental monitoring programs that allow a more accurate response to the effects of human activities on planktonic organisms.

5. Conclusions

This chapter presents studies on zooplankton in the SWAO, highlighting information on the distribution, ecology, and local and global anthropic impacts on this community. The first studies on zooplankton focused on the search for knowledge and distributions of species. More recently, these studies have been related to the assessment of biodiversity, the relationship of communities with oceanographic forcing, studies of natural and anthropic impacts, and the search for indexes that express the environmental quality and how to quantify these impacts. Studies that focus on the metagenomic knowledge of zooplanktonic species have also evolved a lot in recent times.

Although many studies have been conducted, there is still a large gap in the knowledge of the zooplankton community in the SWAO. Through this chapter, we hope, to disseminate information about the zooplankton in this area and encourage researchers to expand their knowledge of this important marine community in this extensive yet little-known region.

6. Acknowledgments

The authors would like to thank Dr. Fernando Morgado for the invitation to participate in the book *Zooplankton Importance in a Changing World*.

References

Acha, E.M., Mianzan, H., Guerrero, R., Carreto, J., Giberto, D., Montoya, N. et al. 2008. An overview of physical and ecological processes in the rio de la plata estuary. Cont. Shelf Res. 28: 1579–1588. https://doi.org/10.1016/j.csr.2007.01.031.

Almeida, L.R., Costa, I.S. and Eskinazi-Sant'Anna, E.M. 2012. Composition and abundance of zooplankton community of an impacted estuarine lagoon in northeast Brazil. Brazil. J. Biol. 72: 13–24. https://doi.org/10.1590/S1519-69842012000100002.

Amelia, T.S.M., Khalik, W.M.A.W.M., Ong, M.C., Shao, Y.T., Pan, H.-J. and Bhubalan, K. 2018. Marine microplastics as vectors of major ocean pollutants and its hazards to the marine ecosystem and humans. Prog Earth Planet Sci. 8: 12. https://doi.org/10.1186/s40645-020-00405-4.

Araujo, H.M.P., Nascimento-Vieira, D.A., Neumann-Leitao, S., Schwamborn, R., Lucas, A.P.O. and Alves, J.P.H. 2008. Zooplankton community dynamics in relation to the seasonal cycle and nutrient inputs in an urban tropical estuary in Brazil. Brazil. J. Biol. 68: 751–62.

Araujo, A.V., Dias, C.O. and Bonecker, S.L.C. 2016. Differences in the structure of copepod assemblages in four tropical estuaries: importance of pollution and the estuary hydrodynamics. Mar. Pollut. Bull 115: 412–420. https://doi.org/10.1016/j.marpolbul.2016.12.047.

Araujo, A.V., Dias, C.O. and Bonecker, S.L.C. 2017. Effects of environmental and water quality parameters on the functioning of copepod assemblages in tropical estuaries. Estuar Coast Shelf Science 194: 150–161. http://dx.doi.org/10.1016/j.ecss.2017.06.014.

Araujo, A.V., Dias, C.O. and Bonecker, S.L.C. 2020. Diversity and functional groups of copepods as a tool for interpreting trophic relationships and ecosystem functioning in estuaries. Mar. Environ. Res. 162: 105190. https://doi.org/10.1016/j.marenvres.2020.105190.

Arienzo, M., Ferrara, L. and Trifuoggi, M. 2021. Research progress in transfer, accumulation and effects of microplastics in the oceans. J. Mar. Sci. Eng. 9: 433. https://doi.org/10.3390/jmse9040433.

Arruda, M.R., Ávila, L.R.M. and Bonecker, S.L.C. 2010. Chaetognatha, spadellidae, *Paraspadella nana* owre, 1963. New Occurrence from the Southwest Atlantic Ocean. CheckList 6: 628–629.

Arruda, W.Z., Campos, E.J.D., Zharkov, V., Soutelino, R.G. and Silveira, I.C.A. 2013. Events of equatorward translation of the vitoria eddy. Cont. Shelf Res. 70: 61–73. https://doi.org/10.1016/j.csr.2013.05.004.

Barbalho, E.S., Barros, J.D. and da Silva, F.M. 2013. Comportamento da produção pesqueira norte-rio-grandense em anos de el niño e la niña. Sociedade e Território 25: 55–66.

Barbosa, C. 2020. Como a acidificação dos oceanos impacta as nossas vidas? IBDMAR/BILOS. Available in http://www.ibdmar.org/2020/09/como-a-acidificacao-dos-oceanos-impacta-as-nossas-vidas/. Accessed in 01/04/2022.

Bianchini, A., da Silva, C.C., Lauer, M.M., Jorge, M.B., Costa, P.G., Marques, J.A. et al. 2016. Avaliação do impacto da lama/pluma samarco sobre os ambientes costeiros e marinhos (ES e BA) com ênfase nas unidades de conservação 1a expedição no navio de pesquisa soloncy moura do CEPSUL/ICMBio. Ministério do Meio Ambiente, Instituto Chico Mendes de Conservação da Biodiversidade-ICMBio. Diretoria de Pesquisa, Avaliação e Monitoramento da Biodiversidade. Avaiable in http://www.icmbio.gov.br/ portal/publicacoes?id=7862:documentos-rio-doce. Accessed in 12/05/2021.

Boltovskoy, D 1981. Atlas del zooplancton del atlantico sudoccidental y metodos de trabajo con el zooplancton marino. Mar del Plata, Argentina: INIDEP.

Boltovskoy, D. 1999. South Atlantic Zooplankton, volumes 1–2. Leiden: Backhuys Publishers.

Boltovskoy, D., Correa, N. and Boltovskoy, A. 2003. Marine zooplanktonic diversity: a view from the south Atlantic diversité du zooplancton marin: un regard sur l'Atlantique Sud. Oceanologica Acta 25: 271–278.

Boltovskoy, D., Correa, N. and Boltovskoy, A. 2005. Diversity and endemism in cold waters of the south Atlantic: contrasting patterns in the plankton and the benthos. Sci. Mar. 69: 17–26. https://doi.org/10.3989/scimar.2005.69s217.

Boltovskoy, D. and Valentin, J.L. 2018. Overview of the history of biological oceanography in the southwestern Atlantic, with emphasis on plankton. pp. 3–34. *In*: Hoffmeyer, M.S., Sabatini, M.E., Brandini, F.P., Calliari, D.L. and Santinelli, N.H. (eds.). Plankton Ecology of the Southwestern Atlantic. Springer, Cham. https://doi.org/10.1007/978-3-319-77869-3_1.

Bonecker, A.C.T. and Castro, M.S. (Org.). 2006. Atlas de larvas de peixes da região central da zona econômica exclusiva brasileira. 1 ed. Rio de Janeiro: Museu Nacional.

Bonecker, A.C.T. and Castro, M.S. 2018. Larval fish assemblages in the foz do amazonas basin. Pan-American J. Aquatic Sci. 13: 114–120.

Bonecker, A.C.T., Castro, M.S., Costa, P.G., Bianchini, A. and Bonecker, S.L.C. 2019. Larval fish assemblages of the coastal area affected by the tailings of the collapsed dam in southeast Brazil. Reg. Stu. Mar. Sci. 32: 100848. https://doi.org/10.1016/j.rsma.2019.100848.

Bonecker, A.C.T., Dias, C.O., Castro, M.S., Carvalho, P.F., Araujo, A.V., Paranhos, R. et al. 2018. Vertical distribution of mesozooplankton and ichthyoplankton communities in the south-western Atlantic ocean (23°14'1"S 40° 42'19"W). Journal of the Marine Biological Association of the United Kingdom 99: 51–65. https://doi.org/10.1017/ S0025315417001989.

Bonecker, A.C.T., Menezes, B.S., Dias Jr., C., Silva, C.A., Ancona, C.M., Dias, C.O. et al. 2022. An integrated study of the plankton community after four years of fundão dam disaster. Sci. Total Environ. 806: 150613. https://doi.org/10.1016/j.scitotenv.2021.150613.

Bonecker, S.L.C. (Org.). 2006. Atlas de zooplâncton da região central da zona econômica exclusiva brasileira. 1 ed. Rio de Janeiro: Museu Nacional.

Bonecker, S.L.C., Araujo, A.V., Carvalho, P.F., Dias, C.O., Fernandes, L.F.L, Migotto, A.E. et al. 2014. Horizontal and vertical distribution of mesozooplankton species richness and composition down to 2,300 m in the southwest Atlantic ocean. Zoologia 31: 445–462. http://dx.doi.org/ 10.1590/S1984-46702014000500005.

Bonecker, S.L.C., Dias, C.O., Fernandes, L. and Araujo, A.V. 2012. Avaliação da comunidade Mesozooplanctônica. pp. 100–145. *In*: PETROBRAS (ed.). Baía de guanabara: síntese do conhecimento ambiental, v.2, Rio de Janeiro.

Botterell, Z.L.R., Beaumont, N., Dorrington, T., Steinke, M., Thompson, R.C. and Lindeque, P.K. 2019. Bioavailability and effects of microplastics on marine zooplankton: a review. Environ. Pollut. 245: 98e110. https://doi.org/10.1016/j.ocecoaman.2018.09.013.

Bradford-Grieve, J.M., Markhaseva, E.L., Rocha, C.E.F. and Abiahy, B. 1999. Copepoda. *In*: Boltovskoy, D. (ed.). South Atlantic Zooplankton – Vol. 2. Leiden: Backhuys Publishers.

Brandini, F.P., Lopes, R.M., Gutseit, K.S., Spach, H.L. and Sassi, R. 1997. Planctonologia na plataforma continental do brasil: diagnose e revisão bibliográfica. MMA, CIRM, FEMAR, 196 p.

Brierley, A.S. and Kingsford, M.J. 2009. Impacts of climate change on marine organisms and ecosystems. Curr. Biol. 19: 602–614. https://doi.org/10.1016/j.cub.2009.05.046.

Calil, P.H.R., Suzuki, N., Baschek, B. and Silveira, I.C.A. 2021. Filaments, fronts and eddies in the cabo frio coastal upwelling system, Brazil. Fluids 6: 54. https://doi.org/10.3390/fluids6020054.

Campelo, R.P.S., Bonou, F.K., de Melo Jr., M., Diaz, X.F.G., Bezerra, L.E.A. and Neumann-Leitão, S. 2019. Zooplankton biomassa round marine protected islands in the tropical Atlantic ocean. J. Sea Res. 154: 101810. https://doi.org/10.1016/j.seares.2019.101810.

Campelo, R.P.S., Lima, C.D.M., Santana, C.S., Silva, A.J., Neumann-Leitão, S., Ferreira, B.P. et al. 2021. Oil spills: the invisible impact on the base of tropical marine food webs. Mar. Pollut. Bull. 167: 112281. https://doi.org/10.1016/j.marpolbull.2021.112281.

Carvalho, P.F. and Bonecker, S.L.C. 2008. Notes on geographic distribution: tunicata, thaliacea, pyrosomatidae, *Pyrosomella verticillata* (Neumann, 1909): first record from the southwest Atlantic ocean. Check List 4: 272–274.

Carvalho, P.F. and Bonecker, S.L.C. 2016. Variação da composição e abundância das espécies da classe appendicularia e seu uso como potenciais bioindicadoras de regiões e massas de água superficiais na área da bacia de campos, rio de janeiro, Brasil. Iheringia Sér Zool 106: e2016022. https://doi.org/ 10.1590/1678-4766e2016022.

Carvalho, P.F., Bonecker, S.L.C. and Nassar, C.A.G. 2016. Analysis of the appendicularia class (subphylum urochordata) as a possible tool for biomonitoring four estuaries of the tropical region. Environ. Monit. Assess. 188. http://dx.doi.org/10.1007/s10661-016-5616-5.

Castro, F. 2008. O plâncton ignorado. agência FAPESP. Available in https://agencia.fapesp.br/plancton-ignorado/9171/. Accessed 01/05/2022.

Castro, R.O., Silva, M.L., Marques, M.R.C. and Araújo, F.V. 2016. Evaluation of microplastics in jurujuba cove, niterói, RJ Brazil, an area of mussels farming. Mar. Poll. Bull. 110: 555–558. https://doi.org/10.1016/j.marpolbul.2016.05.037.

Castro, R.O., da Silva, M.L. and de Araújo, F.V. 2018. Review on microplastic studies in Brazilian aquatic ecosystems. Ocean Coast Management 165: 385–400. https://doi.org/10.1016/j.ocecoaman.2018.09.013.

Chaudhary, C., Richardson, A.J., Schoeman, D.S. and Costello, M.J. 2021. Global warming is causing a more pronounced dip in marine species richness around the equator. Proc. Natl. Acad. Sci. 118: e201509411. https://doi.org/10.1073/pnas.2015094118.

Conceição, L.R., Souza, C.S., Mafalda Jr., P.O., Schwamborn, R. and Neumann-Leitão, S. 2021. Copepods community structure and function under oceanographic influences and antropic impacts from the narrowest continental shelf of southwestern Atlantic. Reg. Stud. Mar. Sci. 47: 10193. https://doi.org/10.1016/j.rsma.2021.101931.

Condon, R.H., Graham, W.M., Duarte, C.M., Pitt, K.A., Lucas, C.H., Haddock, S.H.D. et al. 2012. Questioning the rise of gelatinous zooplankton in the world's oceans. Biosci. J. 62: 160–169. https://doi.org/10.1525/bio.2012.62.2.9.

Cordes, E.E., Jones, D.O.B., Schlacher, T.A., Amon, D.J., Bernardino, A.F., Brooke, S. et al. 2016. Environmental impacts of the deep-water oil and gas industry: a review to guide management strategies. Front. Environ. Sci. 4: 58. https://doi.org/10.3389/fenvs.2016.00058.

Cotta, P.S. 2004. Impacts of offshore oil industryonthe marine biota and mitigating measures; impactos das atividades offshore da industria do petroleo sobre a comunidade biologica marinha e as medidas mitigadoras. Brazil: N. p., Web.

da Costa, K.G., Vallinoto, M. and da Costa, R.M. 2011. Molecular identification of a new cryptic species of *Acartia tonsa* (copepoda, acartiidae) from the northern coast of brazil, based on mitochondrial COI gene sequences. J. Coast Res. Special Issue 64: 359–63. https://www.jstor.org/stable/26482193.

da Cruz Lopes da Rosa, J., Dias, C.O., Suárez-Morales, E., Weber, L.I. and Fischer, L.G. 2021. Record of *caromiobenella* (copepoda, monstrilloida) in Brazil and discovery of the male of *C. brasiliensis*: morphological and molecular evidence. Diversity 13: 241. https://doi.org/10.3390/d13060241.

Dias, C.O. and Araujo, A.V. 2006. Copepoda. pp. 23–101. *In*: Bonecker, S.L.C. (ed.). Atlas da Região Central da Zona Econômica Exclusiva brasileira. Museu Nacional Série de livros n. 21: Rio de Janeiro.

Dias, C.O. and Araujo, A.V. 2019. Diversity, and habitat partitioning of scolecitrichidae species (copepoda: calanoida) down to 1,200 m in the southwestern Atlantic Ocean. An Acad Bras Cienc 91: e20170973. https://doi.org/10.1590/0001-376520192017.

Dias, C.O., Araujo, A.V. and Bonecker, S.L. 2019. Distribution, diversity, and habitat partitioning of scolecitrichidae species (copepoda: calanoida) down to 1,200 m in the southwestern Atlantic ocean. Na. Acad. Bras. Ciênc. 91. https://doi.org/10.1590/0001-3765201920170973.

Dias, C.O., Araujo A.V., Vianna S.C., Loureiro Fernandes L.F., Paranhos R., Suzuki M.S. et al. 2015. Spatial and temporal changes in biomass, production and assemblage structure of mesozooplanktonic copepods in the tropical south-west Atlantic ocean. J. Mar. Biol. Assoc. UK 95: 483–496. https://doi.org/10.1017/S0025315414001866.

Dias, C.O., Bonecker, A.C.T., Castro, M.S., Carvalho, P.F., Paranhos, R. and Bonecker, S.L.C. 2020. Holoplankton and meroplankton of three western Atlantic sedimentary basins. Mar. Biol. Res. 16, 10: 695–713, https://doi.org/10.1080/17451000.2021.1894341.

Dias, C.O., Carvalho, P.F., Bonecker, A.C.T. and Bonecker, S.L.C. 2018. Biomonitoring of the mesoplanktonic community in a polluted tropical bay as a basis for coastal management. Occan Coast Management 161: 189–200. https://doi.org/10.1016/j.ocecoaman.2018.05.007.

Diaz, R.J. and Rosenberg, R. 2008. Spreading dead zones and consequences for marine ecosystems. Science 321: 926–929. https://doi.org/10.1126/science.1156401.

Eskinazi-Sant'anna, E.M. 2000. Zooplankton abundance and biomass in a tropical estuary (pina estuary-northeast Brazil). Trab. Oceanog. Univ. Fed. PE 28: 21–34.

Fernandes, L.F.L., Paiva, T.R.M., Longhini, C.M., Pereira, J.B., Ghisolfi, R.D., Lázaro, G.C.S. et al. 2020. Marine zooplankton dynamics after a major mining dam rupture in the doce river, southeastern Brazil: rapid response to a changing environment. Sci. Total Environ. 736: 139621. https://doi.org/10.1016/j.scitotenv.2020.139621.

Figueiredo, G.M. and Vianna, T.M.P. 2018. Suspended microplastics in a highly polluted bay: abundance, size, and availability for mesozooplankton. Mar. Pollut. Bull. 135: 256–265. https://doi.org/10.1016/j.marpolbul.2018.07.020.

Fonseca, A.L., Newton, A. and Cabral, A. 2021. Local and meso-scale pressures in the eutrophication process of a coastal subtropical system: challenges for effective management. Estuar. Coast. Shelf Res. 250: 107109. https://doi.org/10.1016/j.ecss.2020.107109.

Franco, B.C., Defeo, O., Piola, A.R., Barreiro, M., Yang, H., Ortega, L. et al. 2020. Climate change impacts on the atmospheric circulation, ocean, and fisheries in the southwest south Atlantic ocean: a review. Climatic Change 162: 2359–2377. https://doi: 10.1007/s10584-020-02783-6.

Gasalla, M.A., Velasco, G., Rossi-Wongtschowski, C.L.D.B., Haimovici, M. and Madureira, L.S.P. 2007. Modelo de biomassas do ecossistema da plataforma continental externa, talude, e região oceânica adjacente do sudeste/sul do Brasil, entre 100 e 1000 m. série documentos revizee: Score Sul. São Paulo, Instituto Oceanográfico da USP.

Gilbert, P.M., Fullerton, D., Burkholder, J.M., Cornwell, J. and Kana, T.M. 2011. Ecological stoichiometry, biogeochemical cycling, invasive species, and aquatic food webs: san francisco estuary and comparative systems. Rev. Fish Sci. 19: 358e417. https://doi.org/10.1080/10641262.2011.611916.

Gomes, C.L. 2007. O mesozooplâncton da baía de guanabara: distribuição temporal dos principais grupos e produção de duas espécies de copepoda dominantes. Ph.D. Thesis. Federal University of Rio de Janeiro, Rio de Janeiro.

Gomes, C., Marazzo, A. and Valentin, J.L. 2004. The vertical migration behaviour of two calanoid copepods, *Acartia tonsa* dana, 1849 and *paracalanus parvus* (claus, 1863) in a stratified tropical bay in Brazil. Crustaceana 77: 941–954. http://dx.doi.org/10. 1163/1568540042781784.

Gruber, N. 2011. Warming up, turning sour, losing breath: ocean biogeochemistry under global change. Phil. Trans. R. Soc. A 369: 1980–1996. https://doi.org/10.1098/rsta.2011.0003.

Guenther, M., Lima, I., Mugrabe, G., Tenenbaum, D.R., Gonzalez-Rodriguez, E. and Valentin, J.L. 2012. Small time scale plankton structure variations at the entrance of a tropical eutrophic bay (guanabara bay, Brazil). Braz. J. Oceanogr. 60: 405–414. http:// dx.doi.org/10.1590/S1679-87592012000400001.

Hobday, A.J. and Pecl, G.T. 2014. Identification of global marine hotspots: sentinels for change and vanguards for adaptation action. Rev Fish Biol Fisheries 24: 415–425. https://doi.org/10.1007/s11160-013-9326-6.

Ivar do Sul, J.A., Costa, M.F., Barletta, M. and Cysneiros, F.J.A. 2013. Pelagic microplastics around an archipelago of the equatorial Atlantic. Mar. Pollut. Bull. 75: 305–309. https://doi.org/10.1016/j.marpolbul.2013.07.040.

Ivar do Sul, J.A., Costa, M.F. and Fillmann, G. 2014a. Surface waters are sources of microplastics to insular beaches in the western tropical Atlantic ocean. *In*: Proceedings of the 2nd International Ocean Research Conference (IORC). Barcelona-Spain.

Ivar do Sul, J.A., Costa, M.F. and Fillmann, G. 2014b. Microplastics in the pelagic environment around oceanic Islands of the western tropical Atlantic. Water Air Soil Pollut 225: 2004. https://doi.org/10.1007/s11270-014-2004-z.

Jablonski, S., Martins, A.S., Amaral, A.C.Z., Silva, A.O.Á., Rossi-Wongtschowski, C.L.D.B., Hazin. F.V. et al. 2006. Programa REVIZEE: relatório executivo. Brasília: Ministério do Meio Ambiente.

Larrazábal, M.E., Cavalcanti, E.A.H., Vieira, D.A.N., Oliveira-Koblitz, V.S., Araújo, E.M., Barreto, T.M.S. et al. 2009. Macrozooplâncton na zona econômica exclusiva do nordeste do Brasil (REVIZEE NE II e NE III). pp. 48–102. *In*: Hazin, F.H.V. (ed.). Oceanografia Biológica: Biomassa fitoplanctônica, zooplanctônica, macrozooplâncton, avaliação espacial e temporal do ictioplâncton, estrutura da comunidade de larvas de peixes e distribuição e abundância do ictionêuston. Fortaleza: Martins and Cordeiro.

Levitus, S., Antonov, J. and Boyer, T. 2005. Warming of the world ocean, 1955–2003. Geophys. Res. Letters 32: L0260.

Lima, A.R.A., Costa, M.F. and Barletta, M. 2014. Distribution patterns of microplastics within the plankton of a tropical estuary. Environ. Res. 132: 146–155. https://doi.org/10.1016/j.envres.2014.03.031.

Lima, A.R.A., Barletta, M., Costa, M.F., Ramos, J.A.A., Dantas, D.F., Melo, P.A.M.C. et al. 2015a. Changes in the composition of ichthyoplankton assemblage and plastic debris in mangrove creeks relative to moon phases. J. Fish Biol. 89: 619–640. https://doi.org/10.1111/jfb.1283.

Lima, A.R.A., Barletta, M. and Costa, M.F. 2015b. Seasonal distribution and interactions between plankton and microplastics in a tropical estuary. Estuar Coast Shelf Sci. 165: 213–225. https://doi.org/10.1016/j.ecss.2015.05.018.

Lopes, R.M., Katsuragawa, M., Dias, J.F., Montú, M.A., Muelbert, J.H., Gorri, C. et al. 2006. Zooplankton and ichthyoplankton distribution on thesouthern Brazilian shelf: an overview. Sci. Mar. 70: 189–202. https://scientiamarina.revistas.csic.es/index.php/scientiamarina/article/view/147.

Lopes, R.M. 2007. Marine zooplankton studies in Brazil: a brief evaluation and perspectives. An. Acad. Bras. Ciênc. 79: 369–379. https://doi.org/10.1590/S0001-37652007000300002.

Lutz, V.A., Segura, V., Dogliotti, A.I., Gagliardini, D.A., Bianchi, A.A. and Balestrini, C.F. 2010. Primary production in the argentine Sea during spring estimated by field and satellite models. J. Plankton Res. 32: 181–195. https://doi.org/10.1093/plankt/fbp117.

MacArthur, D.E., Waughray, D. and Stuchtey, M.R. 2016. The new plastics economy: rethinking the future of plastics. *In*: World Economic Forum.

Margalef, R. 1961. Communication of structure in planktonic populations. Limnol. Oceanogr. 6: 124–128.

Marques, S.C., Azeiteiro, U.M., Marques, J.C., Miguel Neto, J. and Pardal, M.A. 2006. Zooplankton and ichthyoplankton communities in a temperate estuary: spatial and temporal patterns. J. Plankton Res. 28: 297–312. https://doi:10.1093/plankt/fbi126.

McGlathery, K.J., Sundbäck, K. and Anderson, I.C. 2007. Eutrophication in shallow coastal bays and lagoons: the role of plants in the coastal filter. Mar. Ecol. Prog. Ser. 348: 1–18. https://doi.org/10.3354/meps07132.

Melo, P.A.M.C., Neumann-Leitão, S., Zanardi-Lamardo, E., Flores-Montes, M.J. and de Melo Jr. M. 2021. Morphological abnormalities in acartia lilljeborgii giesbrecht (1889) (copepoda, calanoida) in a tropical estuary under industrial development. An. Acad. Bras. Ciênc. 93: e20190231. https://doi.org/10.1590/0001-3765202120190231. PMID: 33852671.

Mianzan, H., Lasta, C., Acha, E., Guerrero, R., Macchi, G. and Bremec, C. 2001. The río de la plata estuary, argentina-uruguay. *In*: Seeliger, U. and Kjerfve, B. (eds.). Coastal Marine Ecosystems of Latin America. Ecological Studies (Analysis and Synthesis) 144. Springer, Berlin, Heidelberg. https://doi.org/10.1007/978-3-662-04482-7_14.

MMA - Ministério do Meio Ambiente. 2006. Programa REVIZEE: avaliação do potencial sustentável de recursos vivos na zona econômica exclusiva: relatório executivo/MMa, Secretaria de Qualidade Ambiental, Brasil.

Ministério do Meio Ambiente. 2009. Informe sobre as espécies exóticas invasorasz marinhas no Brasil, p. 440. http://www.mma.gov.br/estruturas/chm/_arquivos/biodiversidade_33_14.pdf, Accessed 08th feb 2022.

Moisan, J.R., Moisan, T.A. and Abbott, M.R. 2002. Modelling the effect of temperature on the maximum growth rates of phytoplankton populations. Ecol. Model. 153: 197–215. https://doi.org/10.1016/S0304-3800(02)00008-X.

Montú, M., Duarte, A.K. and Gloeden, I.M. 1997. Environment and biota of the patos lagoon estuary. Zooplankton. pp. 40–43. *In*: Seeliger, U., Odebrecht, C. and Castello, J.P. (eds.). Subtropical Convergence Environments: The Coast and Sea in the Southwestern Atlantic. Berlin: Springer.

Nascimento, N.R., Vital, G.V., Plaza, A.S. and Souza, G.L.M. 2021. Atividade petrolífera offshore e sua relação com os impactos ambientais nos ecossistemas marinhos. Meio Ambiente (Brasil) 3: 46–63.

Neumann-Leitão, S., Eskinazi Sant'anna, E.M., Gusmão, L.M.O., Nascimento-Vieira, D.A., Paranaguá, M.N. and Schwamborn, R. 2008. Diversity and distribution of the mesozooplankton in the tropical southwestern Atlantic. J. Plankton Res. 30: 795–805. https://doi.org/10.1093/plankt/fbn040.

Neumann-Leitão, S., Melo, P.A.M.C., Schwamborn, R., Diaz, X.F.G., Figueiredo, L.G.P., Silva, A.P. et al. 2018. Zooplankton from a reef system under the influence of the amazon river plume front. Microbiol. 9: 355. https://doi.org/10.3389/fmicb.2018.00355.

Neumann-Leitão, S., de Melo Jr. M., Porto Neto, F.F., Silva, A.P., Diaz, X.F.G., Silva, T.A. et al. 2019. Connectivity between coastal and oceanic zooplankton from rio grande do norte in the tropical western Atlantic. Front. Mar. Sci. https://doi.org/10.3389/fmars.2019.00287.

Nogueira, C.R., Bonecker, A.C.T. and Bonecker, S.L.C. 1989. Zooplâncton da baía de guanabara (RJ-Brasil). composição e variações espaço-temporal. Memórias do III Encontro Brasileiro de Plâncton: 151–156.

Nogueira Jr, M., Rodriguez, C.S., Mianzan, H., Haddad, M.A. and Genzano, G. 2013. Description of a new hydromedusa from the southwestern Atlantic ocean, *bougainvillia pagesi* sp. nov. (Cnidaria, Hydrozoa, Anthoathecata). Mar. Ecol. 34: 113–122. https://doi.org/10.1111/maec.12030.

Nogueira Jr, M., Santos, P.C.B., Aguilar, M.T., Brandini, F. and Miyashita, L.K. 2018. Diversity of gelatinous zooplankton (cnidaria, ctenophora, chaetognatha and tunicata) from a subtropical estuarine system, southeast Brazil. Mar. Biodiv. 1–16. https://doi.org/10.1007/s12526-018-0912-7.

Odebrech, C. and Castello, J.P. 2001. The convergence ecosystem in the southwest Atlantic. pp. 145–165. *In*: Seeliger, U. and Kjerfve, B. (eds.). Subtropical Convergence Environments: The Coast and Sea in the Southwestern Atlantic. Ecological Studies (Analysis and Synthesis), 144, Springer, Berlin, Heidelberg. https://doi.org/10.1007/978-3-662-04482-7_12: 2001.

Odebrecht, C., Secchi, E.R., Abreu, P.C., Muelbert, J.H. and Uiblein, F. 2017. Biota of the patos lagoon estuary and adjacent marine coast: long-term changes induced by natural and human-related factors, Mar. Biol. Res. 13: 3–8. https://doi.org/10.1080/17451000.2016.1258714.

Olivatto, G.P., Martins, M.C.T., Montagner, C.C., Henry, T.B. and Carreira, R.S. 2019. Microplastic contamination in surface waters in guanabara bay, rio de janeiro, Brazil. Mar. Poll. Bull. 139: 157–162. https://doi.org/10.1016/j.marpolbul.2018.12.042.

Orr, J.C., Caldeira, K., Fabry, V., Gattuso, J-P., Haugan, P., Lehodey, P. et al. 2009. Research priorities for ocean acidification, report from the second symposium on the ocean in a high-CO_2 world, Monaco, October 6–9, 2008, convened by SCOR, UNESCO-IOC,IAEA, and IGBP, 25P. Available in http:// ioc3.unesco.org/oanet/HighCO2World.htm. Accessed in 12/10/2021.

PBMC- Painel Brasileiro de Mudanças Climáticas. 2021. Available in http://pbmc.coppe.ufrj. br/index.php/en/news/476-acidificacao-dos-oceanos-um-grave-problema-para-a-vida-no-planeta#:~:text=Vida%20marinha%20em%20risco&text=Estudos%20preliminares%20 apontam%20que%20a,conchas%2C%20levando%20ao%20seudesaparecimento.   Accessed   in 01/10/2022.

Pecl, G.T., Araújo, M.B., Bell, J.D., Blanchard, J., Bonebrake, T.C., Chen, I-C. et al. 2017. Biodiversity redistribution under climate change: impacts on ecosystems and human well-being. Science 355: eaai9214. https://doi.org/10.1126/science.aai9214.

Poloczanska, E.S., Burrows, M.T., Brown, C.J., Molinos, J.G., Halpern, B.S., Hoegh-Guldberg, O. et al. 2016. Responses of marine organisms to climate change across oceans. Front. Mar. Sci. 3: 62. https://doi.org/10.3389/fmars.2016.00062.

Popova, E., Yool, A., Byfield, V., Cochrane, K., Coward, A.C., Salim, S.S. et al. 2016. From global to regional and back again: common climate stressors of marine ecosystems relevant for adaptation across five ocean warming hotspots. Global Change Biol. 22: 2038–2053. https://doi.org/10.1111/ gcb.13247.

REVIZEE. 2022. Avaliação do Potencial Sustentável de Recursos Vivos na Zona Econômica Exclusiva. Available in https://www.marinha.mil.br/secirm/psrm/revizee. Accessed in 12/09/2021.

Robb, C.K. 2014. Assessing the impact of human activities on bristish columbia's estuaries. PLoS ONE 9: e99578. https://doi.org/10.1371/journal.pone.0099578.

Rocha, G.M., Salvador, B., Laino, P.S., Santos, G.H.C., Demoner, L.E., Conceição, L.R. et al. 2022. Responses of marine zooplankton indicators after five years of a dam rupture in the doce river, southeastern Brazil. Sci. Total Environ. 806: 151249. https://doi.org/10.1016/j.scitotenv.2021.151249.

Salvador, B. and Bersano, J.G.F. 2017. Zooplankton variability in the subtropical estuarine system of paranaguá bay, Brazil, in 2012 and 2013. Estuar Coastal Shelf Sci. 199: 1–13. https://doi. org/10.1016/j.ecss.2017.09.019.

Santos, P.V. 2012. Impactos ambientais causados pela perfuração de petróleo, cadernos de graduação, ciências exatas e tecnológicas 1: 53–163. https://periodicos.set.edu.br/cadernoexatas/article/view/297.

Santos, R.V.S., Ramos, S. and Bonecker, A.C.T. 2017. Can we assess the ecological status of estuaries based on larval fish assemblages? Mar. Pollut. Bull. 124: 367–375. 10.1016/j.marpolbul.2017.07.043.

Schutze, M.L.M. and Ramos, J.M. 1999. Variação anual do zooplâncton na baía de guanabara e na região costeira litorânea adjacente (rio de janeiro– Brasil) com especial referência aos copépodes. pp. 61–72. *In*: Silva, S.H.G. and Lavrado, H.P. (eds.). Ecologia dos Ambientes Costeiros do Estado do Rio de Janeiro.

Sinev, A.Y. and Elmoor-Loureiro, L.M.A. 2010. Three new species of chydorid cladocerans of subfamily aloninae (branchipoda: anomopoda: chydoridae) from Brazil. Zootaxa 2390: 1–25.

Soares, J.M.F. 2016. Impacto de alterações climáticas na receita da pesca do camarão-rosa na lagoa dos patos, evidenciando o município de são lourenço do Sul/RS, Brasil. Dissertação em Gerenciamento Costeiro, Instituto de Oceanografia, Universidade Federal do Rio Grande, 97p. Available in https:// sistemas.furg.br/sistemas/sab/arquivos/conteudo_digital/746d0280f41c68fae1fbed5730063d0a.pdf. Accessed in 12/07/2021.

Sodré, C.F.L., Silva, Y.J.A. and Monteiro, I.P. 2016. Acidificação dos oceanos: fenômeno, consequências e necessidade de uma governança ambiental global. Revista Científica do CEDS da UNDB 1. Available in www.undb.edu.br/ceds/revistadoceds. Accessed in 12/07/2021.

Sorrentino, R., Alves, J., Johnssson, R. and Senna, A.R. 2016. A new species of cyphocarididae (crustacea, amphipoda, lysianassoidea) from off the northeastern Brazilian coast. Zootaxa 4161: 345–356. https://doi.org/10.11646/zootaxa.4161.3.3.

Souza, C.A., Gomes, L.F., Nabout, J.C., Velho, L.F.M. and Vieira, L.C.G. 2018. Temporal trends of scientific literature about zooplankton community. Neotrop. Biol., Conserv. 13: 274–286. https://doi.org/10.4013/nbc.2018.134.01.

Steffen, W., Richardson, K., Rockström, J., Cornell, S.E., Fetzer, I., Bennett, E.M. et al. 2015. Sustainability. planetary boundaries: guiding human development on a changing planet. Science 347: 259855. https://doi/org/10.1126/science.1259855.

Sterza, J.M. and Fernandes, L.L. 2006. Zooplankton community of the vitória bay estuarine system (southeastern Brazil): Characterization during a three-year study. Braz. J. Oceanogr. 54: 95–105.

Tabarelli, M., Rocha, C.F.D., Romanowski, H.P., Rocha, O. and Lacerda, L.D. (eds.). 2013. PELD–CNPq, Dez anos do programa de pesquisas ecológicas de longa duração do Brasil: achados, lições e perspectivas. Editora Universitária da Universidade Federal de Pernambuco, Recife.

Teixeira-Amaral, P.T., Amaral, W.J.A., Ortiz, D.O., Agostini, V.O. and Muxagata, E. 2017. The mesozooplankton of the patos lagoon estuary, Brazil: trends in community structure and secondary production. Mar. Biol. Res. 13: 48–61. https://doi.org/10.1080/17451000.2016.1248850.

Troina, G.C., Riekenberg, P., van der Meer, M.T., Botta, S., Dehairs, F. and Secchi, E.R. 2021. Combining isotopic analysis of bulk-skin and individual amino acids to investigate the trophic position and foraging areas of multiple cetacean species in the western south Atlantic. Environ. Res. 201: 111610. https://doi.org/10.1016/j.envres.2021.111610.

Valentin, J.L. 2008. Consequências das mudanças climáticas para o plâncton do Atlântico Sul. SBPC, Reunião Anual, Energia, Ambiente, Tecnologia, Campinas, 6 p. Available in http://www.sbpcnet. org.br/livro/60ra/textos/si-jeanlouisvalentin.pdf. Accessed in 10/13/2021.

Valentin, J.L., Gouvêa, G.V. and Gomes, C.l. 2020. Mesozooplâncton e massas d´água na baía de guanabara: dez anos de monitoramento. Oecol. Aust. 24: 349–364. https://doi.org/10.4257/oeco.2020.2402.09.

Valentin, J.L., Tenenbaum, D.R., Bonecker, A.C.T., Bonecker, S.L.C., Nogueira, C.R., Paranhos, R. et al. 1999. Caractéristiques hydrobiologiques de la baie de guanabara (rio de janeiro, brésil). J. Rech. Océanographique 24: 33–41.

Vieira-Menezes, F.G., Dias, C.O., Cornils, A., Silva, R. and Bonecker, S.L.C. 2021. New Paracalanidae species from the central coast of Brazil: morphological description and molecular evidence. Mar. Biodivers. 51: 54. https://doi.org/10.1007/s12526-021-01188-7.

Yoneda, N.T. 2000. Avaliação e ações prioritárias para a conservação da biodiversidade da zona costeira e marinha - plâncton. Base de dados tropical. Available in: http://www.bdt.org.br/workshop/costa/plancton/inde. Accessed in 11/10/2021.

Xie, S.-P., Deser, C., Vecchi, G.A., Jian, M., Teng, H. and Wittenberg, T. 2009. Global warming pattern formation: sea surface temperature and rainfall. J. Climate 23: 966–986. https://doi.org/10.1175/2009JCLI3329.1.

3.3

Marine Plankton of Mozambique
The Rich (In)Visible World Supporting
Biodiversity Hotspots

Fernando Morgado[1],* and *Luis R. Vieira*[2]

1. Introduction

Mozambique is located in tropical East Africa, which includes the coasts of Somalia, Kenya, and Tanzania, between latitudes 10°20'S and 26°50'S and has an approximate area of 783,000 km², of which about 4,500 km² is marine. Mozambique is covered by a diversity of ecosystems and species of global interest for conservation, being part of the five main phytogeographic zones of Southern Africa, namely: (i) Regional Center of Zambezian Endemism, (ii) Regional Center of Swahilian Endemism (Regional Mosaic Zanzibar-Inhambane), (iii) Swahili-Maputaland Regional Transition Zone, (iv) Maputaland-Tongoland Regional Mosaic, and (v) Center for Afromontane Endemism (White 1983, Burgess and Clarke 2000, Burgess et al. 2006, Van Wyk 1994, 1996, Van Wyk and Smith 2001). In these phytoregions, there are five different phytochories, subdivided into 12 ecoregions (Burgess et al. 2004) that are in different conservation states and represent important biodiversity and endemism hotspots (Van Wyk 1996, Van Wyk and Smith 2001). Diversity hotspots and centers of plant endemism in Mozambique include the Maputaland and Chimanimani regions, coastal forests and the "inselbergs" island hills in northern Mozambique (Burgess et al. 2004). The coastline has a length of about 2,700 km, the third longest coastline in Africa from Rovuma River in the North (parallel 100 27'S) to Ponta do Ouro in the south (parallel 260 52'S) on the borders with the Republic of Tanzania and the Republic of South Africa, north and south, respectively (Hoguane and Pereira 2003). The continental shelf, up to the 200 m isobath, has an area of 104 km². The coast is

[1]   The Center for Environmental and Marine Studies (CESAM) and Department of Biology University of Aveiro, Campus Universitário de Santiago, Aveiro, Portugal.

[2]   Interdisciplinary Center of Marine and Environmental Research (CIIMAR). University of Porto, Terminal de Cruzeiros do Porto de Leixões, Av. General Norton de Matos s/n, 2250–208 Matosinhos, Portugal.

*   Corresponding author: fmorgado@ua.pt

divided into three major natural regions: the coral coast, marsh coast and parabolic dune coast. It is characterized by a diversity of habitats that includes sandy beaches, coastal dunes, coral reefs, estuaries, bays, forests, and mangrove swamps and seagrass beds (Hewawasam 2000, Hoguane and Pereira 2003, Hoguane 2007, TRANSMAP 2008). The morphology of the coast is characterized by low areas, with an altitude up to about 200 m above the mean sea level. Three distinct hydrogeological zones can be identified along the Mozambican coast:

(i) Dune coast, characteristic of the area south of the Save River, where the porous areas deposited by wind agents form a regional phreatic aquifer. Soil permeability decreases from the coast to the interior as soils become rich in clay;

(ii) Alluvial plains that developed along the main rivers, characteristic of the central zone;

(iii) Volcanic lands, which mark the boundary between sea and land, are characteristic of the northern zone (Hoguane 2007).

Mozambique is inserted in the Marine Eco-Region of East Africa, with an extensive coastal area covering 4,600 km and extending from southern Somalia to the Kwazulu-Natal coast of South Africa (Boon 2009, Spalding et al. 2007, WWF 2009). This coastal region constitutes a region with high biodiversity and characterized by a great diversity of habitats, such as mangrove forests, seagrass beds, coral reefs and open sea, and characterized by a very specific type of physical and ecological link between coastal and marine habitats (Spalding et al. 2007). Along the coast, the habitats constitute a diversified matrix supporting a wide, rich and complex range of marine populations, including bacteria, phytoplankton, zooplankton, fishes, mammals and birds, that depend on this diversity for their productivity that include complex trophic networks in interaction and characterized by high productivity (Spalding et al. 2007). The sea around the Southern Indian Ocean is a mixture of water masses from several different origins, consisting of nutrient-rich Agulhas current and southern ocean influenced waters off South Africa, oligotrophic offshore waters in the central areas, with warm and cold eddies mixing in toward tropical waters off Australia (Harris 1972, Saetre and da Silva 1979, 1984, Nehring et al. 1987, Lutjeharms 1976, Lutjeharms and Van Ballegooyen 1988, Grundlingh 1995, Stammer 1997, Schouten et al. 2002a, Biastoch and Krauss 1999, Stramma and Lutjeharms 1997, Schott and McCreary 2001, Chemane et al. 1997, Kostianoy et al. 2003, Ridderinkhof and Ruijter 2003). The marine biological and coastal ecosystems of the Africa and Mozambican east and west coasts are modulated by the inter-dependencies between physical, biogeochemical and ecological components of regional ocean systems and also under their anthropogenic influence (de Ruijter et al. 2002, Groeneveld and Koranteng 2017, Biastoch and Krauß 1999, de Ruijter et al. 2005, Van Aken et al. 2004, Ridderinkhof and de Ruijter 2003, Lutjeharms and Jorge da Silva 1988, Lutjeharms 2006, Weimerskirch et al. 2004, Tew Kai and Marsac 2008, Groeneveld and Koranteng 2017, Sætre and Silva 1982, Hoguane 1996, 1999, Gammelsrød and Hoguane 1996, Brinca et al. 1983, Steen and Hoguane 1990).

The rich (in)visible world of plankton in these areas is crucial for supporting biodiversity hotspots. Planktonic organisms are one of the most abundant life forms on Earth (besides bacteria) and play a crucial trophic role in the marine food chain, they occupy multiple trophic levels in pelagic food webs (Biller et al. 2018, Carradec et al. 2018, Gregory et al. 2019). They are an intermediate trophic level between phytoplankton and fish and an important component of carbon and nutrient cycles in the ocean (Clifford et al. 2017). Plankton range in size from tiny microbes (> 100 nm), which are invisible to the naked eye, to jellyfish (up to 30 m meters long) and comprise a phylogenetically and functionally diverse assemblage of protistan and metazoan consumers, composed of viruses, bacteria, phytoplankton, zooplankton and the pelagic larvae of many marine invertebrates and fishes (Biller et al. 2018, Carradec et al. 2018, Gregory et al. 2019). Microphytoplankton includes cyanobacteria, diatoms and dinoflagellates, the autotrophic picoplankton, that deliver about half of the ocean's primary production in the biosphere (Boyce et al. 2010, Vallina et al. 2014). Most eukaryotic plankton biodiversity belonged to heterotrophic protistan groups, particularly those known to be parasites or symbiotic hosts. The microzooplankton size class (< 200 μm) is functionally dominated by protistan (unicellular eukaryote), metazoan consumers but also includes smaller juvenile stages of metazooplankton species (Biller et al. 2018, Carradec et al. 2018, Gregory et al. 2019). The mesozooplankton size class (0.2–20 mm) consists mainly of true animals but also includes large protists, such as pelagic foraminifera and radiolaria (Moriarty et al. 2013, Clifford et al. 2017). Many zooplanktonic organisms, from single-cell bacteria to large ones, are known to produce light (bioluminescence is the production and emission of visible light by living organisms). Plankton biodiversity is fundamental for the marine ecosystems' functioning due to their role in ocean ecology and biogeochemistry.

2. Plankton Studies Over the Coast of Mozambique

Plankton studies along the coast of Mozambique are scarce. The first records were restricted mainly to specific aspects of some taxonomic groups, such as phytoplankton (Silva 1956) and foraminifera (Braga 1960). Some years later, some studies were carried out in the southern region (island of Inhaca) by Gove and Cuamba (1989), where preliminary observations were made on the seasonality of plankton on the island of Inhaca. Paula (1998) described the seasonal cycle of planktonic communities on the east coast of Inhaca Island, analyzing the abundance and distribution of phytoplankton, the relationship with nutrients and physical parameters, in addition to the abundance, distribution and seasonal fluctuations of zooplankton. Special attention was given to the larval stages of bivalves and decapods due to the importance of a large number of these organisms for local consumption. Subsequently, in the same place, Clark and Paula (2003) developed studies on larval stages of decapods and produced identification keys for three common crab species on the island of Inhaca and restricted distribution in the Western Indian Ocean. The understanding of the relations between productivity and distribution of organisms, phytoplankton, micronekton, zooplankton, fish and marine megafaunas, such as whales and whale sharks and the circulation patterns of the current of the Mozambique Channel and the Banco de Sofala and Delagoa Bight (Ternon et al. 2014), triggered a recent increase of studies off the coast of Mozambique. Several studies in the Mozambique Channel have been developed with the aim of analyzing the composition and distribution of the mesoscale zooplankton community (Lebourges-Dhaussy et al. 2009, Jaquemet 2014). Hydrodynamic processes, availability of nutrients, phytoplankton, zooplankton and abundance and distribution of fish larvae were also investigated in some areas of coastal waters in Mozambique, such as the Sofala Bank during "Dr. Fridtjof Nansen" Survey, in 2007 (Johnsen et al. 2007). Another study carried out in the central region (Sofala Bank) addressed issues related to the dynamics of the vertical distribution of zooplankton (Leal 2009). Recently, several studies have been developed on the nature of ocean circulation, regional ocean-atmosphere interactions and the effect of changing sea surface temperature fields on weather and climate around Southern Africa (Lebourges-Dhaussy et al. 2009, Krakstad et al. 2015, Everett 2017).

3. Marine Plankton of Mozambique

The oligotrophic marine waters of Mozambique are dominated by small plankton with low biomass and driven by nutrient recycling (Johnsen et al. 2007, Tew Kai and Marsac 2008, Lebourges-Dhaussy et al. 2009, Huggett 2014, Krakstad et al. 2015, Groeneveld et al. 2017), as seen in coral reef habitats (O'Donnell et al. 2017). Plankton biogeographic fluctuations are strongly correlated with basin-scale climate indices, whereas long-term declining trends are related to increasing sea surface temperatures (Leal 2009, Raj et al. 2010, Swart et al. 2010, Barlow et al. 2014, Huggett 2014, Lamont et al. 2014). The seasonal range of upper ocean environmental conditions is typically greatest at mid and high latitudes, but substantial seasonal variations of environment and plankton biomass also occur in many tropical regions (Barlow et al. 2007, Tew Kai and Marsac 2009, Sá et al. 2013). In oligotrophic waters, such as in Mozambican marine waters, particularly the vast oceanic gyres, abundances and distribution are uniform over large geographical areas and were primarily driven by factors related to vertical mixing and nutrient delivery (Huggett 2007, Olofsson et al. 2017, Barlow et al. 2017). However, in the Indian Ocean, a complex seasonality is observed related to the effects of monsoon dynamics and freshwater inputs on nutrient delivery (Steen and Hoguane 1990, Hoguane 2007, Leal et al. 2009). Global autotrophic picoplankton distribution abundance is expected to increase in temperate and subpolar oceans as warming shifts their thermal regimes toward the thermal niche of picoautotrophs, although their abundance is much more moderate or even decline in the subtropical and tropical ocean (Tew Kai et al. 2008, Raj et al. 2010, Lamont et al. 2014, Barlow et al. 2014, 2017). As subtropical gyres represent 70% of the ocean surface, the largest biome in the planet, the responses predicted represent a significant change in the Earth system with important implications for global biodiversity and biogeochemical cycling (Quartly and Srokosz 2004, Lamont et al. 2010, Ternon et al. 2014, José et al. 2014). Mesoscale oceanic processes such as localized wind-driven shelf currents and Ekman-driven upwelling are key sources of nutrient input in these areas (de Ruijter et al. 2002, Ridderinkhof and de Ruijter 2003, Schouten et al. 2002a, b, Schouten et al. 2003, Van Aken et al. 2004, Weimerskirch et al. 2004, de Ruijter et al. 2005, Lutjeharms 2006, Tewkai and Marsac 2008, Ridderinkhof et al. 2010, Lutjeharms et al. 2012, Groeneveld and Koranteng 2017). Within oligotrophic waters, such as in Mozambique tropical latitudes, the

Table 1. Most important groups and species.

Planktonic Groups	Species	References
Phytoplankton, nanoplankton and picoplankton: diatoms	*Chaetoceros* spp., *Pseudo-nitzschia* spp., *Proboscia alata, Cerataulina pelagica* and *Thalassionema nitzschioides*	Leal et al. 2009, Raj et al. 2010, Sá et al. 2013, Lamon et al. 2014, Barlow et al. 2014, Barlow et al. 2017, Kelchner 2020
Phytoplankton, nanoplankton and picoplankton: dinoflagellates	*Protoperidinium* spp., *Ceratium* spp. and *Pyrophacus* spp., were also observed in low abundance *Alexandrium* spp., *Dinophysis* spp., *Prorocentrum* spp. and *Akashiwo* spp.	
Phytoplankton, nanoplankton and picoplankton: coccolithophores	*Discosphaera tubifera* and *Emiliania huxleyi*	Huggett 2007, Leal et al. 2009, Huggett 2014, Johnsen et al. 2007, Tew Kai and Marsac 2008, 2009, 2010, Lebourges-Dhaussy et al. 2009, Huggett 2014, Krakstad et al. 2015, Groeneveld et al. 2017, Kelchner 2020
Zooplankton. Holoplankton. Microzooplankton, mesozooplankton, and macrozooplankton	Calanoid, cyclopoid and poecilostomatoid copepods (Paracalanidae, Oithonidae, Oncaeidae, *Clausocalanus* spp., *Oithona* spp., *Paracalanus* spp., *Macrostella* spp, *Acartia* spp., *Oncaea* spp., *Cyclopo*ida spp.), Tunicata (appendicularians and Dolliolids), ostracods, chaetognaths and siphonophores, cladocera, mysids, and Euphausiids	
Meroplankton	Fish eggs and larvae, shrimp larvae, coral larvae, Cirripedia Nauplii and Metanauplii, cyphonaut larvae, decapods larvae, echinoderm larvae and mollusk larvae, and euphausiid nauplii	

phytoplankton community tends to be dominated by picoplankton (> 3 μm), where low surface-to-volume ratio makes them the best competitor for limited nutrients (Leal et al. 2009, Raj et al. 2010, Sá et al. 2013, Lamon et al. 2014, Barlow et al. 2014, Barlow et al. 2017, Kelchner 2020).

Temperature Is the parameter related to differences observed: areas of cooler water dominated by microphytoplankton (diatoms) and regions of warmer water with pico-sized-dominated communities. In shallower areas of the continental shelf dominated by microflagellates, common with other oligotrophic regions, where phytoplankton dominance is not only due to microplankton (Tew Kai et al. 2008, Raj et al. 2010, Lamont et al. 2014, Barlow et al. 2014, 2017). Zooplankton taxonomic composition is typical of tropical and subtropical ecosystems (≤ 1 mm length), dominated by small copepods and other microzooplankton that drive larval fish viability (Jenny and Huggett 2007, Leal et al. 2009, Huggett 2014, Johnsen et al. 2007, Tew Kai and Marsac 2008, Lebourges-Dhaussy et al. 2009, Huggett 2014, Krakstad et al. 2015, Groeneveld et al. 2017, Kelchner 2020). Zooplankton plays a major role in the functioning and productivity of these marine waters and links primary productivity to fisheries abundance (Rohner et al. 2013b, Jaquemet et al. 2014, Pierce et al. 2008, Pereira et al. 2014, Williams et al. 2017, Fordyce 2018, O'Donnell et al. 2017). Planktonic organisms are food for a range of animals, from barnacles and sea squirts to large fish and whales, whereby plankton biomass models are crucial to identify biodiversity hotspots. Marine megafauna (mammals, large fish, and birds) also consume large amounts of plankton and small fish and can balance trophic dynamics (Rohner et al. 2013b, Jaquemet et al. 2014, Pierce et al. 2008, Pereira et al. 2014, Williams et al. 2017, Fordyce 2018, O'Donnell et al. 2017). Current climate change is shifting plankton abundance, composition and distribution because of their impact on physiological response and its effect on plankton food webs (Mackas and Beaugrand 2010). Jellyfish aggregations (Cnidarians, Siphonophora, Medusa, and Ctenophora) are a natural feature of healthy pelagic ecosystems, but evidence is accumulating that the severity and frequency of outbreaks are increasing in many areas related to climate change. The most important registed groups and species of marine plankton research from Mozambique are presented in Table 1.

4. Marine Plankton Vertical and Horizontal Distribution in Mozambique

Plankton high biomass is associated with the highest productive areas of Mozambique, mainly in the southern inshore edge of a strong cyclonic eddy located south of Angoche and in other productive areas of the Sofala Bank and southern inshore regions (Bazaruto and Delagoa Bight) (Johnsen et al. 2007, Tew Kai and Marsac 2008, Lebourges-Dhaussy et al. 2009, Huggett 2014, Krakstad et al. 2015, Groeneveld et al. 2017). Phytoplankton present longitude (coast-offshore, more productive waters were found near the coast) and latitudinal differences (increase in biomass from north to south). Microphytoplankton were associated with cooler southern water masses, while picophytoplankton were related to warmer northern water masses. The highest concentrations of micro-sized phytoplankton are observed both at the surface and sub-surface (at depths between 30 m and 40 m), where higher concentrations of nutrients are also normally recorded. The nano and pico-sized phytoplankton revealed opposing patterns, being most abundant at the surface (Leal et al. 2009, Raj et al. 2010, Sá et al. 2013, Lamon et al. 2014, Barlow et al. 2014, Barlow et al. 2017, Kelchner 2020). Mesozooplankton presents higher biomass inshore compared to offshore and higher biomass in the southern sector of the western channel compared to the oligotrophic northern region. Mesozooplankton is largely concentrated in the upper 100 m, but in shallower areas of the continental shelf, they can occur between 30 m and 40 m. Concentrations of zooplankton biovolumes are higher in relatively cool water (< 20°C) at depths of 100 m but are low at shelf-edge stations to the north off the Mozambique coast, relatively warmer temperatures (> 23°C) at 100 m (Jenny and Huggett 2007, Leal et al. 2009, Huggett 2014, Johnsen et al. 2007, Tew Kai and Marsac 2008, Lebourges-Dhaussy et al. 2009, Huggett 2014, Krakstad et al. 2015, Groeneveld et al. 2017, Kelchner 2020).

5. Plankton Contribution to Density of Higher Trophic Level Organisms

Higher trophic level productivity is linked to cycles in lower food web production at both the phytoplankton-zooplankton scale as well as zooplankton-fish interactions (Johnsen et al. 2007, Huggett 2014, Krakstad et al. 2015, Groeneveld and Koranteng 2017, Kelchner 2020). Plankton account for a large proportion of the energy transfer to areas of high primary production, such as coastal upwelling areas and coral reefs (Fordyce 2018), seagrass meadows (Guidi et al. 2016) that are associated with highly productive fish and fisheries on continental shelves (Ternon et al. 2014, José et al. 2014, Johnsen et al. 2007, Olsen et al. 2009). The Southeast Indian Ocean circulation patterns, and Mozambique channel currents and the Sofala Bank and Delagoa Bight have a profound influence on the biological oceanography of the Mozambican coastal areas, including the productivity and distribution of phytoplankton and micronekton and also of higher trophic levels from zooplankton to fish to megafauna, such as whales and whale sharks (Ternon et al. 2014). The northern part of the coast also has favorable feeding conditions for tuna spawning in surface chlorophyll fronts on the annual migration routes of the Indian Ocean tuna stocks (Druon et al. 2017). Mozambique has several coastal regions known as biodiversity hotspots, Inhambane Province and the parabolic dune coast, having a relatively high density of whale sharks (*Rhincodon typus*) planktivorous reef manta rays (*Manta alfredi*) and giant manta rays (*M. birostris*) (Rohner et al. 2013b) and also, for migrating humpback whales, bottlenose dolphins, humpback dolphins, dugongs, sea turtles, small-eye stingray, guitar sharks, corals, and many reef fishes (Pierce et al. 2008, Pereira et al. 2014, Williams et al. 2017, Fordyce 2018, O'Donnell et al. 2017). These planktivores are considered opportunistic filter feeders, and habitat choice is highly dependent on localized productivity (Stevens 2007). Whale sharks and manta rays are often seen feeding at the surface on small, planktonic organisms such as phytoplankton, copepods, mysids, shrimp krill, small fishes, fish and coral larvae (Stevens 2007, Rohner et al. 2013a). Climate-mediated changes in plankton abundance and composition may affect upper trophic levels and fisheries (Mackas and Beaugrand 2010). In addition, plankton can also be used as biological indicators for pollution, water quality and eutrophication (Jickells et al. 2017, Miloslavich et al. 2018).

6. Final Remarks and Future Perspectives

In the Mozambican context, current societal global challenges associated with the accelerated growth of technological development, the exponential growth of the human population and urbanized areas and inadequate policies have led to severe impacts on marine ecosystems and also to insufficient production of food, fresh water, and consumption materials, leading to an economic, social and environmental crisis (Globe Afrique 2016, World Bank 2016, 2018, World Food Programme 2021, IMF 2016, Ministério do Mar, Águas Interiores e Pescas (MIMAIP) 2018, UNEP 2020, OECD 2018, 2020, 2021). In developing countries such as Mozambique, these aspects are more expressive, with increases in economic and social inequalities (Anon 2018, Africa Sustainable Development Report 2018, Ministério do Mar,

Águas Interiores e Pescas (MIMAIP) 2018), which demands an increase of efficiency on use of biological resources, the resilience of natural ecosystems and sustainable development in order to reduce the impacts on the economic and social development of the human communities (World Bank 2018). Ocean science research is key to sustainable use of the marine environment (Visbeck 2018) based on close communication and cooperation between governments, international institutions, marine science and operational oceanography services and industry (Wisz et al. 2020). An integrated approach for the scientific, teaching and research systems requires a significant improvement in the governance of marine resources and efficient Coastal Zone Management (IOC-UNESCO 2018, OECD 2018, 2020, 2021). Plankton is an essential ocean and biodiversity descriptor for regional-scale assessment of pelagic habitats and support for decision-making policy (Miloslavich et al. 2018a, b, Muller-Karger et al. 2018, Batten et al. 2019). The marine and planktonic research can significantly contribute to the sustainable exploitation and protection of marine biodiversity in the Mozambican Ocean, coastal area, estuaries and bays, allowing, in a regional approach, to predict global impacts of climate change and global marine and ocean management (OECD 2021).

Knowledge of the distribution processes of planktonic communities has been increasingly studied on the coast of Mozambique, mainly in the Mozambique Channel and the Exclusive Economic Zone, not only for their ecological importance but also for their responses to hydroclimatic characteristics and economic importance. However, in a climate change context and associated global warming, ocean acidification and anthropogenic pressures (IPCC 2014), it is crucial to a new narrative for the ocean (Lubchenco and Gaines 2019) to face new frontiers in ocean exploration (Raineault and Flanders 2019) and new technologies for plankton observations and monitoring of ocean in order to improve knowledge of the variability of spatial and temporal distribution patterns of zooplankton in the sea in order to understand the changes and impacts on the structure and general functioning of marine and coastal ecosystems (Kwiatkowski et al. 2018, Benedetti et al. 2019, Buck et al. 2019, Hablützel et al. 2021). Current challenges of climate change impacts, pollution, eutrophication, harmful algal blooms, and distribution of invasive species have led to severe changes in marine food webs (Borja et al. 2020), decrease in plankton biodiversity, spatial and temporal dispersion, fishery stocks and the trophic pathways of organic matter in the food web (IOC-UNESCO 2017, Everett et al. 2017, IOC-UNESCO 2018b, Batten et al. 2019, Lombart et al. 2019). Current and future challenges are to understand and predict how global changes and multiple stressors drivers, especially anthropogenic stressors, affect the structure and function of planktonic organisms (Chiba et al. 2018, Romagnan et al. 2016, Brun et al. 2019, Fennel et al. 2019, Batten et al. 2019) that need future ocean observations to connect climate, fisheries and marine production (Schmidt et al. 2019, Beaugrand et al. 2019). Given their contribution to production in the oceans and estuaries and to the variability of recruitment of fish species (largely determined in its planktonic phase), planktonic research can also contribute with important indicators, in the medium and long term, to the context of the conservation of living marine resources and the correct management of fisheries in this region of Mozambique.

6.1   The Importance of the Research Centers Education and Training Supporting Scientific Networking

The development and increment of marine and planktonic research in Mozambique imply the development and improvement of planktonic scientific research and plankton research centers in the Mozambican Universities with networking efforts of the researchers in order to build a wide net of research centers distributed across different regions (Mackenzie et al. 2019) and with the economic support of private national and international companies in order to combine science, technology and innovation (OECD 2018). Building a national and international network with different responsibilities in the implementation of scientific research on plankton is crucial to promoting socio-economic and technological development (OECD 2020). The research centers and universities must develop capacity building throughout training and mobility programs and projects, supporting scientific networking for science and education that should include experienced marine scientists and also permit implementation of young scientists' grants in order to increase scientific and knowledge production and the dissemination of technologies for zooplankton monitoring and trends in marine plankton diversity and structure, temporal and spatial scales in life cycles and distributions addressing biogeographic marine plankton traits and to contribute toward global biodiversity conservation challenges assessment (Chiba et al. 2018). The definition of new plankton key areas for scientific development is also fundamental as it explores opportunities to integrate attractive and innovative R+D management for the next decades (Chiba et al. 2018, OECD 2020).

Following the current global challenges for plankton research, some priority issues are crucial to Mozambican marines and coastal research development, such as study the marine biodiversity, climatic variability and global changes (Beaugrand 2014), the biomass composition of the oceans (Bar-On and Milo 2019), the global patterns of phytoplankton and zooplankton diversity driven by temperature and environmental variability (Righetti et al. 2019), the genomic of marine plankton species (Carradec et al. 2018), the paradigms in biological carbon cycling in the ocean (Zhang et al. 2018). This knowledge will constitute fundamental tools for the establishment of actions in the field of marine environmental policies and biodiversity conservation:

(i)   Set goals for the preservation of biodiversity and marine resources.

(ii)   Provide scientific support for protection legislation as a way of supporting decision-making.

(iii)   Identify and implement opportunities for cooperation and participation in exchange networks between researchers for the development of projects in the area of marine biology.

(iv)   Improve data collection and monitoring framework to support marine management.

(v)   Improve understanding of marine ecosystem services (climate regulation) to be incorporated into marine management.

(vi) Improve governance and marine and coastal policy framework in order to enable the sustainable management of the marine environment.

(vii) Improve marine citizenship in order to engage the public with the marine environment.

(viii) Improve the health of marine ecosystems and the use and management of marine resources.

References

Africa Sustainable Development Report. 2018. Towards a transformed and resilient continent. African Union, Economic Commission for Africa; African Development Bank and United Nations Development Programme. 135 pp.

Anon. 2018. Mozambique economic update: less poverty, but more inequality [Online]. Available online at: https://www.worldbank.org/en/country/mozambique/publication/mozambique-economic-update-less-poverty-but-more-inequality.

Barlow, R., Lamont, T., Gibberd, M.J., Airs, R., Jacobs, L. and Britz, K. 2017. Phytoplankton communities and acclimation in a cyclonic eddy in the southwest Indian ocean. Deep-Sea Res Pt I 124: 18–30.

Barlow, R., Lamont, T., Morris, T., Sessions, H. and van den Berg, M. 2014. Adaptation of phytoplankton communities to mesoscale eddies in the mozambique channel. Deep-Sea Res Pt Ii 100: 106–118.

Barlow, R., Stuart, V., Lutz, V., Sessions, H., Sathyendranath, S., Platt, T. et al. 2007. Seasonal pigment patterns of surface phytoplankton in the subtropical southern hemisphere. Deep-Sea Res Pt I 54: 1687–1703.

Bar-On, Y.M. and Milo, R. 2019. The biomass composition of the oceans: a blueprint of our blue planet. Cell 179(7): 1,451–1,454. https://doi.org/10.1016/j.cell.2019.11.018.

Batten, S.D., Abu-Alhaija, R., Chiba, S., Edwards, M., Graham, G., Jyothibabu, R. et al. 2019. A global plankton. Diversity Monitoring Program. Front. Mar. Sci. 6: 321. doi: 10.3389/fmars.2019.00321.

Beaugrand, G. 2014. Marine biodiversity, climatic variability and global change. Routledge, London, 486 pp. https://doi.org/10.4324/9780203127483.

Beaugrand, G., Conversi, A., Atkinson, A., Cloern, J., Chiba, S., Fonda-Umani, S. et al. 2019. Prediction of unprecedented biological shifts in the global ocean. Nat. Clim. Change 9: 237–243.

Benedetti, F., Ayata, S-D., Irisson, J-O., Adloff, F. and Guilhaumom, F. 2019. Climate change may have minor impact on zooplankton functional diversity in the Mediterranean Sea. Divers Distrib 25: 568–581.

Biastoch, A. and Krauss, W. 1999. The role of mesoscale eddies in the source regions of the agulhas current. J. Phys. Oceanogr. 29: 2303–2317.

Biller, S.J., Berube, P.M., Dooley, K., Williams, M., Satinsky, B.M., Hackl, T. et al. 2018. Marine microbial metagenomes sampled across space and time. Sci. Data 5: 180176.

Boon, E.K. 2009. Area studies (regional sustainable development review): Africa - Volume I, EOLSS Publication, Oxford, United Kingdom, 335 pp.

Borja, A., Andersen, J.H., Arvanitidis, C.D., Basset, A., Buhl-Mortensen, L., Carvalho, S. et al. 2020. Past and future grand challenges in marine ecosystem ecology. Front. Mar. Sci. 7: 362. doi: 10.3389/fmars.2020.00362.

Boyce, D.G., Lewis, M.R. and Worm, B. 2010. Global phytoplankton decline over the past century. Nature 466: 591–596.

Braga, J.M. 1960. Foraminíferos da costa de moçambique. Etudos, Ensaios e Documentos No. 67, Junta de Investigações do Ultramar, Lisboa, 208 pp.

Brönnimann, P. 1954. Probable occurrence of oligocene on saipan. American Journal of Science 252(11): 673–682. https://doi.org/10.2475/ajs.252.11.673.

Brinca, L., Jorge da Silva, A., Sousa, L. and Sousa e R. Saetre, I.M. 1983. A survey on the fish resources at sofala bank – mozambique. September 1982. Reports on Surveys with the R/V Dr. Fridjtof Nansen. Instituto Nacional de Investigação Pesqueira- Maputo 69 p.

Brun, P., Stamieszkin, K., Visser, A.W., Licandro, P., Payne, M.R. and Kiorboe, T. 2019. Climate change has altered zooplankton-fuelled carbon export in the north Atlantic. Nat. Ecol. Evol. 3: 416–423.

Buck, J.J.H., Bainbridge, S.J., Burger, E., Kraberg, A., Casari, M., Casey, K.S. et al. 2019. Ocean data product integration through innovation—the next level of data interoperability. Front. Mar. Sci. 6: 32.

Burgess, N.D. and Clarke, G.P. (Eds.). 2000. Coastal forests of eastern africa. IUCN Forest Conservation Programme, Gland, Switzerland and Cambridge, England. 443pp.

Burgess, N.D., D'Amico Hales, J., Dinerstein, E., Itoua, I., Newman, K., Olson, D. et al. 2004. Terrestrial eco regions of africa and madagascar: a continental assessment. Island Press, Washington, DC, USA. ISBN: 1-55963-364-6.

Carradec, Q., Pelletier E., Da Silva, C., Alberti, A., Seeleuthner, Y., Blanc-Mathieu, R. et al. 2018. Tara oceans coordinators a global ocean atlas of eukaryotic genes. Nat. Commun, 9: 373.

Chemane, D., Motta, H. and Achimo, M. 1997. Vulnerability of coastal resources to climate changes in mozambique: a call for integrated coastal zone management. Ocean and Coastal Management 37(1): 63–83.

Chiba, S., Batten, S., Martin, C.S., Ivory, S., Miloslavich, P. and Weatherdon, L.V. 2018. Zooplankton monitoring to contribute towards addressing global biodiversity conservation challenges. J. Plankt. Res. 40: 509–518. doi: 10.1093/plankt/fby030.

Clark, P.F. and Paula, J. 2003. Descriptions of ten xanthoidean (crustacea: decapoda: brachyura) first stage zoeas from Inhaca Island, Mozambique. The Raffles Bulletin of Zoology 51(2): 323–378.

Clifford, E.L., Hansell, D.A., Varela, M.M., Nieto-Cid, M., Herndl, G.J. and Sintes, E. 2017. Crustacean zooplankton release copious amounts of dissolved organic matter as taurine in the ocean. Limnol. Oceanogr. 62: 2745–2758. doi: 10.1002/lno.10603.

Costa, D. and Soto, B. 2012. Meio Ambiente em Moçambique: Notas para reflexão sobre a situação actual e os desafios para o futuro.

de Ruijter, W.P.M., Ridderinkhof, H., Lutjeharms, J.R.E., Schouten, M.W. and Veth, C. 2002. Observations of the flow in the mozambique channel. Geophysical Research Letters 29(10): 1502.

de Ruijter, W.P.M., Ridderinkhof, H. and Schouten, M.W. 2005. Variability of the southwest Indian ocean, Philos. Trans. R. Soc. A 363: 63–76.

de Vargas, C., Audic, S., Henry, N., Decelle, J., Mahé, F., Logares, R. et al. 2015. Eukaryotic plankton diversity in the sunlit ocean. Science 348(6237). https://doi.org/10.1126/science.1261605.

Druon, J.-N., Chassot, E., Murua, H. and Lopez, J. 2017. Skipjack tuna availability for purse seine fisheries is driven by suitable feeding habitat dynamics in the Atlantic and Indian oceans. Frontiers in Marine Science, 4.

Everett, B. 2017. Study area, vessels and surveys. pp. 23–35. *In*: book: The RV Dr Fridtjof Nansen in the Western Indian Ocean: Voyages of marine research and capacity development, Chapter: 3, Publisher: FAO, Rome, Italy, Editors: Johan C. Groeneveld, Kwame A. Koranteng.

Fennel, K., Gehlen, M., Brasseur, P., Brown, C., Ciavatta, S., Cossarini, G. et al. 2019. Advancing marine biogeochemical and ecosystem reanalyses and forecasts as tools for monitoring and managing ecosystem health. Front. Mar. Sci. 6: 89. doi: 10.3389/fmars.2019.00089.

Findlay, K., Cockcroft, V.G. and Guissamulo, A.T. 2011. Dugong abundance and distribution in the bazaruto archipelago, mozambique. African Journal of Marine Science 33: 441–452.

Findlay, K., Meyer, M., Elwen, S., Kotze, D., Johnson, R., Truter, P. et al. 2011. Distribution and abundance of humpback whales, megaptera novaeangliae, off the coast of mozambique, 2003. Journal of Cetacean Research and Management (special issue 3): 163–174.

Fordyce, A. 2018. Reef fishes of praia do tofo and praia da barra, inhambane, mozambique. Western Indian Ocean Journal of Marine Science 17: 71–91.

Federico, M., Ibarbalz, Nicolas Henry, Manoela C. Brandão, Séverine Martini, Greta Busseni, Hannah Byrne et al. 2019. Global trends in marine plankton diversity across kingdoms of life. Cell 179(5, 14): 1084–1097.e21.

Gammelsrød T. and Hoguane A.M. 1996. Water masses, currents and tides at the sofala bank, November 1987. Revista de Investigação Pesqueira, 21.

Globe Afrique. 2016. Africa faces some serious environmental problems: greenpeace. Globe Afrique. https://globeafrique.com/africa-faces-some-serious-environmental-problems-greenpeace/.

Gregory, A.C., Zayed, A.A., Conceição-Neto, N., Temperton, B., Bolduc, B., Alberti, A. et al. 2019. Tara oceans coordinators marine DNA viral macro- and microdiversity from pole to pole. Cell 177: 1109–1123.

Groeneveld, J.C. and Koranteng, K.A. (eds.). 2017. The RV Dr fridtjof nansen in the Western Indian Ocean: Voyages of Marine Research and Capacity Development. FAO. Rome, Italy. 258.

Grundlingh, M.L. 1995. Tracking eddies in the southeast atlantic and southwest indian ocean with TOPEX/poseidon. Journal of Geophysical Research 100: 977–986.

Guidi, L., Chaffron, S., Bittner, L., Eveillard, D., Larhlimi, A., Roux, S. et al. 2016. Plankton networks driving carbon export in the oligotrophic ocean. Nature 532: 465.

Hablützel, P.I., Rombouts, I., Dillen, N., Lagaisse, R., Mortelmans, J., Ollevier, A. et al. 2021. Exploring new technologies for plankton observations and monitoring of ocean health. pp. 20–25. *In*: Hazards, E.S., Kappel, S.K., Juniper, S., Seeyave, E. Smith and Visbeck, M. (eds.). Frontiers in Ocean Observing: Documenting Ecosystems, Understanding Environmental Changes, Forecasting, A Supplement to Oceanography 34(4).

Harris, T.F.W. 1972. Sources of the agulhas current in the spring of 1964. Deep-Sea Res. 19: 633–650.

Hewawasam, I. 2000. Advancing knowledge: a key element of the world bank's integrated coastal management strategic agenda in sub-saharan Africa. Ocean and Coastal Management 43: 361–377.

Hoguane, A.M. 1996. Hydrodynamics, Temperature and Salinity in Mangrove Swamps in Mozambique. PhD Thesis; University of Wales, Bangor, UK.

Hoguane, A.M. 1999. Sea level measurement and analysis. *In*: The Western Indian Ocean. National Report: Mozambique.

Hoguane, A.M. and Pereira, M.A.M. 2003. National report: marine biodiversity in mozambique - the known and the unknown. pp. 138–155. *In*: Decker, C., Griffiths,C., Prochazka, K., Ras C. and Whitefield, A. (eds.). Marine Biodiversity in Sub-Saharan Africa: the known and the unknown. Proceedings of the Marine Biodiversity in Sub-Saharan Africa: the Known and the Unknown Cape Town, South Africa 23–26 September 2003.

Hoguane, A.M. 2007. Perfil da zona costeira de moçambique. Revista de Gestão Costeira integrada, Univali 7(1): 69–82.

Huggett, J.A. 2014. Mesoscale distribution and community composition of zooplankton in the mozambique channel. Deep Sea Research Part II: Topical Studies in Oceanography 100: 119–135.

IMF. 2016. World Economic Outlook Data. International Monetary Fund.

IPCC. 2014. Climate change, 2014. Impacts, adaptation, and vulnerability. Summaries, frequently asked questions, and cross-chapter boxes. *In*: Field, C.B., Barros, V.R., Dokken, D.J., Mach, K.J., Mastrandrea, M.D., Bilir, T.E. et al. (eds.). A Contribution of Working Group II to the Fifth Assessment Report of the Intergovernmental Panel on Climate Change. World Meteorological Organization, Geneva, 190 pp.

IOC-UNESCO. 2017. Global Ocean Science Report-The current status of ocean science around the world. *In*: Valdés, L. (ed.). United Nations Educational, Scientific and Cultural Organization.

IOC-UNESCO. 2018. Revised roadmap for the UN decade of ocean science for sustainable development. In IOC/EC-LI/2 Annex 3. Retrieved from https://oceandecade.org/assets/uploads/documents/EC51-2A3-Roadmap_e_1564761714.pdf.

IOC-UNESCO. 2018b. Global ocean observing system- in situ networks. Available at: http://www.goosocean.org/index.php?option=com_contentandview=articleand id=24andItemid=12 3.

Jaquemet, S., Ternon, J.F., Kaehler, S., Thiebot, J.B., Dyer, B., Bemanaja, E. et al. 2014. Contrasted structuring effects of mesoscale features on the seabird community in the mozambique channel. Deep-Sea Research II. Deep-Sea Research II 100(2014): 200–211.

Jickells, T., Buitenhuis, E., Altieri, K., Baker, A., Capone, D., Duce, R. et al. 2017. A reevaluation of the magnitude and impacts of anthropogenic atmospheric nitrogen inputs on the ocean. Glob. Biogeochem. Cycles 31: 289–305.

Johnsen, E., Kraskstad, J., Ostrowski, M., Serigstad, B., Stromme, T., Alvheim, O. et al. 2007. Surveys of the Living Marine Resources of Mozambique: Ecosystem Survey and Special Studies. Report No. 8/20072007409. Bergen, Norway, Institute of Marine Research.

José, Y.S., Aomont, O., Machu, E. and Penven, C.L. 2014. Influence of mesoscale eddies on biological production in the mozambique channel: Deep Sea Research Part II: Topical Studies in Oceanography. Volume 100, Elsevier Ltd. All rights reserved.79–93.

Kelchner, H. 2020. Role of coastal environmental conditions during austral winter on plankton pommunity dynamics and the occurrence of pseudo-nitzschia spp. and domoic acid in Inhambane Province, Mozambique. LSU Master's Theses 135 pp.

Kwiatkowski, L., Aumont, O., Bopp, L. and Ciais, P. 2018. The impact of variable phytoplankton stoichiometry on projections of primary production, food quality, and carbon uptake in the global ocean. Glob Biogeochem Cycles 32: 516–528.

Krakstad, J.-O., Krafft, B., Alvheim, O., Kvalsund, M., Bernardes, I., Chacate, O. et al. 2015. Marine ecosystem survey of mozambique. Cruise report Dr Fridtjof.Nansen. 11 November–02 December 2014. Report No. GCP/INT/003/NOR. FAO–NORAD.

Kostianoy, G., Ginzburg, A.I., Lebedev, S.A., Frankignoulle, M. and Delille, B. 2003. Fronts and mesoscale variability in the southern Indian ocean as inferred from the TOPEX/POSEIDON and ERS-2 Altimetry Data.

Lamont, T., Barlow, R.G., Morris, T. and van den Berg, M.A. 2014. Characterisation of mesoscale features and phytoplankton variability in the mozambique channel. Deep Sea Research Part II: Topical Studies in Oceanography 100: 94–105.

Lamont, T., Roberts, M.J., Barlow, R.G., Morris, T. and van den Berg, M.A. 2010. Circulation patterns in the delagoa bight, mozambique, and the influence of deep ocean eddies. Afr. J. Mar. Sci. 32: 553–562.

Leal, M.C., Sá, C., Nordez, S., Brotas, V. and Paula, J. 2009. Distribution and vertical dynamics of planktonic communities at sofala bank, mozambique. Estuarine, Coastal and Shelf Science 84(4): 605–616.

Lebourges-Dhaussy, A., Huggett, J., Ockuis, S., Roudaut, G., Jossee, E. and Verheye, H. 2009. Zooplankton size and distribution within mesoscale structures in the mozambique channel: a comparative approach using the TAPS acoustic profiler, a multiple net sampler and ZooScan image analysis.Institut de Recherche pour le Développement (IRD), UMR LEMAR 195 (UBO/CNRS/IRD/Ifremer.45.

Lindkvist, E., Wijermans, N., Daw, T.M., Gonzales-Mon, B., Giron-Nava, A., Johnson, A.F. et al. 2020. Navigating complexities: agent-based modeling to support research, governance, and management in small-scale fisheries. Frontiers in Marine Science 6: 733.

Lombard, F., Boss, E., Waite, A.M., Vogt, M., Uitz, J., Stemmann, L. et al. 2019. Globally consistent quantitative observations of planktonic ecosystems. Front. Mar. Sci. 6: 196.

Lubchenco, J. and Gaines, S.D. 2019. A new narrative for the ocean. Science 364(6444): 911. https://doi.org/10.1126/science.aay2241.

Lutjeharms, J.R.E. 1976. The agulhas current system during the northeast monsoon system. Journal of Physical Oceanography 6: 665–670.

Lutjeharms, J.R.E. 2006. The ocean environment off southeastern africa: a review. S Afr. J. Sci. 102: 419–426.

Lutjeharms, J.R.E. and Van Ballegooyen, R.C. 1988. The retroflection of the agulhas current. Journal of Physical 50 8. References Oceanography 18: 1570–1583.

Lutjeharms, J.R.E., Biastoch, A., van der Werf, P.A., Ridderhinkhof, H. and de Ruijter, W.P.M. 2012. On the discontinuous nature of the mozambique current. South African Journal of Science 108(1/2).

Lutjeharms, J.R.E. 2006. The costal oceans of southeastern Africa. pp. 783–834. *In*: Robinson, A.R. and Brink, K.H. (eds.). The Sea Vol 14. Harvard University Press, Cambridge, MA.

Mackas, D.L. and Beaugrand, G. 2010. Comparisons of zooplankton time series. J. Mar. Syst. 79: 286–304.

Mackenzie, B., Celliers, L., Assad, L.P., de, F., Heymans, J.J., Rome, N. et al. 2019. The role of stakeholders in creating societal value from coastal and ocean observations. Front. Mar. Sci. 6(Mar): 137. https://doi.org/10.3389/fmars.2019.00137.

Meyers, G. 2019. Environmental Issues in Africa's Cities. Wiley Online Library, https://doi.org/10.1002/9781118568446.eurs0002.

Miloslavich, P., Bax, N.J., Simmons, S.E., Klein, E., Appeltans, W. and Aburto-Oropeza, O. et al. 2018. Essential ocean variables for global sustained observations of biodiversity and ecosystem changes. Glob. Change Biol. 24: 2416–2433.

Ministério do Mar, Águas Interiores e Pescas, MIMAIP. 2018. Balanco do plano social e economico de 2018. Maputo, Mozambique: Mozambique Ministry of Seas, Inland Waters and Fisheries.

Moriarty, R. and O'Brien, T.D. 2013. Distribution of mesozooplankton biomass in the global ocean. Earth Syst. Sci. Data 5: 45–55.

Muller-Karger, F.E., Miloslavich, P., Bax, N.J., Simmons, S., Costello, M.J., Sousa Pinto, I. et al. 2018. Advancing marine biological observations and data requirements of the complementary essential ocean variables (eovs) and essential biodiversity variables (ebvs) frameworks. Front. Mar. Sci. 5: 211. doi: 10.3389/fmars.2018.00211.

Nehring, D., Hagen, E., da Silva, J.A., Schemainda, R., Wolf, G., Michelchen, N. et al. 1987. Results of oceanological studies in the mozambique channel in february–march 1980. Beitr. Meereskd 56: 51–63.

O'Donnell, J.L., Beldade, R., Mills, S.C., Williams, H.E. and Bernardi, G. 2017. Life history, larval dispersal and connectivity in coral reef fish among the scattered Islands of the mozambique channel. Coral Reefs 36: 223–232.

OECD. 2018. OECD Science, Technology and Innovation Outlook 2018: Adapting to Technological and Societal Disruption. OECD Publishing, Paris, https://doi.org/10.1787/sti_in_outlook-2018-en.

OECD. 2020. Sustainable Ocean for All: Harnessing The Benefits of Sustainable Ocean Economies for Developing Countries, The Development Dimension. OECD Publishing, Paris, https://doi.org/10.1787/bede6513-en.

OECD. 2021. Adapting to a changing climate in the management of coastal zones, OECD environment Policy Papers, No. 24, OECD Publishing, Paris, https://doi.org/10.1787/b21083c5-en.

Olofsson, M., Karlberg, M., Lage, S. and Ploug, H. 2017. Phytoplankton community composition and primary production in the tropical tidal ecosystem, Maputo Bay (the Indian Ocean). J. Sea Res. 125: 18–25.

Pereira, M.A.M., Litulo, C., Santos, R., Leal, M.C., Fernandes, R.S, Tibiriçá, Y. et al. 2014. Mozambique marine ecosystem review. *In*: Costa, H. (ed.). Biodinâmica, Maputo.

Pierce, S.J., Mendez-Jimenez, A., Collins, K., Rosero-Caicedo, M. and Monadjem, A. 2010. Developing a code of conduct for whale shark interactions in Mozambique. Aquat Conserv 20: 782–788.

Pierce, S.J., White, W.T. and Marshall, A.D. 2008. New record of the smalleye stingray, dasyatis microps (myliobatiformes: dasyatidae), from the western Indian Ocean. Zootaxa 1734.

Quartly, G.D. and Srokosz, M.A. 2004. Eddies in the southern mozambique channel. Deep-Sea Res. Pt Ii 51: 69–83.

Raineault, N.A. and Flanders, J. 2019. New frontiers in ocean exploration: the E/V nautilus, NOAA ship okeanos explorer, and R/V falkor 2018 field season. Oceanography 32(1 suplement): 150. https://doi.org/10.5670/oceanog.2019.supplement.01.

Raj, R.P., Peter, B.N. and Pushpadas, D. 2010. Oceanic and atmospheric influences on the variability of phytoplankton bloom in the southwestern Indian ocean. Journal of Marine Systems 82: 217–229.

Ridderinkhof, H. and de Ruijter, W.P.M. 2003. Moored current observations in the mozambique channel, Deep Sea Res. Part II 50: 1933–1955.

Ridderinkhof, H., van der Werf, P.M., Ullgren, J.E., van Aken, H.M., van Leeuwen, P.J. and de Ruijter, W.P.M. 2010. Seasonal and interannual variability in the mozambique channel from moored current observations. Journal of Geophysical Research 115(C6).

Righetti, D., Vogt, M., Gruber, N., Psomas, A. and Zimmermann, N.E. 2019. Global pattern of phytoplankton diversity driven by temperature and environmental variability. Science Advances 5(5). https://doi.org/10.1126/sciadv.aau6253.

Rohner, C.A., Couturier, L.I.E., Richardson, A.J., Pierce, S.J., Prebble, C.E.M., Gibbons, M.J. et al. 2013a. Diet of whale sharks rhincodon typus inferred from stomach content and signature fatty acid analyses. Marine Ecology Progress Series 493: 219–235.

Rohner, C.A., Pierce, S.J., Marshall, A.D., Weeks, S.J., Bennett, M.B. and Richardson, A.J. 2013b. Trends in sightings and environmental influences on a coastal aggregation of manta rays and whale sharks. Marine Ecology Progress Series 482: 153–168.

Rohner, C.A., Richardson, A.J., Jaine, F.R.A., Bennett, M.B., Weeks, S.J., Cliff, G. et al. 2018. Satellite tagging highlights the importance of productive mozambican coastal waters to the ecology and conservation of whale sharks. PeerJ 6: e4161.

Romagnan, J.B., Aldamman, L., Gasparini, S., Nival, P., Aubert, A., Jamet J.L. et al. 2016. High frequency mesozooplankton monitoring: can imaging systems and automated sample analysis help us describe and interpret changes in zooplankton community composition and size structure – an example from a coastal site. J. Mar. Syst. 162: 18–28.

Sá, C., Leal, M.C., Silva, A., Nordez, S., André, E., Paula, J. et al. 2013. Variation of phytoplankton assemblages along the mozambique coast as revealed by HPLC and microscopy. J. Sea Res. 79: 1–11.

Saetre, R. and Silva, R. Paula. 1979. The marine fish resources of mozambique. reports on surveys with R/V Dr. fritjof nansen. Serviços de Investigação Pesqueira. Maputo/Institute of Marine Research, Bergen. 179 p.

Sætre, R. and da Silva, A.J. 1982. Water mass and circulation of the mozambique channel. Revista de investigacaopesqueira. Technical Reprt 3. Instituto de Desenvolvimanto Pesqueiro, Maputo.

Saetre, R. and Da Silva, A.J. 1984. The circulation of the mozambique channel. Deep-Sea Res. 31, 5: 485–508.

Schlüter, M., Baeza, A., Dressler, G., Frank, K., Groeneveld, J., Jager, W. et al. 2017. A framework for mapping and comparing behavioural theories in models of social-ecological systems. Ecological Economics 131: 21–35.

Schlüter, M., Orach, K., Lindkvist, E., Martin, R., Wijermans, N., Bodin, Ö. et al. 2019. Toward a methodology for explaining and theorizing about social-ecological phenomena. Current Opinion in Environmental Sustainability 39: 44–53.

Schmidt, J.O., Bograd, S.J., Arrizabalaga, H., Azevedo, J.L., Barbeaux, S.J., Barth, J.A. et al. 2019. Future ocean observations to connect climate, fisheries and marine ecosystems. Front Mar. Sci. 6(Sept): 550. https://doi.org/10.3389/fmars.2019.00550.

Schott, F. and McCreary, J. 2001. The monsoon circulation of the Indian ocean. Progress in Oceanography 51: 1–123.

Schouten, M.W., de Ruijter, P.M., van Leeuwen, P.J. and Dijkstra, H.A. 2002a. A teleconnection between the equatorial and southern Indian ocean. Geophysical Research Letters 29. doi:10.1029/2001GL014542.

Schouten, M., de Ruijter, W. and Van Leeuwen, P. 2002b. Upstream control of agulhas ring shedding. Journal of Geophysical Research 107. doi:10.1029/ 2001JC000804.

Schouten, M.W., de Ruijter, W.P.M., van Leeuwen, P.J. and Ridderinkhof, H. 2003. Eddies and variability in the mozambique channel. Deep Sea Research Part II: Topical Studies in Oceanography 50: 1987.

Silva, E.S. 1956. Contribuiçao para o estudo do microplancton marinho de moçambique. Jta Invest. Ultramar, Est., Ens. Doc., 28: 1–97, pl. 1–14; Jta Invest. Ultramar, Miss. Biol, marit., Colect. 1(8).

Stammer, D. 1997. Global characteristics of ocean variability estimated from regional TOPEX/POSEIDON altimeter measurements. Journal of Physical Oceanography 27: 1743–1769.

Stevens, J.D. 2007. Whale shark (rhincodon typus) biology and ecology: a review of the primary literature. Fish Res 84: 4–9.

Steen, J.-E. and Hoguane, A.M. 1990. Oceanographic results on expedition carried out by R/V Dr. Fridjof nansen in mozambique waters during april-may 1990. 35p., Relatório No. 13. Maputo, Mozambique.

Spalding, M.D., Fox, H.E., Allen, G.R., Davidson. N., Ferdana. Z.A., Finlayson. M. et al. 2007. Marine ecoregions of the world: a bioregionalization of coastal and shelf areas. Bioscience 57(7): 573–582.

Stramma, L. and Lutjeharms, J.R.E. 1997. The flow field of the subtropical gyre of the south Indian ocean. Journal of Geophysical Research 102: 5513–5530.

Swart, N.C., Lutjeharms, J.R.E., Ridderinkhof, H. and de Ruijter, W.P.M. 2010. Observed characteristics of mozambique channel eddies. Journal of Geophysical Research 115.

Ternon, J.F., Bach, P., Barlow, R., Huggett, J., Jaquemet, S., Marsac, F. et al. 2014. The mozambique channel: from physics to upper trophic levels. Deep Sea Research Part II: Topical Studies in Oceanography 100: 1–9.

Tew Kai, E. and Marsac, F. 2008. Patterns of variability of sea surface chlorophyll in the mozambique channel: a quantitative approach. Journal of Marine Systems 77: 77–88.

Tew-Kai, E. and Marsac, F. 2009. Patterns of variability of sea surface chlorophyll in the mozambique channel: a quantitative approach. Journal of Marine Systems 77: 77–88.

Tew Kai, E. and Marsac, F. 2010. Influence of mesoscale eddies on spatial structuring of top predators' communities in the mozambique channel. Progress in Oceanography 86: 214–223.

TRANSMAP. 2008. Transboundary networks of marine protected areas for conservation and sustainable development: biophysical, socioeconomic and governance assessment in eastern Africa. Final report, Project INCO-CT2004-510862, European Commission.

UNEP. 2020. Africa: the challenge. United National Environment Program. https://www. unenvironment.org/regions/africa#:~:text=Africa%20faces%20serious%20environmental%20 challenges,extreme%20vulnerability%20to%20climate%20change.

Vallina, S.M., Follows, M.J., Dutkiewicz, S., Montoya, J.M., Cermeno, P. and Loreau, M. 2014. Global relationship between phytoplankton diversity and productivity in the ocean. Nat. Commun. 5: 4299.

Visbeck, M. 2018. Ocean science research is key for a sustainable future. Nat. Commun. 9(1): 1–4. https://doi.org/10.1038/s41467-018-03158-3.

Van Aken, H.M., Ridderinkhof, H. and De Ruijter, W.P.M. 2004. North Atlantic deep water in the south–western Indian ocean, Deep Sea Res. Part I 51(6): 755–776.

Van Wyk, A.E. 1994. Maputaland-pondoland region. South Africa, Swaziland and Mozambique. Em: Centers of Plant Diversity. A Guide and Strategy for Their Conservation. Vol. 1. WWF and IUCN, Pretoria.

Van Wyk, A.E. 1996. Biodiversity of the maputaland centre. pp. 198–207. *In*: Van Den Burgt, X.M. and Van Medenbach De Rooy, J.M. (eds.). The Biodiversity of African Plants. Dordrecht, The Netherlands, Kluwer Academic Publishers.

Van Wyk, A.E. and Smith, G.F. 2001. Regions of floristic endemism in southern Africa. A Review with Emphasis on Succulents. Umdaus Press, Pretoria, South Africa.

Weimerskirch, H., Le Corre, M., Jaquemet, S., Potier, M. and Marsac, F. 2004. Foraging strategy of a top predator in tropical waters: great frigatebirds in the mozambique channel. Mar. Ecol. Prog. Ser. 275: 297–308.

Williams, J.L., Pierce, S.J., Rohner, C.A., Fuente, M.M.P.B. and Hamann, M., 2017. Spatial distribution and residency of green and loggerhead Sea turtles using coastal reef habitats in southern mozambique. Frontiers in Marine Science 3.

White, F. 1983. The vegetation of Africa. Natural Resources Research 20, UNESCO, Paris.

Wisz, M.S., Satterthwaite, E.V., Fudge, M., Fischer, M., Polejack, A., John, M.S. et al. 2020. 100 opportunities for more inclusive ocean research: cross-disciplinary research questions for sustainable ocean governance and management. Front. Mar. Sci. 7(576): 1–23. https://doi.org/10.3389/ fmars.2020.00576.

World Bank. 2016. Mozambique. Systematic Country Diagnostic. Washington, DC: World Bank.

World Bank. 2018. Communities livelihoods fisheries: fisheries governance and shared growth in mozambique. [WWW Document]. The World Bank Data. Available at: https://documents. worldbank. org/pt/publication/documents-reports/ documentdetail/403651525888008345/communities-livelihoods-fisheries-fisheries-governance-andshared-growth-in-mozambique.

World Bank. 2020. DataBank on mozambique. Accessed December 5, 2020. https://data.worldbank.org/ country/MZ.

World Health Organization (WHO). 2017. World Bank, 2021. Mozambique. [WWW Document]. The World Bank Data. Available at: https://data. worldbank.org/country/mz (accessed 5.2.21).

World Food Programme. 2021. Mozambique' 'Country Brief'. [WWW Document]. Available at: https:// www.wfp.org/countries/mozambique.

WORLD WILDLIFE FUND. 2009. The Global Conservation Program Achievements and lessons learned from 10 years of support for threats-based conservation at a landscape and seascape scale. Eastern African Marine Ecoregion. Final Closeout Report 2003–2009.

Zhang, C., Dang, H., Azam, F., Benner, R., Legendre, L., Passow, U. et al. 2018. Evolving paradigms in biological carbon cycling in the ocean. National Science Review 5(4): 481–499. https://doi. org/10.1093/nsr/nwy074.

3.4

Temporal Changes in Zooplankton Abundance and Assemblages in Japanese Coastal and Offshore Waters

Taketoshi Kodama[1,*] and *Hiroomi Miyamoto*[2,‡]

1. Introduction

East Asian countries and the marine environments of East Asia have changed extensively since the start of the "Anthropocene" in the early and middle of the 20th century (Yasuhara et al. 2012, Kuwae et al. 2022). The water temperature of the East Asian Seas has increased rapidly (Belkin 2009, Wu et al. 2012), and biogeochemical cycles are changing (Chen and Guo 2020). Although activities on the mainland undoubtedly contribute substantially to these environmental changes, East Asian waters also feature highly active fisheries and aquaculture industries (FAO 2020), and these industrial activities affect the marine environments and ecosystems of East Asian waters.

East Asian waters, especially Japanese waters, are important for both local and global fisheries (Ando et al. 2021). Zooplankton are generally considered secondary or tertiary producers and prey of fish. Thus, zooplankton biomass and assemblages are thought to be directly linked to the production of fisheries. In Japanese waters, time-series data on zooplankton biomass and assemblages have been collected since 1951 (Odate 1994). These data show that changes in zooplankton biomass and assemblages can be tracked in Japanese waters in the Anthropocene.

Japan is surrounded by the Western North Pacific (WNP) and three marginal seas (the East China Sea, the Sea of Japan (SOJ, Japan Sea), and the Sea of Okhotsk, Figure 1). In this review, we focused on the WNP and the SOJ because zooplankton studies of these two regions have been conducted over various decades.

---

[1] Graduate School of Agriculture and Life Science, The University of Tokyo, 1-1-1, Yayoi, Bunkyo, Tokyo, Japan.

[2] Fisheries Resources Institute, Japan Fisheries Research and Education Agency, 25–269, Shimomekuraku-bo, Same, Hachinohe, Aomori, Japan.

‡ Present address: Fisheries Resources Institute, Japan Fisheries Research and Education Agency, Fukuura, Kanazawa, Yokohama, Kanagawa, Japan.

* Corresponding author: takekodama@g.ecc.u-tokyo.ac.jp

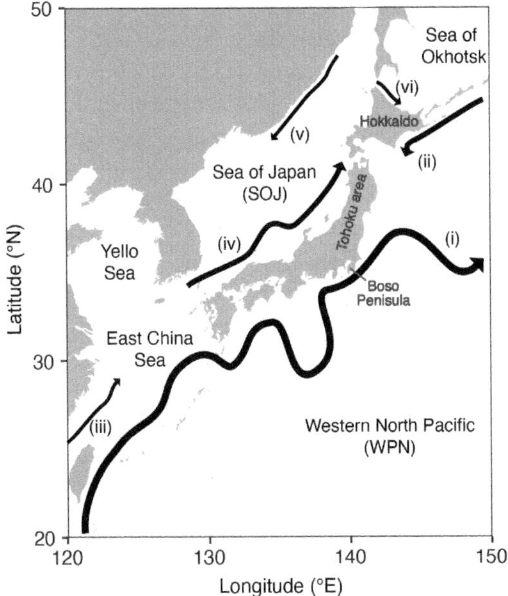

Figure 1. Map of the seas surrounding Japan with a schematic diagram of surface currents: (i) the Kuroshio and Kuroshio Extension, (ii) the Oyashio, (iii) the Taiwan Warm Current, (iv) the Tsushima Warm Current, (v) the Liman Current, and (vi) the Soya Warm Current.

1.1 The Sea of Japan

The SOJ is the marginal sea of the WNP, and it surrounds the Japanese archipelago, the Korean Peninsula, and Russia (Figure 1). The SOJ is connected to the East China Sea, the Sea of Okhotsk, and the WNP, and the straits connected to these seas are narrow and shallow. Therefore, the SOJ is a semi-closed sea; only its surface water is rapidly exchanged with other seas (Yanagi 2002). The residence time of the surface water (< 200 m depth) is a few years, but the residence time of the deep water is approximately 90 years (Yanagi 2002). The SOJ has a deep convection system, western boundary currents, and subarctic circulation (Gamo et al. 2014). Thus, this sea is referred to as a miniature ocean or ocean microcosm (Ichiye 1984, Chen et al. 2017).

One of the unique characteristics of the SOJ is "Japan Sea Proper Water" (JSPW). JSPW comprises the middle and deep layer of the SOJ, and its temperature (0–1°C) and salinity (33.96–34.14) are homogenous in horizontal and vertical (Sudo 1986). The other unique characteristic of the JSPW is its high sensitivity to the effects of anthropogenic activities (Gamo et al. 2014). The dissolved oxygen concentration and pH in the JSPW are decreasing and increasing, respectively (Gamo et al. 2014, Chen et al. 2017). These changes are largely driven by global warming (Gamo 1999, Chen et al. 2017), warm winter air temperatures weaken the formation of JSPW, which prolongs the residence time of JSPW and increases the consumption of dissolved oxygen (Gamo 1999). When winter air temperatures are colder, more oxygen-rich water is supplied to the bottom layer (Senjyu et al.

2002). Also, phosphate and dissolved oxygen concentrations have declined for several decades in the surface water above the deep JSPW (Kodama et al. 2016, Ono 2021). The atmospheric deposition of nitrogenous compounds is increasing the concentrations of nitrogenous nutrients (Kitayama et al. 2012), however, excess nitrogenous nutrient inputs via the Chengjiang River to the East China Sea and horizontal advective transport of Changjiang diluted water are considered the main causes of the decline in phosphate in the SOJ (Kim et al. 2013, Kodama et al. 2016, Kodama et al. 2017).

Several studies on the abundances and assemblages of zooplankton in the SOJ have been conducted in the late 20th century (Iguchi 2004). According to the review (Iguchi 2004), zooplankton ecology in the SOJ differs from that in the WNP in three ways: (1) the vertical migration of cold-water species is restricted, (2) the vertical dispersal of epipelagic carnivorous zooplankton to the deep-sea is facilitated by reduced interspecific competition, and (3) cold-water species have long lifetimes. The first feature is caused by the thermocline during summer. The surface water temperature is elevated during summer because warm and less-saline water flows into the SOJ from the East China Sea (Iguchi 2004). The second feature is caused by the low diversity of zooplankton in the middle and deep layers of the SOJ. Zooplankton that inhabits middle and deep-sea waters from the other seas rarely invade the SOJ because the straits connected to the other seas are shallow, and only warm water passes. The third feature is caused by the physical properties and biological diversity of the JSPW, specifically, the growth of zooplankton is slow in cold water, and predator diversity is low in the JSPW (Iguchi 2004).

Most 20th-century studies on zooplankton in the SOJ have focused on the ecology and biology of single species (Iguchi 2004), studies of long-term variation and changes in community structure are lacking by comparison. Clarifying the dynamics of single species can provide important information for marine biological studies; however, studies of the interactions among zooplankton taxa are necessary to understand comprehensively the marine ecosystems in the rapidly changing SOJ. Long-term variation in zooplankton taxa and changes in community structure have not yet been characterized in the SOJ, yet this information is important given the uniqueness of the plankton ecosystem of the SOJ and the rapid rate at which environments are changing.

1.2 The Western North Pacific

In the WNP, two western boundary currents flow: the warm current is Kuroshio, and the cold one is Oyashio (Figure 1). The Kuroshio originates in the North Equatorial Current and flows off Taiwan, the Nansei Islands, and the Japanese archipelago. After that, the current changes its direction eastward around the Boso Peninsula, which is referred to as the Kuroshio Extension. The current velocity of the Kuroshio sometimes reaches four knots, and it transports warm, saline, and low-nutrient water from the subtropical region to Japanese coastal waters. The Oyahshio is the western boundary current of the western subarctic gyre. The Oyashio flows along the Kuril Islands and changes its direction eastward off eastern Hokkaido. The water of the Oyashio is cold, less saline, and nutrient-rich. Therefore, these two western boundary currents have different hydrographic characters, contributing to the unique pelagic biogeography and biomes of the WNP.

In the WNP, the ecoregions are generally classified into the western subarctic gyre region, Oyashio region, Kuroshio region, and western subtropical gyre region, corresponding to the current systems (Longhurst 2007). Additionally, the transition region is formed in the boundary region between subarctic and subtropical waters (Odate 1994). Zooplankton community structure is changed with the difference of ecoregions (Yoshiki et al. 2013, Miyamoto et al. 2022). The western subarctic gyre is characterized by high-nutrient and low-chlorophyll (HNLC) water (Tsuda et al. 2003). In this region, because phytoplankton production is restricted by iron depletion, which is the essential trace nutrient for phytoplankton growth, and the abundance of phytoplankton is low. Massive blooms of phytoplankton occur in spring in the Oyashio region, and these blooms are induced by the influx of iron-rich waters from the Sea of Okhotsk (Nishioka et al. 2011, Kuroda et al. 2019). In the Kuroshio region and subtropical gyre, primary production is low because of the low concentrations of macronutrients (Saito 2019). However, the high biological production and active nutrient flux in the Kuroshio region are driven by several processes, including the entrainment of coastal water, the island mass effect, and upwelling associated with Kuroshio frontal eddies (Nagai et al. 2009, Kodama et al. 2014, Nagai et al. 2019, Hasegawa et al. 2021). The transition region is defined as the area between the subarctic (= 4°C at 100 m) front and the Kuroshio Extension front (= 15°C at 200 m) of the Tohoku area. The transition region shows a mosaic pattern of interspersed water masses of the Oyashio and Kuroshio waters. In this area, spring phytoplankton blooms occur because of the supply of macronutrients in the winter and the improvement of light availability (Nishibe et al. 2015).

Various small pelagic fish use the Oyashio region and transition regions as feeding and nursery grounds, including Japanese sardine (*Sardinops melanostiscus*), Japanese anchovy (*Engraulis japonicus*), Jack mackerel (*Trachurus japonicus*), mackerels (*Scomber japonicus* and *S. australasicus*), and Pacific saury (*Cololabis saira*), and these fish are main components of Japanese fishery catches (Yatsu et al. 2005). Ecological and biological studies of zooplankton have been actively conducted to understand variability in feeding environments and the availability of prey for small pelagic fish (Tsuda 2013). Wet weight-based zooplankton biomass at depth of 0 m to 150 m in the Oyashio region was highest in May (41 g m^{-2}) and lowest in winter (Odate 1994). These distinct seasonal changes in total biomass are driven by large subarctic calanoid copepods, such as *Neocalanus* and *Eucalanus* spp. (Yamaguchi et al. 2014). The three subarctic *Neocalanus* species, *Neocalanus flemingeri*, *N. plumchrus*, and *N. cristatus* undergo ontogenetic vertical migration during their one- to two-year life span (Kobari and Ikeda 1999, Tsuda et al. 1999, Kobari and Ikeda 2001a, b). During the spring and summer, *Neocalanus* copepods grow from the copepodite I to V stage in the epipelagic layer (0–150 m depth) and then migrate to the mesopelagic layer at the copepodite V stage. After the downward migration, they become adults and reproduce in the following winter and spring. Although *Eucalanus bungii* usually inhabits the mesopelagic layer, this species can also occur in the epipelagic layer during spring blooms during its reproductive period (Tsuda et al. 2004, Shoden et al. 2005). In the Kuroshio Extension and transition regions, the biomass in the epipelagic layer peaks in spring (May) and autumn (October and November) (Odate 1994). The peaks in biomass in the Kuroshio Extension and transition regions are lower than those in the Oyashio region. The relative contributions of *Neocalanus*, *Eucalanus*, and *Metridia* spp. transported from the Oyashio region to the total biomass of subarctic copepods are large during the spring peak in the transition region (Kobari et al. 2008).

The hydrographical conditions in the Oyashio and Kuroshio areas are changing (Ono et al. 2001, Watanabe et al. 2005, Tadokoro et al. 2009). Warming trends have been observed, and the warming of seawater induces changes in biogeochemical processes and flow patterns in these areas (Wang et al. 2016, Nishikawa et al. 2020, Mensah and Ohshima 2021). Analysis of accumulated zooplankton samples has provided new insights into spatiotemporal changes (e.g., long-term variation and geographical patterns) in zooplankton biomass, the abundance of single species, and species composition. The distribution and communities of zooplankton have been continuously monitored using zooplankton net sampling, and these surveys are conducted on cruises in ocean environments and during fisheries stock assessments. These net sampling methods are standardized and are conducted using the same mesh (0.33 mm), ring size (45 cm), and tow depth (150 m) (Takasuka et al. 2017). Some of these samples have been referred to as the "Odate Collection" after Dr. Kazuko Odate (Odate 1994), the researcher who assembled and analyzed these zooplankton samples. The Japan Fisheries Research and Education Agency (FRA) continues to collect and store zooplankton samples from around Japan. More than 50,000 samples are now stored. The Odate Collection is a valuable scientific resource for understanding spatiotemporal changes in pelagic ecosystems in the WNP and is comparable with the north Atlantic continuous plankton recorder (CPR) data and the Eastern North Pacific CalCOFI (California Cooperative Oceanic Fisheries Investigations) data.

1.3 Aims of This Review

The physical and chemical properties of the East Asian Seas are rapidly changing (Chen and Guo 2020). However, a single study of long-term variation in biological processes is not sufficient for characterizing long-term changes in biological production in these seas. In addition, comparisons of SOJ and WNP are expected to provide a deep understanding of zooplankton variations in the East Asian Seas because the marine systems are different between the two areas. Therefore, this review mainly focused on studies from the 21st century to examine long-term variation in zooplankton communities in the SOJ and WNP (mainly off the Tohoku area). Then, we discussed some of the insights they provide regarding patterns of change in zooplankton communities and abundances in the East Asian Seas and the future directions of zooplankton studies.

2. Changes in Zooplankton in the Sea of Japan

2.1 Zooplankton Biomass

The wet weight of zooplankton has historically been used as an indicator of zooplankton biomass near Japan. Many research institutes in Japan have estimated the biomass of zooplankton via sampling at depths of 0–150 m using vertical net tows (mesh size: 0.33–0.34 mm) and then measuring the wet weight of samples. Two studies in the 20th century have reported interannual variation in the wet weight of zooplankton in the SOJ. The first study presents data collected from the central SOJ (named PM-line) (Minami et al. 1999), and the other study reports interannual variation from several larger data sets from the SOJ (Hirota and Hasegawa 1999). The PM-line data were also included in Hirota and Hasegawa (1999), and the interannual variation reported in these two studies was similar. Zooplankton biomass decreased in the late 1980s. Minami et al. (1999) collected data until 1995, and zooplankton biomass increased in the 1990s.

Minami et al. (1999) suggested that changes in zooplankton biomass are driven by changes in water temperature. Minami et al. (1999) also reported that variation in zooplankton biomass undergoes six-year and 20-year cycles. The six-year cycle of zooplankton biomass is synchronized with cycles of water temperature, and the relationship between biomass and temperature is negative over this cycle (Minami et al. 1999). However, the 20-year cycle of zooplankton biomass parallels the 16-year cycle in water temperature, and the relationship between biomass and temperature is positive over this scale (Minami et al. 1999). However, Hirota and Hasegawa (1999) concluded that temperature is not the primary cause of interannual variation in zooplankton biomass in the SOJ. They also reported that variation in zooplankton biomass is not always consistent with the Monsoon Index defined by Hanawa et al. (1988), the thickness of the mixed layer, the pattern of the Tsushima Current, and chlorophyll-*a* concentrations, which led them to conclude that the primary cause of interannual variation in zooplankton biomass in the SOJ remains unclear.

Chiba et al. (2005) used data sets similar to those of Hirota and Hasegawa (1999) and analyzed decadal variation in zooplankton biomass and environmental parameters in spring and winter. They suggested that decadal variation in zooplankton biomass and its underlying causes vary in the northern and southern parts of the SOJ. In the northern part of the SOJ, zooplankton biomass slightly increased after 1977 when the regime shift occurred in the North Pacific, and zooplankton biomass in spring is positively associated with the chlorophyll-*a* concentration (Chiba et al. 2005). In the southern part of the SOJ, zooplankton biomass decreased after 1977, but a significant relationship between the chlorophyll-*a* concentration and zooplankton biomass was not observed (Chiba et al. 2005). Both Hirota and Hasegawa (1999) and Chiba et al. (2005) suggested that bottom-up processes do not explain interannual variation in zooplankton biomass in the southern part of the SOJ.

Data on zooplankton biomass in the SOJ have been collected up to 2019, and a recent study re-analyzed patterns of interannual variation with these new data (Kodama et al. 2022) (Figure 2). Kodama et al. (2022) reported two major findings regarding interannual variation in zooplankton biomass in the SOJ: (1) zooplankton

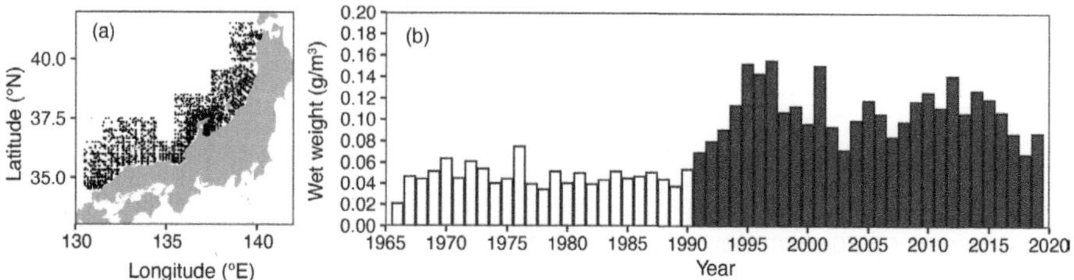

Figure 2. Interannual variation in the wet weight of zooplankton in the southern SOJ from 1966 to 2019. (a) The sampling station, and (b) median values of the wet weights of samples collected with 0.33-mm mesh nets each year. The original data were from Kodama et al. (2022), but much of these data were collected by Hirota and Hasegawa (1999) before 1990 (blank bars).

biomass has not linearly decreased as water temperature has increased over the past four decades, and (2) top-down control (predatory pressure of small pelagic fish) appears to regulate interannual variation in zooplankton biomass.

In Kodama et al. (2022), the southern SOJ was divided into three subareas (including one in the East China Sea), and interannual variation in zooplankton biomass was evaluated using an empirical modeling approach. Consistent with previous studies, zooplankton biomass was low in the 1980s, rapidly increased in the 1990s, and gradually decreased in the 2010s. These changes in zooplankton biomass mirror changes in the biomass of small pelagic fish (Japanese sardine, Japanese Anchovy, and Pacific Round Herring) in the SOJ, and there was a significant negative relationship between zooplankton biomass and small pelagic fish biomass (Kodama et al. 2022). However, comparing differences from the previous year is necessary in the case of time-series data sets. Kodama et al. (2022) found a significant negative relationship between differences in zooplankton biomass from the previous year and differences in small pelagic fish biomass over the same periods (Kodama et al. 2022). This suggests that the biomass of zooplankton decreases when the biomass of small pelagic fish increases in the southern SOJ; thus, small pelagic fish biomass appears to be the primary cause of long-term variation in zooplankton biomass.

Previous studies examining interannual variation in zooplankton biomass in the SOJ have indicated that the factors affecting zooplankton biomass in the northern and southern parts of the SOJ differ. In the northern part of the SOJ, bottom-up processes, such as primary production and temperature, play key roles, but top-down processes play a more important role in the southern part of the SOJ. This indicates that the effects of anthropogenic activities on zooplankton biomass might differ in the northern and southern parts of the SOJ. In the northern region, global warming or nutrient inputs might have more pronounced effects on zooplankton biomass compared with other factors, whereas in the southern region, the impacts of fisheries activities might be disproportionately substantial. More recent observations are needed to confirm a relationship between anthropogenic activities and zooplankton biomass in the northern SOJ. However, the northern part of the SOJ is located outside the exclusive economic zone of Japan and Korea; thus, recent changes in zooplankton biomass remain unclear.

2.2 Zooplankton Community

Here, we reviewed studies of zooplankton community structure in the SOJ. We reviewed studies reporting the biomass of zooplankton communities and discussed potential changes in community structure.

2.2.1 Copepods

In the SOJ, copepods are the most numerically dominant zooplankton group. For example, copepods account for 18.6–77.5% (annual mean 58.2%) of the total biomass of zooplankton in the vicinity of the Tsushima Strait, which is located on the southwestern edge of the SOJ (Hirakawa et al. 1995). The wet-weight-based abundance of copepods was 48% of the total wet weight of mesozooplankton in the Yamato Rise in the central SOJ (Hirakawa et al. 1999) and 28.4% of the total wet weight of mesozooplankton in Toyama Bay in the southern central SOJ (Hirakawa et al. 1992). These studies (Hirakawa et al. 1992, Hirakawa et al. 1995, Hirakawa et al. 1999) have reported variation in copepods among months. Seasonal variation in copepod abundances has not been investigated in recent years, and data on copepod abundances in winter and autumn are particularly lacking; nevertheless, all data indicate that copepods are the most dominant zooplankton group in the SOJ (Kodama et al. 2018a, b).

Interannual variation in copepods in the SOJ has not been a major focus of studies conducted in the 20th century. Two studies have provided "snapshots" of copepod assemblages (Iguchi and Tsujimoto 1997, Iguchi et al. 1999), and these studies have categorized copepods into warm water, cold water, and cosmopolitan species. Kodama et al. (2018b) reported variation in zooplankton assemblages in the central coastal SOJ (from Wakasa Bay to Toyama Bay) in spring (Figure 3). Kodama et al. (2018b) mainly focused on patterns of spatial variation, but they also reported information on interannual variation. In the spring, *Corycaeus affinis* (warm-water species) and *Oithona atlantica* (cold-water species) were the dominant copepods. *C. affinis* and *O. atlantica* were the dominant copepods in the western and eastern regions, respectively, and the horizontal advective transport of warm water (coastal branch of the Tsushima Warm Current) increases the abundance of warm-water species in coastal SOJ (Kodama et al. 2018b). A west-east gradient of zooplankton assemblages was observed every year in May from 1999 to 2013, and these relationships were not linear (Kodama et al. 2018b), suggesting that interannual variation was low compared with spatial variation.

The copepod assemblages in the spring of the coastal SOJ were also investigated in 1995 (Iguchi and Tsujimoto 1997, Iguchi et al. 1999). In Wakasa Bay, *Corycaeus* spp. were the most dominant, and *Calanus sinicus* was the second most dominant species (Iguchi et al. 1999). The copepod assemblages in Wakasa Bay were the same as those in Kodama et al. (2018b). In Toyama Bay, which is located in the western part of the study area of Kodama et al. (2018b), *Corycaeus* spp. were the most dominant, and *O. atlantica* was the second most dominant species at five of the six stations in May (Iguchi and Tsujimoto 1997). These findings are inconsistent with those of Kodama et al. (2018b). According to Kodama et al. (2018b), *C. affinis*

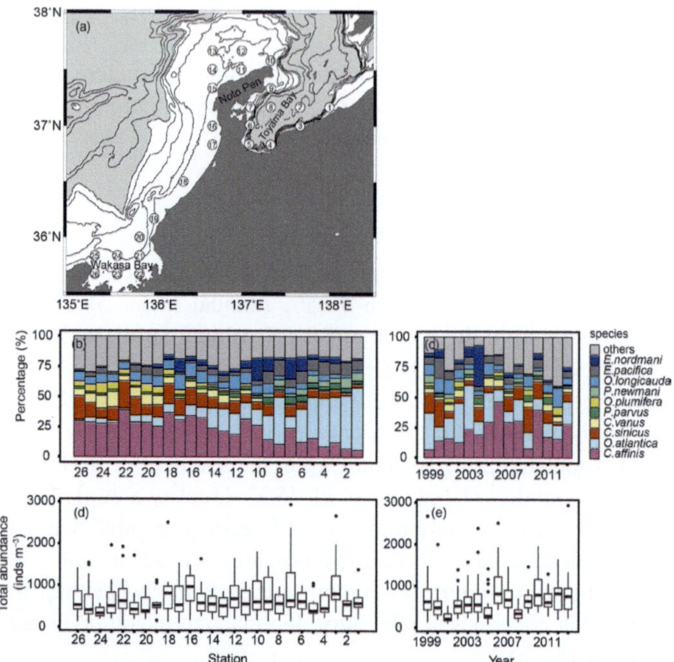

Figure 3. Interannual and spatial variation in zooplankton community structure and abundance in May in the southern coastal area of the SOJ from 1999 to 2013. (a) The map shows sampling sites (stations from 1 to 26). (b) Spatial and (c) interannual variation in zooplankton community structure; zooplankton were sampled using 0.33 mm mesh nets. (d) Spatial and (e) interannual variation in the abundance of zooplankton. The boxplots show the median (horizontal lines within boxes), upper and lower quartiles (boxes), quartile deviation (bars) and outliers (circles). The original figures are from Kodama et al. (2018b) with CCBY 4.0 License.

is a warm-water species, and the sea surface temperature in Toyama Bay is less than 16°C. However, the sea surface temperature in Toyama Bay was 16–18°C in May 1995 (Iguchi and Tsujimoto 1997). These differences in water temperature might explain the differences in the copepod assemblages between Iguchi and Tsujimoto (1997) and Kodama et al. (2018b).

Changes in the copepod assemblages in the SOJ are largely driven by variations in temperature. However, in Kodama et al. (2022), significant decadal variation in copepod assemblages was observed, yet no variation in the range of temperatures was observed over decades. Spring (April and May) zooplankton assemblages (mainly copepod assemblages) were investigated in the 1970s (western and center SOJ, offshore), 1980s (center and eastern SOJ, offshore), 2000s, and 2010s (eastern SOJ, offshore and coastal). The proportions of calanoid *Mesocalanus* and *Neocalanus* were significantly higher in the 1970s than in the other three decades, whereas the proportion of the cyclopoid *Oithona* was lower in the 1970s than in the other three decades (Kodama et al. 2022). The proportion of the calanoid *Pseudocalanus* was higher in the 1980s than in the 2000s and the 2010s, and the proportion of *Oithona* was lower in the 1980s than in the 2000s and the 2010s (Kodama et al. 2022). Kodama et al. (2022) suggested that this variation among decades might be associated with geographical differences; however, alternative causes yet to be discovered might also contribute to explaining these patterns. Hence, the causes of interannual variation in copepod assemblages in the SOJ remain unclear; additional studies are needed to clarify the causes driving this variation.

2.2.2 Cladocerans

Marine cladocerans have been reported to make significant contributions to total zooplankton biomass in the SOJ (Kim and Onbé 1989, Kim et al. 1993, Hirakawa et al. 1995, Onbé and Ikeda 1995). The marine cladocerans are reported only eight species: *Penilia avirostris*, *Evadne nordmanni*, *E. spinifera*, *Pseudevadne tergestina*, *Pleopis polyphemoide*, *Pl. schmackeri*, *Podon leuckartii*, and *Po. intermedius* (Egloff et al. 1997), and all of these species, with the exception of *Po. intermedius*, have been documented in the SOJ (Onbé and Ikeda 1995). The life cycles of cladocerans differ from those of copepods; marine cladocerans can reproduce both parthenogenetically (i.e., asexually) and gamogenetically (i.e., sexually). Under favorable conditions, they reproduce parthenogenetically, but resting eggs are produced by females under unfavorable conditions (Egloff et al. 1997). These resting eggs sink to the ocean floor, and their development resumes when the temperature of the sea bottom becomes optimal (Onbe 1974). Therefore, marine cladocerans require neritic environments to complete their life cycles; *Ps. tergestina* and *Pl. schmackeri* are usually present in offshore waters (Longhurst and Seibert 1972).

The abundances of marine cladocerans in the SOJ were first reported in 1966–1974 during the Cooperative Study of the Kuroshio and Adjacent Regions (CSK) (Kim and Onbé 1989, Kim et al. 1993). These samples were collected at a depth of 0–150 m using 0.33-mm mesh tow nets. The abundances of *Pl. schmackeri* and *E. spinifera* were estimated by Kim and Onbé (1989) and Kim et al. (1993). The abundance of *Pl. schmackeri* was high in August and usually was ≤ 30 individuals (inds) m^{-3} in the Tsushima Strait (max: 79.2 inds m^{-3}) (Kim and Onbé 1989). The abundance of *Evadne spinifera* was only ≤ 11 inds m^{-3} in the SOJ, according to the CSK (Kim et al. 1993).

Onbé and Ikeda (1995) studied annual variation in marine cladoceran abundances in Toyama Bay from February 1990 to January 1991. The samples in Onbé and Ikeda (1995) were collected at a depth of 0–500 m using 0.33-mm mesh plankton nets. Marine cladocerans were observed in the SOJ from March to December, and their abundance was highest in August (approximately 60 inds m^{-3}) (Onbé and Ikeda 1995). During this period, *Pe. avirostris* was the dominant marine cladoceran (53 inds m^{-3}). The abundance of cladocerans in the family *Podonidae* (except *Pe. avirostris*) was always <10 inds m^{-3}, the abundances of *E. nordmanni*, *E. spinifera*, and *Ps. tergestina* were highest in April (5.6 inds m^{-3}), July (8.0 inds m^{-3}), and June (5.8 inds m^{-3}), respectively (Onbé and Ikeda 1995).

Kodama et al. (2021) reported marine cladoceran abundances in the SOJ using massive data sets from 1997 to 2019. Sampling was conducted with nets of several mesh sizes (0.06 mm, 0.1 mm, and 0.33 mm), and significant differences in abundances were observed between samples taken with 0.33-mm mesh nets compared with samples taken using nets of the other mesh sizes. The major cladoceran

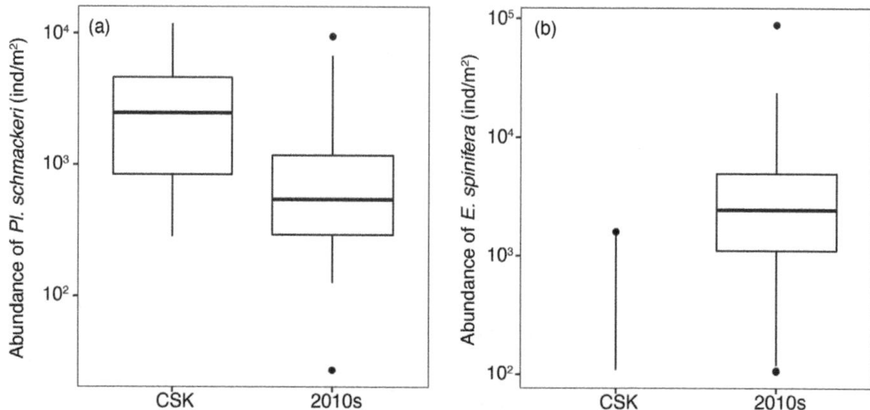

Figure 4. Comparison of the abundances of (a) *Pleopis schmackeri* and (b) *Evadne spinifera* between the CSK (Cooperative Study of the Kuroshio and Adjacent Regions) period (1966–1974) and the 2010s (2011–2017). The data were from Kim and Onbé (1989) (for *Pl. schmackeri* during the CSK period), Kim et al. (1993) (for *E. spinifera* during the CSK period), and Kodama et al. (2021) (for the 2010s). To facilitate comparison, the data in Kodama et al. (2021) were limited to samples collected using 0.33 mm mesh nets towed from a depth of 200 m. Only the maximum values of *E. spinifera* during the CSK period were reported by Kim et al. (1993).

species detected using the 0.33-mm mesh samples were *Pe. avirostris*, *E. nordmanni*, *E. spinifera*, *Ps. tergestina*, and *Pl. schmackeri*, and the abundances of these five major species were sometimes greater than 10^4 inds m^{-2} (Kodama et al. 2021), peaks in the abundances of *Pe. avirostris*, *E. spinifera*, *Ps. tergestina*, and *Pl. schmackeri* were observed in August or September, and that of *E. nordmanni* was in April (Kodama et al. 2021). An empirical statistical model (a generalized additive model) indicated that the abundances of the major marine cladocerans, with the exception of *E. nordmanni*, increase with warming (Kodama et al. 2021).

Kodama et al. (2021) did not evaluate interannual variation in marine cladocerans because of high variation in sampling effort over the years; thus, patterns of variation in the abundance of marine cladocerans in this region have not yet been reported. The units of abundance in Kodama et al. (2021) (inds m^{-2}) were different from the units used in previous studies (inds m^{-3}) (Kim and Onbé 1989, Kim et al. 1993, Onbé and Ikeda 1995), but conversions are possible using the tow depths. The abundance of *Pl. schmackeri* in the water column was $< 1.2 \times 10^4$ inds m^{-2}, and usually $< 4.5 \times 10^3$ inds m^{-2} in 1968 (Kim and Onbé 1989), and that of *E. spinifera* was $\leq 1.6 \times 10^3$ inds m^{-2} from 1965 to 1974 (Kim et al. 1993) (Figure 4). In 1990 (Onbé and Ikeda 1995), the highest abundance of *Pe. avirostris* observed was approximately 2.6×10^4 inds m^{-2}, and the highest abundance of the family *Podonidae* observed was $< 5.0 \times 10^3$ inds m^{-3}. The abundance of *Pl. schmackeri* in 1968 was at the same level or higher than its abundance in the 2010s from July to September (median: 5.4×10^2 inds m^{-2}, data points with an abundance of 0 were removed; Figure 4) (Kodama et al. 2021). The abundance of *E. spinifera* in the 2010s from July to September was 2.4×10^3 inds m^{-2} (data points with an abundance of 0 were removed), and this was higher than the abundance of *E. spinifera* from 1965 to 1974 in the SOJ. These findings indicate that the abundance of *E. spinifera* might have increased over the past 50 years.

Three hypotheses have been proposed to explain the increase in the abundance of *E. spinifera* in the SOJ: (1) warming trends, (2) a decrease in predatory pressure in the 21st century, and (3) an increase in the abundance of *E. spinifera* in source water, such as Kuroshio water. The water temperature of the SOJ increased by more than 1°C from 1982 to 2006 (Belkin 2009). The abundance of major marine cladocerans, with the exception of *E. nordmanni*, increased in the warm water in the SOJ (Kodama et al. 2021). Therefore, spring and summer are the most optimal periods for the growth of *E. spinifera* in the SOJ. Predatory pressure also requires consideration because marine cladocerans are preyed upon by pelagic fish and their larvae (Robert et al. 2014, Landry et al. 2018). In Kodama et al. (2022), zooplankton biomass (according to wet-weight values) was lower in the SOJ in 1990 than at any point in the 21st century because of predatory pressure. These findings indicate that the low marine cladoceran abundances in Onbé and Ikeda (1995) might be explained by predatory pressure from small pelagic fish. We have no reliable estimates of small pelagic fish biomass before 1974, thus, we are unable to reject the hypothesis that predatory pressure is the main cause of differences in the abundance of *E. spinifera* between the CSK period and recent years. However, none of the aforementioned hypotheses can explain the similarity in the abundance of *Pl. schmackeri* in 1968 and the 2010s. Therefore, the possibility that increases in the abundance of *E. spinifera* in the source water contribute to this pattern requires consideration. The abundance of *E. spinifera* in the Kuroshio water during the CSK period was not measured, but several tens of individuals per 1 m² of *E. spinifera* have been documented in Kuroshio water (Kodama et al. 2021).

In summary, changes in the abundance of marine cladocerans in the SOJ have varied extensively over previous decades. The abundance of *E. spinifera* was higher in the SOJ in the 2010s than in the 1960s, and no significant differences were observed in the abundance of *Pl. schmackeri* over this same period. Because only seven marine cladocerans are present in the SOJ, studies of changes in cladoceran assemblages and biomass at the species level are more tractable compared with such studies in copepods. Given that many cladocerans prefer warm water, marine cladocerans could be used as indicators of warming in the SOJ.

2.2.3 Jellyfish and Salps

Jellyfish biomass is increasing in coastal oceans worldwide (Condon et al. 2013). Although Condon et al. (2013) did not find that jellyfish biomass is increasing or fluctuating, increases in the biomass of jellyfish have been observed in East Asian coastal waters and on a global scale over the past three decades (Dong et al. 2010). In the Chinese waters adjacent to the SOJ, blooms of three species of jellyfish (*Aurelia aurita*, *Cyanea nozakii*, and *Nemopilema nomurai*) have been documented (Dong et al. 2010). Blooms of *N. nomurai* have occurred several times in the 21st century in the SOJ (Uye 2008).

Nemopilema nomurai is one of the largest jellyfish in the world, and its bell diameter is over 1 m long (Omori and Kitamura 2004). Blooms of *N. nomurai* occurred approximately once every 40 years in the 20th century (Uye 2008). The first bloom was observed in 1920 (Kishinouye 1922), the second one was in 1958 (Shimomura 1959), and the third one was in 1995 (Kubota et al. 1996). Shimomura (1959) noted that a bloom of *N. nomurai* occurred in 1940 in the SOJ. The bloom in 1958 was described in detail by Shimomura (1959). Specifically, Shimomura (1959) noted that *N. nomurai* was generally not a rare species and that several individuals would be observed every year in the SOJ; however, in 1958, individuals were observed every 4 m to 5 m in the horizontal direction. Shimomura (1959) also reported that *N. nomurai* is often observed in the Yellow Sea, which is surrounded by China and the Korean Peninsula.

Blooms of *N. nomurai* have been considered rare in the 20th century; however, blooms of this species have been observed frequently in the 21st century (Kawahara et al. 2006, Uye 2008, Kitajima et al. 2015), including 2002, 2003, 2005, 2006, and 2009. Uye (2008) indicated that the 2008 bloom was the largest ever documented and estimated that $3–5 \times 10^8$ medusae flowed from the East China Sea to the SOJ. Since 2010, blooms of this species have not been reported, but approximately 50 inds of *N. nomurai* were collected in a set-net in 2016 (Kawano et al. 2020). This suggests that the frequency of *N. nomurai* blooms in the SOJ has increased in recent years.

Many studies have attempted to determine the causes of *N. nomurai* blooms in East Asian waters, including the SOJ. Shimomura (1959) suggested that *N. nomurai* originated in the Tsushima Strait, which is the gate of the SOJ, and that the stocks of the Yellow Sea and the SOJ are different. Recent studies have revealed a link between the biomass of *N. nomurai* in the northern East China Sea and the SOJ (Uye 2008, Kitajima et al. 2020), and numerical modeling studies have indicated that *N. nomurai* in the SOJ is derived from Chinese coastal areas (Reizen and Isobe 2006). *N. nomurai* does not reproduce in the SOJ because they are unable to survive the cold winter temperatures (Kitajima et al. 2015).

The environmental conditions in Chinese coastal areas, especially in the Changjiang Riverine, might contribute to the increased frequency of *N. nomurai* outbreaks (Yoon et al. 2008, Dong et al. 2010, Kawahara et al. 2013, Xu et al. 2013, Yoon et al. 2014, Sun et al. 2015). Yoon et al. (2008) reported that *N. nomurai* is only present in Changjiang diluted water, and the abundance of *N. nomurai* is positively associated with the abundance of small and medium-sized zooplankton. Xu et al. (2013) reported that eutrophication and the warming of seawater in late spring and early summer are the key factors contributing to *N. nomurai* outbreaks. Kawahara et al. (2013) reported that podocysts of *N. nomurai* can survive at least six years, and excystment occurs under warm, less saline and hypoxic conditions. Thus, when these conditions prevail in the East China Sea estuary, blooms of *N. nomuraii* occur (Kawahara et al. 2013).

Anthropogenic activities also have substantial effects on jellyfish blooms in the East China Sea (Uye 2008, Dong et al. 2010). Both Dong et al. (2010) and Uye (2008) noted four major ways that anthropogenic activities affect jellyfish blooms in the East China Sea: (1) eutrophication, (2) overfishing, (3) habitat modification, and (4) climate change (warming). Overfishing and habitat modification are considered more important causes of jellyfish blooms in the East China, Bohai, and Yellow Seas than in the South China Sea (Dong et al. 2010). In these seas, the intensity of mariculture activities has increased rapidly, and this has coincided with decreases in fishery catches (Dong et al. 2010).

Temporal variation in *N. nomurai* biomass has not yet been estimated in the SOJ, but blooms of *N. nomurai* have increased. The effects of *N. nomurai* blooms on ecosystems in the SOJ remain unclear (Iguchi et al. 2017). *Nmopilema nomurai* preys on macrozooplankton and the carbon requirement for the growth and respiration of *N. nomurai* in the SOJ is 0.1–0.33 mgC m^{-3} d^{-1} (Iguchi et al. 2017). In bloom years, the total daily carbon requirement of *N. nomurai* in the SOJ is equivalent to that of the common squid; however, the amount of microzooplankton consumed by *N. nomurai* is not sufficiently large to have a significant effect on macrozooplankton biomass (Iguchi et al. 2017). *N. nomurai* blooms in the SOJ mainly affect human activities, such as fisheries (Uye 2008).

Other salps and jellyfish also significantly contribute to biological production in the SOJ. Kuroda et al. (2000) and Shimomura (1959) reported outbreaks of various gelatinous planktons (jellyfish and salps) aside from *N. nomurai* in the SOJ, including the common salp, *Salpa fusifolmis*; moon jellyfish, *Aurelia aurita*; and northern sea nettle, *Chrysaora melanaster*. Blooms of *A. aurita* and *C. melanaster* have been observed on local scales nearly every year between 1995 and 2000 (Kuroda et al. 2000). By contrast, blooms of *S. fusifolmis* are rare; the *S. fusifolmis* bloom in 2000 was the first bloom of this species documented in the SOJ since 1965, and blooms of this species were frequently recorded from 1950 to 1965 (Kuroda et al. 2000). In the spring of 2004, Iguchi and Kidokoro (2006) reported outbreaks of the salp *Thetys vagina*, the largest salp in the world, in the SOJ. This provided the first record of *T. vagina* outbreaks in the SOJ. Iguchi and Kidokoro (2006) also reported that warm water and high chlorophyll-*a* concentrations might induce *T. vagina* outbreaks.

Many studies have indicated that large gelatinous plankton, such as jellyfish and salps, make large contributions to biological production in the SOJ. However, Iguchi and Kidokoro (2006) noted that quantitative estimates of salp abundance are difficult to generate because they show large variations in body size; furthermore, the effect of sampling bias requires consideration. As mentioned in Kuroda et al. (2000), robust approaches for monitoring and estimating the biomass of large gelatinous plankton need to be developed to detect changes in ecosystems of the SOJ.

2.2.4 Other Zooplankton

Chaetognaths have been used as indicators of water mass in the 20th century in the SOJ (Nagai et al. 2006, 2008, Nagai et al. 2012). The four dominant chaetognath species in the SOJ are *Sagitta elegans*, *S. enflata*, *S. minima*, and *S. nagae* (Nagai et al. 2006, 2008, Nagai et al. 2012). *Sagitta elegans* is a cold-water species and is usually present in the northern part of the SOJ; the other three species are present in the southern part of the SOJ.

Both Nagai et al. (2006) and Nagai et al. (2008) characterized variation in chaetognath community structure from 1972 to 2002. They found that temperature controls the distribution of dominant species present in warm water (*S. enflata*, *S. minima*, and *S. nagae*) and community structure. Interannual variation in the abundance of the three dominant warm-water chaetognath species was affected by water temperature (Nagai et al. 2006). Positive relationships between temperature were generally observed in every season in these three species. These findings suggest that the abundance of warm-water species might increase in the SOJ as warming continues. In the East China Sea, the abundance of the cosmopolitan chaetognath *Pterosagitta draco* increased from 1975 to 2005 (Aoyama et al. 2008). Nagai et al. (2008) showed a schematic diagram of the factors controlling chaetognath assemblages; however, direct evidence of a link between climate indexes and chaetognath assemblages has not yet been obtained.

Chaetognath abundances are thought to increase with temperature in the SOJ. However, no studies using 21st-century datasets have examined the abundance of chaetognaths in the SOJ. Given that chaetognaths are carnivorous, increases in their abundance would likely increase predatory pressure on secondary producers (Nagai et al. 2012). Analyses of recent observations of chaetognath abundances and community structure are needed to clarify the effect of global warming on their top-down effects.

Appendicularia are the second most dominant zooplankton during the summer in the SOJ (Kodama et al. 2018a). The most dominant species during the summer in the SOJ is *Oikopleura longicuada*, and the abundance of *O. longicuada* is positively associated with the abundance of *Oncaea* and *Microsetella* (Kodama et al. 2018a). These findings suggest that increases in the abundance of Appendicularia species can increase production at higher trophic levels in epipelagic ecosystems in the SOJ (Kodama et al. 2018a). The numerical abundance of *O. longicuada* increased with warming, which suggests that the warming trend in the SOJ might increase the relative contribution of *O. longicuada* to total production in the SOJ (Kodama et al. 2018a). However, few 20th-century studies on appendicularians have been conducted in the SOJ; two studies (Tomita et al. 1999, Tomita et al. 2003) have examined appendicularians in Toyama Bay. In Tomita et al. (1999), appendicularian assemblages and biomass were studied over a single year from 1990 to 1991, and *O. longicuada* was the dominant appendicularian during the summer. This finding is consistent with the results of Kodama et al. (2018a). The maximum biomass of *O. longicuada* in 1990 was 1,650 inds m^{-3}, which is consistent with the results of Kodama et al. (2018a) (mean ± SD: 271 ± 298 inds m^{-3} from 0 to 200 m in depth). Given the limitation in 20th-century observations, long-term patterns in the abundance and assemblages of appendicularians in the SOJ remain unclear.

3. Changes in Zooplankton in the Western North Pacific

3.1 *Zooplankton Biomass off the Tohoku Area*

Odate (1994) investigated long-term changes in zooplankton biomass in the epipelagic layer of the Tohoku area from 1951 to 1990 (Figure 5). During this period, decadal fluctuations in biomass in the Oyashio region were observed, and there were three biomass peaks: from 1956 to 1962, from 1965 to 1989, and from 1984 to 1989. Sugimoto and Tadokoro (1998) observed a fluctuating pattern similar to that observed in the different data sets from Odate (1994). Interannual variability was lower in the Kuroshio Extension and transition regions than in the Oyashio region;

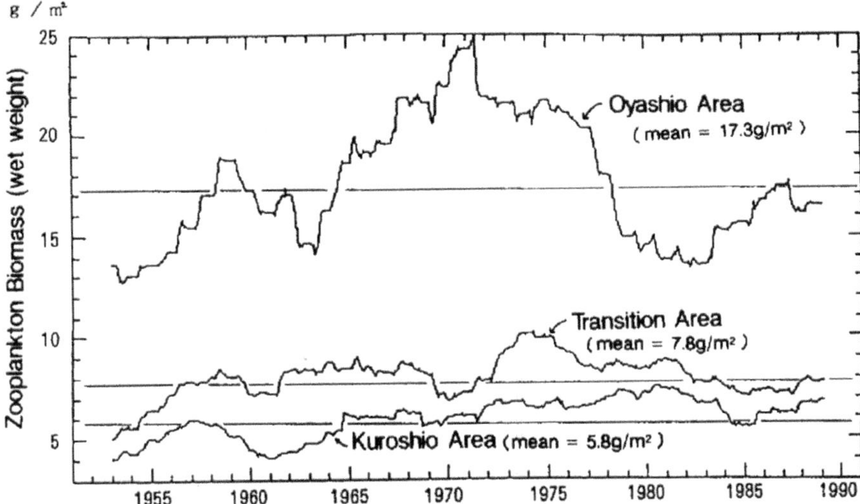

Figure 5. Long-term variability in zooplankton biomass indicated by the 48-month running mean from 1951 to 1990. Horizontal lines in the figure show the mean value of zooplankton biomass. The original figure is from Odate (1994), and the Japan Fisheries Research and Education Agency (FRA) has the copyright.

however, biomass peaks in the transition zone were observed between 1957 and 1969 and around 1974, and those in the Kuroshio Extension region were observed in the late 1950s and early 1980s. Tadokoro (2004) provided an update to the data of Odate (1994) and found a distinct biomass peak around 1995 that surpassed the previous peaks. Chiba et al. (2006) and Chiba et al. (2008) demonstrated that a distinct shift from high to low zooplankton biomass occurred in 1976/1977, and the inverse pattern was observed in 1988/1989. These shifts were consistent with shifts in climatic regimes in response to the Pacific oscillation index (PDO).

3.2 Copepod Communities

Long-term changes in zooplankton communities have been examined using data on copepod species composition and abundance. Chiba et al. (2006) studied the abundance and phenology of pelagic copepod communities in the subarctic WNP. In Chiba et al. (2006), the copepod community was divided into five subgroups. The abundance of the spring-summer subgroup (copepods mainly observed in spring to summer) decreased, and the abundance peak of this subgroup was delayed one month during the cold regime from the mid-1970s to the 1980s, however, the abundance of this group increased again in the 1990s. Possible geographical shifts are discussed in Chiba et al. (2009). Chiba et al. (2009) showed that the abundance of subarctic copepods was low, but it was higher south of the Oyashio fronts in the cold regime from 1981 to 1999, according to the data presented by Chiba et al. (2006). The abundance of subtropical copepods in the transition region also increased during this regime. Chiba et al. (2006) and Chiba et al. (2009) focused on the effects of climate systems, such as the PDO and Aleutian Low, on primary productivity, but

Chiba et al. (2013) mentioned that the increase in the abundance of subtropical copepods after 1981 might be related to the weakening of the Kuroshio Extension. They suggested that the offshore advection of zooplankton transported from south of Honshu reduced in weaken, while the zooplankton transported offshore (east of 150°E) under the strengthened current environment.

Miyamoto et al. (2017) investigated long-term variation in copepod communities in the spring and the relationship with the Kuroshio on this variation. They showed that the abundance of warm-water species, such as *Clausocalanus* and *Oncaea*, increased in the north front of the Kuroshio, but typical coastal copepods, such as *Paracalanus parvus* s.l., were abundant in the coastal area when the Kuroshio axis was located near the coast. Sogawa et al. (2017) and Sogawa et al. (2019) also noted that the transport of copepods by the Kuroshio (Extension) Current increases the abundance of copepods in the Kuroshio area. These findings indicate that the Kuroshio and Kuroshio Extension flow patterns affect zooplankton biomass and communities in the WNP.

3.3 Dominant Copepod Species

Because *Neocalanus* and *Eucalanus* spp. are the only dominant zooplankton in the Oyashio area and subarctic zone, changes in the biomass of dominant species might reflect changes in total biomass. Kobari et al. (2007) investigated the abundance of *E. bungii* in the Oyashio area. The monthly abundance of *E. bungii* from 1960 to 1990 decreased after the 1976/1977 regime shift and increased after the 1988/1989 regime shift (Kobari et al. 2007). The period of maximum abundance was shorter (from May to June) from the late 1970s to the early 1980s than in the other years (from April to July) (Kobari et al. 2007).

Tadokoro et al. (2005) investigated long-term changes in the abundance of three subarctic *Neocalanus* species, which were the most dominant species in mesozooplankton communities in the subarctic area of the North Pacific. Although interannual variation in the abundance of *N. flemingeri* was comparatively stable, interannual variation was more pronounced in *N. plumchrus* and *N. cristatus* (Tadokoro et al. 2005). The abundance of these two species increased after the 1988/1989 regime shift. The authors suggested that the abundance of these species was reduced extensively via predation by the Japanese sardine, which increased in abundance in the 1980s. Tadokoro et al. (2009) also noted that interannual variation in the phosphate concentration fluctuated in response to the 18.6-yr nodal tidal cycle, and this was consistent with long-term variation in the abundance of *N. plumchrus* (Figure 6). These findings indicated that variation in zooplankton abundance and biomass in the Oyashio area and transition region was also driven by changes in the nutrient supply associated with astronomical cycles.

Recently, interannual changes in the Oyashio region have been observed using CPR. Chiba et al. (2015) investigated interannual changes in copepod body size from 2000 to 2011 and found that the abundance of subarctic copepods increased after 2007 when the Oyashio area and subarctic regions were in a warm phase (PDO negative). The mean body size in the community increased. Although the increase in mean body size in warmer water was counterintuitive, given that large copepods

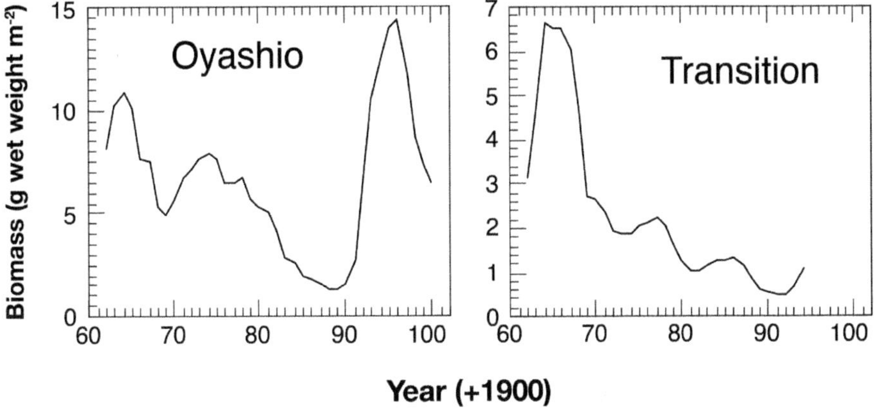

Figure 6. Five-year running mean of spring–summer (April–September) mean biomass of *Neocalanus plumchrus* from the 1960s to early 2000s. The original data was from Tadokoro et al. (2009) and provided by K. Tadokoro.

are generally distributed in cold water; the increase in abundance of these subarctic copepods likely increased in the warmer region because of the high primary production.

Yoshiki et al. (2013) investigated the interannual change in the abundance of the population structure of the subarctic copepods collected by the CPR in the Oyashio area and the western subarctic gyre. They showed that the timing of *Neocalanus* copepod occurrence in the epipelagic layer was stable interannually, whereas the timing of the occurrence of *E. bungii* exhibited a different interannual pattern. The occurrence of *E. bungii* in this layer was strongly affected by phenological variation in phytoplankton blooms (Yoshiki et al. 2013). Yoshiki et al. (2015) also investigated copepod communities in June and July using data collected by the CPR in the same region and found that warm-water species tended to be distributed at higher latitudes because of high temperatures after 2011. However, the warm-water effect was not observed in subarctic copepods, suggesting that the effect of warm water differed among species.

3.4 Other Zooplankton Communities

Long-term variation in zooplankton has not been reported in the WNP, with the exception of copepods. In the WNP, blooms of the salp *Thalia democratica* have been reported (Tsuda and Nemoto 1992, Ishak et al. 2022). Tsuda and Nemoto (1992) found that pelagic tunicates were distributed in the warm-core ring and its adjacent water, and blooms of *T. democratica* occurred in the front of the warm-core ring. Ishak et al. (2022) reported that *T. democratica* blooms occur in water masses with high nutrient concentrations and a high N/P ratio, indicating that this species would thrive in the HNLC water mass from the subarctic region. Ishak et al. (2022) also reported doliolid blooms (*Dolioletta gegenbauri*) east of 180° and north of the subarctic boundary, which was associated with the high chlorophyll-*a* region. Blooms of this same species have been reported in Kuroshio-Oyashio mixed water (Takahashi et al. 2015). These findings suggest that gelatinous zooplankton are key species, but long-term variations in their abundances and the frequency of their blooms have not yet been examined.

4. Synthesis and Future Directions

Given that zooplankton is the prey of many fish, variation in zooplankton biomass and community structure likely affects the production of fisheries. There are numerous approaches for monitoring zooplankton biomass and community structure. Continuous monitoring efforts are needed to enhance our understanding of the status of zooplankton populations and the feeding environments of fish. According to studies of zooplankton biomass and community structure in the SOJ and WNP, long-term changes in mesozooplankton biomass and communities have not changed linearly with increases in temperature, which indicates that top-down effects are important in the SOJ. Patterns of decadal variation in the WNP differ from those in the SOJ. This suggests that the warming trend is not a major factor currently affecting zooplankton biomass in the WNP and its marginal seas; however, increases in temperature have been shown to affect the survival of zooplankton (Tang et al. 2019). Given that the temperature continues to increase, zooplankton populations need to be monitored so that changes in marine ecosystems can be promptly identified. Marine heat waves have been documented in the WNP, and these have affected marine ecosystems in the WNP (Takagi et al. 2022). The effects of global warming thus require attention.

In both the SOJ and WNP, currents play a major role in the transportation of zooplankton. This indicates that zooplankton and physical oceanographic properties are strongly linked. Global warming can also affect the paths of currents (Nishikawa et al. 2020). The effects of currents on secondary producers in Japanese waters require further study to clarify patterns of future change. There is a special need for studies of long-term variation in zooplankton in the East China Sea, including upstream of the Kuroshio area and the Sea of Okhotsk, because no studies to date have characterized variation in zooplankton in these regions.

Strong evidence demonstrating a link between primary production and secondary production is lacking. Although many studies have suggested that "bottom-up" processes explain variation in abundance and community structure, there is no direct evidence indicating that bottom-up processes are the major factor driving interannual variation in zooplankton in Japanese waters. The relationship between primary and secondary production is assumed to be non-linear because there are temporal lags in the growth of phytoplankton and zooplankton. In addition, data on phytoplankton biomass in the ocean can be obtained on a weekly basis using satellites, but data on zooplankton biomass cannot be obtained on a weekly basis in offshore areas. More monitoring efforts are needed to clarify the relationship between primary and secondary production.

Further development of monitoring methodologies is also needed. There is a special need for approaches that facilitate the detection of zooplankton assemblages, especially copepods, and for estimating the biomass of noncrustacean zooplankton.

Microscopy has long been the classical method for monitoring zooplankton assemblages, but DNA barcoding has become a powerful monitoring tool in recent years. DNA barcoding analysis was developed for the identification of zooplankton species during the Census of Marine Zooplankton project (Bucklin et al. 2010), and the metabarcoding approach is now considered adequate for the identification of copepods (Hirai et al. 2015, Laakmann et al. 2020). In Japanese waters, metabarcoding analysis of size-fractioned copepod community samples can provide new insights into the relative contributions of small copepods to secondary production; this approach could also possibly be used as a monitoring tool (Hirai et al. 2021).

Although a powerful approach for monitoring changes in zooplankton community structure, metabarcoding data are not free of bias. Metabarcoding-based estimates of zooplankton assemblages often differ from morphological identification-based estimates (Ershova et al. 2021). In the SOJ, Kodama et al. (2020) investigated PCR bias in metabarcoding analysis using the 18S RNA V9 regions and found significant positive relationships in numerical abundances and abundances estimated via metabarcoding analysis, but the relationships varied among species. The same approach (primer sets, PCR conditions, and sample storage) must consistently be used when quantifying temporal variation in zooplankton assemblages. The metabarcoding approach cannot be effectively applied to old samples stored in formaldehyde. Shiozaki et al. (2021) reported that the results of DNA metabarcoding analysis did not change after samples were stored for approximately 1.5 years in formaldehyde solution, but whether this is the case for long periods remains unclear.

The metabarcoding approach can also be applied to other zooplankton taxa, especially noncrustacean zooplankton (Bucklin et al. 2021). However, the problem of sampling bias needs to be resolved before this approach can be applied to some noncrustacean zooplankton, such as jellyfish and salps. The sampling method also requires consideration. Videos with remotely operated vehicles (ROV) have been used to investigate gelatinous plankton (William et al. 2003, Takahashi et al. 2015, Choy et al. 2017), however, the robustness of quantitative biomass estimates of gelatinous plankton obtained via videos has been questioned given that the volume of water that can be observed via video is limited compared with the amount of water that can be sampled via classical sampling approaches such as plankton nets. Visual surveys, including quantitative estimates of zooplankton biomass, have not yet been used to monitor zooplankton populations.

Although visual surveys have been used to estimate zooplankton biomass, they could potentially be made even more useful through the development of statistical approaches. Visual surveys with modern statistical approaches have been used to estimate the biomass of marine mammals and their habitats (Kanaji and Gerrodette 2020, Kanaji et al. 2021). The application of modern statistical approaches is not only necessary for the use of visual surveys but also for maximizing the utility of classical data sets. Empirical modeling approaches such as generalized linear and generalized additive models have been commonly used in ecological studies in the 21st century (Zuur and Ieno 2016), yet these approaches have rarely been used to study variation in zooplankton biomass.

Thus, the summary of the review is that zooplankton biomass and community structure in the East Asian Seas are changing, but the causes are not simple. Some studies indicate a warming of temperature is significant, but ocean currents or predation are important in some studies. Anthropogenic impacts are not ignorable but not dominant for changes in zooplankton biomass and community structure. We need more observations, development of technologies, and applications of modern statistical analyses to detect the anthropogenic impacts of Japanese waters on the zooplankton communities before the changes are over the ecosystem resilience.

5. Acknowledgments

We have no conflict of interest to declare. The details of the Odate Collection and a figure were provided by Dr. K. Tadokoro, Japan Fisheries Research and Education Agency. This work was financially supported by grants from the Japan Fisheries Research and Education Agency and JSPS KAKENHI Grant Number 23H02285 (for TK) and 19K06215 (for HM).

References

Ando, K., Lin, X.P., Villanoy, C., Danchenkov, M., Lee, J.H., He, H.J. et al. 2021. Half-century of scientific advancements since the cooperative study of the Kuroshio and adjacent regions (CSK) programme-need for a new Kuroshio research. Progress in Oceanography 193: 102513. https://doi.org/10.1016/j.pocean.2021.102513.

Aoyama, M., Goto, H., Kamiya, H., Kaneko, I., Kawae, S., Kodama, H. et al. 2008. Marine biogeochemical response to a rapid warming in the main stream of the Kuroshio in the western north pacific. Fisheries Oceanography 17(3): 206–218. https://doi.org/10.1111/j.1365-2419.2008.00473.x.

Belkin, I.M. 2009. Rapid warming of large marine ecosystems. Progress in Oceanography 81(1–4): 207–213. https://doi.org/10.1016/j.pocean.2009.04.011.

Bucklin, A., Nishida, S., Schnack-Schiel, S., Wiebe, P.H., Lindsay, D., Machida, R.J. et al. 2010. A census of zooplankton of the global ocean. Life in the World's Oceans: 247–265. https://doi.org/10.1002/9781444325508.ch13.

Bucklin, A., Peijnenburg, K.T.C.A., Kosobokova, K.N., O'Brien, T.D., Blanco-Bercial, L., Cornils, A. et al. 2021. Toward a global reference database of COI barcodes for marine zooplankton. Marine Biology 168(6): 78. https://doi.org/10.1007/s00227-021-03887-y.

Chen, C.-T.A. and Guo, X. 2020. Changing Asia-Pacific Marginal Seas. Singapore, Springer Singapore. https://doi.org/10.1007/978-981-15-4886-4.

Chen, C.-T.A., Lui, H.-K., Hsieh, C.-H., Yanagi, T., Kosugi, N., Ishii, M. et al. 2017. Deep oceans may acidify faster than anticipated due to global warming. Nature Climate Change 7(12): 890–894. https://doi.org/10.1038/s41558-017-0003-y.

Chiba, S., Aita, M.N., Tadokoro, K., Saino, T., Sugisaki, H. and Nakata, K. 2008. From climate regime shifts to lower-trophic level phenology: Synthesis of recent progress in retrospective studies of the western North Pacific. Progress in Oceanography 77(2–3): 112–126. https://doi.org/10.1016/j.pocean.2008.03.004.

Chiba, S., Batten, S.D., Yoshiki, T., Sasaki, Y., Sasaoka, K., Sugisaki, H. et al. 2015. Temperature and zooplankton size structure: climate control and basin-scale comparison in the north pacific. Ecology and Evolution 5(4): 968–978. https://doi.org/10.1002/ece3.1408.

Chiba, S., Di Lorenzo, E., Davis, A., Keister, J.E., Taguchi, B., Sasai, Y. et al. 2013. Large-scale climate control of zooplankton transport and biogeography in the Kuroshio-Oyashio extension region. Geophysical Research Letters 40(19): 5182–5187. https://doi.org/10.1002/grl.50999.

Chiba, S., Hirota, Y., Hasegawa, S. and Saino, T. 2005. North-south contrasts in decadal scale variations in lower trophic-level ecosystems in the Japan Sea. Fisheries Oceanography 14(6): 401–412. https://doi.org/10.1111/j.1365-2419.2005.00355.x.

Chiba, S., Sugisaki, H., Nonaka, M. and Saino, T. 2009. Geographical shift of zooplankton communities and decadal dynamics of the Kuroshio-Oyashio currents in the Western North Pacific. Global Change Biology 15(7): 1846–1858. https://doi.org/10.1111/j.1365-2486.2009.01890.x.

Chiba, S., Tadokoro, K., Sugisaki, H. and Saino, T. 2006. Effects of decadal climate change on zooplankton over the last 50 years in the Western Subarctic North Pacific. Global Change Biology 12(5): 907–920. https://doi.org/10.1111/j.1365-2483.2006.01136.x.

Choy, C.A., Haddock, S.H.D. and Robison, B.H. 2017. Deep pelagic food web structure as revealed by *in situ* feeding observations. Proceedings of the Royal Society B: Biological Sciences 284(1868): 20172116. https://doi.org/10.1098/rspb.2017.2116.

Condon, R.H., Duarte, C.M., Pitt, K.A., Robinson, K.L., Lucas, C.H., Sutherland, K.R. et al. 2013. Recurrent jellyfish blooms are a consequence of global oscillations. Proc. Natl. Acad. Sci. USA 110(3): 1000–1005. https://doi.org/10.1073/pnas.1210920110.

Dong, Z., Liu, D. and Keesing, J.K. 2010. Jellyfish blooms in China: dominant species, causes and consequences. Marine Pollution Bulletin 60(7): 954–963. https://doi.org/10.1016/j.marpolbul.2010.04.022.

Egloff, D.A., Fofonoff, P.W. and Onbé, T. 1997. Reproductive biology of marine cladocerans. Advances in Marine Biology. Blaxter, J.H.S. and Southward, A.J., Academic Press. 31: 79–167. https://doi.org/10.1016/S0065-2881(08)60222-9.

Ershova, E.A., Wangensteen, O.S., Descoteaux, R., Barth-Jensen, C. and Præbel, K. 2021. Metabarcoding as a quantitative tool for estimating biodiversity and relative biomass of marine zooplankton. ICES Journal of Marine Science 78(9): 3342–3355. https://doi.org/10.1093/icesjms/fsab171.

FAO. 2020. The State of World Fisheries and Aquaculture 2020. Sustainability in Action. Rome.

Gamo, T. 1999. Global warming may have slowed down the deep conveyor belt of a marginal sea of the northwestern Pacific: Japan Sea. Geophysical Research Letters 26(20): 3137–3140. https://doi.org/10.1029/1999gl002341.

Gamo, T., Nakayama, N., Takahata, N., Sano, Y., Zhang, J., Yamazaki, E. et al. 2014. The Sea of Japan and its unique chemistry revealed by time-series observations over the last 30 years. Monographs on Environment, Earth and Planets 2(1): 1–22. https://doi.org/10.5047/meep.2014.00201.0001.

Hanawa, K., Watanabe, T., Iwasaka, N., Suga, T. and Toba, Y. 1988. Surface thermal conditions in the western north Pacific during the ENSO events. Journal of the Meteorological Society of Japan 66(3): 445–456. https://doi.org/10.2151/jmsj1965.66.3_445.

Hasegawa, D., Matsuno, T., Tsutsumi, E., Senjyu, T., Endoh, T., Tanaka, T. et al. 2021. How a small reef in the Kuroshio cultivates the ocean. Geophysical Research Letters 48(7): e2020GL092063. https://doi.org/10.1029/2020gl092063.

Hirai, J., Kuriyama, M., Ichikawa, T., Hidaka, K. and Tsuda, A. 2015. A metagenetic approach for revealing community structure of marine planktonic copepods. Mol. Ecol. Resour. 15(1): 68–80. https://doi.org/10.1111/1755-0998.12294.

Hirai, J., Yamazaki, K., Hidaka, K., Nagai, S., Shimizu, Y. and Ichikawa, T. 2021. Characterization of diversity and community structure of small planktonic copepods in the Kuroshio region off Japan using a metabarcoding approach. Marine Ecology Progress Series 657: 25–41. https://doi.org/10.3354/meps13539.

Hirakawa, K., Imamura, A. and Ikeda, T. 1992. Seasonal variability in abundance and composition of zooplankton in Toyama Bay, southern Japan Sea. Bulletin of Japan Sea Regional Fisheries Research Laboratory 42: 1–15.

Hirakawa, K., Kawano, M., Nishihama, S. and Ueno, S. 1995. Seasonal variability in abundance and composition of zooplankton in the vicinity of the Tsushima straits, southern Japan Sea. Bulletin of Japan Sea Regional Fisheries Research Laboratory 45: 25–38.

Hirakawa, K., Morita, A. and Nagata, H. 1999. Seasonal variability of the zooplankton assemblage and its relationship with oceanographic structures at Yamato tai, central Japan Sea (In Japanese with English Abstract). Bulletin of Japan Sea Regional Fisheries Research Laboratory 49: 37–56.

Hirota, Y. and Hasegawa, S. 1999. The zooplankton biomass in the Sea of Japan. Fisheries Oceanography 8(4): 274–283. https://doi.org/10.1046/j.1365-2419.1999.00116.x.

Ichiye, T. 1984. Some problems of circulation and hydrography of the Japan Sea and the Tsushima Current. Ocean Hydrodynamics of the Japan and East China Seas. Ichiye, T., Elsevier. 39: 15–54. https://doi.org/10.1016/s0422-9894(08)70289-7.

Iguchi, N. 2004. Spatial/temporal variations in zooplankton biomass and ecological characteristics of major species in the southern part of the Japan Sea: a review. Progress in Oceanography 61(2–4): 213–225. https://doi.org/10.1016/j.pocean.2004.06.007.

Iguchi, N., Iwatani, H., Sugimoto, K., Kitajima, S., Honda, N. and Katoh, O. 2017. Biomass, body elemental composition, and carbon requirement of *Nemopilema nomurai* (scyphozoa: rhizostomeae) in the southwestern Japan Sea. Plankton and Benthos Research 12(2): 104–114. https://doi.org/10.3800/pbr.12.104.

Iguchi, N. and Kidokoro, H. 2006. Horizontal distribution of *Thetys vagina* tilesius (tunicata, thalicaea) in the Japan Sea during spring 2004. Journal of Plankton Research 28(6): 537–541. https://doi.org/10.1093/plankt/fbi138.

Iguchi, N. and Tsujimoto, R. 1997. Seasonal changes in the copepoda assemblage as food for larval anchovy in Toyama Bay, southern Japan Sea (In Japanese with English abstract). Bulletin of the Japan Sea National Fisheries Research Institute 47: 79–94.

Iguchi, N., Wada, Y. and Hirakawa, K. 1999. Seasonal changes in the copepoda assemblage as food for larval anchovy in western Wakasa Bay, southern Japan Sea (In Japanese with English abstract). Bulletin of the Japan Sea National Fisheries Research Institute 49: 69–80.

Ishak, N.H.A., Motoki, K., Miyamoto, H., Fuji, T., Taniuchi, Y., Kakehi, S. et al. 2022. Basin-scale distribution of salps and doliolids in the transition region of the North Pacific Ocean in summer: drivers of bloom occurrence and effect on the pelagic ecosystem. Progress in Oceanography 204: 102793. https://doi.org/10.1016/j.pocean.2022.102793.

Kanaji, Y. and Gerrodette, T. 2020. Estimating abundance of Risso's dolphins using a hierarchical Bayesian habitat model: a framework for monitoring stocks of animals inhabiting a dynamic ocean environment. Deep Sea Research Part II: Topical Studies in Oceanography 175: 104699. https://doi.org/10.1016/j.dsr2.2019.104699.

Kanaji, Y., Maeda, H., Okamura, H., Punt, A.E. and Branch, T. 2021. Multiple-model stock assessment frameworks for precautionary management and conservation on fishery-targeted coastal dolphin populations off Japan. Journal of Applied Ecology 58(11): 2479–2492. https://doi.org/10.1111/1365-2664.13982.

Kawahara, M., Ohtsu, K. and Uye, S.-I. 2013. Bloom or non-bloom in the giant jellyfish *Nemopilema nomurai* (scyphozoa: rhizostomeae): roles of dormant podocysts. Journal of Plankton Research 35(1): 213–217. https://doi.org/10.1093/plankt/fbs074.

Kawahara, M., Uye, S.I., Ohtsu, K. and Iizumi, H. 2006. Unusual population explosion of the giant jellyfish *Nemopilema nomurai* (scyphozoa: rhizostomeae) in east asian waters. Marine Ecology Progress Series 307: 161–173. https://doi.org/10.3354/meps307161.

Kawano, M., Sonoyama, T., Hori, S., Ogimoto, K., Kunimori, T. and Uchida, Y. 2020. Noteworthy Phenomena on the marine organisms in the southwestern Japan Sea off Yamaguchi prefecture during 2014–2018. Bulletin of Yamaguchi Prefectural Fisheries Research Center 17: 9–31.

Kim, S.K., Chang, K.I., Kim, B. and Cho, Y.K. 2013. Contribution of ocean current to the increase in N abundance in the Northwestern Pacific marginal seas. Geophysical Research Letters 40(1): 143–148. https://doi.org/10.1029/2012GL054545.

Kim, S.W. and Onbé, T. 1989. Distribution and zoogeography of the marine cladoceran *Podon schmackeri* in the Northwestern Pacific. Marine Biology 102(2): 203–210. https://doi.org/10.1007/bf00428281.

Kim, S.W., Onbé, T. and Yoo, K.I. 1993. Distribution of the marine cladoceran *Evadne spinifera* in waters adjacent to Korean peninsula. The Journal of the Oceanographic Society of Korea 28(1): 47–51.

Kishinouye, K. 1922. Echizen kurage (*Nemopilema nomurai*). Dobutsugaku Zasshi 34: 343–345.

Kitajima, S., Hasegawa, T., Nishiuchi, K., Kiyomoto, Y., Taneda, T. and Yamada, H. 2020. Temporal fluctuations in abundance and size of the giant jellyfish *Nemopilema nomurai* medusae in the northern east China Sea, 2006–2017. Marine Biology 167(6): 75. https://doi.org/10.1007/s00227-020-03682-1.

Kitajima, S., Iguchi, N., Honda, N., Watanabe, T. and Katoh, O. 2015. Distribution of *Nemopilema nomurai* in the southwestern Sea of Japan related to meandering of the Tsushima Warm Current. Journal of Oceanography 71(3): 287–296. https://doi.org/10.1007/s10872-015-0288-2.

Kitayama, K., Seto, S., Sato, M. and Hara, H. 2012. Increases of wet deposition at remote sites in Japan from 1991 to 2009. Journal of Atmospheric Chemistry 69(1): 33–46. https://doi.org/10.1007/s10874-012-9228-3.

Kobari, T. and Ikeda, T. 1999. Vertical distribution, population structure and life cycle of *Neocalanus cristatus* (crustacea: copepoda) in the Oyashio region, with notes on its regional variations. Marine Biology 134(4): 683–696. https://doi.org/10.1007/s002270050584.

Kobari, T. and Ikeda, T. 2001a. Life cycle of *Neocalanus flemingeri* (crustacea: copepoda) in the Oyashio region, western subarctic pacific, with notes on its regional variations. Marine Ecology Progress Series 209: 243–255. https://doi.org/10.3354/meps209243.

Kobari, T. and Ikeda, T. 2001b. Ontogenetic vertical migration and life cycle of *Neocalanus plumchrus* (crustacea: copepoda) in the Oyashio region, with notes on regional variations in body sizes. Journal of Plankton Research 23(3): 287–302. https://doi.org/10.1093/plankt/23.3.287.

Kobari, T., Moku, M. and Takahashi, K. 2008. Seasonal appearance of expatriated boreal copepods in the Oyashio–Kuroshio mixed region. ICES Journal of Marine Science 65(3): 469–476. https://doi.org/10.1093/icesjms/fsm194.

Kobari, T., Tadokoro, K., Sugisaki, H. and Itoh, H. 2007. Response of *Eucalanus bungii* to oceanographic conditions in the western subarctic Pacific ocean: retrospective analysis of the odate collections. Deep Sea Research Part II: Topical Studies in Oceanography 54(23–26): 2748–2759. https://doi.org/10.1016/j.dsr2.2007.08.005.

Kodama, T., Hirai, J., Tawa, A., Ishihara, T. and Ohshimo, S. 2020. Feeding habits of the Pacific bluefin tuna (*Thunnus orientalis*) larvae in two nursery grounds based on morphological and metagenomic analyses. Deep-Sea Research Part II-Topical Studies in Oceanography 175: 104745. https://doi.org/10.1016/j.dsr2.2020.104745.

Kodama, T., Igeta, Y. and Iguchi, N. 2022. Long-term variation in mesozooplankton biomass caused by top-down effects: a case study in the coastal Sea of Japan. Geophysical Research Letters: e2022GL099037. https://doi.org/10.1029/2022gl099037.

Kodama, T., Igeta, Y., Kuga, M. and Abe, S. 2016. Long-term decrease in phosphate concentrations in the surface layer of the southern Japan Sea. Journal of Geophysical Research-Oceans 121(10): 7845–7856. https://doi.org/10.1002/2016jc012168.

Kodama, T., Iguchi, N., Tomita, M., Morimoto, H., Ota, T. and Ohshimo, S. 2018a. Appendicularians in the southwestern Sea of Japan during the summer: abundance and role as secondary producers. Journal of Plankton Research 40(3): 269–283. https://doi.org/10.1093/plankt/fby015.

Kodama, T., Morimoto, A., Takikawa, T., Ito, M., Igeta, Y., Abe, S. et al. 2017. Presence of high nitrate to phosphate ratio subsurface water in the Tsushima strait during summer. Journal of Oceanography 73(6): 759–769. https://doi.org/10.1007/s10872-017-0430-4.

Kodama, T., Ohshimo, S., Tanaka, H., Ashida, H., Kameda, T., Tanabe, T. et al. 2021. Abundance and habitats of marine cladocerans in the Sea of Japan over two decades. Progress in Oceanography 194: 102561. https://doi.org/10.1016/j.pocean.2021.102561.

Kodama, T., Shimizu, Y., Ichikawa, T., Hiroe, Y., Kusaka, A., Morita, H. et al. 2014. Seasonal and spatial contrast in the surface layer nutrient content around the Kuroshio along 138°E, observed between 2002 and 2013. Journal of Oceanography 70(6): 489–503. https://doi.org/10.1007/s10872-014-0245-5.

Kodama, T., Wagawa, T., Iguchi, N., Takada, Y., Takahashi, T., Fukudome, K.I. et al. 2018b. Spatial variations in zooplankton community structure along the Japanese coastline in the Japan Sea: influence of the coastal current. Ocean Science 14(3): 355–369. https://doi.org/10.5194/os-14-355-2018.

Kubota, S., Nakamura, T. and Yasuda, T. 1996. First record of a schyphomedusa, *Stomolophus nomurai* (kishinouye) (cnidaria; schyphozoa) found in Asou Bay, Tsushima Island, Nagasaki Prefecture, Japan (in Japanese). NankiSeibutsu 38(1): 55–56.

Kuroda, H., Toya, Y., Watanabe, T., Nishioka, J., Hasegawa, D., Taniuchi, Y. et al. 2019. Influence of coastal Oyashio water on massive spring diatom blooms in the Oyashio area of the north Pacific ocean. Progress in Oceanography 175: 328–344. https://doi.org/10.1016/j.pocean.2019.05.004.

Kuroda, K., Morimoto, H. and Iguchi, N. 2000. Mass occurrence of salps and jellyfishes in the Japan Sea in 2000 (in Japanese). Bulletin of the Japanese Society of Fisheries Oceanography 64(4): 311–315.

Kuwae, M., Tsugeki, N.K., Amano, A., Agusa, T., Suzuki, Y., Tsutsumi, J. et al. 2022. Human-induced marine degradation in anoxic coastal sediments of Beppu Bay, Japan, as an Anthropocene marker in east Asia. Anthropocene 37: 100318. https://doi.org/10.1016/j.ancene.2021.100318.

Laakmann, S., Blanco-Bercial, L. and Cornils, A. 2020. The crossover from microscopy to genes in marine diversity: from species to assemblages in marine pelagic copepods. Philosophical Transactions of the Royal Society B: Biological Sciences 375(1814): 20190446. https://doi.org/10.1098/rstb.2019.0446.

Landry, M.R., Beckley, L.E. and Muhling, B.A. 2018. Climate sensitivities and uncertainties in food-web pathways supporting larval bluefin tuna in subtropical oligotrophic oceans. ICES Journal of Marine Science 76(2): 359–369. https://doi.org/10.1093/icesjms/fsy184.

Longhurst, A.R. and Seibert, D.L.R. 1972. Oceanic distribution of *evadne* in the eastern Pacific (cladocera). Crustaceana 22(3): 239–248.

Longhurst, R. 2007. Ecological Geography of the Sea. Academic Press.

Mensah, V. and Ohshima, K.I. 2021. Weakened overturning and tide control the properties of Oyashio intermediate water, a key water mass in the north Pacific. Scientific Reports 11(1): 14526. https://doi.org/10.1038/s41598-021-93901-6.

Minami, H., Kawae, S., Jifuku, J. and Nagai, N. 1999. Long-term variability of oceanic conditions along the PM line in the Japan Sea (in Japanese). Umi to Sora 74(4): 141–146.

Miyamoto, H., Itoh, H. and Okazaki, Y. 2017. Temporal and spatial changes in the copepod community during the 1974–1998 spring seasons in the Kuroshio region; a time period of profound changes in pelagic fish populations. Deep Sea Research Part I: Oceanographic Research Papers 128: 131–140. https://doi.org/10.1016/j.dsr.2017.07.007.

Miyamoto, H., Takahashi, K., Kuroda, H., Watanabe, T., Taniuchi, Y., Kuwata, A. et al. 2022. Copepod community structure in the transition region of the north Pacific ocean: water mixing as a key driver of secondary production enhancement in subarctic and subtropical waters. Progress in Oceanography: 102865. https://doi.org/10.1016/j.pocean.2022.102865.

Nagai, N., Tadokoro, K., Kuroda, K. and Sugimoto, T. 2006. Occurrence characteristics of chaetognath species along the PM transect in the Japan Sea during 1972–2002. Journal of Oceanography 62(5): 597–606. https://doi.org/10.1007/s10872-006-0079-x.

Nagai, N., Tadokoro, K., Kuroda, K. and Sugimoto, T. 2008. Chaetognath species-specific responses to climate regime shifts in the Tsushima Warm Current of the Japan Sea. Plankton and Benthos Research 3(2): 86–95. https://doi.org/10.3800/pbr.3.86.

Nagai, N., Takahashi, T. and Kuroda, K. 2012. Review on the fauna and the distributional ecology of pelagic chaetoganaths in the Japan Sea (In Japanese with English abstract). Bulletin of the Plankton Society of Japan 59(2): 95–97. https://doi.org/10.24763/bpsj.59.2_95.

Nagai, T., Clayton, S. and Uchiyama, Y. 2019. Multiscale routes to supply nutrients through the Kuroshio nutrient stream. pp. 105–125. *In*: Nagai, T., Saito, H., Suzuki, K. and Takahashi, M. (eds.). Kuroshio Current: Physical, Biogeochemical, and Ecosystem Dynamics. John Wiley and Sons. https://doi.org/10.1002/9781119428428.ch6.

Nagai, T., Tandon, A., Yamazaki, H. and Doubell, M.J. 2009. Evidence of enhanced turbulent dissipation in the frontogenetic Kuroshio front thermocline. Geophysical Research Letters 36(12). https://doi.org/10.1029/2009gl038832.

Nishibe, Y., Takahashi, K., Shiozaki, T., Kakehi, S., Saito, H. and Furuya, K. 2015. Size-fractionated primary production in the Kuroshio extension and adjacent regions in spring. Journal of Oceanography 71(1): 27–40. https://doi.org/10.1007/s10872-014-0258-0.

Nishikawa, H., Nishikawa, S., Ishizaki, H., Wakamatsu, T. and Ishikawa, Y. 2020. Detection of the Oyashio and Kuroshio fronts under the projected climate change in the 21st century. Progress in Earth and Planetary Science 7(1): 29. https://doi.org/10.1186/s40645-020-00342-2.

Nishioka, J., Ono, T., Saito, H., Sakaoka, K. and Yoshimura, T. 2011. Oceanic iron supply mechanisms which support the spring diatom bloom in the Oyashio region, western Subarctic Pacific. Journal of Geophysical Research-Oceans 116: C02021. https://doi.org/10.1029/2010jc006321.

Odate, K. 1994. Zooplankton biomass and its long-term variation in the western North Pacific ocean, tohoku Sea area, Japan. Bulletin of Tohoku Regional Fisheries Research Laboratory 56: 115–173.

Omori, M. and Kitamura, M. 2004. Taxonomic review of three Japanese species of edible jellyfish (Scyphozoa: Rhizostomeae). Plankton Biology and Ecology 51: 36–51.

Onbe, T. 1974. Studies on the ecology of marine cladocerans. Journal of the Faculty of Fisheries and Animal Husbandry, Hiroshima University 13(1): 83–179. https://doi.org/10.15027/41211.

Onbé, T. and Ikeda, T. 1995. Marine cladocerans in Toyama Bay, southern Japan Sea - seasonal occurrence and day-night vertical distributions. Journal of Plankton Research 17(3): 595–609. https://doi.org/10.1093/plankt/17.3.595.

Ono, T. 2021. Long-term trends of oxygen concentration in the waters in bank and shelves of the southern Japan Sea. Journal of Oceanography 77(4): 659–684. https://doi.org/10.1007/s10872-021-00599-1.

Ono, T., Midorikawa, T., Watanabe, Y. W., Tadokoro, K. and Saino, T. 2001. Temporal increases of phosphate and apparent oxygen utilization in the subsurface waters of western Subarctic Pacific from 1968 to 1998. Geophysical Research Letters 28(17): 3285–3288. https://doi.org/10.1029/2001gl012948.

Reizen, N. and Isobe, A. 2006. Numerical tracer experiments representing behavior of the giant jellyfish, *Nemopilema nomurai*, in the yellow and East China Seas (in Japanese with English abstract). Oceanography in Japan 15(5): 425–436.

Robert, D., Murphy, H.M., Jenkins, G.P. and Fortier, L. 2014. Poor taxonomical knowledge of larval fish prey preference is impeding our ability to assess the existence of a "critical period" driving year-class strength. Ices Journal of Marine Science 71(8): 2042–2052. https://doi.org/10.1093/icesjms/fst198.

Saito, H. 2019. The Kuroshio. pp. 1–11. *In*: Nagai, T., Saito, H., Suzuki, K. and Takahashi, M. (eds.). Kuroshio Current: Physical, Biogeochemical, and Ecosystem Dynamics. John Wiley and Sons. https://doi.org/10.1002/9781119428428.ch1.

Senjyu, T., Aramaki, T., Otosaka, S., Togawa, O., Danchenkov, M., Karasev, E. et al. 2002. Renewal of the bottom water after the winter 2000–2001 may spin-up the thermohaline circulation in the Japan Sea. Geophysical Research Letters 29(7). https://doi.org/10.1029/2001gl014093.

Shimomura, T. 1959. On the unprecedented flourishing of "Echizen-Kurage", *Stomolophus nomurai* (KISHINOUYE), in the Tsushima Warm Current regions in autumn. Bulletin of Japan Sea Regional Fisheries Research Laboratory 7: 85–107.

Shiozaki, T., Itoh, F., Hirose, Y., Onodera, J., Kuwata, A. and Harada, N. 2021. A DNA metabarcoding approach for recovering plankton communities from archived samples fixed in formalin. PLOS ONE 16(2): e0245936. https://doi.org/10.1371/journal.pone.0245936.

Shoden, S., Ikeda, T. and Yamaguchi, A. 2005. Vertical distribution, population structure and lifecycle of *Eucalanus bungii* (copepoda: calanoida) in the Oyashio region, with notes on its regional variations. Marine Biology 146(3): 497–511. https://doi.org/10.1007/s00227-004-1450-3.

Sogawa, S., Hidaka, K., Kamimura, Y., Takahashi, M., Saito, H., Okazaki, Y. et al. 2019. Environmental characteristics of spawning and nursery grounds of Japanese sardine and mackerels in the Kuroshio and Kuroshio extension area. Fisheries Oceanography 28(4): 454–467. https://doi.org/10.1111/fog.12423.

Sogawa, S., Kidachi, T., Nagayama, M., Ichikawa, T., Hidaka, K., Ono, T. et al. 2017. Short-term variation in copepod community and physical environment in the waters adjacent to the Kuroshio current. Journal of Oceanography 73(5): 603–622. https://doi.org/10.1007/s10872-017-0420-6.

Sudo, H. 1986. A note on the Japan Sea proper water. Progress in Oceanography 17(3–4): 313–336. https://doi.org/10.1016/0079-6611(86)90052-2.

Sugimoto, T. and Tadokoro, K. 1998. Interdecadal variations of plankton biomass and physical environment in the north Pacific. Fisheries Oceanography 7(3–4): 289–299. https://doi.org/10.1046/j.1365-2419.1998.00070.x.

Sun, S., Zhang, F., Li, C., Wang, S., Wang, M., Tao, Z. et al. 2015. Breeding places, population dynamics, and distribution of the giant jellyfish *Nemopilema nomurai* (Scyphozoa: Rhizostomeae) in the Yellow Sea and the East China Sea. Hydrobiologia 754(1): 59–74. https://doi.org/10.1007/s10750-015-2266-5.

Tadokoro, K. 2004. Regime shift and variations in Oyashio ecosystems (in Japanese). Ocean Current and Biological Resouces. Sugimoto, T. Tokyo, Seizando: 208–216.

Tadokoro, K., Chiba, S., Ono, T., Midorikawa, T. and Saino, T. 2005. Interannual variation in *Neocalanus* biomass in the Oyashio waters of the western north Pacific. Fisheries Oceanography 14(3): 210–222. https://doi.org/10.1111/j.1365-2419.2005.00333.x.

Tadokoro, K., Ono, T., Yasuda, I., Osafune, S., Shiomoto, A. and Sugisaki, H. 2009. Possible mechanisms of decadal-scale variation in PO4 concentration in the western north Pacific. Geophysical Research Letters 36: L08606. https://doi.org/10.1029/2009gl037327.

Takagi, S., Kuroda, H., Hasegawa, N., Watanabe, T., Unuma, T., Taniuchi, Y. et al. 2022. Controlling factors of large-scale harmful algal blooms with *Karenia selliformis* after record-breaking marine heatwaves. Frontiers in Marine Science 9. https://doi.org/10.3389/fmars.2022.939393.

Takahashi, K., Ichikawa, T., Fukugama, C., Yamane, M., Kakehi, S., Okazaki, Y. et al. 2015. *In situ* observations of a doliolid bloom in a warm water filament using a video plankton recorder: bloom development, fate, and effect on biogeochemical cycles and planktonic food webs. Limnology and Oceanography 60(5): 1763–1780. https://doi.org/https://doi.org/10.1002/lno.10133.

Takasuka, A., Tadokoro, K., Okazaki, Y., Ichikawa, T., Sugisaki, H., Kuroda, H. et al. 2017. *In situ* filtering rate variability in egg and larval surveys off the Pacific coast of Japan: Do plankton nets clog or over-filter in the sea? Deep Sea Research Part I: Oceanographic Research Papers 120: 132–137. https://doi.org/10.1016/j.dsr.2016.12.017.

Tang, K.W., Ivory, J.A., Shimode, S., Nishibe, Y. and Takahashi, K. 2019. Dead heat: copepod carcass occurrence along the Japanese coasts and implications for a warming ocean. ICES Journal of Marine Science 76(6): 1825–1835. https://doi.org/10.1093/icesjms/fsz017.

Tomita, M., Ikeda, T. and Shiga, N. 1999. Production of *Oikopleura longicauda* (tunicata : appendicularia) in Toyama Bay, southern Japan Sea. Journal of Plankton Research 21(12): 2421–2430. https://doi.org/10.1093/plankt/21.12.2421.

Tomita, M., Shiga, N. and Ikeda, T. 2003. Seasonal occurrence and vertical distribution of appendicularians in Toyama Bay, southern Japan Sea. Journal of Plankton Research 25(6): 579–589. https://doi.org/10.1093/plankt/25.6.579.

Tsuda, A. 2013. Dynamics of zooplankton and lower trophic organisms in the subarctic Pacific (in Japanese with English abstract). Oceanography in Japan 22(3): 85–96. https://doi.org/10.5928/kaiyou.22.3_85.

Tsuda, A. and Nemoto, T. 1992. Distribution and growth of salps in a Kuroshio warm-core ring during summer 1987. Deep Sea Research Part A. Oceanographic Research Papers 39: S219-S229. https://doi.org/10.1016/S0198-0149(11)80013-0.

Tsuda, A., Saito, H. and Kasai, H. 1999. Life histories of *Neocalanus flemingeri* and *Neocalanus plumchrus* (calanoida: copepoda) in the western subarctic pacific. Marine Biology 135(3): 533–544. https://doi.org/10.1007/s002270050654.

Tsuda, A., Saito, H. and Kasai, H. 2004. Life histories of *Eucalanus bungii* and *Neocalanus cristatus* (copepoda: calanoida) in the western subarctic Pacific ocean. Fisheries Oceanography 13(s1): 10–20. https://doi.org/10.1111/j.1365-2419.2004.00315.x.

Tsuda, A., Takeda, S., Saito, H., Nishioka, J., Nojiri, Y., Kudo, I. et al. 2003. A mesoscale iron enrichment in the western subarctic Pacific induces a large centric diatom bloom. Science 300(5621): 958–961. https://doi.org/10.1126/science.1082000.

Uye, S.-I. 2008. Blooms of the giant jellyfish *Nemopilema nomurai*: a threat to the fisheries sustainability of the east asian marginal Seas. Plankton and Benthos Research 3(Supplement): 125–131. https://doi.org/10.3800/pbr.3.125.

Wang, Y.-L., Wu, C.-R. and Chao, S.-Y. 2016. Warming and weakening trends of the Kuroshio during 1993–2013. Geophysical Research Letters 43(17): 9200–9207. https://doi.org/10.1002/2016gl069432.

Watanabe, Y.W., Ishida, H., Nakano, T. and Nagai, N. 2005. Spatiotemporal decreases of nutrients and chlorophyll-*a* in the surface mixed layer of the western north Pacific from 1971 to 2000. Journal of Oceanography 61(6): 1011–1016. https://doi.org/10.1007/s10872-006-0017-y.

William, M.G., Daniel, L.M. and Jonathan, C.M. 2003. *In situ* quantification and analysis of large jellyfish using a novel video profiler. Marine Ecology Progress Series 254: 129–140.

Wu, L.X., Cai, W.J., Zhang, L.P., Nakamura, H., Timmermann, A., Joyce, T. et al. 2012. Enhanced warming over the global subtropical western boundary currents. Nature Climate Change 2(3): 161–166. https://doi.org/10.1038/Nclimate1353.

Xu, Y., Ishizaka, J., Yamaguchi, H., Siswanto, E. and Wang, S. 2013. Relationships of interannual variability in SST and phytoplankton blooms with giantjellyfish (*Nemopilema nomurai*) outbreaks in the yellow Sea and east China Sea. Journal of Oceanography 69(5): 511–526. https://doi.org/10.1007/s10872-013-0189-1.

Yamaguchi, A., Matsuno, K., Abe, Y., Arima, D. and Ohgi, K. 2014. Seasonal changes in zooplankton abundance, biomass, size structure and dominant copepods in the Oyashio region analysed by an optical plankton counter. Deep Sea Research Part I: Oceanographic Research Papers 91: 115–124. https://doi.org/https://doi.org/10.1016/j.dsr.2014.06.003.

Yanagi, T. 2002. Water, salt, phosphorus and nitrogen budgets of the Japan Sea. Journal of Oceanography 58(6): 797–804. https://doi.org/10.1023/A:1022815027968.

Yasuhara, M., Hunt, G., Breitburg, D., Tsujimoto, A. and Katsuki, K. 2012. Human-induced marine ecological degradation: micropaleontological perspectives. Ecology and Evolution 2(12): 3242–3268. https://doi.org/10.1002/ece3.425.

Yatsu, A., Watanabe, T., Ishida, M., Sugisaki, H. and Jacobson, L.D. 2005. Environmental effects on recruitment and productivity of Japanese sardine *Sardinops melanostictus* and chub mackerel *Scomber japonicus* with recommendations for management. Fisheries Oceanography 14(4): 263–278. https://doi.org/10.1111/j.1365-2419.2005.00335.x.

Yoon, W.D., Lee, H.E., Han, C., Chang, S.-J. and Lee, K. 2014. Abundance and distribution of *Nemopilema nomurai* (Scyphozoa, Rhizostomeae), in korean waters in 2005–2013. Ocean Science Journal 49(3): 183–192. https://doi.org/10.1007/s12601-014-0018-5.

Yoon, W.D., Yang, J.Y., Shim, M.B. and Kang, H.K. 2008. Physical processes influencing the occurrence of the giant jellyfish *Nemopilema nomurai* (Scyphozoa: Rhizostomeae) around Jeju Island, Korea. Journal of Plankton Research 30(3): 251–260. https://doi.org/10.1093/plankt/fbm102.

Yoshiki, T.M., Chiba, S., Sasaki, Y., Sugisaki, H., Ichikawa, T. and Batten, S. 2015. Northerly shift of warm-water copepods in the western subarctic north Pacific: continuous plankton recorder samples (2001–2013). Fisheries Oceanography 24(5): 414–429. https://doi.org/10.1111/fog.12119.

Yoshiki, T.M., Chiba, S., Sugisaki, H., Sasaoka, K., Ono, T. and Batten, S. 2013. Interannual and regional variations in abundance patterns and developmental timing in mesozooplankton of the western north Pacific ocean based on continuous plankton recorder during 2001–2009. Journal of Plankton Research 35(5): 993–1008. https://doi.org/10.1093/plankt/fbt047.

Zuur, A.F. and Ieno, E.N. 2016. A protocol for conducting and presenting results of regression-type analyses. Methods in Ecology and Evolution 7(6): 636–645. https://doi.org/10.1111/2041–210X.12577.

3.5

Zooplankton Species Diversity, Distribution and Long-Term Change of the Community Along the Bulgarian Black Sea Coast

A Brief Review

Kremena Stefanova, * *Elitsa Stefanova, Ivelina Zlateva, Valentina Doncheva, Natalya Slabakova* and *Nadezhda Valcheva*

1. Introduction

Zooplankton is an integral component of holistic ecosystem assessments in line with the intermediate trophic link between phytoplankton and fish. As essential elements of the pelagic food web of aquatic ecosystems, they are subject to both top-down predation and bottom-up controls (Jernberg et al. 2017). Additionally, zooplankton are the main drivers of marine biological pumps with important contribution to the vertical transport of carbon (Steinberg and Landry 2017). Zooplankton community structure changes and distribution can influence the biogeochemical cycles and energy flows in aquatic ecosystems (Beaugrand et al. 2003, Brun et al. 2019). Zooplankton are often seen as reliable indicators of environmental variations caused by climate change or pollution (Hays et al. 2005, Batchelder et al. 2013, Ndah et al. 2022). Understanding the seasonal changes in zooplankton community structure and the distribution pattern adds valuable insights to the overall knowledge of the ecosystem dynamics and the potential impacts of climate change.

The Black Sea is a unique sea, the largest land-locked basin in the world relatively rich in biodiversity, heritage and natural resources. It is an estuarine basin, characterized by a positive net freshwater balance, mainly due to the outflow of some of the largest European rivers, such as the Danube and Dnieper, and a high rate of precipitation (Kara et al. 2008, Volkov and Landerer 2015). The resulting salinity of about 18 psu in the upper layer forms a strong stratification all over the

Institute of Oceanology–Bulgarian Academy of Sciences, Varna Bulgaria.

* Corresponding author: stefanova@io-bas.bg

basin. The significant volume of freshwater river inflows is also responsible for the higher organic matter production (Kovalev 1993). Moreover, anoxic conditions below 200 m in the Black Sea, with high levels of hydrogen sulfide, result in an azoic environment below the epipelagic zone (Sorokin 1982). Another main characteristic of the Black Sea is the Cold Intermediate Layer (CIL) formed at the depth of the winter convection (Özsoy and Ünlüata 1997). The reduced events of CIL formation in recent years are related to the amplified response to climate change in the Black Sea (Gunduz et al. 2020). In addition, sea surface temperature observations have helped to detect recent warming in the Black Sea as a response to climate change (Ginzburg et al. 2004, Shapiro et al. 2010, Mulet et al. 2018). Considering the latter, the characteristics of the zooplankton community in the Black Sea have been driven by a multitude of factors that have formed their specific structure. Furthermore, the basin was exposed to significant pressure mainly due to human-induced factors, such as eutrophication and hypoxia, overfishing, and the introduction of alien species, in addition to the effects of climate change. The combination of these stressors is considered to be the main cause of the degradation of the Black Sea marine ecosystem since the early 1970s (Akoglu 2023).

The northwestern and western regions are more eutrophic than the eastern one due to the Danube runoff (Humborg et al. 1997, Yunev et al. 2005), while the south-western region is influenced by the Mediterranean waters penetrating through the Bosphorus (Kovalev et al. 1999). The northern part of the Bulgarian Black Sea zone is under the influence of Danube discharge and part of the contamination/eutrophication comes from the shore and near-shore regions. Three areas along the Bulgarian coast have been identified as exposed to maximum anthropogenic pressure: Varna and Burgas bays and Kamchia River mouth are under the direct or indirect influence of industrial and municipal discharges, diffuse sources, port operations and tourism development (Shtereva et al. 2015). The main stressors along the Bulgarian coast are pollution (nutrient enrichment leading to eutrophication, contaminants and marine litter), fishing and physical disturbance, invasive species, tourism and climate change. The main drivers implying an impact on the Bulgarian Black Sea water quality (WQ) are industry, urbanization, tourism, marine traffic, and agriculture.

The research on the ecological significance and biological characteristics of the zooplankton community of the Bulgarian Black Sea coast started at the beginning of the 20th century, and the zooplankton are still actively studied. Many case studies are focused on the assessment of the current state and trends of plankton fauna relative to Black Sea environmental changes. The structural alterations have proved to be linked to direct drivers, such as eutrophication, alien species introduction, excessive fishing, and climate changes (Stefanova et al. 2012, Kamburska et al. 2003).

Understanding the processes that shape the patterns of community structure is a fundamental goal in ecological research and makes a valuable contribution to the knowledge of ecosystem functioning, productivity and well-being.

Considering the latter, two different mesozooplankton abundance datasets were utilized in the present review and analyzed to demonstrate the spatiotemporal variation of the zooplankton community on the Bulgarian Black Sea coast and how it relates to selected environmental drivers.

2. Material and Methods

2.1 Study Location

The study area was restricted to the Bulgarian Exclusive Economic Zone (Figure 1) locked between min. lat: 41°58' 41.5" N (41.9782), min. long 27°26' 58.5" E (27.4496°) and max. lat 43° 44' 52" N (43.7478°), max. long 31° 20' 43" E (31.3453°). Sixty-four surveys (1,791 samples) were conducted in the frame of multidisciplinary oceanographic cruises within various projects and programs during the spring-summer-autumn period starting from 2000 to 2020. During the expeditions, the main pelagic habitats under consideration were defined as coastal (< 30 m), shelf (30–200 m) and open sea (> 200 m), according to the Initial Assessment Report of Bulgaria (Moncheva et al. 2013). Time-series analyses were applied to the highest resolution data, collected over the years (almost 600 samples in 20 years), gathered at the Cape Galata transect located in front of the Varna area.

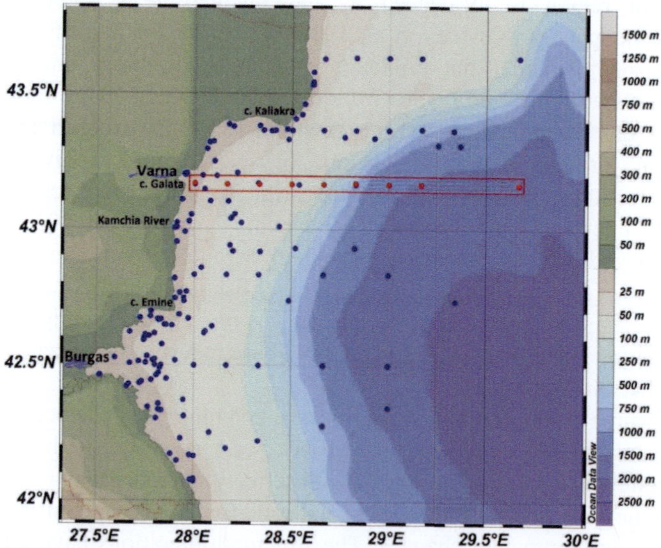

Figure 1. Study area. Map of the sampling stations corresponding to the sampling programs in 2000–2020 period. The red box presents the sampling transect in front of c. Galata and the points – the locations where time series data were gathered.

2.2 Sampling Protocol

At each station, zooplankton samples were collected using a closing Juday net (150 μm mesh size, 36 cm mouth opening in diameter = 0,1 m² mouth surface area). The net was vertically hauled from a depth of two meters above the bottom to the sea surface at integral sampling layers or in discrete layers depending on water stratification, thermocline depth and sigma-theta values (Alexandrov et al. 2014).

Before sample preservation, the gelatinous species (*Aurelia aurita, Pleirobrachia pileus, Mnemiopsis leidyi,* and *Beroe ovata*) were removed, rinsed, measured and counted on board (Shiganova et al. 2020). The samples were preserved in a final 4% formalin solution buffered to pH 8–8.2 with disodium tetraborate (borax) ($Na_2B_4O_3 \cdot 10\ H_2O$) (Alexandrov et al. 2014).

In the laboratory, the samples were elaborated according to Alexandrov et al. (2014). Species were identified with the use of Morduhay-Boltovskoy (1968, 1969, and 1972), and recent taxonomy was checked with the Word Register of Marine Species (http://www.marinespecies.org). Biomass values as wet weight were estimated based on the individual weight given per taxa and size class in Petipa, 1957. Here, we analyze the temporal variability of all plankton taxa with a focus on the holoplanktonic fauna and meroplankton.

2.3 Statistical Analysis

The data collected in the period 2000–2020, along with *in situ* environmental data (Chlorolphyl a, temperature and salinity) were used to study seasonality and more recent structure of the zooplankton communities and a long-term dataset collected in the period 1966–2020 to help investigate the mesozooplankton community response to regime shifts.

Indicator species analysis (ISA) along with breakpoint analysis (for identification of temperature regime shifts) were applied to long-term mesozooplankton abundance data (collected in coastal and shelf areas) to detect structural changes in the community as a result of climatic changes and ISA was applied to 2000–2020 data set to identify indicator species by season (spring, summer and autumn) with an integration of sampling sites to coastal, shelf and open sea regions.

Non-metric multidimensional scaling (NMDS) of 2000–2020 abundance data by sampling sites was overlayed with selected abiotic and limitation factors to facilitate an ordination plot with fitted environmental variables (vectors) by seasons (factors) to identify underlying associations of mesozooplankton abundance variations with the environment.

All analyses and graphic representations were performed using the statistics and programming software R 4.1.2, package 'vegan', 'indicspecies', 'ggplot2', 'segmented', and MATLAB programing environment version 2020a.

3. Results and Discussion

3.1 Species Diversity

In order to detect structural changes in the community composition relative to the abiotic environment, pressures and global climate change in the marine ecosystem, high-resolution data is required. However, the abiotic changes and the biological responses of the pelagic ecosystems can be very complex due to the synergistic links between climate and human activities (Harley et al. 2006). The influence of

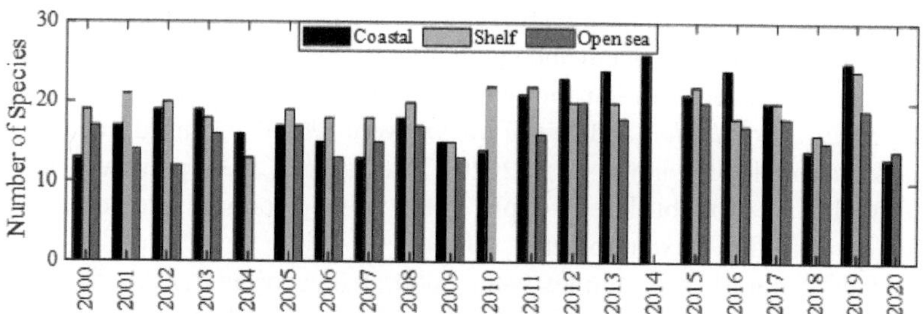

Figure 2. Number of mesozooplankton species in the period 2000–2020.

zooplankton on the production of phytoplankton (grazing) and consumers of a higher trophic level is related to the taxonomic composition and quantitative parameters of the community, which in turn depends on the physio-chemical factors of the marine environment.

The plankton fauna of the Black Sea is poorer compared to the Mediterranean Sea due mainly to the low salinity levels (Kovalev et al. 1999). Black Sea zooplankton was distinguished into three ecogeographical groups–Mediterranean invaders, Pontic relicts and freshwater species (Kovalev et al. 1999). Zooplankton communities in the Mediterranean and Black Seas consist mainly of species of Atlantic origin, but typical marine stenohaline organisms (Radiolaria, Spihonophora, Pteropoda, and Salpae) inhabiting the Mediterranean Sea (salinity: S = 36–38‰) are unable to survive in the Black Sea (salinity: S = 17–20‰) (Kovalev 2001). Transitional waters at Bosphorus and Danube-Black Sea are characterized by rich species diversity. Approximately 80 Mediterranean holoplankton species occur in the Black Sea, mostly in the Bosphorus area and more than 150 plankton species are distributed in the northwestern shelf with half of them brackish and freshwater species inhabiting estuarine areas and bays which, however, could not survive in the Black Sea ecosystem for a long time (Kovalev et al. 1999).

In the last two decades the mesozooplankton community along the Bulgarian coast, featured by increased species diversity (Figure 2), is represented by species/taxa belonging to nine phyla, Ciliophora, Cnidaria, Ctenophora Rotifera, Annelida, Mollusca, Arthropoda, Chaetognatha, and Chordata. Arthropoda is the most diverse, with 24 species/taxa (75% of the total number of species/taxa) of which class Copepoda (12 species) is the richest, followed by Branchiopoda (eight species) and larval forms of benthic organisms (10 species/taxa). In the studied period (2000–2020), the species richness was relatively high, in the range between 13 to 26 species/taxa, and for the last eight years, it shows relative homogeneity (20 ± 6) if compared to the previous years. The highest number of species was traditionally recorded in the coastal and shelf habitats (average 19) in the upper homogenous layer, while the open sea was characterized by lower but stable diversity (17 ± 3). Clearly, the period 2000–2009 showed lower species richness in coastal habitats in comparison with shelf. Most probably, the zooplankton community has been negatively affected by a multitude of factors including *M. leidyi* predation, cold waves and the related phytoplankton assemblages' changes.

Keynote groups in terms of their important ecology role as copepods, Diplostraca (= Cladocera) and Meroplankton could be classified as the most important

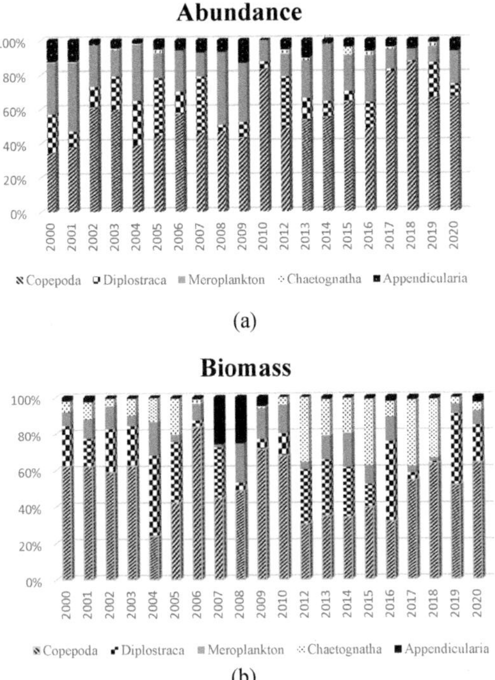

Figure 3. Zooplankton community structure by taxa and groups in the period 2000–2020 (a) abundance and (b) biomass.

component of the plankton fauna (Figures 3a, b). In general, genera Acartia, Paracalanus, Oithona, and Centropages were mostly presented inshore and above the thermocline, while Pseudocalanus and Calanus at offshore and below the thermocline (Figure not shown). Recently found (2009) cyclopoid species *Oithona davisae* in Bulgarian waters (Mihneva and Stefanova 2011, 2013) have become more frequent and abundant. Meroplankton was more diverse in coastal and shelf waters (Polychaeta, Cirripedia, Bivalvia, Gastropoda, Isopoda, Cumacea, and Mysida) in comparison with the open sea where the larvae of Polychaeta and Mollusks are the most frequent representatives. Diplostraca species (*Evadne spinifera, Pseudevadne tergestina, Penilia avirostris* and *Pleopis polyphemoides*) co-dominated the zooplankton assemblages, especially in summer. The most widely distributed zooplankton group was also the most abundant one, such as Copepoda. For the entire study period, copepods prevailed with more than 50% with an increasing trend after 2009 (> 65%) (Figure 3a). Chaetognaths contributed significantly to biomass only (12–31%); as a whole their share in total abundance did not exceed 3%.

3.2 Spatio-Temporal Distribution

The pelagic habitats formed by the simultaneous action of various environmental factors and their alterations in space (horizontally or vertically) form gradients. Gradients in environmental variables may promote heterogeneity in the distribution, abundance or composition of a zooplankton community (Seda and Devetter 2000).

3.2.1 Horizontal Distribution Pattern

It is evident that mesozooplankton abundance and biomass gradually decrease starting from coastal (min 461 and max 44,740 ind/m³ and min 12.3 and max 1,021.2 mg/m³) through shelf waters (min 455 and max 28,180 ind/m³ and min 19.27 and max 379.89 mg/m³) to open sea (min 157 and max 3,127 ind/m³ and min 4.49 and max 330.11 mg/m³) with more frequent peak values observed in coastal waters again during the spring and summer. This tendency may reflect suitable abiotic conditions and biotic interactions, forming the environmental niche for most abundant mesozooplankton species or the eventual presence of limitation factors related to specific spatial extent. Higher abundance variations at the coastal stations in comparison with the offshore ones suggested that zooplankton community and assemblages could be more influenced and disturbed by the anthropogenic and climatic effects as compared with open-water zooplankton community in terms of abundance and species composition (Figures 4a, b).

Copepoda group abundance ranged from min 206 to max 30,708 ind/m³, Diplostraca from 0 individuals presented per sampling site to max 12,740 ind/m³ and

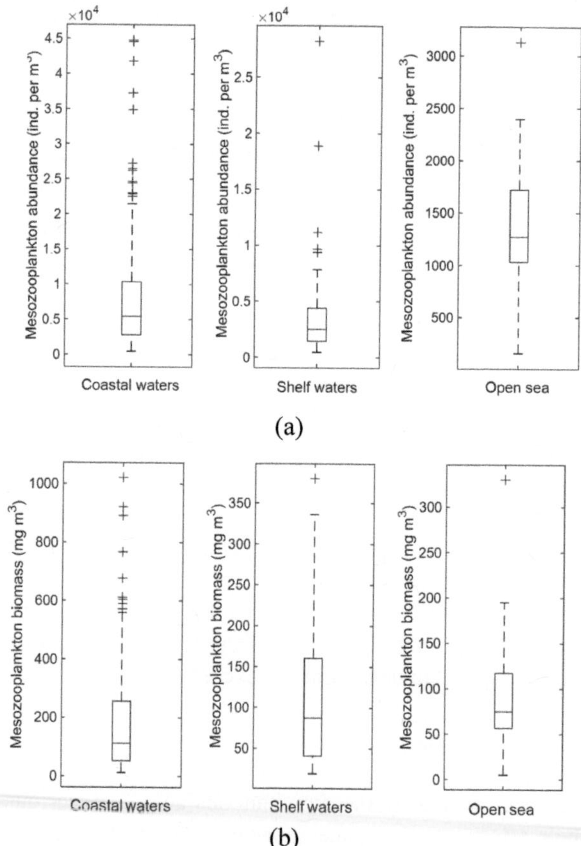

Figure 4. Box-plot of mesozooplankton abundance (a) and biomass (b) by sites grouping to coastal, shelf waters and open sea.

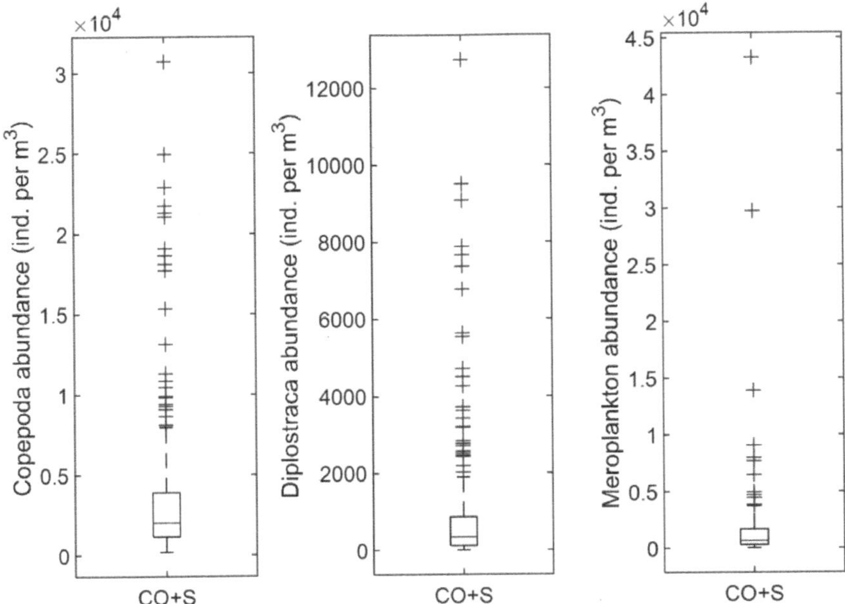

Figure 5. Box-plot of zooplankton groups abundance by site grouping to coastal and shelf waters, considering similar environmental conditions and seasonality.

Meroplankton from min 19 to max 43,273 ind/m³, with numerous outliers presented for all groups, which reflects the seasonal dynamics of species abundances, generally expected to be higher and reach peak values in the spring and summer seasons (Figure 5).

The spatial heterogeneity of plankton communities is a well-known phenomenon. The uneven distribution of plankton is mostly explained as a response to a complex of environmental factors acting at various scales of different biotic and abiotic components. The effects of advective physical forces resulting from river inflows, wind and temperature regime, food availability and trophic conditions, predation or competition have all been cited as key factors responsible for the horizontal patch of zooplankton community (Seda and Devetter 2000). Similar aggregations (patchy), due to the hydrodynamic and trophic features of the region, are often observed off the coast of Bulgaria. Thus, the mesozooplankton community shows heterogeneity between the pelagic habitats, mainly on the coast and shelf, while in the open sea, the variety range is less pronounced. As an example of a spatial distribution pattern here is presented the autumn model of zooplankton community displacement (Figure 6). Spatial variability in the quantitative parameters differs substantially between the coast and inner shelf compared to offshore sometimes more than 30 folds. Another gradient projection was to axes from north to south as a reflection (partly) of favorable trophic conditions due to the Danube runoff mainly in spring (Humborg et al. 1997, Yunev et al. 2005) and local drivers' impact.

Biological seasonality is best expressed in temperate latitudes with four seasons present. Pelagic communities, as dynamic systems, undergo significant changes in time and space, especially well expressed in seasons. Zooplankton seasonal changes

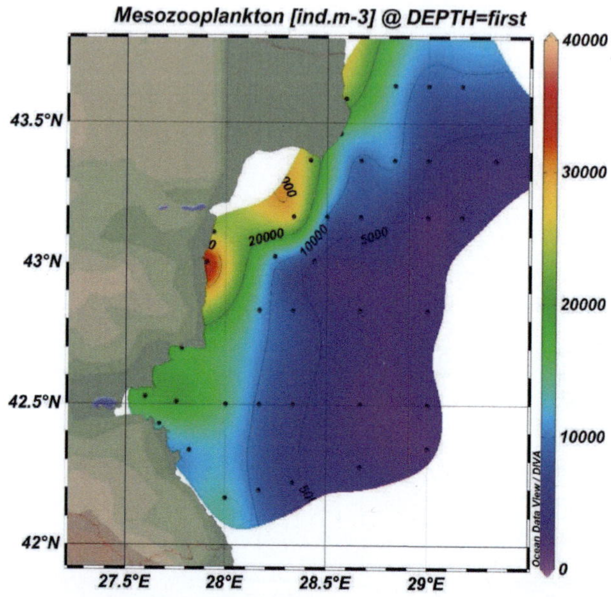

Figure 6. Mesozooplankton abundance [ind/m³] spatial distribution in October 2017.

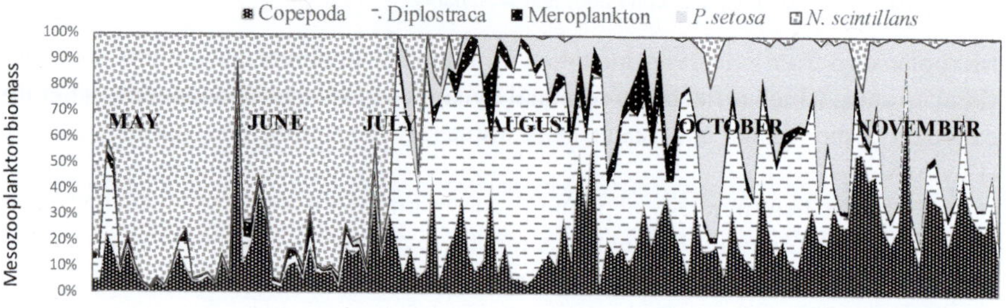

Figure 7. Example of seasonal succession of zooplankton community from May to November 2012.

reflected both species composition and abundance and biomass. In the shallow waters in front of the Bulgarian coast, two peaks in numerical abundance are recorded – a large one in spring and a smaller one in autumn. Favorable trophic and temperature conditions determine high species diversity and biomass as well. Open sea habitat is characterized by one peak usually in summer. Huge seasonal fluctuations of mesozooplankton quantity are observed in coastal habitats, between 7–30 times higher than the winter ones, while 2–3 times only in the open sea.

Marine ecosystems are strongly affected by climatic factors at various temporal scales. For example, the ecosystem's seasonal cycle is related to seasonal variability of sea temperature, water circulation, wind, insolation and other factors. Investigations in different regions of the World Ocean, including the Mediterranean and the Black Sea, reveal climate influence on marine ecosystems and also on interannual and decadal scales (Conversi et al. 2010, Oguz et al. 2006).

Seasonal succession of various key species and groups are very well demonstrated in Figure 7. Development of *N. scintillans* in spring and *M. leidyi* in summer reflected indirectly on mesozooplankton biomass since both negatively correlated with the biomass of planktonic fauna (SoE Report 2019). *N. scintillans* dominated in spring (May–June), with an average of 78% of the total, thermophilic representatives of Cladocera (= Diplostraca) prevailed in summer (45%), *P. setosa*, mainly in the fall (48%) when is the peak of reproduction processes. Copepods are presented all year round with a slight excess to the summer-autumn period.

3.2.2 Vertical Distribution

The Black Sea has a wide spectrum of hydrographic features (sharp seasonal and permanent stratification layers with changing conditions vertically in salinity, temperature, density, dissolved oxygen, hydrogen sulfur, currents, and aerobic and anaerobic zones). Zooplankton distribution in depth is very heterogeneous, quantitatively and qualitatively; nevertheless, some features have been established as similar in the water column distribution of the biomass. Vertical distribution and diel migration of the zooplankton occurred within the oxygenated zone of the water column. The lower distribution boundary is restricted by the upper limit of the hydrogen sulfide (H_2S) layer at depths corresponding to sigma theta 16.2 (Mutlu 2020). Its extent is spread between 170–200 m depth in the area of the Black Sea Rim Current and from 85 m to 100 m in the center of the cyclonic eddies bounded by it (Nikitin 1949).

The vertical pattern reflects the decrease in food availability due to light-limited primary production in deeper waters, and the decrease in temperature from the surface to the meso- and bathypelagic layers (Rutherford et al. 1999), but despite the low diversity of the large cold water species as *Pseudocalanus elongatus*, *Calanus euxinus* in the zooplankton assemblages are of crucial importance for transportation of energy and organic matter to deeper layers. The latter also explained the high zooplankton biomass values in depth. While the majority of copepods remained in the upper mixed layer and performed limited vertical migration, a few copepod species (two mentioned above) were either closely associated with the thermocline or oxygen minimum zone. In spring and autumn *P. elongatus* and *C. euxunus* contributed with 50–60% to copepod biomass respectively representing 40–50% of total mesozooplankton biomass. The homogeneous layer above the thermocline during spring, summer and autumn water body stratification determined a complex of thermophilic organisms from Copepoda and Diplostraca (= Cladocera), while eurythermic are found in the thermocline and the layer with depth up to 100 m. With the cooling of the surface water in autumn, thermophilic forms gradually decreased and were replaced by eurythermic ones, which reached a depth of 50–60 m. Figure 8 presents the vertical distribution pattern of Copepoda with two peaks coinciding with the uppermost layer and derived by *A. clausi*, *A. tonsa*, *P. parvus*, and deep layer *P. elongatus* within layer with temperature < 8°C (the Cold Intermediate Layer – CIL), and *C. euxinus* throughout the oxygenated layer between 16.2–15.8 sigma theta.

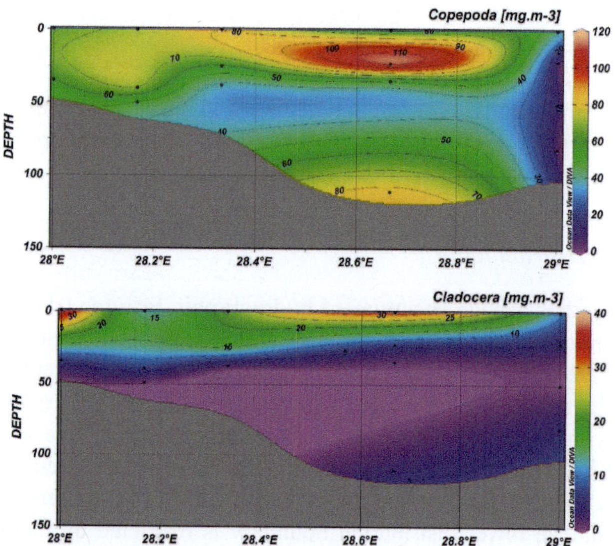

Figure 8. Vertical distribution of zooplankton biomass in October 2017 (a) Copepoda biomass [mg/m³] and (b) Cladocera [mg/m³].

3.3 Long-Term Community Dynamics

Understanding the effect of multiple pressures, environmental variability and global climate changes in marine ecosystems requires high-resolution long-term data to explain short-term fluctuations in the qualitative composition and quantitative parameters of the metazoan. This variability can be induced by both: anthropogenic and climatic factors or it can be simply linked to natural system development stages. The Black Sea ecosystem evolution has undergone different structural changes related to three different pressure regimes (Akoglu et al. 2014). The first period (1966–1973) is associated with the phase of early eutrophication, the second period (1974–1993) is related to coinciding events of intensive eutrophication, invasion of *M. leidyi*, overfishing and loss of biodiversity, the last one (1994–2007) corresponds to *B. ovata* introduction, *M. leidyi/B. ovata* interaction and the post-eutrophication process related to changes in phytoplankton community and climatic signals. In the reference period (before the beginning of the 70s), the trophic capacity of the system allowed full utilization and regeneration of biogenic elements. The community was relatively stable with natural seasonal succession and low primary productivity (Akoglu et al. 2014). Respectively, the Black Sea ecosystem alterations later have been directly or indirectly linked to changes that occurred both in the temperature and phytoplankton community. As phytoplankton comprises the base of the marine food web, changes in phytoplankton production and community composition may have profound consequences for the higher trophic levels (Edwards and Richardson 2004). Due to their short life cycles and quick response to changes in their environment, phytoplankton species are sensitive to ecosystem change (Hays et al. 2005). Together with temperature anomalies the years after 2000 mark a new phase in the phytoplankton assembly featured by a decrease in the total biomass (chlorophyll a) on

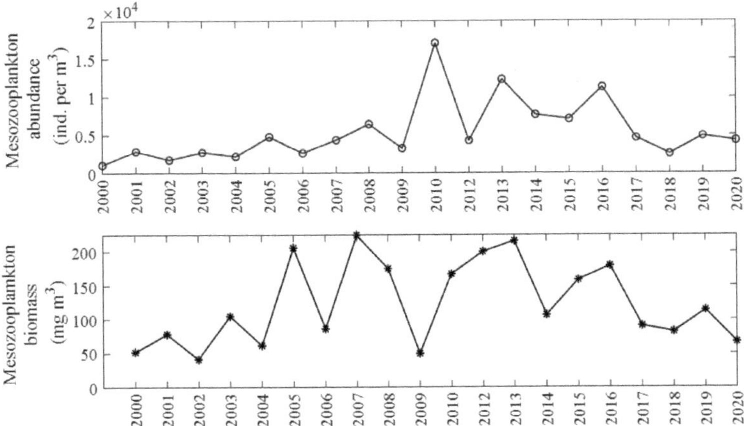

Figure 9. Mesozooplankton long-term dynamic of abundance (upper panel) and biomass (lower panel) at c. Galata transect in the period 2000–2020.

the account of a striking increase of the abundance of small-sized species, "other" than the typical diatom/dinoflagellate composite, dominated mainly by microflagellates and *Emiliania huxleyi* in particular (Moncheva et al. 2006). Changes in plankton fauna species composition over this period have also been associated with increased occurrence of small-size taxa (*A. clausi, P. parvus, P. polyphemoides, E. tergestina, O. dioica,* and benthic larvae) similar to the phytoplankton community changes and as a possible result of climate alterations. It is important to note that total mesozooplankton abundance and biomass were observed in low values in the period 2000–2005 (2,493 ind/m^3 ± 2,141, respectively 95.21 mg/m^3 ± 64.9), almost 3 and 2 folds below records in 2006–2020 (Figure 9a, b). The largest contribution to the total abundance in 2010 (Figure 9a) was due to the extremely high abundance of *O. davisae* with maxima in July–August which coincided with the highest water temperature over the observed period. Summer 2010 was extraordinarily warm in Eastern Europe including the Black Sea coastal area heat wave (Barriopedro et al. 2011).

To understand the natural mechanisms behind the occurrence or abundance of zooplankton species a reduced set of indicator species, as an alternative to sampling the entire community, has been recognized as an advantageous instrument in long-term environmental monitoring for conservation or ecological management. Species are chosen as indicators if they reflect the biotic or abiotic characteristics of the environment, provide evidence for the influence of environmental variations or changes or are useful for predicting the diversity of other species, taxa or communities within an area.

Indicator species analysis (ISA) instantly integrates relative species abundance and the frequency of occurrence to yield a maximum indicator value for each of the studied species, forming a given community.

Here we demonstrated two different approaches to studying indicator species toward two different ecological concepts. It is of utmost importance for ecologists and biologists to identify indicator species reflecting seasonal dynamics and also to validate and explore indicator species' presence to different spatial extents.

Furthermore, in light of the global climatic changes it is essential also to detect and study possible environmental shifts or limitation factors that may lead to structural changes in the entire community and ecosystem and thus give the prevalence of some species over others at a macro-ecological level, which may lead to further loss of biodiversity, occurrence of new species or structural changes in indicator species complex.

Marine ecosystems are strongly affected by climatic factors at various temporal scales. For example, the ecosystem's seasonal cycle is related to seasonal variability of sea temperature, water circulation, wind, insolation and other factors. Investigations in different regions of the World Ocean, including the Mediterranean and the Black Sea, reveal climate influence on marine ecosystems and also on interannual and decadal scales (Oguz et al. 2006).

Seasonal succession of various key species and groups are very well demonstrated in Figure 7. Development of *N. scintillans* in spring and *M. leidyi* in summer reflected indirectly on mesozooplankton biomass, since both negatively correlated with the biomass of planktonic fauna (Stefanova et al. 2019). *N. scintillans* dominated in spring (May–June), with an average of 78% of the total, thermophilic representatives of Cladocera (= Diplostraca) prevailed in summer (45%), *P. setosa*, mainly in the fall (48%) when is the peak of reproduction processes. Copepods are presented all year round with a slight excess to the summer-autumn period.

3.4 Key Zooplankton Species

Species abundance data collected in the period 2000–2020 was analyzed to identify indicator species complex in terms of seasonal dynamics in coastal, shelf and open sea waters (Table 1). The set was preliminarily reduced by the exclusion of rare species to nine species which were expected to represent some of the community indicator species (*Calanus euxinus, Pseudocalanus elongatus, Paracalanus parvus, Acartia clausi, Centropages ponticus, Penilia avirostris, Noctiluca scintillans, Aurelia aurita* and *Mnemiopsis leidyi*). *N. scintillans, A. aurita* and *M. leidyi* were included in studying the mesozooplankton abundance limitation factors as season-specific.

Zooplankton indicator species reflect rather a seasonal pattern of the zooplankton community than distinguishing habitats. The populations of dinoflagellate *N. scinitllans* characterized the spring community, warm-water species *C. ponticus, P. avirostris* and *M. leidyi* represented summer assemblages and eurythermal *P. parvus* with cold water *P. elongatus* belongs to autumn metazoan assemblage.

Results of ISA confirmed *N. scintillans* importance. *Noctiluca* is a particular component of the Black Sea mesozooplankton (Konsulov 1984, Sorokin 1982, Zaika 2005). The species characterized the three habitats but the most pronounced presence is in the coastal area. *N. scintillans* exhibited positive anomalies in spring (2,384 ± 5,456 ind/m³), especially during 2015 when the maximum was recorded at 36,531 in front of the Varna area. It often forms massive blooms when its concentration can reach millions of cells per square meter, exceeding the total abundance of mesozooplankton metazoans (Konsulov and Kamburska 1998). After 1976, during the eutrophication period of the Black Sea ecosystem, the abundance

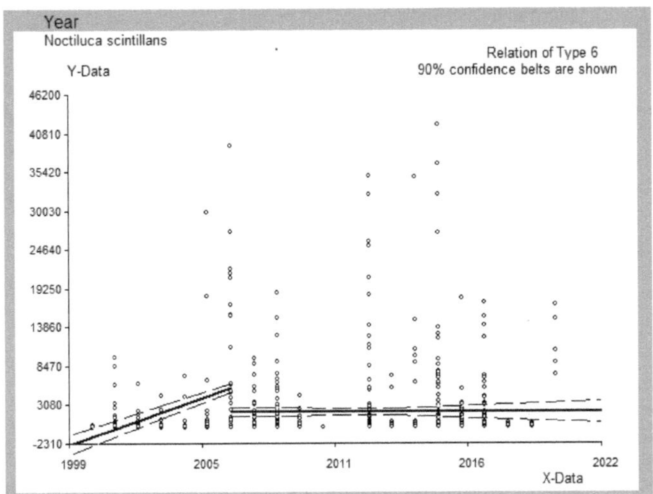

Figure 10. Breakpoint analysis, segmented regression model fit of mean *N. scintillans* abundance [ind/m³] for the 2000–2020 period along the c. Galata transect.

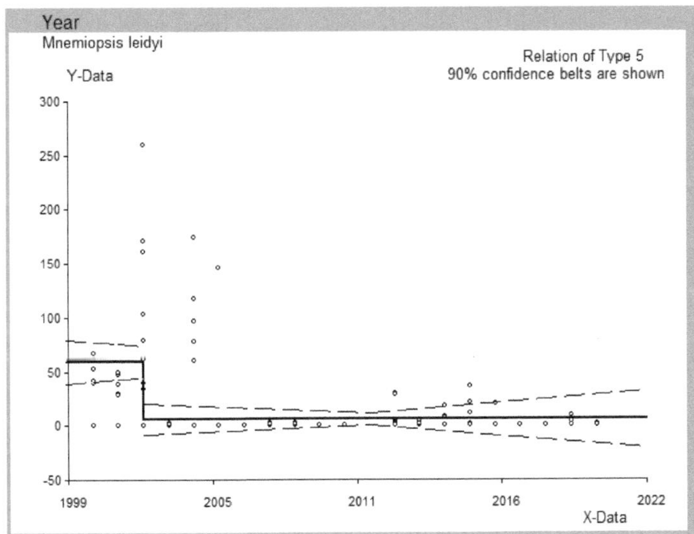

Figure 11. Breakpoint analysis, segmented regression model fit of mean *Mnemiopsis leidyi* abundance [ind/m³] for the 2000–2020 period along the c. Galata transect.

of the large dinoflagellate *N. scintillans* increased substantially especially in coastal and shelf stations. The breakpoint analysis of *N. scintillans* abundance resulted in an increasing trend within the cold period 2000–2005 (Figure 10) and stable presence but in lower abundances after 2006 with the temperature increase, excluding 2015 when the maximum was observed.

One of the most dramatic events in the Black Sea ecosystem development was an invasion of the ctenophore *M. leidyi* (after 1985), which caused a degradation of the pelagic ecosystem. Abundances of *M. leidyi* started to decline after penetration and

Table 1. Multilevel pattern analysis – Indicator species period 2000–2020; site grouping: Coastal, Shelf and Open sea by seasons.

Coastal Waters *(Total number of species: 9; selected number of species: 6; number of species associated with 1 group: 5; Number of species associated with 2 groups: 1)*		
Group	**Species Name**	**Stat., p-value (significance α=0.05)** Signif. codes: 0 '***' 0.001 '**' 0.01 '*' 0.05 '.' 0.1 ' ' 1
Spring	*Noctiluca scintillans*	0.33 0.0019**
Summer	*Centropages ponticus* *Mnemiopsis leidyi*	0.471 0.0001 *** 0.302 0.0019 **
Autumn	*Paracalanus parvus* *Pseudocalanus elongatus*	0.425 0.0001 *** 0.323 0.0029 **
Autumn+Summer	*Penilia avirostris*	0.286 0.0049 **
Shelf Waters *(Total number of species: 9; selected number of species: 5; number of species associated with 1 group: 4; number of species associated with 2 groups: 1)*		
Spring	*Noctiluca scintillans*	0.276 0.0443 *
Summer	*Penilia avirostris* *Mnemiopsis leidyi*	0.323 0.0039 ** 0.322 0.0143 *
Autumn	*Paracalanus parvus*	0.433 4e-04 ***
Autumn+Summer	*Centropages ponticus*	0.301 0.03 *
Open Sea *(Total number of species: 9; selected number of species: 5; number of species associated with 1 group: 5; number of species associated with 2 groups: 0)*		
Spring	*Pseudocalanus elongatus* *Noctiluca scintillans*	0.464 0.0028 ** 0.371 0.0345 *
Summer	*Penilia avirostris* *Centropages ponticus* *Mnemiopsis leidyi*	0.555 0.0004 *** 0.462 0.0019 ** 0.370 0.0313 *

mass development of the ctenophore *B. ovata*, a predator of *M. leidyi* (Vereshchaka et al. 2019). Before the invasion of *B. ovata*, which now controls the population of *M. leidyi*, the *M. leidyi*–linked signal was almost as strong as the temperature signal (Vereshchaka et al. 2019). In the period of the uncontrolled bloom, *M. leidyi* explained a significant part of the mesozooplankton community variance at coastal, shelf and open sea habitats (Table 1). A breakpoint of Mnemiopsis was distinguished in 2004 (Figure 11), almost 10 years after the second ctenophore introduction.

The second ISA approach, introduced in the present research was rather related to the detection of breakpoints or temperature regime shifts that may have led to an ecosystem response such as structural changes within the community and respectively changes in indicator species complex. The sea surface temperature (SST) is being used very often as a proxy for describing climatic variability. It indicates a relatively mild cooling phase during 1960–1980 and a subsequent more pronounced cooling phase during 1980–1993 (SoE report 2019). Similar variations were also observed in the warm seasons meaning cold intermediate layer (CIL) temperature. They were followed by a warmer phase during 1993–2014. The climate-induced temperature

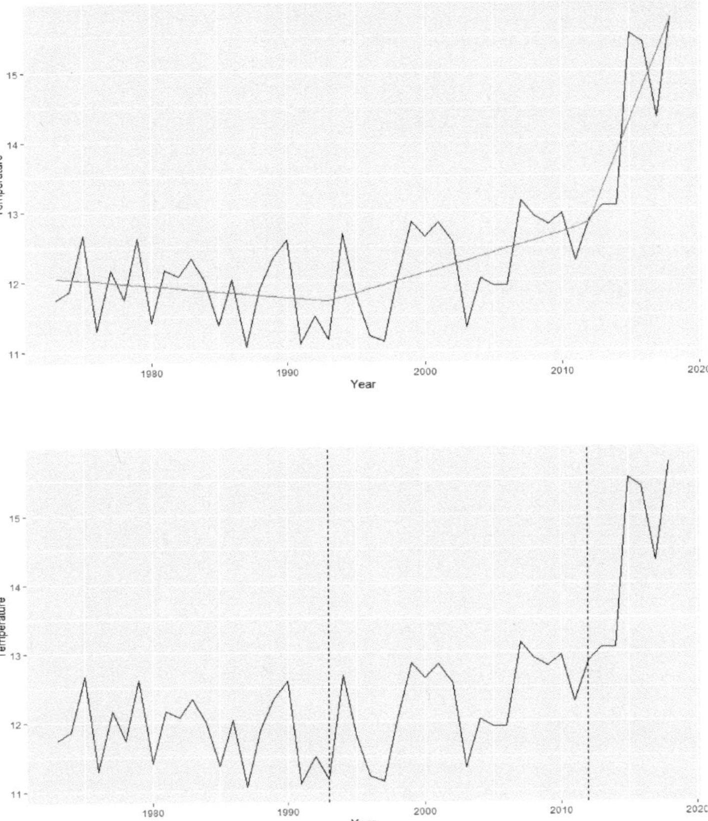

Figure 12. Breakpoint analysis, segmented regression model fit of mean air temperature in Varna bay, measured in the period 1960–2018. (A) trend of mean air temperature and (B) identified periods after the ISA approach applied.

changes are related to the strengthening of the NAO; its positive phase results in colder, drier, and more severe winters in contrary to the wetter, warmer, and milder winters observed over northwestern Europe and the Eastern North Atlantic Ocean (Oguz et al. 2006). The subsequent warming trend starting from 1993 up to 2001 increased the SST and CIL temperatures. Afterwards, both SST and CIL temperatures undergo a decadal scale oscillation between the minimum in 2005–2006 and the maximum in 2010–2011 (SoE report 2019). According to de Lima et al. (2021) an overall warming of the basin has been observed especially in the period 2005–2018 with a decreasing trend in deeper layers.

For the provisions of the ISA, a long-term mesozooplankton dataset was taken into consideration – abundance data collected over the period 1966–2020. Prior to running the breakpoint analysis to the available annual mean temperature time-series (SST measurements over the period 2000–2020 and air temperature measurements (NIMH) in the period 1960–2018 in Varna Bay), all published case studies were carefully examined to support the proper selection of initial breakpoints values in a combination with iterative search applied to temperature time-series (Crawley).

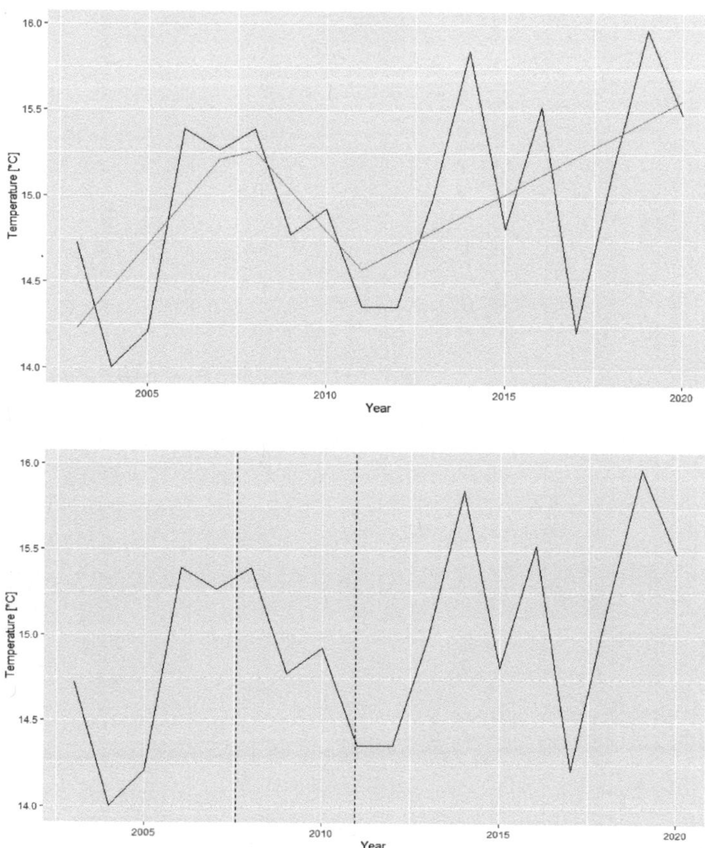

Figure 13. Breakpoint analysis, segmented regression model fit of mean SST temperature in Varna bay, measured in the period 2000–2020.

Table 2. Multilevel pattern analysis – Indicator species under regime shifts in the sub-periods 1966–1993, 1994–2007, 2008–2011, 2003–2020, site grouping: Coastal waters.

Indicator Species Under Regime Shifts (1966–1993, 1994–2007, 2008–2011, 2003–2020)		
(Total number of species: 7; selected number of species: 5; number of species associated with 1 group: 4; number of species associated with 2 groups: 0; number of species associated with 3 groups: 0)		
Group	**Species Name**	**Stat., p-value (significance α=0.05)** Signif. codes: 0 '***' 0.001 '**' 0.01 '*' 0.05 '.' 0.1 ' ' 1
P1 (1966–1993)	*Noctiluca scintillans*	0.162 0.0493 *
P3 (2008–2011)	*Acartia.clausi*	0.45 2e-04 ***
P4 (2012–2020)	*Paracalanus parvus* *Centropages ponticus*	0.40 5e-04 *** 0.31 9e-03 **

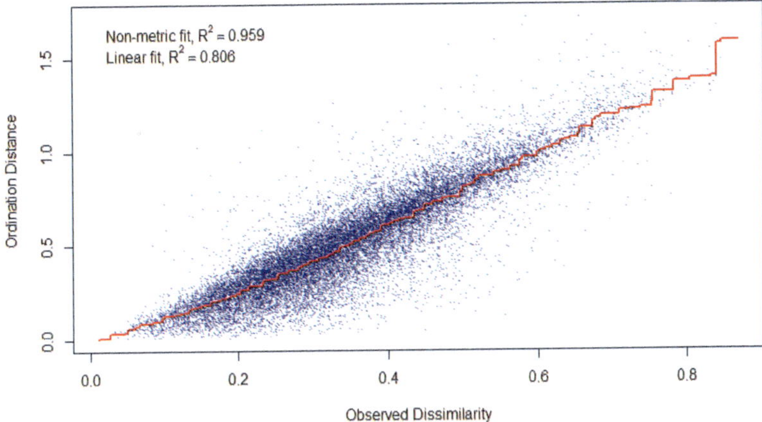

Figure 14. Shepard plot for nonmetric multidimensional scaling (NMDS) results. The dashed line indicates a perfect linear relationship between calculated and ordination distances. NMDS plot (figure shows only statistically significant fits of the overlaid environmental gradients).

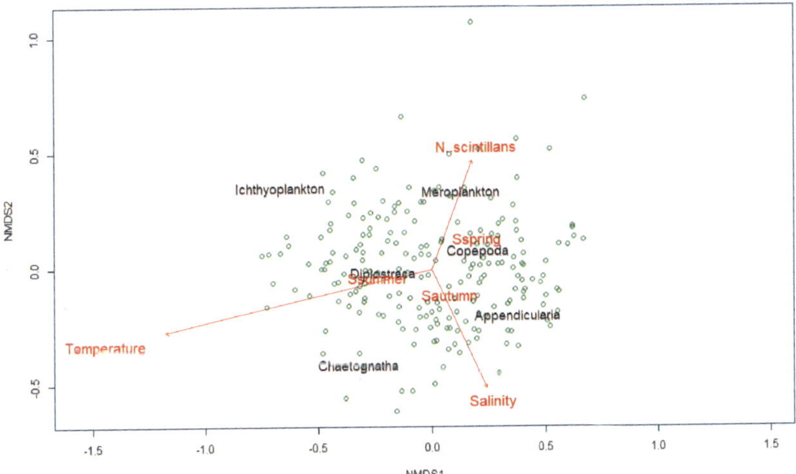

Figure 15. NMDS plot of mesozooplankton taxonomic groups abundance data (centroids) with fitted environmental variables (vectors) and seasons centroids (factors).

The estimated breakpoints delivered by the segmented regression model fit were as follows: for Varna Bay air temperature time-series regime shifts were identified in 1993 and 2012 and for SST time-series in 2008 and 2011 (Figures 12a, b, 13a, b).

Multiple R^2 for air temperature model fit was estimated to be 0.74 (adjusted R^2–0.71) and respectively for SST model fit multiple R^2: 0.3706, adjusted R^2: 0.11. Despite the low R^2 value of the overall SST model fit (which was expected as the number of observations was small n = 18) the results were considered realistic in regards to biomonitoring observations in the studied period and abundance data were divided into four sub-periods, consistent with the breakpoint analysis results and considered as different temperature regimes: 1966–1993, 1994–2007, 2008–2011 and 2012–2020. ISA results (Table 2) showed different species indicative for the

Table 3. Statistical outcome of goodness-of-fit of environmental data to NMDS plot.

Global Multidimensional Scaling Using MonoMDS		
***VECTORS**		
Group	*NMDS$_1$, NMDS$_2$*	*r^2* Pr(>r) (significance α=0.05)* Signif. codes: 0 '***' 0.001 '**' 0.01 '*' 0.05 '.' 0.1 ' ' 1
Chl.a	0.20987 0.97773	0.0229 0.106
Temperature	−0.97493 −0.22253	0.4274 0.001 ***
Salinity	.42206 −0.90657	0.0929 0.001 ***
Noctiluca scintillans	0.34957 0.93691	0.0732 0.002 **
***FACTORS**		
Season: autumn	0.0798 −0.1125	Season 0.276 0.001 ***
Season: spring	0.1966 0.1282	
Season: summer	−0.2367 −0.0357	

* The envfit output contains information on the length of the segments for each variable. The segments are scaled to the r^2 value so that the environmental variables with a longer segment are more strongly correlated with the data than those with a shorter segment.

defined periods with *N. scintillans*, assigned with the higher indicator value for the period 1966–1993, *A. clausi* for 2008–2011 and *P. parvus* and *C. ponticus* in the period 2012–2020. The latter completely matches the structural changes, observed in the community as a result of the temperature regime shifts.

A non-metric multidimensional scaling (NMDS) was used to show 2000–2020 mesozooplankton taxonomic groups abundance data in two dimensions. NMDS plot generally highlights the similarities between samples or translated to the present case study; the more similar are two samples (points), the more similar they are in terms of their mesozooplankton community structure and abundance. Additionally, there are possibilities to overlay underlying environmental variables that are covarying with mesozooplankton community structure and possibly driving changes in the community. The selected variables were related to the abiotic environment *in situ* data for temperature and salinity, biotic interactions Chlorophyl a as a proxy to species feeding ecology (competition for food and food availability), and *N. scintillans* abundance as a limitation factor.

Shepard diagram (Figure 14) showed that reduced scaling visualization or goodness-of-fit is accurate, and Figure 15 show the NMDS plot of mesozooplankton taxonomic groups abundance data (centroids) with fitted environmental variables (vectors) and seasons centroids (factors) (statistical outcome of goodness-of-fit of environmental data to NMDS plot detailed in Table 3).

It is clear that longer gradients are expected to have more significant effects on abundance data, i.e., temperature can potentially have a high impact on Ichtyoplankton, Chaetognata and Diplostraca abundance variations in the summer seasons, salinity may have an association with Chaetognata and Appendicularia groups abundance in autumn seasons and *N. scintillans* may have an impact on Copepoda and Meroplankton as a limitation factor in the spring seasons.

4. Concluding Remarks

In general, the present case study aimed at identifying the prevalent and more general trends in the ecosystem and the zooplankton community in the Bulgarian Black Sea waters with a particular focus on the structural changes and community response to anthropogenic pressures and environmental variations within the context of the global warming. Different hypotheses were tested in terms of identification of indicator species, related to seasonality and regime shifts. The results showed clearly that the zooplankton community has undergone structural changes, which may be considered indicative of the ecosystem as a whole and regular monitoring can be implemented as global climate change footprint analysis and early warning for primary producers' structural changes. Considering the latter, further complex case studies may contribute to the actual assessment of the zooplankton's role as a "bottom-up" and "top-down" control for the entire ecosystem.

5. Acknowledgments

This work has been carried out partly in the framework of the National Science Program 'Environmental Protection and Reduction of Risks of Adverse Events and Natural Disasters', approved by the Resolution of the Council of Ministers № 577/17.08.2018 and supported by the Ministry of Education and Science (MES) of Bulgaria (Agreement № Д01-271/09.12.2022) and Advancing Black Sea Research and Innovation to Co-Develop Blue Growth within Resilient Ecosystems (BRIDGE – BS) project, funded by the European Union's Horizon 2020 research and innovation program under grant agreement No. 101000240.

References

Akoglu, E., Salihoglu, B., Libralato, S., Oguz, T. and Solidoro Cosimo. 2014. An indicator-based evaluation of black Sea food web dynamics during 1960–2000. Journal of Marine Systems 134: 113–125.

Akoglu, E. 2023. Ecological indicators reveal historical regime shifts in the Black Sea ecosystem. PeerJ 11: e15649. https://doi.org/10.7717/peerj.15649.

Alexandrov, B., Arashkevich, E., Gubanova, A. and Korshenko, Al. 2014. Black Sea monitoring guidelines mesozooplankton. In the frame of the EU/UNDP Project: Improving Environmental Monitoring in the Black Sea – EMBLAS. 31 pp.

Arashkevich, E.G., Stefanova, K., Bandelj, V., Siokou, I., Terbiyik Kurt, T., Ak Örek, Y. et al. 2014. Mesozooplankton in the open black Sea: Regional and seasonal characteristics. Journal of Marine Systems 135: 81–96.

Batchelder, H., Kendra, L. Daly, CabelL S. Davis, Rubao, J., Mark D. Ohman, William T. Peterson et al. 2005. Climate change and marine plankton. Trends in Ecology and Evolution 20: 337–344.

Barriopedro, D., Fischer, E.M., Luterbacher, J., Trigo, R.M. and García-Herrera, R. 2011. The hot summer of 2010: redrawing the temperature record map of Europe. Science 332(6026): 220–224.

Beaugrand, G., Brander, K.M., Lindley, J.A., Souissi, S. and Reid, P.C. 2003. Plankton effect on cod recruitment in the north Sea. Nature 426: 661–664. doi: 10.1038/nature02164.

Brun, P., Stamieszkin, K., Visser, A.W., Licandro, P., Payne, M.R. and Kiørboe, T. 2019. Climate change has altered zooplankton-fuelled carbon export in the north Atlantic. Nat. Ecol. Evol. 3: 416–423. doi: 10.1038/s41559-018-0780-3.

de Lima, C.Z., Buzan, J.R., Moore, F.C., Baldos, U.L.C., Huber, M. and Hertel, T.W. 2021. Heat stress on agricultural workers exacerbates crop impacts of climate change. Environmental Research Letters 16: 044020.

Edwards, M. and Richardson, A. 2004. Impact of climate change on marine pelagic phenology and trophic mismatch. Nature 430: 881–884. https://doi.org/10.1038/nature02808.

Ginzburg, A.I., Kostianoy, A.G. and Sheremet, N.A. 2004. Seasonal and interannual variability of the black Sea surface temperature as revealed from satellite data (1982–2000). J. Mar. Syst. 52: 33–50. doi: 10.1016/j.jmarsys.2004. 05.002.

Gunduz, M., Özsoy, E. and Hordoir, R. 2020. A model of black Sea circulation with strait exchange (2008–2018). Geosci. Model Dev. 13: 121–138. doi: 10.5194/ gmd-13-121-2020.

Humborg, C., Ittekkot, V., Cociasu, A. and Bodungen, B. 1997. Effect of danube river dam on black Sea biogeochemistry and ecosystem structure. Nature 386: 385–388.

Hays, G.C., Richardson, A.J. and Robinson, C. 2005. Climate change and marine plankton. Trends in Ecology and Evolution 20: 337–344.

Jernberg, S., Lehtiniemi, M. and Uusitalo, L. 2017. Evaluating zooplankton indicators using signal detection theory. Ecol. Indic. 77: 14–22.

Kara, A.B., Wallcraft, A.J., Hurlburt, H.E. and Stanev, E.V. 2008. Air–sea fluxes and river discharges in the black Sea with a focus on the danube and bosphorus. J. Mar. Syst. 74: 74–95. doi: 10.1016/j. jmarsys.2007.11.010.

Kamburska, L., Doncheva, V. and Stefanova, K. 2003. On the recent changes of zooplankton community structure along the bulgarian black Sea coast—a post-invasion effect of exotic ctenophores interactions. pp. 69–84. Proceeding of First Intern. Conf. on Environmental Research and Assessment (Bucharest, Romania, March 23–27, 2002). Docendi Publishing House, Bucharest, Romania.

Konsulov, A. and Kamburska, L. 1998. Ecological determination of the new ctenophora – *Beroe ovata* invasion in the black Sea. Океанология, ИО-БАН, Варна, т. 2: 195–198.

Kovalev, A.V. 1993. Zooplankton. 1. Mesozooplankton. pp. 144–164. *In*: Kovalev, A.V. and Finenko, Z.Z. (eds.). Plankton of the Black Sea. Naukova Dumka, Kiev (in Russian).

Kovalev, A.V., Skryabin, V.A., Zagorodnyaya, Y.A., Bingel, F., Kideyş, A.E., Niermann, U. et al. 1999. The black Sea zooplankton: composition, spatial/temporal distribution and history of investigations Tr. J. of Zoology 23: 195–209.

Mihneva, V. and Stefanova, K. 2011. Species diversity. abundance and biomass dynamic of mesozooplankton along the bulgariam black Sea (2008–2010). Proceeding of Union of Scientists –Varna. Series "Marine sciences" 2011: 97–104.

Mihneva, V. and Stefanova, K. 2013. The non-native copepod oithona davisae (Ferrari, F.D. and Orsi. 1984) in the western black Sea: seasonal and annual abundance variability. BioInvasions Records 2(2): 119–124. doi: http://dx.doi.org/10.3391/bir.2013.2.2.04.

Moncheva, S., Todorova, V., Shivarov, A., Dzhurova, B., Prodanov, B., Doncheva, V. et al. 2013. Initial assessment of the marine environment - Bulgaria in art. 8, 9, 10 of MSFD 2008/56/EC and NOOSMV. 476 pp.

Moncheva, S. 2006. Phytoplankton- technical report-control of eutrophication, hazardous substancesand related measures for rehabilitating the black Sea ecosystem: Phase 2: Leg I PIMS 30651, Istanbul, Turkey The state of phytoplankton. Available from: https://www.researchgate.net/publication/312372456_ The_state_of_phytoplankton [accessed Dec 19 2023].

Mordukhay-Boltovskoy, F.D. (ed.). 1968. The identification book of the black Sea and the Sea of azov fauna.- kiev: naukova dumka publ., T. 1 (Protozoa, Porifera, Coelenterata, Ctenophora, Nemertini, Nemathelminthes, Annelida, Tentaculata), 423 pp. (in Russian).

Mordukhay-Boltovskoy, F.D. (ed.). 1969. The identification book of the black Sea and the Sea of azov fauna.- Kiev: naukova dumka Publ., T. 2 (Artropoda: Cladocera, Calanoida, Cyclopoida, Monstrilloida, Harpacticoida, Ostracoda, Cirripedia, Malacostraca, Decapoda), 536 pp. (in Russian).

Mordukhay-Boltovskoy, F.D. (ed.). 1972. The identification book of the Black Sea and the Sea of Azov Fauna.-Kiev: Naukova Dumka Publ., T. 3 (Arthropoda, Mollusca, Echinodermata, Chaetognatha, Chordata: Tunicata, Ascidiacea, Appendicularia), 340 pp. (in Russian).

Mulet, S., Buongiorno Nardelli, B., Good, S., Pisano, A., Greiner, E., Monier, M. et al. 2018. Ocean temperature and salinity. *In*: Copernicus Marine Service Ocean State Report. J. Oper. Oceanogr. 11(suppl. 1): s13–s16. doi: 10.1080/1755876X.2018.1489208.

Mutlu, E. 2020. Offshore diel vertical distribution of meso/macro-holo/mero zooplankton in the southern black Sea. Journal of Applied Biological Sciences, E-ISSN: 2146–0108 14(3): 249–267.

Ndah, A., Meunier, C.L., Kirstein, I., G¨obel, J., R¨onn, L. and Boersma, M. 2022. A systematic study of zooplankton-based indices of marine ecological change and water quality: application to the European marine strategy framework Directive (MSFD). Ecological Indicators 135. https://doi.org/10.1016/j.ecolind.2022.108587.

Oguz, T. 2006. Black Sea ecosystem response to climatic teleconnections. Oceanography, Special ssue features: Black Sea Oceanography 18(2): 122–133.

Özsoy, E. and Ünlüata, Ü. 1997. Oceanography of the black Sea: a review of some recent results. Earth Sci. Rev. 42: 231–272. doi: 10.1016/s0012-8252(97)81859–4.

Petipa, T.S. 1957. On the mean weight of the principle forms of zooplankton in the black Sea. Tr. Sevast. Biol. Station 9: 39–57 (in Russian).

Seda, J. and Devetter M. 2000. Zooplankton community structure along a trophic gradient in a canyon-shaped dam reservoir. J. Plankton Res. 22: 1829–1840.

Shapiro, G.I., Aleynik, D.L. and Mee, L.D. 2010. Long term trends in the seasurface temperature of the black Sea. Ocean Sci. 6: 491–501. doi: 10.5194/os-6-491-2010.

Sorokin, Y.I. 2002. The Black Sea: Ecology and Oceanography. Backhuys Publishers, Leiden, The Netherlands, 875 pp.

Shtereva, G., Velikova, V. and Doncheva, V. 2015. Human impact on marine water nutrients enrichment. J. Environ. Prot. Ecol. 16: 40–48.

Stefanova, K., Stefanova, E., Doncheva, V., Mgeladze, M., Khalvashi, M., Arashkevich, E. et al. 2019. Mesozooplankton. *In*: Krutov, A. (ed.). BSC, 2019. State of the Environment of the Black Sea (2009–2014/5). Istanbul, Turkey: Publications of the Commission on the Protection of the Black Sea against Pollution (BSC).

Steinberg, D.K. and Landry, M.R. 2017. Zooplankton and the ocean carbon cycle. Annu. Rev. Mar. Sci. 9: 413–444. doi: 10.1146/annurev-marine-010814–015924.

Stefanova, E., Stefanova, K. and Kozuharov, D. 2012. Mesozooplankton assemblages in northwestern part of black Sea. Acta Zool. Bulg. 64(4): 403–412. ISSN 0324–0770.

Shiganova, T.A., Anninsky, B., Finenko, G.A., Kamburska, L., Mutlu, E., Mihneva, V. et al. 2020. Black Sea monitoring guidelines. Macroplankton (gelatinous plankton). In the Frame of the EU/UNDP Project: Improving Environmental Monitoring in the Black Sea – EMBLAS.

Volkov, D.L. and Landerer, F.W. 2015. Internal and external forcing of sea level variability in the black Sea. Clim. Dyn. 45: 2633–2646. doi: 10.1007/s00382- 015-2498-0.

Yunev, O.A., Moncheva, S. and Carstensen, J. 2005. Long-term variability of vertical chlorophyll a and nitrate profiles in the open black Sea: eutrophication and climate change. Mar. Ecol. Prog. Ser. 294: 95–107.

Vereshchaka, A.L, Anokhina, L.L., Lukasheva, T.A. and Lunina A.A. 2019. Long-term studies reveal major environmental factors driving zooplankton dynamics and periodicities in the Black Sea coastal zooplankton, PeerJ 2019 Sep 27(7): e7588. doi: 10.7717/peerj.7588. eCollection 2019.

3.6

A Review of Planktonic Studies in Africa's Contribution to Maintenance of Marine Biodiversity, Sustainable Consumption and Biological Production

Fernando Morgado[1],* and *Luis R. Vieira*[2]

1. Introduction

Africa is the world's second-largest continent, with about 30.3 million km^2, and the second-most populous continent, with 1.4 billion people, which represents about 18% of the world's human population (IPBES 2018, World Population Prospects 2022a). Predictions for future African population growth indicate an increase of almost 1.3 billion people by 2050, corresponding to double the current population (Neumann et al. 2015, Global Platform for Sustainable Cities World Bank 2018, IPBES 2018), which implies also an expected increase of urbanization and population settlement on coastal zones by 2060 (Neumann et al. 2015, Global Platform for Sustainable Cities World Bank 2018). This increasing human settlement and their activities on the coasts of the continent, such as harbors, fisheries activities, industries, and other development infrastructures have contributed, during the last decades, to multiple anthropogenic pressures, that are associated with its climatic and environmental conditions and hence increasingly pressure and threaten the integrity of the coastal and marine ecosystems (Diop et al. 2011, Africa Sustainable Development Report 2018, UNEP 2018, Meyers 2019). The African geographic and particular climatic conditions, such as the pressure systems, winds and the Intertropical Convergence Zone, surface wind patterns that influence rainfall and temperature regimes on the continent also pose serious environmental vulnerabilities to climate change and

[1] The Center for Environmental and Marine Studies (CESAM) and Department of Biology University of Aveiro, Campus Universitário de Santiago, Aveiro, Portugal.

[2] Interdisciplinary Center of Marine and Environmental Research (CIIMAR). University of Porto, Terminal de Cruzeiros do Porto de Leixões, Av. General Norton de Matos s/n, 2250–208 Matosinhos, Portugal.

* Corresponding author: fmorgado@ua.pt

extreme events impacts, reflected in an increase on the number and frequency of natural disasters, droughts, water pollution and degradation of coastal and marine environment and habitats (UNEP-DTIE 2010, WBCSD 2010, United Nations 2018, UNPD 2018, Lamere 2013, Globe Afrique 2016, IPBES 2018, Matata and Adan 2018, Meyers 2019). These direct and indirect drivers associated with inappropriate economic policies, technologies and governance systems, as well as socio-political and cultural pressures, will affect marine biodiversity and food production, fresh water, sustainable resources consumption, poverty, and increase in economic and social inequalities (Diop et al. 2011, Popova et al. 2016), and health problems (UNEP-DTIE 2010, WBCSD 2010, Lamere 2013, Grant 2015, Globe Afrique 2016, Global Platform for Sustainable Cities World Bank 2018, IPBES 2018, Meyers 2019, Okonjo-Iweala 2020, UNEP 2020).

Worldwide human survival depends on biological resources (Pecl et al. 2017) but human activities are globally threatening ecosystems and species with the consequent reduction of biodiversity and profound impacts on the economic and social development of human communities (OECD 2019, 2020, Cepic et al. 2022). Today's challenges and demands for development and competitiveness, in an increasingly global economy, converge in a broader vision of efficient use of marine resources, ensuring the resilience of marine ecosystems toward sustainable development (IPBES 2018, UNEP 2018, OECD 2020). These aspects take on greater expression in the poorest and developing countries, such as African countries due to the large extent of the Ecoregions and the political and socio-economic particularities add enormous complexity to the maintenance of biodiversity and the balance and sustainability of ecosystems in these regions (Lamere 2013, Meyers 2019, Globe Afrique 2016, UNEP 2020).

The coastal and marine environment of Africa has a wide marine biodiversity and diversity of habitats that include biodiversity hotspots and productive fishing grounds regions, crucial for the ecological and socio-economic activities, goods and services of the populations (Diop et al. 2011, Africa Sustainable Development Report 2018, IPBES 2018, Cepic et al. 2022). Africa continent includes five major large marine ecosystems, the Agulhas current LME and the Somali current LME, the Benguela current large marine ecosystem, the Canary current large marine ecosystem and the Guinea current large marine ecosystem (the Canary and Benguela currents are two of the world's major coastal upwelling systems) (SCLME/SWIOFP 2013a, b), that are areas of great marine productivity, representing significant contributions to the African countries economies and are central to a diverse livelihood strategies (Africa Sustainable Development Report 2018, IPBES 2018, IOC-UNESCO 2020). The development of studies to evaluate marine ecosystems' condition and biodiversity loss under the cumulative impact of human uses and climate change, at different scales, remains a global knowledge gap for a better understanding of the impacts on ecosystems and human well-being (Borja 2014, Pecl et al. 2017, Ehler et al. 2019), which requires policy and management actions in order to ensure a sustainable marine resources management (Diop et al. 2011, Satia 2016, Globe Afrique 2016, Meyers 2019, UNEP 2020). In the African context, this knowledge constitutes essential management tools for smarter use of marine and coastal resources, the reduction of bycatch and the support of sustainable fisheries (Climate Smart Oceans

2016, Barlow et al. 2018, Hazen et al. 2018) to promote sustainable production, consumption practices and for food security, the creation of new models for obtaining income (Schlüter et al. 2017, 2019, Lindkvist et al. 2020) and fundamental to decision-support for integrated ocean management (Barlow et al. 2018, Matata and Adan 2018, Lombard et al. 2019).

2. Plankton Studies' Contribution to the Maintenance of Marine Biodiversity, Sustainable Consumption and Biological Production

Plankton communities are very important in the trophic dynamics and the structure of marine ecosystems and exhibit quick responses to environmental changes through changes in their abundance distribution and species composition, which makes them suitable as indicators of ecosystem health and biodiversity over both short and longer time scales (Beaugrand 2014, 2019). The planktonic research allows exploring new technologies for ocean observations, monitoring and health production (Hablützel et al. 2021), significantly contributing to the maintenance of marine biodiversity, understanding climatic variability and global change (Beaugrand 2014, 2019), global patterns of phytoplankton and zooplankton diversity driven by temperature and environmental variability (Righetti et al. 2019) and realizing the evolving paradigms in biological carbon cycling in the ocean (Zhang et al. 2018). The planktonic communities can also be important indicators of the trophic structure of communities in estuaries, bays and coastal and marine areas and allow a more efficient assessment of the processes taking place in water bodies, contributing to the maintenance of marine biodiversity, sustainable consumption and organic production (McQuatters-Gollop et al. 2019). The knowledge of plankton ecology and trophic dynamics allows the establishment of interfaces for understanding the processes by which plankton organisms interact with the environmental and biological components of marine ecosystems and coastal areas (Greenwood et al. 2019) and helps to understand the biomass composition of the oceans toward a sustainable and biological marine resources consumption (Bar-On and Milo 2019).

3. A Review of Plankton Studies on Marine Ecosystems in African Countries

Plankton on the coast of Africa and the Indian Ocean has been studied since the middle of the 20th century and research on plankton has shown significant development to date. The works were focused on the plankton of the Indian Ocean, including coastal waters, the phytoplankton and zooplankton composition, abundance and diversity of oceanic and estuarine species for certain coastal areas in Kenya, Tanzania, Senegal and Ivory Coast, coast of Benin, Nigerian coast, coast of Namibia (Benguela current), Seychelles coastal waters, Mauritania, Mauritius, Cameroon, Ghana, Kenya, Mozambique, Namibia, and South Africa. The works also have focused on the dynamics of plankton relations with environmental conditions and the effects on fresh water supply in communities. In regions of importance for fishing, some works

Table 1. Review of global and thematic planktonic studies in African countries (annex).

Plankton of the Indian Ocean, West Africa, and SE Atlantic	Kramp 1955, Hendey 1958, Lebour 1959, Furnestin 1966, Bé and Hutson 1977, Thiriot 1977, John 1985, Madhupratap 1986, Saraswathy 1986, Postel 1990, Nicol et al. 1992, Gibbons et al. 1995, Postel et al. 1995, Sedletskaya 1999, Conway et al. 2003, Gallienne et al. 2004, Gallienne and Smythe-Wright 2005, Lathuilière et al. 2008, Seeyave et al. 2009, Demarcq and Somoue 2015, Sreeush et al. 2018.	Ecology of planktonic foraminifera and biogeographic patterns of life and fossil assemblages, larval decapod Crustacea, manual for fish eggs and larvae, zooplankton standing stock and diversity, Epipelagic mesozooplankton dynamics, Pleuromamma in the Indian Ocean, phytoplankton and primary productivity, marine diatoms, marine medusae, Chaetognathes diversity and distribution, zoogeography and diversity of euphausiids, growth and shrinkage of Antarctic krill Euphausia superba, horizontal and vertical distribution patterns of fish larvae, meso-zooplankton response to coastal upwelling seasonal and intraseasonal surface chlorophyll-a variability, nitrogen nutrition in assemblages dominated by Pseudo-nitzschia spp., Alexandrium catenella and Dinophysis acuminate, guide to the coastal and surface zooplankton of the south-western Indian Ocean, biological production in upwelling areas
Plankton from Morocco	Boucher 1982, Ouldessaib et al. 1998, Dolan et al. 2002, Youssara et al. 2004, John et al. 2004, Somoue et al. 2005, Berraho 2007, El Khalki and Moncef 2007, Badsi et al. 2010, Elghrib et al. 2012, Zaafa et al. 2012, Badsi et al. 2012, Zizah et al. 2012, Salah 2013, Natij et al. 2014, Anabalón et al. 2014, Anhichem et al. 2017, Berraho et al. 2016, 2019, Aouititen et al. 2019, Amazzal et al. 2020, Rhomad et al. 2021.	Faunal composition and demographic structure of copepods population in the north-west African coastal upwellings, copepods population in the estuaries, microzooplankton diversity: relationships of tintinnid ciliates with resources, competitors and predators, spatio-temporal variations of mesozooplankton, zonal structures in fish larval abundance and diversity, ecological factors affecting the distribution of zooplankton community, variability of spatial and temporal distribution of marine zooplankton communities in relation with environmental parameters, phytoplankton diversity and community composition along estuaries, phytoplankton distribution in upwelling areas, phytoplankton abundance and diversity in the coastal waters, spatial-temporal structure of zooplankton in upwelling filaments, the structure of planktonic communities under variable coastal upwelling conditions, water and sediment quality, abundance and structure of copepod communities along the Atlantic coast, spatio-temporal evolution of zooplankton abundances between Cap Boujdour and Cap Blanc, predicting jellyfish strandings in the Moroccan north-west mediterranean coastline, spatial relationships between the environment and ichthyoplankton of small pelagics, copepod community along the Mediterranean coast of Morocco, biodiversity and spatio-temporal variability of copepods community, importance of hydrological parameters in the distribution of planktonic eggs and larvae in an upwelling zone, modeling investigation of the nutrients and phytoplankton dynamics in the Moroccan Atlantic coast
Plankton from Egypt	Abdel-Aziz Aboul-Ezz 2003, Zakaria et al. 2007, Zakaria 2006, 2015, Zaghloul et al. 2020.	Zooplankton community of the Egyptian Mediterranean coast, influence of salinity variations on zooplankton community. The zooplankton community in Egyptian Mediterranean waters, the impact of climate variability and anthropogenic activities on zooplankton community in the neritic waters, and marine gelatinous zooplankton of the Egyptian waters

Table 1 contd. ...

...Table 1 contd.

Plankton from Algeria	Seridji 2000, Riandey et al. 2005, Hafferssas and Seridji 2010, Khames and Hafferssas 2018, 2019, Kherchouche and Hafferssas 2019, 2020, Mellak and Hafferssas 2022, Khames et al. 2022.	Copepod Diversity and community structure in the Algerian Basin, zooplankton distribution related to the hydrodynamic features in the Algerian Basin, relationships between the hydrodynamics and changes in copepod structure, biodiversity and abundance of gelatinous zooplankton along the Algerian coast, abundance and species composition of gelatinous zooplankton in Habibas islands and Sidi Fredj, species composition and distribution of medusae (Cnidaria: Medusozoa) along the Algerian coast, copepod communities of the Algerian coast, gelatinous zooplankton versus Algerian environmental factors.
Plankton from Tunisia	Daly Yahia et al. 2003, 2004	Distribution et écologie des Méduses (Cnidaria) du golfe de Tunis, the spatial and temporal structure of planktonic copepods in the Bay of Tunis
Plankton from Senegal and Guiné-Bissau	Gaudy and Seguin 1964, Seguin 1966, An 1971, Voituriez and Dandonneau 1974, Séret 1983, Casanova et al. 1982, Gaertner 1988, Diouf 1990, Arkhipov et al. 2015, Ndour et al. 2018	Annual distribution of pelagic copepods in the waters of Dakar, zooplankton of the South Coast of the Cape Verde peninsula, Atlantic Ocean phytoplankton south of the Gulf of Guinea, biology of surface plankton in the Bay of Dakar, relations between the thermal structure, primary production and the regeneration of nutrient salts, zooplankton of Senegal, zooplankton communities in the Bay of Dakar and their relationship with hydrography, biomass and chemical and faunal composition of zooplankton, species composition and features of ichthyoplankton distribution, composition, distribution and abundance of zooplankton and ichthyoplankton
Plankton from Cape Verde	Touré 1971, Andersen et al. 1997, Alamo et al. 2000, Hernández et al. 2000, Lindley et al. 2001, 2004, Vinogradov et al. 2004, Séguin 2010, Hanel et al. 2010, Ramos et al. 2012, Teuber et al. 2013, Denda et al. 2014, Karstensen et al. 2015, 2016, Fiedler et al. 2016, Fischer et al. 2016, Rodrigues et al. 2016, Hoving et al. 2020, Kiko et al. 2020, Morais et al. 2021, Stenvers et al. 2021, Lüskow et al, 2022	Quantitative and qualitative variation of zooplankton in the Cape Verde region, macroplankton and micronekton in the northeast tropical Atlantic: abundance, community composition and vertical distribution in relation to different trophic environments, pelagic polychaetes of the cape verde islands, pelagic decapods (larvae and adults) from the Cape Islands, pelagic amphipods from the Cape Verde Islands, Phyllosoma larvae (Decapoda: Palinuridea) of the Cape Verde Islands, zooplankton distribution patterns at two seamounts, planktonic nudibranch molluscs in the Cape Verde archipelago, copepods in relation to the oxygen minimum zone, zooplankton community near the island of São Vicente, insight on pelagic copepod respiration, larval fish abundance, composition and distribution at Senghor Seamount (Cape Verde Islands), variation of the chlorophyll a related to sea surface temperature, analysis of zooplankton samples using the zooscan approach, open ocean dead zones wind and geostrophic currents, distribution, upwelling and isolation in oxygen-depleted anticyclonic modewater eddies and implications for nitrate cycling, oxygen utilization and downward carbon flux in an oxygen-depleted eddy, phytoplankton community structure, dynamics and their relationship to water quality in five Santiago Island reservoirs, bathypelagic particle flux signatures from a suboxic eddy: production, sedimentation and preservation, zooplankton-mediated fluxes, associations and role in the biological carbon pump of Pyrosoma atlanticum (Tunicata and Thaliacea), *in situ* observations show vertical community structure of pelagic fauna, distribution and biomass of gelatinous zooplankton in relation to an oxygen minimum zone and a shallow seamount

Plankton from the Seychelles archipelago	Bijoux et al. 2003, Dilmahamod 2014, Dilmahamod et al. 2016, Kim et al. 2022.	Marine biodiversity in the Seychelles archipelago, thermocline ridge and large-scale climate modes, primary productivity and the annual cycle of chlorophyll-a, chlorophyll-a variability and analysis of a coupled biophysical model, mesozooplankton community variability
Plankton from Ghana	Mensah 1966, 1974, Anang 1979, Biney 1990, Wiafe and Frid 1997, Wiafe 1997, Wiafe and Frid 1999, 2001, Sywula et al. 2002.	Coastal plankton, marine copepod distribution, zooplankton composition and distribution, seasonal cycle cf the phytoplankton, zooplankton biodiversity, hydrographic forcing and zooplankton temporal variation, temporal variability within coastal zooplankton communities, genetic subdivision of the upwelling copepod communities
Plankton from Kenya	Wickstead 1961, 1962, Fenaux 1973, Fleminger et al. 1982, De Decker 1984, Baidya and Choudhury 1984, Reay and Kimaro 1984, Grove et al. 1986, Kimaro 1986, Revis 1988, Revis and Okemwa 1988, Kimaro and Jaccarini 1989, Okemwa 1989, 1889, 1992, Okemwa and Revis 1986, Demeulenaere 1991, Mwaluma 1993, 1997, Mwaluma et al. 1993, Oldewage 1995, Mwaluma et al. 2003, Osore 1992, 1994, Osore et al. 1997, Kasyi 1994, Kitheka et al. 1996, Fondo 2003.	Zooplankton distribution and diversity, zooplankton, composition, distribution and abundance, zooplankton community structure and seasonal variation, zooplankton annual cycle, zooplankton distribution and effects cf monsoons, copedods species distribution and description, appendicularians distribution and description, early life-history stages of fish and prawns, rainfall effects and tidal rhythms, nutrient levels, phytoplankton biomass and zooplankton composition, biomass and abundance, water-circulation dynamics, water column nutrients and plankton productivity, Zooplankton from mangroves, seagrass and coral reef systems
Plankton from Mozambique	Leal et al. 2009, Raj et al. 2010, Sá et al. 2013, Lamon et al. 2014, Barlow et al. 2014, Barlow et al. 2017, Kelchner 2020.	Phytoplankton, nanoplankton and picoplankton: diatoms, dinoflagellates and occolithophores
	Huggett 2007, Leal et al. 2009, Huggett 2014, Johnsen et al. 2007, Tew Kai and Marsac 2008, 2009, 2010, Lebourges-Dhaussy et al. 2009, Huggett 2014, Krakstad et al. 2015, Groeneveld et al. 2017, Kelchner 2020.	Zooplankton. Holoplankton. Microzooplankton, mesozooplankton and macrozooplankton and meroplankton

Table 1 contd. ...

...Table 1 contd.

Plankton from Mauritius	Bhikajee 2003, Lim and Tan 1981.	The marine biodiversity of mauritius, larval development of Brachyura
Plankton from Ivory Coast	Boden 1961, Amosse 1970, Carmouze and Caumette 1985, Reyssac 1970, Bainbridge 1972, Binet 1977, Yacouba and Nestor 2003, Etilé et al. 2009, Lebrato and Jones 2009, Ndour et al. 2018.	Marine diatoms, organic pollution and biomasses of phytoplancton and heterotrophs bacteria, phytoplankton and primary production, zooplankton diversity and distribution, Euphausiace diversity and distribution, mass deposition event of Pyrosoma atlanicum carcasses, marine biodiversity in Côte d'Ivoire, spatiotemporal variations of the zooplankton abundance and composition
Plankton from Sierra Leone	Leigh 1973, Findlay 1978, Aleem 1979, Longhurst 1983, Conteh 2001, Ndomahina 2002, John 2003, Lamin 2011.	Studies on the Zooplankton of estuaries, marine biology of Sierra Leone, marine microplankton, benthic-pelagic coupling and export of organic carbon, plankton composition, abundance and diversity around the sewage outfall discharge, assessment of the coastal and marine biodiversity, fish larvae from the Dampier seamount and Sierra Leone rise, plankton distribution and abundance in relation to some environmental factors and fish availability on the continental shelf
Plankton from Tanzania	Okera 1974, Bryceson 1977, Mgaya 2003.	Zooplankton diversity and distribution, phytoplankton diversity and distribution. Marine biodiversity in Tanzania
Plankton from Cameroon	Folack 1988, 1989, Gabche 2003.	Marine biodiversity in Cameroon, phytoplankton diversity and distribution, Phytoplankton and Chlorophyll a
Plankton from Nigeria	Nwankwo 1991, Nwankwo 1997, Kadiri 1999, Kadiri 2006a, Kadiri 2006b, Davies 2009, Kadiri 2011, Abdul et al. 2016.	Phytoplankton distribution, Phytoplankton and physicochemical attributes, harmful algae, spatiotemporal distribution, abundance and species composition of zooplankton, dinoflagellates composition, effects of environmental parameters on zooplankton assemblages in coastal estuaries

Plankton from Mauritania	John et al. 1991, John and Zelck 1997, Bory et al. 2001, Helmke et al. 2005, Bouimetarhan et al. 2009, Romero et al. 1999, Romero et al. 2002, 2005, 2008, 2009a, 2009b, 2010, 2015, 2016, 2019, 2020, Romero and Fischer 2017.	Distribution and drift of horse mackerel larvae (genus Trachurus), features, boundaries and connecting mechanisms of the Mauritanian province exemplified by oceanic fish larvae particle fluxes, productivity regimes and upwelling, northwest African upwelling and its effect on offshore organic carbon export to the deep sea, Dinoflagellate cyst distribution regimes and upwelling, marine diatoms as indicators, diatom species composition and upwelling, marine planktonic diatoms production, siliceous phytoplankton fluxes, biogeography of major diatom taxa, submillennial-to-millennial variability of diatom production, effects of hydrographic and climatic forcing on diatom production and export, seasonal and interannual dynamics in diatom production, eddies as trigger for diatom productivity, fluxes of diatoms, coccolithophorids, calcareous and organic-walled dinoflagellate cysts, planktonic foraminifera and pteropods and the species-specific composition, flux variability of phyto- and zooplankton communities and upwelling
Plankton from Angola (Benguela current)	Silva 1953a, b, 1955, 1958, 2003, 2004, Silva and Boersma 1977, Caron 1978, Packard 1979, Chapman and Shannon 1985, Stuart 1986, Borchers and Hutchings 1986, Armstrong et al. 1987, Stuart and Pillar 1988, Verheye and Hutchings 1988, Olivar and Barange 1990, Barange 1990, Bailey 1991, Hutchings et al. 1991, John et al. 1991, Hutchings et al. 1991, Brown et al. 1991, Verheye 1991, Pagès and Gili 1991, Barange et al. 1991, 1992, Barange and Stuart 1991, OberhäNsli et al. 1992, Barange and Pillar 1992, Pillar et al. 1992, Timonin et al. 1992, Verheye et al. 1991, 1992, 1998, 2001, Timonin et al. 1992, Chapman et al. 1994, Timonin 1995, Shannon and	Diatom of marine plancton of Angola, "Red water" by *Exuviella baltica* L. with simultanious mortality of fishes in coastal waters, dinoflagelate of Angola's marine plancton, Angola's marine microplanton, Cretaceous Planktonic Foraminifers DSDP Leg 39 (South Atlantic), Cretaceous planktonic foraminifers from dsdp leg 40, Southeastern Atlantic Ocean, respiration and respiratory electron transport activity in plankton from the northwest African upwelling area, the Benguela ecosystem: chemistry and related processes, feeding and metabolism of Euphausia lucens (Euphausiacea) in the southern Benguela current, starvation tolerance, development time and egg production of Calanoides carinatus in the southern Benguela Current, short-term variability during an anchor station study in the southern Benguela upwelling system. Abundance, distribution and estimated production of mesozooplankton with special reference to Calanoides carinatus, horizontal and vertical distribution of zooplankton biomass in the southern Benguela Diatom of marine plancton of Angola, "Red water" by *Exuviella baltica* L. with simultanious mortality of fishes in coastal waters, dinoflagelate of Angola's marine plancton, Angola's marine microplanton, Cretaceous Planktonic Foraminifers DSDP Leg 39 (South Atlantic), Cretaceous planktonic foraminifers from dsdp leg 40, Southeastern Atlantic Ocean, respiration and respiratory electron transport activity in plankton from the northwest African upwelling area, the Benguela ecosystem: chemistry and related processes, feeding and metabolism of Euphausia lucens (Euphausiacea) in the southern Benguela current, starvation tolerance, development time and egg production of Calanoides carinatus in the southern Benguela Current, short-term variability during an anchor station study in the southern Benguela upwelling system. Abundance, distribution and estimated production of mesozooplankton with special reference to Calanoides carinatus, horizontal and vertical distribution of zooplankton biomass in the southern Benguela ecosystem,

Table 1 contd. ...

...Table 1 contd.

	Nelson 1996, Gibbons and Hutchings 1996, Gibbons 1997, Richardson and Verheye 1999, Silva et al. 2000, Boyer et al. 2000, Lass et al. 2000, O'Toole et al. 2001, Ekau et al. 2001, Mayfield et al. 2001, Shannon et al. 2003, Shannon et al. 2003, John et al. 2004, Shannon et al. 2004a, b, Heymans et al. 2004, Ekau and Verheye 2005, Verheye and Ekau 2005, Wasmund et al. 2005, Hagen et al. 2005, Verheye et al. 2005, Moloney et al. 2005, Shannon et al. 2006, Shannon et al. 2006, Fawcett et al. 2007, Rangel and Silva 2007, Auel and Verheye 2007, Shannon et al. 2008, Mohrholz et al. 2008, Heymans et al. 2009, Huggett et al. 2009, Hutchings et al. 2009; Werner et al. 2012, Blanco et al. 2013, Huenerlage and Buchholz 2013, Schukat et al. 2013, Osma et al. 2014, Schukat et al. 2014, Fernández-Urruzola et al. 2014, Verheye 2016, Louw et al. 2016, Louw et al. 2017, Marshall et al. 2022.	vertical migration and habitat partitioning of six euphausiid species in the northern Benguela upwelling system, microplankton ETS measurements as a means of assessing respiration in the Benguela ecosystem, hydrographic and current measurements in the area of the Angola-Benguela front, organic carbon flux and development of oxygen deficiency on the modern Benguela continential shelf, cross-front hydrography and fish larval distribution at the Angola-Benguela frontal zone, effects of large-scale advective processes on gelatinous zooplankton populations in the northern Benguela ecosystem, variability of zooplankton in the region of the Angola-Benguela front, phytoplaton community on the Angolan coast, zooplankton dynamics in the northern Benguela ecosystem, with special reference to the copepod Calanoides carinatus, life history and population maintenance strategies of Calanoides carinatus (Copepoda: Calanoida) in the southern Benguela ecosystem, zooplankton dynamics in the northern Benguela ecosystem, with special reference to the copepod Calanoides carinatus, southwestern Africa Northern Benguela current region, long-term trends in the abundance and community structure of coastal zooplankton in the southern Benguela system, maintenance mechanisms of plankton populations in frontal zones in the Benguela and Angola Current systems, bio-geographic zooplankton patterns in the vicinity of the Angola–Benguela-Frontal-Zone, sustainability of the Benguela, mesozooplankton dynamics in the Benguela ecosystem, with emphasis on the herbivorous copepods, the Benguela; large scale features and processes and system variability, influence of oceanographic fronts and low oxygen on the distribution of ichthyoplankton in the Benguela and southern Angola currents, winter ichthyoplankton in the northern Benguela upwelling and Angola-Benguela ront regions, community structure and trophic ecology of euphausiids in the Benguela ecosystem, diet and feeding of Euphausia hanseni and Nematoscelis megalops (Euphausiacea) in the northern Benguela current, cross-shelf circulation, growth rates of copepods in the southern Benguela upwelling system: the interplay between body size and food, Integrated management of the Benguela current region. A framework for future development, zonation and maintenance mechanisms of Nyctiphanes capensis and Euphausia hanseni (Euphausiacea) in the northern Benguela upwelling system, distribution patterns, abundance and population dynamics of the euphausiids Nyctiphanes capensis and Euphausia hanseni in the northern Benguela upwelling system, major pelagic borders of the Benguela upwelling system according to euphausiid species distribution, dynamics of diurnal distribution of plankton of the northern part of Benguela upwelling system, chlorophyll and phytoplankton primary production in relation to hydrographic structures in the Angola and Benguela current system,

		changes in the northern Benguela ecosystem over three decades: 1970s, 1980s and 1990s, trophic flows in the southern Benguela during the 1980s and 1990s; the seasonal variability of the northern benguela undercurrent and its relation to the oxygen budget, ecosystem states, ecosystem changes and changes in ecosystem functioning of the southern and northern Benguela ecosystem, ecosystem approach to fisheries management in the southern Benguela, modelling stock dynamics in the southern Benguela ecosystem for the period 1978–2002, comparing the Benguela and Humboldt marine upwelling ecosystems with indicators derived from inter-calibrated models, fish eggs and larvae collected in the region of the northern Benguela and Angola currents, metabolic adaptations and reduced respiration of the copepod Calanoides carinatus during diapause at depth in the Angola-Benguela front and northern Benguela upwelling regions, oceanographic and faunistic structures across an Angola Current intrusion into northern Namibian waters, thermal constraints on the respiration and excretion rates of krill, Euphausia hanseni and Nematoscelis megalops, in the northern Benguela upwelling system off Namibia, Pseudo-nitzschia spp. and Prorocentrum micans blooms in Luanda Bay, Angola, shellfish poisoning (ASP) toxins in plankton and molluscs, Benguela: Predicting a Large Marine Ecosystem, Policy options for the northern Benguela ecosystem using a multispecies, multifleet ecosystem model, Hypoxia tolerance in the copepod Calanoides carinatus and the effect of an intermediate oxygen minimum layer on copepod vertical distribution in the northern Benguela Current upwelling system and the Angola-Benguela Front, Copepod biomass, size composition and production in the southern Benguela; spatiotemporal patterns of variation, and comparison with other eastern boundary upwelling systems, contrasting wind patterns and toxigenic phytoplankton in the southern Benguela upwelling system, krill of the northern Benguela Current and the Angola-Benguela frontal zone compared: physiological performance and short-term; the Benguela Current: an ecosystem of four components, distribution of zooplankton biomass and potential metabolic activities across the northern Benguela upwelling system, complex trophic interactions of calanoid copepods in the Benguela upwelling system, comparing internal and external drivers in the southern Benguela and the southern and northern Humboldt upwelling ecosystems, energetics and carbon budgets of dominant calanoid copepods in the northern Benguela upwelling system, starvation in Euphausia hanseni water-body preferences of dominant calanoid copepod species in the Angola-Benguela frontal zone, life strategies, energetics and growth characteristics of Calanoides carinatus (Copepoda) in the Angola-Benguela frontal region, short-term patterns of vertical particle flux in northern Benguela: a comparison between sinking POC and respiratory carbon consumption, plankton productivity of the Benguela Current large marine ecosystem, seasonal and inter-annual phytoplankton dynamics and forcing mechanisms, annual patterns, distribution and long-term trends of Pseudo-nitzschia species in the northern Benguela upwelling system, the Angola Gyre is a hotspot of dinitrogen fixation in the South Atlantic Ocean

Table 1 contd. ...

...Table 1 contd.

Plankton from Namibia	Kruger 1980, Belyanina and Stejker 1988, Kosobokova et al. 1988, Olivar and Barange 1989, Olivar et al. 1991, Pillar et al. 1991, Karaseva and Shiganov 1993, Carola 1994, Brierley et al. 2001, Voges 2003, Postel et al. 2007, Hansen et al. 2014, Benavides et al. 2014.	Phytoplankton composition and distribution, Ichthyoplankton species composition, physiological and biochemical characteristics of Calanoides carinatus in waters on an upwelling, ichthyoplankton distribution and the Benguela upwelling, vertical distribution of fish eggs and larvae, marine planktonic copepoda, euphausids structure and trophic ecology, zooplankton biomass variability off Angola and Namibia investigated by a lowered ADCP and net sampling, acoustic observations of jellyfish in the Namibian Benguela, marine biodiversity in Namibia, a succession of micro- and nanoplankton groups in aging upwelled water, microbial uptake and regeneration of inorganic nitrogen off the coastal Namibian upwelling system
Plankton from South Africa	Hill 1966, Grindley 1977, 1981, Wooldridge and Melville-Smith 1979, Wooldridge and Melville-Smith 1979, Bremner 1980, Kruger 1980, Coetzee 1981, 1985, Wooldridge and Bailey 1982, Whitfield 1985, Webb and Wooldridge 1990, Jerling and Wooldridge 1991, 1995a, 1995b, Schlacher and Wooldridge 1996, 1998, Gibbons and Hutchings 1996, Timonin 1997, Pakhomov and Perissinotto 1997, Wooldridge 1999, 2000, 2005, Perissinotto et al. 2000, Wooldridge and Callahan 2000, Froneman 2001, Cooper 2001, Nozais et al. 2001, Kibirige, 2002, Kibirige and Perissinotto 2003, Griffiths 2003, 2004, Froneman and Vorwerk 2003, Kibirige and	A contribution to the ecology of the Umlalazi Estuary, zooplankton dversity and distribution, estuarine zooplankton community structure and dynamics, marine biodiversity in South Africa, estuarine plankton, feeding ecology of the mysids, Copepod succession in two South African estuaries, zooplankton distribution in relation to environmental conditions in the Swartvlei system, southern Cape, zooplankton and some environmental conditions in the Bot River estuary, estuarine zooplankton community structure and biomass, predation impacts of fishes on estuarine zooplankton, population dynamics and production of calanoid copepods, physical parameters of the diatomaceous mud belt, checklist of marine phytoplankton, diurnal vertical migrations of zooplankton in the upwelling area off the western coast of South Africa, zooplankton diversity and community structure around southern Africa, with special attention to the Benguela upwelling system, Trophic pathways at the cross-roads (is detritus the pivotal point in all estuarine food webs?). Geomorphological variability among microtidal estuaries from the wave dominated South African coast, the structure and trophic role of the zooplankton community of the Mpenjati Estuary, plankton distribution and abundance, feeding ecology of mysid species, diel horizontal migration of Mesopodopsis slabbery, mesozooplankton community structure and grazing impact in the region of the subtropical convergence, relationships between zoo- and phytoplankton in a warm-temperate semi-permanently closed estuary, zooplankton in the feeding ecology of estuarine fishes, freshwater inputs and estuarine zooplankton, annual cycle of microalgal biomass in a South African temporarily-open estuary: nutrient versus light limitation, ecological responses to reductions in freshwater supply and quality in South Africa's estuaries: lessons for management and conservation; The zooplankton community of the Mpenjati Estuary, the effects of a single freshwater release into the Kromme Estuary,

Perissinotto 2003, Stretch and Zietsman 2004, Thomas et al. 2005, Kibirige et al. 2006, Perissinotto et al. 2007, Montoya-Maya and Strydom 2009, Wooldridge and Deyzel 2009, Carrasco et al. 2007, 2013, Whitfield et al. 2012, Deale et al. 2013, Whitfield and Baliwe 2013, Schukat et al. 2013, Jones et al. 2015, Jones et al. 2016, Tagliarolo and Scharler 2018, Lemley et al. 2018, Tagliarolo et al. 2019.	a comparative study of zooplankton dynamics in two subtropical temporarily open/closed estuaries, a South African temporarily open/closed system, Copepods succession in estuarine environment, phytoplankton biomass and size structure in two South African eutrophic, temporarily open/closed estuaries, the hydrodynamics of Mhlanga and Mdloti Estuaries: flows, residence times, water levels and mouth dynamics, energetics and carbon budgets of dominant calanoid copepods in the northern Benguela upwelling system, effects of silt loading on the feeding and mortality of the mysid Mesopodopsis africanain the St. Lucia Estuary, zooplankton composition, abundance and distribution in selected south and west coast estuaries, A review of the ecology and management of temporarily open/closed estuaries in South Africa, with particular emphasis on river flow and mouth state as primary drivers of these systems, temperature and salinity as abiotic drivers of zooplankton community dynamics in the Great Berg Estuary, grazing by Pyrosoma atlanticum (Tunicata, Thaliacea) in the south Indian Ocean, a century of science in South African estuaries; bibliography and review of research trends, Turbidity effects on feeding and mortality of the copepod Acartiella natalensis in the St Lucia Estuary, recovery dynamics of zooplankton following mouth-breaching in the temporarily open/closed Mdloti Estuary, turbidity effects on the feeding, respiration and mortality of the copepod Pseudodiaptomus stuhlmanni in the St Lucia Estuary, impact of a flood event on the zooplankton of an estuarine lake, spatial and temporal variability of carbon budgets of shallow South African subtropical estuaries, triggers of phytoplankton bloom dynamics in permanently eutrophic waters of a South African estuary triggers of phytoplankton bloom dynamics in permanently eutrophic waters of a South African estuary, zooplankton metabolism in South African estuaries; does habitat type influence ecological strategies?

studied the relationship between plankton and fisheries analyzing various estimates of primary productivity in several African countries. A review of planktonic studies in African countries is given in Table 1.

4. Plankton Research and Biodiversity and Sustainability of Marine Resources: Challenges in the African Context

Biodiversity and sustainability of biological resources in the African context, in a climate change scenario, are crucial priority goals and challenges toward the Millennium Development Goals (UNEP-WCMC 2016) and for achieving sustainable and efficient consumption and production patterns for the development of African populations (CBD/COP 2018, African Union 2015a, UNEP-WCMC 2016, Globe Afrique 2016, IBPES 2018, Meyers 2019, Okonjo-Iweala 2020, UNEP 2020). In order to contribute to the maintenance of marine biodiversity, sustainable consumption and biological production, it is necessary to understand the fluctuations and sustainable use of marine stocks, the proper management of fisheries and the conservation of species and ecosystems (IOC-UNESCO 2020). Marine biodiversity conservation and management is fundamental for maintaining interactions and functions on all levels from individuals to ecosystem, including changes related to natural and anthropogenic environmental pressures (Beaugrand et al. 2014, 2019, Chiba et al. 2018, McQuatters-Gollop et al. 2019) and also contribute to a better understanding how climate change and technological economic growth is affecting people's ecosystems and livelihoods, to enhance a more holistic understanding of people's needs and their relationship to climate change in agriculture, food security in fisheries are crucial (Schlüter et al. 2017, 2019, Lindkvist et al. 2020). In this context, these challenges require management and conservation commitments and plans that involve the different dimensions of the biodiversity management process, including the scientific, political and social components (UN 2015, Africa Sustainable Development Report 2018).

These global challenges have led to several global and regional programs and plans, involving different countries, institutions and stakeholders (Agenda 2063 of the African Union, the Sustainable Development Goals, the Africa Regional Roadmap, the Aichi Biodiversity Targets and other globally agreed goals) in order to develop research projects and capacity building to implement inter-regional governance policies and actions (African Union 2015a, 2015b, IPBES 2018). The African 10-Year Framework Programme on SCP The African 10-Year Framework Programme on Sustainable Consumption and Production (African-10YFP) has been development with the collaboration of UNEP and UN-DESA in close consultation with the Secretariats of the African Ministerial Conference on Environment (AMCEN) and the Secretariat of the African Roundtable on Sustainable Consumption and Production (ARSCP) which have been established and supported by UNEP (African Union 2015a, 2015b, IBPES 2018). This program is one of the main activities identified in the context of African-10YFP regional monitoring to support African countries in developing their sustainable consumption and production programs (Regional African Roadmap for the 10YFP Report 2014, Africa Sustainable Development

Report 2018). African countries have also developed sharing plans and programs to improve knowledge of African water spaces and identify transboundary concerns, the "Planning and elaboration of Transboundary Diagnostic Analysis, Development" program and the "Strategic Action Programs" to identify environmental issues and problems and promote scientific and technical assessment (African Union 2015a, 2015b, Africa Sustainable Development Report 2018, IPBES 2018). Most African countries currently participate in some global conventions and agreements that encourage regional cooperation and coordination among countries sharing common resources along their coastlines, such as the four African Regional Seas Conventions and Action Plans (RSCAPs) including the Barcelona Convention (Mediterranean), the Abidjan Convention (West Africa), the Nairobi Convention (Eastern Africa and the Island States) and PERSGA (Red Sea and Gulf of Aden) (African Union 2015a, 2015b, IPBES 2018). Recently, in a more global action, the United Nations platform "Transforming our world: Sustainable Development Agenda 2030" based on 17 Sustainable Development Goals (SDGs) constitute a common vision for humanity and a contract between world leaders and peoples (UN 2015). This resolution stemmed from the Millennium Development Goals between 2000 and 2015, which aimed to combat poverty and ensure basic living conditions for populations in developing countries through measures of Sustainable Consumption and Production and Resource Efficiency to contribute to progress toward Sustainable Development Goals (SDGs) (UN 2015).

Monitoring of coastal and marine plankton communities has been taking place in African countries for the past 100 years constituted comprehensive studies, on a regional and global scale, to characterize marine plankton biodiversity, interpret the processes by which planktonic organisms interact with the diverse set of scales of environmental and biological components of marine ecosystem and responses to anthropogenic and climate change pressures. These works were carried through diverse local, regional and major international projects which constituted a global network of regional studies for the construction of extensive plankton databases and monitoring. Current challenges of globalized markets and production processes, the anthropogenic pressures and climate change, increased demands and competitiveness for marine resources and a more global economy, require in the African context, a converge to a broader vision of efficient use of biological resources, ensuring the resilience of natural ecosystems and sustainable development in order to maintain marine environment sustainability in African countries and their communities (Lamere 2013, Globe Afrique 2016, Africa Sustainable Development Report 2018, IPBES 2028, Meyers 2019, Okonjo-Iweala 2020, UNEP 2020). The combined effects of anthropogenic pollution and climate change have been significantly reducing biological production, especially in oceans, with severe degradation of marine habitats and consequences to ecosystem functioning (Beaugrand 2014, 2019). These questions are currently a global problem in marine ecosystems worldwide (Muller-Karger et al. 2018, Lombard et al. 2019, OECD 2021b) and global marine research strategies and policies, in the medium and long term, are focused on the marine and coastal areas living resources conservation and management in order to achieve sustainable development (Chiba et al. 2018, McQuatters-Gollop et al. 2015, Tweddle et al. 2018). In the fulfilment of Aichi Target 15, the United Nation's

Sustainable Development Goals, delivering the Essential Ocean Variables and Essential Biodiversity Variables that the Global Ocean Observing System and Group on Earth Observation's Biodiversity Observation Network, strategic targets have been defined related to planktonic research to advancing understanding of plankton dynamics and informing policy and management decisions (African Union 2015b, UNEP-WCMC 2016, CBD/SBSTTA 2017, UNEP and International Trade Center 2017, CBD/COP 2018, Miloslavich et al. 2018). This context requires diverse information on the distribution and behavior of planktonic organisms' abundance and distribution and how environmental factors interact to control the population (Beaugrand 2014, 2019, Bar-On and Milo 2019, Righetti et al. 2019, Beaugrand et al. 2019). Plankton plays a key role in food webs in the nutrient cycling through the excretion of various forms of nitrogen and phosphorus that represents the link that transmits energy synthesized by phytoplankton to consumers at higher trophic levels (Zhang et al. 2019, McQuatters-Gollop et al. 2019). The dynamic character of planktonic communities gives a rapid response to the physico-chemical changes of the marine environment and brings profound structural changes at all trophic levels of the ecosystem (Canu et al. 2015, Beaugrand 2014, 2019). Synergies with global networks exploiting satellite, filed data and other plankton sensors could be explored, realizing the survey's capacity to validate earth observation data and to ground-truth emerging plankton observing platforms that are required for a fully integrated ocean observing system, at different scales, in different regions of the world, that can understand global ocean dynamics to inform sustainable marine decision-making (Chiba et al. 2018, Muller-Karger et al. 2018, Batten et al. 2019, Lombard et al. 2019, OECD 2021b).

Considering the emerging challenges of the 21st century, marine research is extremely important for sustainable development strategies and policies (CBD/SBSTTA 2017), in order to make interfaces with other domains of knowledge and to create new interdisciplinary and transdisciplinary fields for the conservation and management of medium and long-term marine living resources and coastal areas (Chiba et al. 2018, McQuatters-Gollop et al. 2015, Batten et al. 2019, Lombard et al. 2019, OECD 2021b). Research strategies may be focused on structural studies, such as the role of plankton in the global functioning of marine and estuarine ecosystems, production in the oceans and estuaries, variability in recruitment of fish species and also through monitoring water bodies (Ibarbalz et al. 2019, United Nations 2017). On the other hand, knowledge of biological factors such as food availability, predation and competition, growth rates, mortality, behavior, histology and histochemistry of ecophysiological processes (Batten et al. 2019, Lombard et al. 2019), constitute also central themes for obtaining information that allows the development of models on the reproductive potential of plankton associated with physical-chemical factors and extrinsic biological factors (Beaugrand et al. 2019). This knowledge is essential for the context of plankton ecology, and its resilience to multiple stress factors, pollution and global climate change (Racault et al. 2014, Beaugrand et al. 2019).

References

Abdel-Aziz, N. and Aboul-Ezz, S. 2003. Zooplankton community of the egyptian mediterranean coast. Egypt J. Aquat. Biol. Fish 7: 91–108.

Abdul, W.O., Adekoya, E.O., Ademolu, K.O., Omoniyi, I.T., Odulate, D.O., Akindokun, T.E. et al. 2016. The effects of environmental parameters on zooplankton assemblages in tropical coastal estuary, south-west, Nigeria. Egypt. J. Aquat. Res. 42(3): 281–287.

Africa Sustainable Development Report. 2018. Towards a transformed and resilient continent. African union, 2014. Progress Report of the Commission on the Africa 2063 Agenda. https://au.int/sites/default/files/newsevents/.../12582-wd-agenda_2063_e_0.p.

African Union. 2015a. Economic commission for africa; african development bank and united nations development programme. 135 pp.

African Union. 2015b. Agenda 2063: The africa we want. Addis ababa, ethiopia: african union commission. Retrieved from http://www.un.org/en/africa/osaa/pdf/au/agenda2063.pdf.

Alamo, M.A.F., Martin, F.H., Tejera, E. and Leon, M.E. 2004. poliquetos pelagicos de las islas de cabo verde. resultados de la campana TFMCBM/98, Proyecto Macaronesia 2000. Rev. Acad. Canar. Ciena, XV (Nums. 3-4): 87–97.

Aleem, A.A. 1979. Marine microplankton from sierra leone, west Africa. Ind. J. Mar. Sc. 8: 291–295.

Amazzal, A., Ait-talborjt, E., Hermas, J. and Hafidi, N. 2020. Importance of hydrological parameters in the distribution of planktonic eggs and larvae in an upwelling zone (Imessouane Bay, Moroccan Atlantic Coast). Caspian J. Environ. Sci. 18: 1–12.

Amosse, A. 1970. Diatomées marines et saum âtres du sénégal et de la côte d'ivoire. Bull. de l'IFAN., T.23, sér. A, n°2: 289–311 + 3 pl.

An, C. 1971. Atlantic ocean phytoplankton south of the gulf of guinea on profiles along 11- and 14-degrees. S. Oceanology 6: 896–901.

Anabalón, V., Arístegui, J., Morales, C.E., Andrade, I., Benavides, M., Correa-Ramírez, M.A. et al. 2014. The structure of planktonic communities under variable coastal upwelling conditions off cape ghir (31°N) in the Canary Current System (NW Africa). Progr. Oceanography 120: 320–339.

Anang, E.R. 1979. The seasonal cycle of the phytoplankton in the coastal waters of ghana. Hydrobiologia 62(1): 33–45.

Andersen, V., Sardou, J. and Gasser, B. 1997. Macroplankton and micronekton in the northeast tropical Atlantic: abundance, community composition and vertical distribution in relation to different trophic environments, Deep-Sea Res. Pt. I 44: 193–222.

Anhichem, M., Dellal, M., Chfiri, R., Yahyaoui, A. and Benbrahim, S. 2017. Etude de la qualité des eaux et des sédiments de la baie de dakhla au maroc. Qualité physico-chimique et contamination métallique, 142, Bulletin de la Société zoologique de France 185–202.

Aouititen, M., Bekkali, R., Nachite, D., Luan, X. and Mrhraoui, M. 2019. Predicting jellyfish strandings in the Moroccan North-West Mediterranean Coastline. European Scientific Journal ESJ 15: 72–72.

Arkhipov, A.G., Mamedov, A.A., Simonova, T.A. and Shnar, V.N. 2015. Species composition and features of ichthyoplankton distribution in the waters of senegal and guinea-bissau. J. Ichthyol. 55: 346–354.

Armstrong, D.A., Mitchell-Innes, B.A., Verheye-Dua, F., Waldron, H. and Hutchings, L. 1987. Physical and biological features across an upwelling front in the southern Benguela. *In*: Payne, A.I.L., Gulland, J.A. and Brink, K.H. (eds.). The Benguela and Comparable Ecosystems. S. Afr. J. Mar. Sci. 5: 171–190.

Auel, H. and Verheye, H.M. 2007. Hypoxia tolerance in the copepod calanoides carinatus and the effect of an intermediate oxygen minimum layer on copepod vertical distribution in the northern Benguela Current upwelling system and the Angola–Benguela. Front, J. Exp. Mar. Biol. Ecol. 352: 234–243.

Badsi, H., Ali, H.O., Loudiki, M., Hafa, M.E., Chakli, R. and Aamiri, A. 2010. Ecological factors affecting the distribution of zooplankton community in the massa lagoon (Southern Morocco). Afr. J. Environ. Sci. Technol. 4(11): 751–762.

Badsi, H., Ali, H.O., Loudiki, M. and Aamiri, A. 2012. Phytoplankton diversity and community composition along the salinity gradient of the massa estuary. Am. J. Hum. Ecol. 1: 58–64.

Baidya, A.U. and Choudhury, A. 1984. Distribution and abundance of zooplankton in a tidal creek of Sagar Island, Sundarbans, West Bengal. Envir. and Ecol. 2(4).

Bailey, G.W. 1991. Organic carbon flux and development of oxygen deficiency on the modern Benguela continental shelf south of 22° S: Spatial and temporal variability. pp. 171–183. *In*: Tyson, R.V. and Pearson, T.H. (eds.). Modern and Ancient Continental Shelf Anoxia. London: Geological Society.

Bainbridge, V. 1972. The zooplankton of the gulf of guinea. Bulletin of Marine Ecology 8: 61–97. Modern and Ancient Continental Shelf Anoxia. Geol. Soc. Spec. Publ. 58: 171–183.

Barlow, R., Lamont, T., Gibberd, M.J., Airs, R., Jacobs, L. and Britz, K. 2017. Phytoplankton communities and acclimation in a cyclonic eddy in the southwest Indian Ocean. Deep-Sea Res Pt I 124: 18–30.

Barlow, R., Lamont, T., Morris, T., Sessions, H. and van den Berg, M. 2014. Adaptation of phytoplankton communities to mesoscale eddies in the mozambique channel. Deep-Sea Res Pt Ii 100: 106–118.

Barlow, J., França, F., Gardner, T., Hicks, C., Lennox, G.D., Berenguer, E. et al. 2018. The future of hyperdiverse tropical ecosystems. Nature 559: 517–526. doi:10.1038/s41586-018-0301-1.

Barange, M. 1990. Vertical migration and habitat partitioning of six euphausiid species in the northern benguela upwelling system. J. Plankton Res. 12: 1223–1237.

Barange, M. and Stuart, V. 1991. Distribution patterns, abundance and population dynamics of the euphausiids nyctiphanes capensis and euphausia hanseni in the northern Benguela upwelling system. Mar. Biol. 109: 93–101.

Barange, M., Gibbons, M. and Carola, M. 1991. Diet and feeding of euphausia hanseni and nematoscelis megalops (euphausiacea) in the northern benguela current: ecological significance of vertical space partitioning. Mar. Ecol. Prog. Ser. 73: 173–181.

Barange, M., Pillar, S.C. and Hutchings, L. 1992. Major pelagic borders of the benguela upwelling system according to euphausiid species distribution. *In*: Payne, A.I.L., Brink, K.H., Mann, K.H. and Hilborn, R. (eds.). Benguela Trophic Functioning. S. Afr. J. Mar. Sci. 12: 3–17.

Barange, M. and Pillar, S. 1992. Cross-shelf circulation, zonation and maintenance mechanisms of nyctiphanes capensis and euphausia hanseni (euphausiacea) in the northern benguela upwelling system. Cont. Shelf Res. 12: 1027–1042.

Bar-On, Y.M. and R. Milo. 2019. The biomass composition of the oceans: a blueprint of our blue planet. Cell 179(7): 1,451–1,454.

Batten, S.D., Abu-Alhaija, R., Chiba, S., Edwards, M., Graham, G., Jyothibabu, R. et al. 2019. A global plankton diversity monitoring program. Frontiers in Marine Science 6: 321. doi: 10.3389/fmars.2019.00321.

Bé, A.W.H. and Hutson, W.H. 1977. Ecology of planktonic foraminifera and biogeographic patterns of life and fossil assemblages in the Indian ocean. Micropaleontology 23: 369–414. 10.2307/1485406.

Beaugrand, G. 2014. Marine Biodiversity, Climatic Variability and Global Change. Routledge, London, 486 pp. https://doi.org/10.4324/9780203127483.

Beaugrand, G., Conversi, A., Atkinson, A., Cloern, J., Chiba, S., Fonda-Umani, S. et al. 2019. Prediction of unprecedented biological shifts in the global ocean. Nat. Clim. Change 9: 237–243.

Belyanina, T.N. and Stejker, T.N. 1988. Contribution to the studies of ichthyoplankton from the benguela upwelling. Oceanology (MOSC>) 28: 663–666.

Benavides, M., Santana, Y., Wasmund, N. and Arístegui, J. 2014. Microbial uptake and regeneration of inorganic nitrogen off the coastal namibian upwelling system. J. Mar. Syst. 140: 51–57.

Berraho, A. 2007. Relations spatialisées entre milieu et ichtyoplancton des petits pélagiques de la côte atlantique marocaine (zones central et sud) (Ph.D. thesis). Mohammed V Univ., Rabat, Morocco.

Berraho, A., Abdelouahab, H., Charib, S., Essarraj, S., Larissi, J., Abdellaoui, B. et al. 2016. Copepod community along the mediterranean coast of morocco (southwestern alboran Sea) during spring Medit. Mar. Sci. 17(3): 661–665. 10.12681/mms.1733.

Berraho, A., Abdelouahab, H., Larissi, J., Baibai, T., Charib, S., Idrissi, M. et al. 2019. Biodiversity and spatio-temporal variability of copepods community in dakhla bay (southern Moroccan coast). Reg. Stud. Mar. Sci. 28: 100437.

Bijoux, J.P., Adam, P-A., Alcindor, R., Bristol, R., Decommarmond, A., Mortimer, J.A. et al. 2003. Marine biodiversity in the seychelles archipelago – the known and the unknown, national Report. pp. 206–227. *In*: Cynthia Decker, Charles Griffiths, Kim Prochazka, Carmen Ras and Alan Whitfield (eds.). Marine Biodiversity in Sub-Saharan Africa: The Known and the Unknown. Proceedings of the Marine Biodiversity in Sub-Saharan Africa: The Known and the Unknown, 310 pp.

Bhikajee, M. 2003. The marine biodiversity of mauritius, national report. pp. 188–205. *In*: Cynthia Decker, Charles Griffiths, Kim Prochazka, Carmen Ras and Alan Whitfield (eds.). Marine Biodiversity in

Sub-Saharan Africa: The Known and the Unknown. Proceedings of the Marine Biodiversity in Sub-Saharan Africa: The Known and the Unknown, 310 pp.

Binet, D. 1977. Grand traits de d'ecologie de principaux taxons du zooplankton ivoirien. Cahiers Orstom Series Oceanographie 15: 89–109.

Biney, 1990. A review of some characteristics of freshwater and coastal ecosystem in ghana. Hydrobiologia 208: 45–52.

Blanco, J., Livramento, F. and Menezes Rangel, I. 2013. Amnesic shellfish poisoning (ASP) toxins in plankton and molluscs from luanda bay, angola. Toxicon February–March 2010 55(2–3): 541–546.

Boden, B.P. 1961. Euphausiacea (crustacea) from tropical west Africa. pp. 251–262. In Atlantide Report No. 6. Danish Science Press Limited, Copenhagen.

Borchers, P. and Hutchings, L. 1986. Starvation tolerance, development time and egg production of Calanoides carinatus in the southern Benguela Current. J. Plankt. Res. 8: 855–874.

Borja, A. 2014. Grand challenges in marine ecosystems ecology. Front. Mar. Sci. 12: 1. doi: 10.3389/fmars.2014.00001.

Boucher, J. 1982. Peuplement de copépodes des upwellings côtiers nord-ouest africains. I. Composition faunistique et structure démographique. Oceanol. Acta 5(1): 49–62.

Bouimetarhan, I., Marret, F., Dupont, L. and Zonneveld, K.A.F. 2009. Dinoflagellate cyst distribution in marine surface sediments off west Africa (17–6°N) in relation to sea-surface conditions, 784 freshwater input and seasonal coastal upwelling. Mar. Micropaleontol. 71: 113–130.

Bory, A., Jeandel, C., Leblond, N., Vangriesheim, A., Khripounoff, A., Beaufort, L. et al. 2001. Downward particle fluxes within different productivity regimes off the mauritanian upwelling zone (EUMELI program). Deep-Sea Res. I 48: 2251–2282.

Boyer, D., Cole, J. and Bartoholomae, C. 2000. Southwestern Africa: Northern Benguela current region. Mar. Pollut. Bull. 41(1–6): 123–140.

Bremner, J.M. 1980. Physical parameters of the diatomaceous mud belt off south west Africa. Mar. Geol. 34: M67–M76 doi:10.1029/2011JC007479. Matching ASCAT and QuikSCAT winds. J. Geophys. Res. 117(C2): C02011.

Brown, P.C., Painting, S.J. and Cochrane, K.L. 1991. Estimates of phytoplankton and bacterial biomass and production in the northern and southern Benguela ecosystems. Afr. J. Mar. Sci. 11(1): 537–564.

Brierley, A.S., Axelsen, B.E., Buecher, E., Sparks, C.A.J., Boyer, H. and Gibbons, M. 2001. Acoustic observations of jellyfish in the namibian benguela. Marine Ecology. Progress Series 210: 55–66.

Bryceson, I. 1977. An ecological study of the phytoplankton off the coastal waters of dar es salaam. PhD thesis, University of Dar es Salaam. 560 pp.

CBD/SBSTTA/22/5. 2017. Updated scientific assessment of progress towards selected aichi biodiversity targets and options to accelerate progress. Retrieved from: https://www.cbd.int/doc/c/c75f/06b1/6fc465496044698feacc47ba/sbstta-22-05-en.pdf.

CBD/COP/14/9. 2018. Long-term strategic directions to the 2050 Vision for Biodiversity, approaches to living in harmony with nature and preparation for the post-2020 global biodiversity framework. Retrieved from: https://www.cbd.int/doc/c/0b54/1750/607267ea9109b52b750314a0/cop-14-09-en.pdf.

Chapman, P. and Shannon, L.V. 1985. The benguela ecosystem. 2. chemistry and related processes. pp. 183–251. In: Barnes, M. (ed.). Oceanography and Marine Biology. An Annual Review 23. Aberdeen; University Press.

Chapman, P., Mitchell-Innes, B.A. and Walker, R. 1994. Microplankton ETS measurements as a means of assessing respiration in the benguela ecosystem. Afr. J. Mar. Sci. 14: 297–312.

Carmouze, J.P. and Caumette, P. 1985. Les effets de la pollution organique sur les biomasses et activités du phytoplancton et des bactéries hétérotrophes dans la lagune Ebrié (Côte d'Ivoire). Rev. Hydrobiol. Trop. 18: 183–212.

Carola, M. 1994. Checklist of the marine planktonic copepoda of southern Africa and their worldwide geographic distribution. South African Journal of Marine Science 14: 225–253.

Carradec, Q., Pelletier, E., Da Silva, C., Alberti, A., Seeleuthner, Y., Blanc-Mathieu, R. et al. 2018. A global ocean atlas of eukaryotic genes. Nature Communications 9: 373.

Carrasco, N.K., Perissinotto, R. and Miranda, N.A. 2007. Effects of silt loading on the feeding and mortality of the mysid mesopodopsis africanain the St. Lucia Estuary, South Africa. J. Exp. Mar. Biol. Ecol. 352: 152–164.

Carrasco, N.K., Perissinotto, R. and Jones, S. 2013. Turbidity effects on feeding and mortality of the copepod acartiella natalensis (connell and grindley, 1974) in the St Lucia Estuary, South Africa. J. Exp. Mar. Biol. Ecol. 446: 45–51.

Caron, M. 1978. Cretaceous planktonic foraminifers from dsdp leg 40, southeastern atlantic ocean. Environmental Science.

Casanova, B., Casanova, J.P., Ducret, F. and Rampal, J. 1982. Biomasse et composition chimique et faunistique du zooplancton du secteur sénégambien (champagne CINECA de la «Thalassa», août 1973. Rapp. P. V. Réun. Const. Int. Explor. Mer. 189: 266–269.

Cepic, M., Bechtold, U. and Wilfing, H. 2022. Modelling human influences on biodiversity at a global scale– a human ecology perspective. Ecological Modelling 465: 109854.

Chiba, S., Batten, S., Martin, C.S., Ivory, S., Miloslavich, P. and Weatherdon, L.V. 2018. Zooplankton monitoring to contribute towards addressing global biodiversity conservation challenges. J. Plankton Res. 40: 509–518. doi: 10.1093/plankt/fby030.

Climate Smart Oceans. 2016. Toward sustainable growth of ocean economies un africa under a changing climate, http://climatesmartoceans.org/wp-content/uploads/2016/08/Policy-brief-basse-def-en.pdf.

Conteh, F.S. 2001. Plankton Composition, Abundance and Diversity Around the Sewage Outfall Discharge Point at Government Wharf. B.Sc. (Gen), USL, 12pp.

Conway, D.V., White, R.G., Hugues-Dit-Ciles, J., Gallienne, C.P. and Robins, D.B. 2003. Guide to the coastal and surface zooplankton of the south-western Indian ocean. Occasional Publication of the Marine Biological Association of the United Kingdom, No. 15, Plymouth, UK. ISSN 0260–2784.

Cooper, J.A.G. 2001. Geomorphological variability among microtidal estuaries from the wavedominated South African coast. Geomorphology 40: 99–122.

Coetzee, D.J. 1981. Zooplankton distribution in relation to environmental conditions in the Swartvlei system, southern cape. Journal of the Limnological Society of Southern Africa 7: 5–12.

Coetzee, D.J. 1985. Zooplankton and some environmental conditions in the bot river estuary. Transactions of the Royal Society of South Africa 45: 363–378.

Da Silva, A.J., Postel, L. and Postel, A. 2000. Bio-geographic zooplankton patterns in the vicinity of the angola–benguela-frontal-zone August/September 2000. African Journal of Marine Science.

Da Silva, A.J. 2004. Verteilung des zooplanktons im bereich der angola–benguela-frontal-zone und seine bedeutung für die ernährungswichtige Schildmakrele (*Trachurus* spp.) im August/September 2000. PhD thesis, University Rostock, Germany, 90 pp.

Davies, O.A. 2009. Spatio-temporal distribution, abundance and species composition of zooplankton of woji-okpoka creek, port harcourt, Nigeria. Res. J. Appl. Sci. Eng. Technol. 1(2): 14–34.

Deale, M., Perissinotto, R. and Carrasco, N.K. 2013. Recovery dynamics of zooplankton following mouth-breaching in the temporarily open/closed Mdloti Estuary, South Africa. Afr. J. Aquat Sci. 38: 67–78.

De Decker, A.H.B. 1984. Near-surface copepod distribution in the south-western Indian and south-eastern Atlantic Ocean. Ann. S. Afr. J. Mar. Sci. 12: 341–354.

Demarcq, H. and Somoue, L. 2015. Phytoplankton and primary productivity off northwest Africa. pp. 161–174. *In*: Valdés, L. and Déniz-González, I. (eds.). Oceanographic and Biological Features in the Canary Current Large Marine Ecosystem. IOC-UNESCO, Paris. IOC Technical Series, No. 115.

Demeulenaere, B. 1991. Marine Phytal Harpacticoida from Kenya: With Empasis on Seagrass Dwelling Species. MSc. Thesis in Fundamental and Applied Marine Ecology. 49 pp.

Denda, A. and Christiansen, B. 2014. Zooplankton distribution patterns at two seamounts in the subtropical and tropical NE Atlantic. Mar. Ecol. 35(2): 159–179.

de Vargas, C., Audic, S., Henry, N., Decelle, J., Mahé, F., Logares, R. et al. 2015. Eukaryotic plankton diversity in the sunlit ocean. Science 348(6237).

Dilmahamod, A.F. 2014. Links between The Seychelles-Chagos Thermocline Ridge and Large-Scale Climate Modes and Primary Productivity; and the Annual Cycle of Chlorophyll-a. Master's Thesis, University of Cape Town. http://hdl.handle.net/11427/921.

Dilmahamod, A.F., Hermes, J.C. and Reason, C.J.C. 2016. Chlorophyll-a variability in the seychelles-chagos thermocline ridge: analysis of a coupled biophysical model. J. Mar. Syst. 154: 220–232. 10.1016/j.jmarsys.2015.10.011.

Diop, S., Arthurton, R., Scheren, P., Kitheka, J., Koranteng, K. and Payet, R. 2011. The Coastal and marine environment of western and eastern africa: challenges to sustainable management and socioeconomic development. pp. 315–335. *In*: Wolanski, E. and McLusky, D.S. (eds.). Treatise on Estuarine and Coastal Science, Vol 11. Waltham: Academic Press.

Diouf, P.S. 1990. Le zooplancton du sénégal. In : Actes du groupe de travail sur "l'impact des fluctuations environnementales sur la dynamique des stocks pélagiques côtiers en Afrique de l'Ouest" organisé à Dakar du 12 au 17 décembre 1988, CRODT, 20 p.

Dolan, J.R., Claustre, H., Carlotti, F., Plounevez, S. and Moutin, T. 2002. Microzooplankton diversity: relationships of tintinnid ciliates with resources, competitors and predators from the Atlantic coast of morocco to the eastern Mediterranean. Deep-Sea Res. Pt. I 49: 1217–1232.

Ehler, C., Zaucha, J. and Gee, K. 2019. Maritime/marine spatial planning at the interface of research and practice. pp. 1–22. *In*: Zaucha, J. and Gee, K. (eds.). Maritime Spatial Planning, (Cham: Palgrave Macmillan).

Ekau, W., Hendricks, A., Kadler, S., Koch, V. and LOICK, N. 2001. Winter ichthyoplankton in the northern benguela upwelling and angola-benguela front regions. S. Afr. J. Sci. 97: 259–265.

Ekau, W. and Verheye, H.M. 2005. Influence of oceanographic fronts and low oxygen on the distribution of ichthyoplankton in the benguela and southern angola currents. Afr. J. mar. Sci. 27(3): 629–639.

Etilé, R.N., Kouassi, M.A., Aka, M.N., Pagano, M., N'douba, V. and Kouassi, N.J. 2009. Spatio-temporal variations of the zooplankton abundance and composition in a west African tropical coastal lagoon (Grand-Lahou, Côte d'Ivoire). Hydrobiologia 624(1): 171–189.

Elghrib, H., Somoue, L., Elkhiati, N., Berraho, A., Makaoui, A., Bourhim, N. et al. 2012. Distribution du phytoplancton dans les zones d'upwelling de la côte atlantique marocaine située entre les latitudes 32°30'N et 24°N. C. R. Biologie 335: 541–554.

El Khalki, A. and Moncef, M. 2007. Etude du peuplement de copépodes de l'estuaire de l'oum er rbia (côte atlantique du maroc): effets des marées et des lâchers de barrages. Leban. Sci. J. 8(1): 3–18.

Fawcett, A., Pitcher, G., Bernard, S., Cembella, A. and Kudela, R. 2007. Contrasting wind patterns and toxigenic phytoplankton in the southern Benguela upwelling system. Marine Ecology Progress Series 348: 19–31.

Fiedler, B., Grundle, D., Schütte, F., Karstensen, J., Löscher, C.R., Hauss, H. et al. 2016. Oxygen utilization and downward carbon flux in an oxygen-depleted eddy in the eastern tropical north Atlantic, Biogeosciences Discuss. doi:10.5194/bg-2016-23.

Fischer, G., Karstensen, J., Romero, O., Baumann, K.-H., Donner, B., Hefter, J. et al. 2016. Bathypelagic particle flux signatures from a suboxic eddy in the oligotrophic tropical north Atlantic: production, sedimentation and preservation. Biogeosciences Discuss. 12: 18253–18313.

Fenaux, R. 1973. Appendicularians from the Indian ocean, red Sea and the Persian Gulf. pp. 409–414. *In*: Zeitchel, B. and Gerlach, S.A. (eds.). The Biology of the Indian Ocean. Ecological Studies Vol. 3. Jacobs, J., Lange, O.L., Olson, J.S. and Wieser, W. (eds.). Chapman & Hall Ltd.

Fernández-Urruzola, I., Osma, N., Packard, T.T., Gómez, M. and Postel, L. 2014. Distribution of zooplankton biomass and potential metabolic activities across the northern benguela upwelling system. Journal of Marine Systems 140, Part B: 138–149.

Findlay, I.W.O. 1978. The Marine biology of the sierra leone river estuary 1. The physical environmental Bull. Inst. Mar. Biol. Oceangr. (3): 48–64.

Fleminger, A., Othman, B.H.R. and Greenwood, J.G. 1982. The Labidocera pectinata group: an Indo-west pacific lineage of planktonic copepods with descriptions of two new species. Journal of Plankton Research 4(2): 245–270.

Folack, J. 1989. Etude préliminaire du phytoplancton d'une zone côtière d'exploitation crevetticole (kribi – cameroun, golfe de guinée – atlantique centre est. Cameroon Journal of Biological and Biochemical Sciences 2(1): 51–65.

Folack, J. 1988 Estimation et degradation de la chlorophyll dans une zone creveticole: kribi-cameroon (golfe de guinée). Cameroon Journal of Biological and Biochemcial Sciences 1(2): 35–43.

Fondo, E. 2003. Marine biodiversity in kenya– the known and the unknown, National Report. pp. 169–187. *In*: Cynthia Decker, Charles Griffiths, Kim Prochazka, Carmen Ras and Alan Whitfield (eds.). Marine Biodiversity in Sub-Saharan Africa: The Known and the Unknown. Proceedings of the Marine Biodiversity in Sub-Saharan Africa: The Known and the Unknown, 310 pp.

Froneman, P.W. 2001. Feeding ecology of the mysid, mesopodopsis wooldridgei, in a temperate estuary along the eastern seaboard of South Africa. Journal of Plankton Research 23: 999–1008.

Froneman, P.W. and Vorwerk, P. 2003. Predation impacts of juvenile gilchristella aestuaria (clupeidae) and atherina breviceps (atherinidae) on the zooplankton in the temperate kariega estuary, south Africa. African Journal of Aquatic Science 28: 35–41.

Froneman, P.W. 2004. Zooplankton community structure and biomass in a southern African temporarily open/closed estuary. Estuarine, Coastal and Shelf Science 60: 125–132.

Furnestin, M.L. 1966. Chaetognathes des eaux africaines. pp. 105–135. In Atlantide Reports No. 9. Danish Science Press Limited.

Gabche, C.E. 2003. Marine biodiversity in cameroon – the known and the unknown. National Report. pp. 64–74. *In*: Cynthia Decker, Charles Griffiths, Kim Prochazka, Carmen Ras and Alan Whitfield (eds.). Marine Biodiversity in Sub-Saharan Africa: The Known and the Unknown. Proceedings of the Marine Biodiversity in Sub-Saharan Africa: The Known and the Unknown, 310 pp.

Gallienne, C.P. and Smythe-Wright, D. 2005. Epipelagic mesozooplankton dynamics around the mascarene plateau and basin, southwestern Indian ocean. Philos. Trans. R. Soc. A Math. Phys. Eng. Sci. 363(1826): 191–202. 10.1098/rsta.2004.1487.

Gallienne, C.P., Conway, D.V.P., Robinson, J., Naya, N., William, J.S., Lynch, T. et al. 2004. Epipelagic mesozooplankton distribution and abundance over the mascarene plateau and basin, south western Indian ocean. J. Mar. Biol. Assoc. U. K., 84: 1–8: 10.1017/S0025315404008835h.

Gaertner, M. 1988. Analyse des communautés zooplanctoniques de la baie de dakar et leurs relations avec l'hydrographie, Inv. Pesq. 52(1): 17–36.

Gaudy, R. and Seguin, G. 1964. Note sur la répartition annuelle des copépodes pélagiques des eaux de dakar, Rec. Trav. St. Mar. End. 34(50): 216–218.

Gibbons, M.J., Barange, M. and Hutchings, L. 1995. Zoogeography and diversity of euphausiids around southern Africa. Marine Biology 123: 257–268.

Gibbons, M.J. (ed.). 1997. An introduction to the zooplankton of the benguela current region. Ministry of Fisheries and Marine Resources, Windhoek, Namibia.

Gibbons, M.J. and Hutchings, L. 1996. Zooplankton diversity and community structure around southern Africa, with special attention to the Benguela upwelling system. S. Afr. J. Sci. 92(2): 63–77.

Globe Afrique. 2016. Africa faces some serious environmental problems: greenpeace. Globe Afrique. https://globeafrique.com/africa-faces-some-serious-environmental-problems-greenpeace/.

Global Platform for Sustainable Cities, World Bank. 2018. urban sustainability framework. 1st ed. Washington, DC: World Bank.

Griffiths, C.L. 2003. Marine biodiversity in south Africa – the known and the unknown, national report. pp 124–137. *In*: Cynthia Decker, Charles Griffiths, Kim Prochazka, Carmen Ras and Alan Whitfield (eds.). Marine Biodiversity in Sub-Saharan Africa: The Known and the Unknown. Proceedings of the Marine Biodiversity in Sub-Saharan Africa: The Known and the Unknown, 310 pp.

Grindley, J.R. 1977. The zooplankton of langebaan lagoon and saldanha bay. Transactions of the Royal Society of South Africa 42: 341–370.

Grindley, J.R. 1981. Estuarine plankton. pp 117–146. *In*: Grindley, J.R. (ed.). Estuarine Ecology, with Particular Reference to Southern Africa. Cape Town: AA Balkema.

Groeneveld, J.C. and Koranteng, K.A. (eds.). 2017. The R.V. Dr fridtjof nansen in the western Indian ocean: Voyages of Marine Research and Capacity Development. FAO. Rome, Italy. 258.

Grove, S.J., Little, M.C. and Reay, P.J. 1986. Tudor creek mombasa: the early life-history stages of fish and prawns 1985. Report of Overseas Development Administration Research Project R3888.

Hablützel, P.I., Rombouts, I., Dillen, N., Lagaisse, R., Mortelmans, J., Ollevier, A. et al. 2021. Exploring new technologies for plankton observations and monitoring of ocean health. pp. 20–25. *In*: Hazards. E.S., Kappel, S.K., Juniper, S., Seeyave, E. Smith and Visbeck, M. (eds.). Frontiers in Ocean Observing: Documenting Ecosystems, Understanding Environmental Changes, Forecasting A Supplement to Oceanography 34(4).

Hafferssas, A. and Seridji, R. 2010. Relationships between the hydrodynamics and changes in copepod structure on the Algerian coast. Zool. Stud. 49: 353–366.

Hafferssas, A., Seridji, R. and Dallot, S. 2010. Mesozooplankton abundance and biomass related to hydrological structure along the Algerian coasts (S.W. Mediterranean Sea). Crustaceana 83: 1281–1302. https://doi.org/10.1163/001121610X526979.

Hagen, W., Ekau, W. and Verheye, H.M. 2005. Metabolic adaptations and reduced respiration of the copepod calanoides carinatus during diapause at depth in the angola-benguela front and northern Benguela upwelling regions. Afr. J. mar. Sci. 27(3): 653–657.

Hanel, R., John, H-C., Meyer-Klaeden, O. and Piatkowski, U. 2010. Larval fish abundance, composition and distribution at senghor seamount (Cape Verde Islands). Journal of Plankton Research. ff10.1093/plankt/FBQ076ff. ffhal-00606280.

Hansen, A., Ohde, T. and Wasmund, N. 2014. Succession of micro- and nanoplankton groups in aging upwelled water off Namibia. J. Mar. Syst. 31: 123–297.

Hazen, E.L., Scales, K.L., Maxwell, S.M., Briscoe, D.K., Welch, H., Bograd, S.J. et al. 2018. A dynamic ocean management tool to reduce bycatch and support sustainable fisheries. Sci. Adv. 4: eaar3001. doi: 10.1126/sciadv.aar3001.

Helmke, P., Romero, O. and Fischer, G. 2005. Northwest african upwelling and its effect on offshore organic carbon export to the deep sea. Global Biogeochemical Cycles 19: GB4015. doi: 10.1029/2004GB002265.

Hendey, N.I. 1958. Marine diatoms from some west african ports J. R. Microsc Soc. Series 3, 77(1–2): 28–85.

Hernández, F., Jiménez, S., Fernández-Alamo, M.A., Tejera, E. and Arbelo, E. 2000. Sobre la presencia de moluscos nudibranquios planctónicos en el archipiélago de Cabo Verde. Rev. Acad Canar. Cienc. XII(3 y 4): 49–54.

Heymans, J.J., Shannon, L.J. and Jarre, A. 2004. Changes in the northern Benguela ecosystem over three decades: 1970s, 1980s and 1990s. Ecol. Mod. 172: 175–195.

Heymans, J.J., Sumaila, U.R. and Christensen, V. 2009. Policy options for the northern benguela ecosystem using a multispecies, multifleet ecosystem model. Prog. Oceanogr. 83: 417–425.

Hill, B.J. 1966. A contribution to the ecology of the Umlalazi estuary. Zoologica Africana 2: 1–24.

Hoving, H.J.T., Neitzel, P., Hauss, H., Christiansen, S., Kiko, R., Robison, B.H. et al. 2020. *In situ* observations show vertical community structure of pelagic fauna in the eastern tropical North Atlantic off Cape Verde. Scientific Reports 10: 21798.

Huenerlage, K. and Buchholz, F. 2013. Krill of the northern benguela current and the angola-benguela frontal zone compared: physiological performance and short-term starvation in Euphausia hanseni. Journal of Plankton Research 35(2): 337–351.

Huggett, J.A. 2014. Mesoscale distribution and community composition of zooplankton in the mozambique channel. Deep Sea Research Part II: Topical Studies in Oceanography 100: 119–135.

Huggett, J., Verheye, H., Escribano, R. and Fairweather, T. 2009. Copepod biomass, size composition and production in the southern benguela: spatio-temporal patterns of variation, and comparison with other eastern boundary upwelling systems. Prog. Oceanogr. 83(1–4): 197–207.

Hutchings, L., Pillar, S.C. and Verheye, H.M. 1991. Estimates of standing stock, production and consumption of meso- and macrozooplankton in the Benguela ecosystem. Afr. J. Mar. Sci. 11(1): 499–512.

Hutchings, L., van der Lingen, C.D., Shannon, L.J., Crawford, R.J.M., Verheye, H.M.S., Bartholomae, C.H. et al. 2009. The Benguela Current: An ecosystem of four components. Progress in Oceanography 83: 15–32.

Ibarbalz, F.M., Henry, N., Brandão, M.C., Martini, S., Busseni, G., Byrne, H. et al. 2019. Global trends in marine plankton diversity across kingdoms of life. Cell 179(5, 14): 1084–1097.e21

IPBES. 2018. Summary for policymakers of the regional assessment report on biodiversity and ecosystem services for Africa of the Intergovernmental Science-Policy Platform on Biodiversity and Ecosystem Services. Archer, E., Dziba, L.E., Mulongoy, K.J., Maoela, M.A., Walters, M., Biggs, R. et al. (eds.). IPBES Secretariat, Bonn, Germany.

Jerling, H.L. and Wooldridge, T.H. 1991. Population dynamics and estimates of production for the calanoid copepod pseudodiaptomus hessei in a warm temperate estuary. Estuarine, Coastal and Shelf Science 33: 121–135.

Jerling, H.L. and Wooldridge, TH. 1995a. Plankton distribution and abundance in the sundays river estuary, south Africa, with comments on potential feeding interactions. South African Journal of Marine Science 15: 169–184.

Jerling, H.L. and Wooldridge, T.H. 1995b. Feeding of two mysid species on plankton in a temperate south African estuary. Journal of Experimental Marine Biology and Ecology 188: 243–259.

John, H.-Ch. 1985. Horizontal and vertical distribution patterns of fish larvae off NW Africa in relation to the environment. pp. 489–512. *In*: Bas, C., Margalef, R. and Rubiés, P. (eds.). Simposio internacional sobre las areas de afloramiento, mas importantes del oeste africano (Cabo Blanco y Benguela) Vol. 1. Inst. In. Pesqueras, Barcelona.

John, H.-Ch., Klenz, B. and Hermes, R. 1991. Distribution and drift of horse mackerel larvae (genus trachurus) off Mauritania January to April 1983. ICES, C.M. 1991/H: 3: 22 pp.

John, H.-Ch. and Zelck, C. 1997. Features, boundaries and connecting mechanisms of the mauritanian province exemplified by oceanic fish larvae. Helgoländer Meeresunters. 5: 213–240.

John, H.-Ch., Mohrholz, V. and Lutjeharms, J.R.E. 2001. Cross-front hydrography and fish larval distribution at the Angola-Benguela frontal zone. J. Mar. Syst. 28: 91–111.

John, H.-Ch. 2003. Fish larvae from the dampier seamount and sierra leone rise (tropical east Atlantic). Mitt. Hamb. Zool. Mus. Inst. 100: 139–152.

John, H.-Ch., Knoll, M. and Müller, T. 2004. Zonal structures in fish larval abundance and diversity off morocco. Mitt. Hamb. Zool. Mus. Inst. 101: 249–273.

John, H.C., Mohrholz, V., Lutjeharms, J.R.E., Weeks, S., Cloete, R., Kreiner, A. et al. 2004. Oceanographic and faunistic structures across an Angola Current intrusion into northern Namibian waters. Journal of Marine Systems 46: 1–22.

Jones, S., Carrasco, N.K. and Perissinotto, R. 2015. Turbidity effects on the feeding, respiration and mortality of the copepod pseudodiaptomus stuhlmanni in the st lucia estuary, south Africa. J. Exp. Mar. Biol. Ecol. 469: 63–68.

Jones, S., Perissinotto, R., Carrasco, N.K. and Vosloo, A. 2016. Impact of a flood event on the zooplankton of an estuarine lake. Mar Biol Res. 12: 158–167.

Johnsen, E., Kraskstad, J., Ostrowski, M., Serigstad, B., Stromme, T., Alvheim, O. et al. 2007. Surveys of the living marine resources of mozambique: Ecosystem survey and special studies. Report No. 8/20072007409. Bergen, Norway, Institute of Marine Research.

Kadiri, M.O. 1999. Phytoplankton distribution in the coastal waters of Nigeria. Nigerian Journal of Botany 12: 51–62.

Kadiri, M.O. 2006a. Phytoplankton flora and physico-chemical attributes of some waters in the eastern Niger delta area of Nigeria. Nigerian J. Botany 19: 188200.

Kadiri, M.O. 2006b. Phytoplankton survey in the western Niger delta, Nigeria. Afr. J. Environ. Pollut. 96(Health 5): 48–58.

Kadiri, M.O. 2011. Notes on harmful algae from nigerian coastal waters. Acta Botanica Hungarica 53(98): 137–143.

Karaseva, E.M. and Shiganov, T.A. 1993. Species composition and seasonal abundance dynamics of ichthyoplankton over the shelf of Namibia in 1988–1989 [Vidovoj sostav I sezonnaya dinamika chislennosti ikhtioplanktona na shel'fe Namibii v 988-1989]. Okeanologiya/Oceanology (MOSC.) 33(2): 242–247.

Karstensen, J., Fiedler, B., Schütte, F., Brandt, P., Körtzinger, A., Fischer, G. et al. 2015. Open ocean dead zones in the tropical north Atlantic ocean. Biogeosciences 12: 2597–2605. doi:10.5194/bg-12-2597-2015.

Karstensen, J., Schütte, F., Pietri, A., Krahmann, G., Fiedler, B., Grundle, D. et al. 2016. Upwelling and isolation in oxygendepleted anticyclonic modewater eddies and implications for nitrate cycling. Biogeosciences Discuss. doi:10.5194/bg-2016-34.

Kasyi, J.N. 1994. Relationship between Nutrient Levels, Phytoplankton Biomass and Zooplankton Composition, Biomass and Abundance in Tudor Creek. M.Sc. Thesis, University of Nairobi, 99 pp.

Khames, G.E.Y. and Hafferssas, A. 2018. Biodiversity and abundance of gelatinous zooplankton along the Algerian coast. Rev d'Ecologie (La Terre la Vie) 73: 211–226.

Khames, E.G.Y. and Hafferssas, A. 2019. Abundance and species composition of gelatinous zooplankton in habibas Islands and sidi fredj (western Mediterranean Sea). Cah Biol. Mar. 60: 143–152. https://doi.org/10.21411/CBM.A.B67AE980.

Khames, G.E.Y., Alioua, Z., Kherchouche, A. and Hafferssas, A. 2022. Gelatinous Zooplankton Versus Algerian Environmental Factors. Thalassas (in press).

Kelchner, H. 2020. Role of Coastal Environmental Conditions During Austral Winter on Plankton Pommunity Dynamics and the Occurrence of Pseudo-Nitzschia spp. and Domoic Acid in Inhambane Province, Mozambique. LSU Master's Theses. 135 pp.

Kherchouche, A. and Hafferssas, A. 2019. Species composition and distribution of medusae (cnidaria: medusozoa) along the Algerian coast between 2°E and 7°E (SW Mediterranean Sea). Mediterr. Mar. Sci. 21: 52–61. https://doi.org/10.12681/MMS.20849.

Kherchouche, A. and Hafferssas, A. 2020. Species composition and distribution of medusae (cnidaria: medusozoa) along the Algerian coast between 2°E and 7°E (SW Mediterranean Sea). Mediterr. Mar. Sci. 21: 52–61 Page 21/27 38.

Kibirige, I. 2002. The Structure and Trophic Role of the Zooplankton Community of the Mpenjati Estuary, A Subtropical and Temporarily-Open System on the Kwazulu-Natal Coast. PhD thesis. University of Durban–Westville, South Africa.

Kibirige, I., Perissinotto, R. and Nozais, C. 2002. Alternative food sources of zooplankton in a temporarily-open estuary: evidence from δ13C and δ15N. Journal of Plankton Research 24(10): 1089–1095.

Kibirige, I. and Perissinotto, R. 2003. The zooplankton community of the mpenjati estuary, a south African temporarily open/closed system. Estuarine, Coastal and Shelf Science 58: 727–741.

Kibirige, I., Perissinotto, R. and Thwala, X. 2006. A comparative study of zooplankton dynamics in two subtropical temporarily open/closed estuaries, South Africa. Mar. Biol. 148: 1307–1324.

Kiørboe, T. and Hirst, A.G. 2014. Shifts in mass scaling of respiration, feeding, and growth rates across life-form transitions in marine pelagic organisms. Am. Nat. 183: E118–E130.

Kiko, R., Brandt, P., Christiansen, S., Faustmann, J., Kriest, I., Rodrigues, E. et al. 2020. Zooplankton-mediated fluxes in the eastern tropical north Atlantic. Front. Mar. Sci. 7: 358.

Kim, M., Kang, J-H., Rho, T., Kang, H-W., Kang, D-J., Park, J-H. et al. 2022. Mesozooplankton community variability in the seychelles–chagos thermocline ridge in the western Indian ocean. Journal of Marine Systems 225: 103649.

Kimaro, M.M. 1986. The Composition, Distribution and Abundance in Near-Surface Zooplankton of Tudor Creek, Mombasa. MSc Thesis. University of Nairobi, 95 pp.

Kimaro, M.M. and Jaccarini, V. 1989. The cycle of near-surface zooplankton abundance in tudor creek, Mombasa, Kenya. J. Sci. Tech. Ser. B. 10: 7–30.

Kitheka, J.U., Ohowa, B.O., Mwashote, B.M., Shimbira, W.S., Mwaluma J.M. and Kazungu, J. 1996. Water-circulation dynamics, water column nutrients and plankton productivity in a wellflushed tropical bay in kenya. Journ. Sea. Res. 35: 257–268.

Kosobokova, K.N., Drits, A.V. and Krylov, P.I. 1988. Physiological and biochemical characteristics of Calanoides carinatus in waters on an upwelling off the coast of namibia. Oceanology 28(3): 375–379.

Krakstad, J.-O., Krafft, B., Alvheim, O., Kvalsund, M., Bernardes, I., Chacate, O. et al. 2015. Marine ecosystem survey of mozambique. Cruise report Dr Fridtjof Nansen. 11 November–02 December 2014. Report No. GCP/INT/003/NOR. FAO–NORAD.

Kramp, P.L. 1955. The medusae of the tropical west coast of Africa. pp. 239–324. *In*: Bruun, A. Fr. (ed.). Atlantide Report, No. 3. Danish science Press Limited, Copenhagen.

Kruger, I. 1980. A checklist of south west African marine phyto-plankton, with some phytogeographical relations. Fisheries Bulletin of South Africa 13: 31–40.

Kruger, I. 1980. A checklist of southwest African marine phytoplankton, with some phytogeographical relations. Fisheries Bulletin of South Africa 13: 31–53.

Lakkis, S. 2013. Le zooplancton marin du Liban (Méditerranée Orientale) Biologie, biodiversité, biogéographie.

Lamere, C. 2013. UNEP highlights environmental impacts on health in Africa, new security beat. https://www.newsecuritybeat.org/2013/03/unep-highlights-environmental-impacts-health-africa/.

Lamin, P.A. 2011. Plankton Distribution and Abundance in Relation to Some Environmental Factors and Fish Availability on the Continental Shelf of Sierra Leone. M.Sc. Thesis.

Lamont, T., Coetzee, J., Shillington, F., Veitch, J., Currie, J.C. and Monteiro, P.M.S. 2009. The benguela current: an ecosystem of four components. Prog. Oceanogr. 83(1–4): 15–32.

Lamont, T., Barlow, R.G., Morris, T. and van den Berg, M.A. 2014. Characterization of mesoscale features and phytoplankton variability in the mozambique channel. Deep Sea Research Part II: Topical Studies in Oceanography 100: 94–105.

Lass, H.U., Schmidt, M., Mohrholz, V. and Nausch, G. 2000. Hydrographic and current measurements in the area of the angola-benguela front. J. phys. Oceanogr. 30: 2589–2609.

Lathuilière, C., Echevin, V. and Lévy, M. 2008. Seasonal and intraseasonal surface chlorophyll-a variability along the northwest African coast. J Geophys. Res. 113: C05007. 10.1029/2007 jc004433.

Leal, M.C., Sá, C., Nordez, S., Brotas, V. and Paula, J. 2009. Distribution and vertical dynamics of planktonic communities at sofala bank, Mozambique. Estuarine, Coastal and Shelf Science 84(4): 605–616.

Lebourges-Dhaussy, A., Huggett, J., Ockuis, S., Roudaut, G., Jossee, E. and Verheye, H. 2009. Zooplankton size and distribution within mesoscale structures in the mozambique channel: a comparative approach using the TAPS acoustic profiler, a multiple net sampler and ZooScan image analysis Institut de Recherche pour le Développement (IRD), UMR LEMAR 195 (UBO/CNRS/IRD/Ifremer.45).

Lebour, M.V. 1959. The larval decapod crustacea of tropical west Africa. Atlantidae Rep. 5: 119–143.

Lebrato, M. and Jones, D.O.B. 2009. Mass deposition event of pyrosoma atlanticum carcasses off Ivory coast (west Africa). Limnol. Oceanogr. 54: 1197–1209.

Leigh, E.A.M. 1973. Studies on the Zooplankton of the Sierra Leone Estuary. M.Sc. Thesis, USL, 223pp.

Lemley, D.A., Adams, J.B. and Strydom, N.A. 2018. Triggers of phytoplankton bloom dynamics in permanently eutrophic waters of a south African estuary triggers of phytoplankton bloom dynamics in permanently eutrophic waters of a south African estuary. Afric. J. Aquatic Sci. 5914.

Lim, -S.S.L. and Tan, -L.W.H. 1981. Larval development of the hairy crab, pilumnus vespertilio (fabricius) (brachyura, xanthidae) in the laboratory and comparisons with larvae of pilumnus dasypodus kingsley and pilumnus sayi rathbun. -Dep. Zool., Natl. Univ. Singapore, Singapore -CRUSTACEANA. 1981. 41(1): 71–88

Lindkvist, E., Wijermans, N., Daw, T.M., Gonzales-Mon, B., Giron-Nava, A., Johnson, A.F. et al. 2020. Navigating complexities: agent-based modeling to support research, governance, and management in small-scale fisheries. Frontiers in Marine Science 6: 733.

Lindley, J.A., Hernández, F., Tejera, E. and Jiménez, S. 2001. Decápodos pelágicos (larvas y adultos) de las islas de cabo verde (campaña TFMCBM/98). Rev. Acad. Canar. Cienc. XIII(4): 87–99.

Lindley, J.A., Hernández, F., Tejera, E. and Correia, S.M. 2004. Phyllosoma larvae (decapoda: palinuridea) of the cape verde Islands. Journal of Plankton Research 26(2): 235–240.

Lindley, J.A., Hernández, F., Tejera, E. and Jiménez, S. 2002. Decápodos pelágicos (larvas y adultos) de las Islas de cabo verde (campaña TFMCBM/98). Rev. Acad. Canar. Cienc. XIII: 87–99.

Loick, N., Ekau, W. and Verheye, H.M. 2005. Water-body preferences of dominant calanoid copepod species in the angola-benguela frontal zone. Afr. J. Mar. Sci. 27(3): 597–608.

Lombard, A.T., Clifford-Holmes, J.K., Vermeulen, E.A., Witteveen, M. and Snow, B. 2019. The advantages of system dynamics models in decision-support for integrated ocean management. *In*: Pillay, N.S. (ed.). Proceedings of the Sixth Annual System Dynamics Conference in South Africa, (Johannesburg: Eskom SOC and University of Witwatersrand).

Lombard, F., Boss, E., Waite, A.M., Vogt, M., Uitz, J., Stemmann, L. et al. 2019. Globally consistent quantitative observations of planktonic ecosystems. Front. Mar. Sci. 6: 196.

Longhurst, A.R. 1983. Benthic-pelagic coupling and export of organic carbon from a tropical Atlantic continental shelf - Sierra Leone. Estuary Coast. Shelf. Sci. 17: 261–285.

Louw, D.C., van der Plas, A.K., Mohrholz, V., Wasmund, N., Junker, T. and Eggert, A. 2016. Seasonal and inter-annual phytoplankton dynamics and forcing mechanisms in the northern Benguela upwelling system. J. Mar. Syst. 157: 124–134.

Louw, D.C., Doucette, G.J. and Voges, E. 2017. Annual patterns, distribution and long-term trends of pseudo-nitzschia species in the northern Benguela upwelling system. Journal of Plankton Research 39(1): 35–47.

Lüskow, F., Christiansen, B., Chi, X., Silva, P., Neitzel, P., Brooks, M.E. et al. 2022. Distribution and biomass of gelatinous zooplankton in relation to an oxygen minimum zone and a shallow seamount in the eastern tropical north Atlantic ocean. Marine Environmental Research 175: 105566.

Madhupratap, M. 1986. Zooplankton standing stock and diversity along an oceanic track in the western Indian ocean. Mahasagar. Bull. of the Nat. Inst. of Oceano. 16: 463–467.

Marshall, T., Granger, J., Casciotti, K.L., Dähnke, K., Emeis, K.-C., Marconi, D. et al. 2022. The Angola Gyre is a hotspot of dinitrogen fixation in the South Atlantic Ocean. Communications Earth & Environment 3: 151.

Matata, A.C. and Adan, A. 2018. Causes of climate change and its impact in the multi-sectoral areas in africa-need for enhanced adaptation policies. Current Journal of Applied Science and Technology 27: 1–10.

Mayfield, S., Elliott, B. and Giddey, C.J. 2001. Analysis of fish eggs and larvae collected in the region of the northern benguela and Angola currents: preliminary results. S. Afr. J. Sci. 97: 270.

McQuatters-Gollop, A., Edwards, M., Helaouët, P., Johns, D.G., Owens, N.J.P., Raitsos, D.E. et al. 2015. The continuous plankton recorder survey: how can long-term phytoplankton datasets deliver good environmental status? Estuarine, Coastal and Shelf Science 162: 88–97.

McQuatters-Gollop, A., Atkinson, A., Aubert, A., Bedford, J., Best, M., Bresnan, E. et al. 2019. Plankton lifeforms as a biodiversity indicator for regional-scale assessment of pelagic habitats for policy. Ecological Indicators 101: 913–925.

Mellak, L. and Hafferssas, A. 2022. Copepod communities of the algerian coast (southwestern Mediterranean Sea, 2°–7°E) and their interaction with hydro-climatic variability. Crustaceana 95(4): 457–482.

Mensah, M.A. 1966. Zooplankton occurrence over the shelf of ghana. pp. 241–254. *In*: Proceedings of the Symposium on the Oceanography and Fisheries Resources of the Tropical Atlantic. Abidjan, La Côte d'Ivoire.

Mensah, M.A. 1974. The occurrence of the marine copepod *Calanoides carinatus* (krøyer) in ghanaian waters. Ghana Journal of Science 14: 147–166.

Meyers, G. 2019. Environmental issues in africa's cities. Wiley Online Library, https://doi. org/10.1002/9781118568446.eurs0002.

Miloslavich, P., Bax, N.J., Simmons, S.E., Klein, E., Appeltans, W., Aburto-Oropeza, O. et al. 2018. Essential ocean variables for global sustained observations of biodiversity and ecosystem changes. Glob. Change Biol. 24: 2416–2433.

Mgaya, Y.D. 2003. Marine biodiversity in tanzania– the known and the unknown, national report. pp. 156–168. *In*: Cynthia Decker, Charles Griffiths, Kim Prochazka, Carmen Ras and Alan Whitfield (eds.). Marine Biodiversity in Sub-Saharan Africa: The Known and the Unknown. Proceedings of the Marine Biodiversity in Sub-Saharan Africa: The Known and the Unknown, 310 pp.

Mohrholz, V., Bartholomae, C., van der Plas, A. and Lass, H. 2008. The seasonal variability of the northern benguela undercurrent and its relation to the oxygen budget on the shelf. Cont. Shelf Res. 28: 424–441.

Moloney, C.L., Jarre, A., Arancibia, H., Bozec, Y.M., Neira, S., Roux, J-P. et al. 2005. Comparing the Benguela and humboldt marine upwelling ecosystems with indicators derived from inter-calibrated models. ICES J. Mar. Sci. 62: 493–502.

Montoya-Maya, P.H. and Strydom, N.A. 2009. Zooplankton composition, abundance and distribution in selected south and west coast estuaries in South Africa. Afr. J. Aquat. Sci. 34(2): 147–157.

Morais, M., Penha, A.M., Novais, M.H., Landim, L., Victória, S.S., Morales, E.A. et al. 2021. Some observations on phytoplankton community structure, dynamics and their relationship to water quality in five santiago Island Reservoirs, Cape Verde. Water 13: 2888.

Muller-Karger, F.E., Miloslavich, P., Bax, N.J., Simmons, S., Costello, M.J., Sousa Pinto, I. et al. 2018. Advancing marine biological observations and data requirements of the complementary essential ocean variables (eovs) and essential biodiversity variables (ebvs) frameworks. Front. Mar. Sci. 5: 211. doi: 10.3389/fmars.2018.00211.

Mwaluma, J. 1993. Zooplankton studies in a mangrove, seagrass and coral reef system, gazi bay, kenya. pp. 77–82. *In*: Hemminga, M.A. (ed.). Second Semi-annual Report on the STD-3 Project Interlinkages between Eastern African Coastal Ecosystems Contract no. T5 3-CT92-0114, Netherlands Inst. of Ecology (CEMO) Yerseke, The Netherlands.

Mwaluma, J.M. 1997. Distribution and Abundance of Zooplankton off the Kenya Coast During The Monsoons. M.Sc. Thesis University of Nairobi. 90 pp.

Mwaluma, J., Osore, M.K. and Okemwa, E. 1993. Zooplankton studies in a mangrove creek system. pp. 74–80. *In*: Woitchik, A.F. (ed.). Dynamics and assessment of Kenyan mangrove ecosystems, No. T5 2-0240-C (GDF), Final Report. ANCH, Vrije Universiteit Brussel.

Mwaluma, J., Osore, M., Kamau, J. and Wawiye, P. 2003. Composition, abundance and seasonality of zooplankton in mida creek, kenya. Western Indian Ocean J. Mar. Sci. 2(2): 147–155.

Natij, L., Damsiri, Z., Khalil, K., Loudiki, M., Ettahiri, O. and Elkalay, K. 2014. Phytoplankton abundance and diversity in the coastal waters of oualidia lagoon, south Moroccan Atlantic in relation to environmental variables. Int. J. Adv. Res. 6: 1022–1032.

Ndomahina, E.T. 2002. An Assessment of the Coastal and Marine Biodiversity of Sierra Leone Consultancy of The Sierra Leone Maritime Administration. 100pp.

Ndour, I., Berraho, A., Ettahiri, O. and Sambec, B. 2018. Composition, distribution and abundance of zooplankton and ichthyoplankton along the senegal-guinea maritime zone (west Africa). The Egyptian Journal of Aquatic Research 44(2): 109–24.

Neumann, B., Vafeidis, A.T., Zimmermann, J. and Nicholls, R.J. 2015. Future coastal population growth and exposure to sea-level rise and coastal flooding-a global assessment. PLoS ONE 10: 6. 10.1371/journal.pone.0131375.

Nicol, S., Stolp, M., Cochran, T., Geijsel, P. and Marshall, J. 1992. Growth and shrinkage of Antarctic krill Euphausia superba from the Indian Ocean sector of the Southern Ocean during summer. Marine Ecology Progress Series 89: 175–181.

Nozais, C., Perissinotto, R. and Mundree, S. 2001. Annual cycle of microalgal biomass in a south African temporarily-open estuary: nutrient versus light limitation. Mar. Ecol. Prog. Ser. 223: 39–48.

Nwankwo, D.I. 1991. A survey of the dinoflagellates of Nigeria. Armoured dinoflagellates of Lagos Lagoon and associated Tidal creeks. Nigerian Journal of Botany 4: 49–60.

Nwankwo, D.I. 1997. A first list of dinoflagellates (pyrrhophyta) from nigerian coastal waters (creek, 119 estuaries lagoons). Pol. Arch. Hydrobiol 44: 313–321.

OberhäNsli, H., BéNier, C., Meinecke, G., Schmidt, H., Schneider, R. and Wefer, G. 1992. Planktonic foraminifers as tracers of ocean currents in the eastern south Atlantic. Paleoceanography 7(5): 607–632.

O'Brien, T.D., Lorenzoni, L., Isensee, K. and Valdes, L. 2017. What is marine ecological time series telling us about the ocean? A status report. IOC Tech. Ser. 129: 1–297.

OECD. 2019. Rethinking innovation for a sustainable ocean economy, OECD publishing, Paris, https://doi.org/10.1787/9789264311053-en.

OECD. 2020. Reframing financing and investment for a sustainable ocean economy, OECD environment policy papers, No. 22, OECD Publishing, Paris, https://doi.org/10.1787/c59ce972-en.

OECD. 2021b. A new era of digitalisation for ocean sustainability? Prospects, benefits, challenges, OECD science, Technology and Industry Policy Papers, No. 111, OECD Publishing, Paris, https://doi.org/10.1787/a4734a65-en.

OECD. 2020. Experimental ocean-based industry database, directorate for science, technology and innovation, OECD, Paris.

Okemwa, E.N. 1989. Analysis of six 24 hours series of zooplankton sampling across a tropical creek, the Port Reitz, Mombasa, Kenya. Trop. Zool. 2: 123–138.

Okemwa, E.N. 1990. A Study of the Pelagic Copepods in a Tropical Marine Creek, Tudor, Mombasa, Kenya With a Special Reference to their Community Structure, Biomass And Productivity. PhD Thesis. Vrije University, Brussels. 225 pp.

Okemwa, E.N. and Revis, N. 1986. Planktonic copepods from coastal and inshore waters of tudor creek, Mombasa. Kenya. J. Sci. Ser. B 7: 27–34.

Okemwa, E.N. 1992. A long-term sampling study of copepods across a tropical creek in mombasa, kenya. E. Afr. Agric. For. J. 57(4): 199–215.

Okera, W. 1974. Zooplankton of the inshore waters of dar es salaam (tanzania, S.E. Africa) with observation to artificial light. Mar. Biol. 26: 13–15.

Okonjo-Iweala, N. 2020. Africa can play a leading role in the fight against climate change. Brookings Institution. https://www.brookings.edu/research/africa-can-play-a-leading-role-in-the-fight-against-climate-change/.

Oldewage, W.H. 1995. Redescription of alebion carchariae kroyer, 1863 (copepoda, siphonostomatoida) from Isurus oxyrhinchus in south Africa. Crustaceana 68(1): 38–42.

Olivar, M.P. and Barange, M. 1989. Vertical distribution of fish eggs and larvae in the northern benguela region. Rapp. P. –V. Reun. CIEM 191: 454.

Olivar, M. and Barange M. 1990. Zooplankton of the northern benguela region in a quiescent upwelling period, J. Plankton Res. 17: 1023–1044.

Olivar, M.P., Rubies, P. and Salat, J. 1991. Horizontal and vertical distribution patterns of ichthyoplankton under intense upwelling regimes off namibia. South African Journal of Marine Science 12: 71–82.

Osma, N., Fernández-Urruzola, I., Packard, T.T., Postel, L., Gómez, M. and Pollehene, F. 2014. Short-term patterns of vertical particle flux in northern benguela: a comparison between sinking POC and respiratory carbon consumption. J. Mar. Syst. 140: 150–162.

Osore, M.K. 1992. A note on zooplankton distribution and diversity in a tropical mangrove creek system, gazi, kenya. Hydrobiologia 247: 119–120.

Osore, M.K. 1994. A Study of Gazi Bay and The Adjacent Waters: Community Structure and Seasonal Variation. M.Sc. Thesis. Vrije Universiteit Brussels. 104 pp.

Osore, M.K., Tackx, M.L.M. and Daro, M.H. 1997. The effect of rainfall and tidal rhythm on the community structure and abundance of the zooplankton of gazi bay, kenya. Hydrobiologia 356: 117–126.

O'Toole, M.J., Shannon, L.V., de Barros Neto, V. and Malan, D.E. 2001. Integrated management of the Benguela current region—a framework for future development. pp. 228–251. *In*: von Bodungen, B. and Turner, R.K. (eds.). Science and Integrated Coastal Management. Dahlem University Press, Berlin.

Ouldessaib, E.T., Khalki, M. El and Moncef, M. 1998. Etude qualitative et quantitative du peuplement de copépodes de la lagune de oualidia (côte atlantique du maroc). Mar. Life 8(1–2): 35–43.

Packard, T.T. 1979. Respiration and respiratory electron transport activity in plankton from the northwest African upwelling area. J. Mar. Res. 37(4): 711–742.

Pagès, F. and Gili, J.M. 1991. Effects of large-scale advective processes on gelatinous zooplankton populations in the northern benguela ecosystem. Marine Ecology Progress Series 75(2/3): 205–215.

Pakhomov, E.A. and Perissinotto, R. 1997. Mesozooplankton community structure and grazing impact in the region of the subtropical convergence south of Africa. J. Plankton Res. 19(6): 675–691. 10.1093/plankt/19.6.675.

Pecl, G.T., Araujo, M.B., Bell, J., Blanchard, J., Bonebrake, T.C., Chen, I. et al. 2017. Biodiversity redistribution under climate change: Impacts on ecosystems and human well-being. Science 355(6332): 1–9.

Perissinotto, R., Walker, D.R., Webb, P., Wooldridge, T.H. and Bally, R. 2000. Relationships between zoo- and phytoplankton in a warm-temperate semi-permanently closed estuary, south Africa. Estuar. Coast. Shelf Sci. 51: 1–11.

Perissinotto, R., Mayzaud, P., Nichols, P.D. and Labat, J.P. 2007. Grazing by pyrosoma atlanticum (tunicata, thaliacea) in the south Indian Ocean. Mar. Ecol. Prog. Ser. 330: 1–11.

Pillar, S.C., Stuart, V., Barange, M. and GIBBONS, M.J. 1992. Community structure and trophic ecology of euphausiids in the benguela ecosystem. *In*: Payne, A.I.L., Brink, K.H., Mann, K.H. and Hilborn, R. (eds.). Benguela Trophic Functioning. S. Afr. J. Mar. Sci. 12: 393–409.

Popova, E., Yool, A., Byfield, V., Cochrane, K., Coward, A.C., Salim, S.S. et al. 2016. From global to regional and back again: common climate stressors of marine ecosystems relevant for adaptation across five ocean warming hotspots. Glob. Chang. Biol. 22: 2038–2053.

Postel, L. 1990. The meso-zooplankton response to coastal upwelling off west Africa with particular regard to biomass. Meereswissenschaftliche Berichte, Warnemünde 1: 127.

Postel, L., Arndt, E.A. and Brenning, U. 1995. Rostock zooplankton studies off west Africa. Helgoländer Meeresuntersuchungen 49: 829–847.

Postel, L., da Silva, A.J., Mohrholz, V. and Lass, H.U. 2007. Zooplankton biomass variability off angola and namibia investigated by a lowered ADCP and net sampling. J. Mar. Syst. 68(1–2): 143–166.

Racault, M.-F., Sathyendranath, S. and Platt, T. 2014. Impact of missing data on the estimation of ecological indicators from satellite ocean-colour time-series. Remote Sens. Environ. 152: 15–28.

Raleigh C. and Jordan L. 2010. Climate change and migration: emerging patterns in the developing world. pp. 103–131. *In*: Mearns, R., Norton, A. (eds.). Chapter 4 in "Social Dimensions of Climate Change: Equity and Vulnerability in a Warming World". Washington DC, USA: World Bank.

Ramos, V., P-Marrero, J., Llinás, O., Cianca, A. and Morales, J. 2012. Variation of the chlorophyll-a related to sea surface temperature, wind and geostrophic currents in the cape verde region using satellite data. INDP 37.

Rangel, I.M. and Silva, S. 2007. Pseudo-nitzschia spp. and prorocentrum micans blooms in luanda bay, Angola. Harmful Algae News 33.

Raj, R.P., Peter, B.N. and Pushpadas, D. 2010. Oceanic and atmospheric influences on the variability of phytoplankton bloom in the southwestern Indian ocean. Journal of Marine Systems 82: 217–229.

Reay, P.J. and Kimaro, M.M. 1984. Surface zooplankton studies in port mombasa during the N. east monsoon. Kenyan J. Sci. and Tech. Ser. B 5: 27–48.

Revis, N. 1988. Preliminary observations on the copepods of Tudor creek, Mombasa, Kenya. Hydrobiologia. 168: 343–350.

Revis, N. and Okemwa, E.N. 1988. Additional records of species of copepods and their distribution in the coastal and inshore waters of kenya. Kenya Journal of Sciences Series B 9: 123–127.

Revis, N. 1988. Preliminary observations on the copepods of tudor creek, mombasa, Kenya. Hydrobiologia. 168: 343–350.

Reyssac, J. 1970. Phytoplancton et production primaire au large de la côte d'Ivoire. Bull. de l'IFAN, 23, Sér. A, n°4: 869–981.

Rhomad, H., Khalila, K., Neves, R., Bougadir, B. and Elkalay, K. 2021. Modeling investigation of the nutrients and phytoplankton dynamics in the moroccan Atlantic coast: a case study of Agadir coast. Ecological Modelling 447: 109510.

Riandey, V., Champalbert, G., Carlotti, F., Taupier-Letage, I. and Thibault-Botha, D. 2005. Zooplankton distribution related to the hydrodynamic features in the Algerian Basin (western Mediterranean Sea) in summer 1997. Deep Sea Research Part I: Oceanographic Research Papers 52: 2029–2048.

Richardson, A.J. and Verheye, H.M. 1999. Growth rates of copepods in the southern benguela upwelling system: the interplay between body size and food. Limnol. Oceanogr. 44(2): 382–392.

Righetti, D., Vogt, M., Gruber, N., Psomas, A. and Zimmermann. N.E. 2019. Global pattern of phytoplankton diversity driven by temperature and environmental variability. Science Advances 5(5).

Rodrigues, E., Kiko, R., Brehmer, P., Machu, E. and Silva, O. 2016 Analysis of zooplankton samples from AWA scientific research on board of R/V Thalassa 2014 using the zooscan approach., ICAWA S2_74.

Romero, O.E., Lange, C.B., Fischer, G., Treppke, U.F. and Wefer, G. 1999. Variability in export production documented by downward fluxes and species composition of marine planktonic diatoms: observations from the tropical and equatorial Atlantic. pp. 365–392. *In*: Fischer, G. and Wefer, G. (eds.). Use of Proxies in Paleoceanography, Examples from the South Atlantic, Springer Verlag, Berlin, Heidelberg.

Romero, O.E., Lange, C.B. and Wefer, G. 2002. Interannual variability (1988–1991) of siliceous phytoplankton fluxes off northwest Africa. J. Plank. Res. 24: 1035–1046.

Romero, O.E., Dupont, L., Wyputta, U., Jahns, S. and Wefer, G. 2003. Temporal variability of fluxes of eolian-transported freshwater diatoms, phytoliths, and pollen grains off cape blanc as reflection of land-atmosphere-ocean interactions in northwest Africa. J. Geophys. Res.-Oceans 108: 3153.

Romero, O.E., Armand, L.K., Crosta, X. and Pichon, J.-J. 2005. The biogeography of major diatom taxa in southern ocean surface sediments: 3. Tropical/Subtropical species, Palaeogeogr. Palaeocl. 223: 49–65.

Romero, O.E., Kim, J. and Donner, B. 2008. Submillennial-to-millennial variability of diatom production off Mauritania, NW Africa, during the last glacial cycle. Paleoceanography 23: PA3218.

Romero, O.E., Rixen, T. and Herunadi, B. 2009a. Effects of hydrographic and climatic forcing on diatom production and export in the tropical southeastern Indian ocean. Mar. Ecol. Prog. Ser. 384: 69–82.

Romero, O.E., Thunell, R.C., Astor, Y. and Varela, R. 2009b. Seasonal and interannual dynamics in diatom production in the cariaco basin, Venezuela. Deep-Sea Res. Pt. I 56: 571–581.

Romero, O.E., Crosta, X., Kim, J.-H., Pichevin, L. and Crespin, J. 2015. Rapid longitudinal migrations of the filament front off namibia (SE Atlantic) during the past 70 kyr. Global Planet. Changes 125: 1–12.

Romero, O.E. and Armand, L.K. 2010. Marine diatoms as indicators of modern changes in oceanographic conditions. pp. 373–400. *In*: Smol, J.P. and Stoermer, E.F. (eds.). The Diatoms, Applications for the Environmental and Earth Sciences (Second Edition), Cambridge University Press, Cambridge.

Romero, O.E., Fischer, G., Karstensen, J. and Cermeño, P. 2016. Eddies as trigger for diatom productivity in the open-ocean northeast Atlantic. Prog. Oceanogr. 147: 38–48.

Romero, O.E. and Fischer, G. 2017. Shift in the species composition of the diatom community in the eutrophic mauritanian coastal upwelling: results from a multi-year sediment trap experiment (2003–2010). Prog. Oceanogr. 159: 31–44.

Romero, O.E., Fischer, G., Baumann, K.-H., Zonneveld, K.A.F. and Donner, B. 2019. Fluxes of diatoms, coccolithophorids, calcareous and organic-walled dinoflagellate cysts, planktonic foraminifera and pteropods and the species-specific composition of the assemblages collected at the mooring site CBeu, PANGAEA, https://doi.pangaea.de/10.1594/PANGAEA.904390, dataset in review.

Romero, O.E., Baumann, K.-H., Zonneveld, K.A.F., Donner, B., Hefter, J., Hamady, B. et al. 2020. Flux variability of phyto- and zooplankton communities in the mauritanian coastal upwelling between 2003 and 2008. Biogeosciences 17: 187–214.

Sá, C., Leal, M.C., Silva, A., Nordez, S., André, E., Paula, J. et al. 2013. Variation of phytoplankton assemblages along the mozambique coast as revealed by HPLC and microscopy. J. Sea Res. 79: 1–11.

Salah, S. 2013. Structure spatio-temporelle du zooplancton dans les deux filaments d'upwelling cap ghir et cap juby. Effet de la dynamique des filaments sur la dispersion des copepods. (Ph.D. thesis), Hassan II Univ., Casablanca, Morocco, 277 pp.

Saraswathy, M. 1986. Pleuromamma (copepoda-calanoida) in the Indian ocean mahasagar-bull. Natl. Inst. Oceanogr. 19(3): 185–201.

Schlacher, T.A. and Wooldridge, T.H. 1998. Trophic pathways at the cross-roads—is detritus the pivotal point in all estuarine food webs? ECSA, 29 Conf. Proceedings, University of Port Elizabeth, p. 73.

Schlüter, M., Baeza, A., Dressler, G., Frank, K., Groeneveld, J., Jager, W. et al. 2017. A framework for mapping and comparing behavioural theories in models of social-ecological systems. Ecological Economics 131: 21–35.

Schlüter, M., Orach, K., Lindkvist, E., Martin, R., Wijermans, N., Bodin, Ö. et al. 2019. Toward a methodology for explaining and theorizing about social-ecological phenomena. Current Opinion in Environmental Sustainability 39: 44–53.

Schukat, A., Teuber, L., Hagen, W., Wasmund, N. and Auel, H. 2013. Energetics and carbon budgets of dominant calanoid copepods in the northern benguela upwelling system. J. Exp. Mar. Biol. Ecol. 442: 1–9.

Schukat, A., Auel, H., Teuber, L., Lahajnar, N. and Hagen, W. 2014. Complex trophic interactions of calanoid copepods in the benguela upwelling system. J. Sea Res. 85: 186–19.

Sedletskaya, V.A. 1999. Manuel d'identification des œufs et des larves des espèces les plus abondantes de poissons habitant au large des côtes de l'Afrique du Nord-Est. Document technique, 26pp.

Seeyave, S., Probyn, T.A., Pitcher, G.C., Lucas, M.I. and Purdie, D.A. 2009. Nitrogen nutrition in assemblages dominated by pseudo-nitzschia spp., alexandrium catenella and dinophysis acuminata off the west coast of south Africa. Mar. Ecol. Prog. Ser. 379: 91–107.

Seguin, G. 1966. Contribution à l'étude de la biologie du plancton de surface de la baie de dakar. Etude quantitative et observation écologique au cours d'un cycle annuel. Bull. IFAN, sér. A 38(1): 1–90.

Séguin, F. 2010. Zooplankton Community Near The Island of São Vicente in The Cape Verde Archipelago: Insight on Pelagic Copepod Respiration, MS thesis, University of Bremen, Bremen, 77 pp.

Seridji, H. 2000. Copepod Diversity and community structure in the algerian basin. Crustaceana 73: 1–23.

Séret, C. 1983. Zooplancton de la côte sud de la presqu'île du cap vert (sénégal). pp. 157–188. *In*: Etude de l'environnement côtier au sud du Cap Vert (Sénégal) Rapp. Prov. CRODT.

Shannon, L.V. and Nelson, G. 1996. The benguela: large-scale features and processes and system variability. pp. 163–210. *In*: Wefer, G., Berger, W.H., Siedler, G. and Webb, D.J. (eds.). The South Atlantic: Present and Past Circulation. Berlin; Springer.

Shannon, L.V. and O'Toole, M.J. 2003. Sustainability of the benguela: ex Africa semper aliquid novi. pp. 227–253. *In*: Hempel, G. and Sherman, K. (eds.). Large Marine Ecosystems of the World–Trends in Exploitation, Protection and Research. Elsevier, Amsterdam, The Netherlands.

Shannon, V., Hempel, G., Melanotte-Rizzoli, P., Moloney, C. and Woods, J. 2006. Benguela: Predicting a Large Marine Ecosystem. Elsevier Science.

Shannon, L.J., Moloney, C.L., Jarre, A. and Field, J.G. 2003. Trophic flows in the southern benguela during the 1980s and 1990s. J. Mar. Sys. 39(1–2): 83–116.

Shannon, L.J., Cochrane, K.L., Moloney, C.L. and Fréon, P. 2004a. Ecosystem approach to fisheries management in the southern Benguela: a workshop overview. Afr. J. Mar. Sci. 26: 1–8.

Shannon, L.J., Christensen, V. and Walters, C. 2004b. Modelling stock dynamics in the southern benguela ecosystem for the period 1978–2002. *In*: Shannon, L.J., Cochrane K.L. and Pillar, S.C. (eds.). Ecosystem Approaches to Fisheries in the Southern Benguela. Afr. J. Mar. Sci. 26: 179–196.

Shannon, L.J. and van der Lingen, C. 2006. Ecosystem states, ecosystem changes and changes in ecosystem functioning of the southern and northern benguela ecosystem. EAF Project Regional Meeting, Cape Town (SA). Document 3a. 15 pp.

Shannon, L.J., Neira, S. and Taylor, M. 2008. Comparing internal and external drivers in the southern benguela and the southern and northern Humboldt upwelling ecosystems. Afr. J. Mar. Sci. 30(1): 63–84.

Silva, E.S. 1953a. Diatomáceas of marine planctoon of angola. Works of Maritime Biologic Mission 4: 11–72.

Silva, E.S. 1953b. Red water by exuviella baltica L. with simultanious mortality of fishes in coastal waters of. Works of Maritime Biologic Mission 4: 75–90.

Silva, E.S. 1955. Dinoflagelated of angola's marine planctoon. Anais, Estudos Biol. Marit. X, Tomo II: 109–191.

Silva, E.S. 1958. New contribution to the study of angola's marine microplantoon. Works of Maritime Biologic Mission 4: 75–90.

Silva, I.P. and Boersma, A. 1977. Cretaceous planktonic foraminifers DSDP leg 39 (south Atlantic). Environmental Science.

Silva, S.C.P. 2003. Contribution to the study of phytoplatoon community on the angolan coast. Graduation Thesis in Biology. Departament of Biology of Sciences Faculty, Agostinho Neto University. Luanda: 79 pp.

Somoue, L., Elkhiati, N., Ramdani, M. and Lam, T. 2005. Abundance and structure of copepod communities along the Atlantic coast of southern Morocco. Acta Adriat. 46(1): 63–76.

Spalding, M.D., Fox, H.E., Allen, G.R., Davidson, N., Ferdana, Z.A., Finlayson, M. et al. 2007. Marine ecoregions of the world: a bioregionalization of coastal and shelf areas. Bioscience 57(7): 573–582.

Sreeush, M.G., Valsala, V., Pentakota, S., Prasad, K.V.S.R. and Murtugudde, R. 2018. Biological production in the Indian ocean upwelling zones-Part 1: refined estimation via the use of a variable compensation depth in ocean carbon models. Biogeosciences 15(7).

Steinberg, D.K. and Landry, M.R. 2017. Zooplankton and the ocean carbon cycle. Annual Review of Marine Science 9(1): 413–444.

Stenvers, V.I., Hauss, H., Osborn, K.J., Neitzel, P., Merten, V., Scheer, S. et al. 2021. Distribution, associations and role in the biological carbon pump of pyrosoma atlanticum (tunicata, thaliacea) off cabo verde, NE Atlantic. Sci. Rep. 11(1): 9231.

Stuart, V. 1986. Feeding and metabolism of euphausia lucens (euphausiacea) in the southern Benguela current. Mar. Ecol. Prog. Ser. 30: 117–125.

Stuart, V. and Pillar, S. 1988. Growth and production of euphausia lucens in the southern Benguela current. J. Plankton Res. 10: 1099–1112.

Sywula, T., Wiafe, G., Sell, J. and Glazewska, I. 2002. Genetic subdivision of the upwelling copepod calanoides carinatus (krøyer, 1849) off the continental shelf of ghana. Journal of Plankton Research 24(5): 523–525.

Tagliarolo, M., Porri, F. and Scharler, U.M. 2018. Temperature-induced variability in metabolic activity of ecologically important estuarine macrobenthos. Mar. Biol. 165: 23.

Tagliarolo, M. and Scharler, U.M. 2018. Spatial and temporal variability of carbon budgets of shallow south African subtropical estuaries. Sci. Total Environ. 626: 915–926.

Tagliarolo, M., Porri, F., Garvie, C.D., Lechman, K. and Scharler, U.M. 2019. Zooplankton metabolism in south African estuaries: does habitat type influence ecological strategies? Journal of Plankton Research 41(4): 535–548.

Teuber, L., Schukat, A., Hagen, W. and Auel, H. 2013. Distribution and ecophysiology of calanoid copepods in relation to the oxygen minimum zone in the eastern tropical Atlantic. PloS One 8: e77590.

Tew-Kai, E. and Marsac, F. 2008. Patterns of variability of sea surface chlorophyll in the mozambique channel: a quantitative approach. Journal of Marine Systems 77: 77–88.

Tew-Kai, E. and Marsac, F. 2009. Patterns of variability of sea surface chlorophyll in the mozambique channel: a quantitative approach. Journal of Marine Systems 77: 77–88.

Tew-Kai, E. and Marsac, F. 2010. Influence of mesoscale eddies on spatial structuring of top predators' communities in the mozambique channel. Progress in Oceanography 86: 214–223.

Thiriot, A. 1977. Peuplements zooplanctoniques dans les regions de remontees d'eaus du littoral atlantique africain. Documentation Scientifique Centre Rechercher Oceanographie. Abidjan 8: 1–72.

Thomas, C.M., Perissinotto, R. and Kibirige, I. 2005. Phytoplankton biomass and size structure in two south African eutrophic, temporarily open/closed estuaries. Estuar Coast Shelf Sci. 65: 223–238.

Timonin, A.G., Arashkevich, E.G., Drits, A.V. and Semenova, T.N. 1992. Zooplankton dynamics in the northern benguela ecosystem, with special reference to the copepod Calanoides carinatus. *In*: Payne, A.I.L., Brink, K.H., Mann, K.H. and Hilborn, R. (eds.). Benguela Trophic Functioning. S. Afr. J. Mar. Sci. 12: 545–560.

Timonin, A.G. 1995. Dynamics of diurnal distribution of plankton of the northern part of benguela upwelling system. Okeanologia 35(4): 556–561.

Timonin, A.G. 1997. Diurnal vertical migrations of zooplankton in the upwelling area off the western coast of south Africa. Oceanology 37(1): 83–88.

Touré, D. 1971. Variation quantitative et qualitative du zooplancton dans la région du cap-vert de septembre 1970 à août 1971. Doc. Scient. Cent. Rech. Océanogr. Dakar-Thiaroye, 39, 25 p.

UN. 2015. Millennium Development Goals Report 2015. New York, NY: United Nations.

UN. 2017. HABITAT III regional report: africa. Retrieved from http://habitat3.org/wp-content/uploads/Habitat-III-Regional-Report-Africa.pdf.

United Nations. 2018. The World's Cities in 2018. United Nations editions, Data Booklet, 31 pp.

UNEP. 2018. The Adaptation Gap Report 2018. United Nations Environment Programme (UNEP), Nairobi, Kenya. 85 pp.

UNEP. 2014a. Design and Development of Integrated Indicators for the Sustainable Development Goals. Report: Senior Expert Meeting, 3–5 December 2014, Gland, Switzerland.

UNEP. 2014b. Measuring Success: Indicators for the Regional Seas Conventions and Action Plans. Nairobi: UNEP.

UNEP-DTIE. 2010. Sustainable Consumption and Production for Development: Background Paper, UNEP-DTIE, France.

UNEP-WCMC. 2016. The State of Biodiversity in Africa: A Mid-Term Review of Progress Toward The Aichi Biodiversity Targets. Cambridge, United Kingdom, UNEP-WCMC. Retrieved from https://wedocs.unep.org/rest/bitstreams/32269/retrieve.

UNEP and International Trade Center. 2017. Guidelines for providing product sustainability information global guidance on making effective environmental, social and economic claims, to empower and enable consumer choice. Retrieved from: http://www.oneplanetnetwork.org/resource/guidelines-providing-product-sustainability-information.

UNEP. 2020. Africa: The challenge. United National Environment Program. https://www.unenvironment.org/regions/africa#:~:text=Africa%20faces%20serious%20environmental%20challenges,extreme%20vulnerability%20to%20climate%20change.

UNEP-WCMC. 2017. Post-2020 global biodiversity framework. Retrieved from: https://www.cbd.int/doc/strategic-plan/Post2020/WCMC.pdf.

UNPD. (United nations, department of economic and social affairs, population division). World Urbanization Prospects: The 2018 Revision, Online Edition. https://population.un.org/wup/Download/Files/WUP2018-F05-Total_Population.xls.

Vavilova, V.V. 1990. Phytoplankton of the Benguela upwelling system off namibia in the austral autumn of 1985. Oceanology 30: 472–477.

Verheye, H.M. and Hutchings, L. 1988. Horizontal and vertical distribution of zooplankton biomass in the southern Benguela, May 1983. Afr. J. Mar. Sci. 6(1): 255–265.

Verheye, H.M. 1991. Short-term variability during an anchor station study in the southern benguela upwelling system. Abundance, distribution and estimated production of mesozooplankton with special reference to Calanoides carinatus (Krøyer, 1849). Prog. Oceanogr. 28(1–2): 91–119.

Verheye, H.M., Hutchings, L. and Peterson, W.T. 1991. Life history and population maintenance strategies of calanoides carinatus (copepoda: calanoida) in the southern Benguela ecosystem. S. Afr. J. Mar. Sci. 11: 179–191.

Verheye, H.M., Hutchings, L., Huggett, J.A. and Painting, S.J. 1992. Mesozooplankton dynamics in the benguela ecosystem, with emphasis on the herbivorous copepods. Afr. J. Mar. Sci. 12(1): 561–584.

Verheye, H.M., Hutchings, L., Huggett, J.A. and Painting, S.J. 1992. Mesozooplankton dynamics in the benguela ecosystem, with emphasis on the herbivorous copepods. *In*: Payne, A.I.L., Brink, K.H., Mann, K.H. and Hilborn, R. (eds.). Benguela Trophic Functioning. S. Afr. J. Mar. Sci. 12: 561–584.

Verheye, H.M., Richardson, A.J., Hutchings, L., Marska, G. and Gianakouras, D. 1998. Long-term trends in the abundance and community structure of coastal zooplankton in the southern benguela system, 1951–1996. Afr. J. Mar. Sci. 19(1): 317–332.

Verheye, H.M., Rogers, C., Maritz, B., Hashoongo, V., Arendse, L.M., Gianakouras, D. et al. 2001. Variability of zooplankton in the region of the angola-benguela front during winter 1999. S. Afr. J. Sci. 97: 257–258.

Verheye, H.M., Rogers, C., Maritz, B., Hashoongo, V., Arendse, L.M., Gianakouras, D. et al. 2001. Variability of zooplankton in the region of the angola-benguela front during winter 1999. S. Afr. J. Sci. 97: 257–258.

Verheye, H.M., Hagen, W., Auel, H., Ekau, W., Loick, N., Rheenen, I. et al. 2005. Life strategies, energetics and growth characteristics of calanoides carinatus (copepoda) in the angola-benguela frontal region. Afr. J. Mar. Sci. 27(3): 641–651.

Verheye, H.M. and Ekau, W. 2005. Maintenance mechanisms of plankton populations in frontal zones in the benguela and angola current systems: a preface, Afr. J. Mar. Sci. 27: 611–615.

Verheye, H.M., Lamont, T., Huggett, J.A., Kreiner, A. and Hampton, I. 2016. Plankton productivity of the benguela current large marine ecosystem (BCLME). Environmental Development 17, Supplement 1: 75–92.

Vinogradov, G., Hernández, F., Tejera, E. and León, M.E. 2004. Pelagic amphipods from the cape verde Islands (TFMCBM/98 cruise, Macaronesia 2000-Project). VIERAEA 32: 7–27.

Voges, L. 2003. Marine biodiversity in namibia–the known and the unknown. National Report. pp. 98–123. *In*: Cynthia Decker, Charles Griffiths, Kim Prochazka, Carmen Ras and Alan Whitfield (eds.). Marine Biodiversity in Sub-Saharan Africa: The Known and the Unknown. Proceedings of the Marine Biodiversity in Sub-Saharan Africa: The Known and the Unknown, 310 pp.

Voituriez, B. and Dandonneau, Y. 1974. Relations entre la structure thermique la production primaire et la régénération des sels nutritifs dans le dôme de guinée. Cah. O.R.S.T.O.M., sér. Océanogr. 12(4): 241–255.

Wasmund, N., Lass, H.U. and Nausch, G. 2005. Distribution of nutrients, chlorophyll and phytoplankton primary production in relation to hydrographic structures in the angola and benguela current system (SE Atlantic). African Journal of Marine Science 27: 177–190.

WBCSD. 2010. Vision 2050, World Business Council for Sustainable Development, Geneva, Switzerland.

Webb, P. and Wooldridge, T.H. 1990. Diel horizontal migration of mesopodopsis slabberi (crustacea: mysidacea) in algoa bay, southern Africa. Marine Ecology Progress Series 62: 73–77.

Werner, T., Huenerlage, K., Verheye, H. and Buchholz, F. 2012. Thermal constraints on the respiration and excretion rates of krill, euphausia hanseni and nematoscelis megalops, in the northern benguela upwelling system off Namibia. Afr. J. Mar. Sci. 34: 391–399.

Whitfield, A.K. 1985. The role of zooplankton in the feeding ecology of fish fry from southern African estuaries. South African Journal of Zoology 20: 166–171.

Whitfield, A.K., Bate, G.C., Adams, J.B., Cowley, P.D., Froneman, P.W., Gama, P.T. et al. 2012. A review of the ecology and management of temporarily open/closed estuaries in south Africa, with particular emphasis on river flow and mouth state as primary drivers of these systems. Afr. J. Mar. Sci. 34: 163–180.

Whitfield, A.K. and Baliwe, N.G. 2013. A century of science in south african estuaries: Bibliography and review of research trends. South African Institute for Aquatic Biodiversity (SAIAB).

Wiafe, G. and Frid, C.L.J. 1997. Hydrographic forcing and zooplankton temporal variation off the coast of ghana, west Africa. pp. 149–59. *In*: Evans, S.M., Vanderpuye, C.J. and Armah, A.K. (eds.). The Coastal Zone of West Africa: Problems and Management. Proceedings of the International Seminar on the Coastal Zone of West Africa. Penshaw Press, U.K.

Wiafe, G. 1997. Zooplankton biodiversity in the gulf of guinea. *In*: Evans, S.M., Vanderpuye, C.J. and Armah, A.K. (eds.). The Coastal Zone of West Africa: Problems and Management. Proceedings of the International Seminar on the Coastal Zone of West Africa. Penshaw Press, U.K. pp 243.

Wiafe, G. and Frid, C.L.J. 1999. Short-term temporal variability within coastal zooplankton communities. Journal of the Ghana Science Association 2(3): 122–128.

Wiafe, G. and Frid, C.L.J. 2001. Marine zooplankton of west Africa (with CD-ROM). Marine Biodiversity Capacity Building in the West African Sub-region. Darwin Initiative Report 5, Ref. 162/7/451.

Wickstead, J.H. 1961. Plankton of the northern Kenyan banks. Nat. Lon. 192: 890–891.

Wickstead, J.H. 1962. Plankton from the coastal stripes of Kenya and Tanzania area of the Indian Ocean. Nat. Lon. 196: 1224–1225.

World Population Prospects. 2022a. population.un.org. United Nations Department of Economic and Social Affairs, Population Division. World Population Prospects - Population Division-United Nations.

World Population Prospects. 2022b.: Demographic indicators by region, subregion and country, annually for 1950–2100 (XSLX). population.un.org (Total Population, as of 1 July) (thousands). United Nations Department of Economic and Social Affairs, Population Division.

Wooldridge, T.H. and Melville-Smith, R. 1979. Copepod succession in two south African estuaries. Journal of Plankton Research 1: 329–341.

Wooldridge, T.H. and Bailey, C. 1982. Euryhaline zooplankton of the sundays estuary with notes on trophic relationships. South African Journal of Zoology 17: 151–163.

Wooldridge, T.H. 1999. Estuarine zooplankton community structure and dynamics. pp. 141–166. *In*: Allanson, B.R. and Baird, D. (eds.). Estuaries of South Africa. Cambridge: Cambridge University Press.

Wooldridge, T.H. and Callahan, R. 2000. The effects of a single freshwater release into the kromme Estuary. 3: Estuarine zooplankton response. Water SA. 26: 311–318.

Wooldridge, T.H. 2005. Invertebrates of the olifants estuary. Unpublished Report. Port Elizabeth: Department of Zoology, Nelson Mandela Metropolitan University.

Wooldridge, T.H. and Deyzel, S.H.P. 2009. Temperature and salinity as abiotic drivers of zooplankton community dynamics in the great berg estuary. Trans R Soc South Afr. 64: 219–237.

Whitfield (eds.). 2003. Marine biodiversity in sub-saharan africa: the known and the unknown. Proceedings of the Marine Biodiversity in Sub-Saharan Africa: The Known and the Unknown, 310 pp.

Yacouba, S. and Nestor, N'G.Y. 2003. Marine biodiversity in côte d'ivoire– the known and the unknown, National Report. pp. 6–25. *In*: Cynthia Decker, Charles Griffiths, Kim Prochazka, Carmen Ras and Alan.

Yahia, M.N.D., Goy, J. and Yahia-Kéfi, O.D. 2003. Distribution et écologie des méduses (cnidaria) du golfe de tunis (méditerranée sud occidentale). Oceanol. Acta 26: 645–655.

Yahia, M.N.D., Souissi, S. and Yahia-Kéfi, O.D. 2004. Spatial and temporal structure of planktonic copepods in the bay of tunis (southwestern Mediterranean Sea). Zool Stud. 43: 366–375.

Youssara, F., Gaudy, R., Moukrim, A. and Moncef, M. 2004. Variations spatio-temporelles du mésozooplancton de la région d'Agadir (Maroc) entre mai 1999 et décembre 2000. Mar. Life 14(1–2): 3–18.

Zaafa, A., Ettahiri, O., Elkhiati, N., Blahen, M., Berraho, A., Somoue, L. et al. 2012. Variability of spatial and temporal distribution of marine zooplankton communities in relation with environmental parameters in tangier and M'Diq (gibraltar strait) regions. J. Mater. Environ. Sci. 3(2): 262–269.

Zaghloul, W., Madkour, F. and Mohammad, S. 2020. Marine gelatinous zooplankton of the egyptian waters: a review. Egypt J. Aquac. 10: 35–45.

Zakaria, H.Y. 2006. The zooplankton community in egyptian mediterranean waters: a review. Acta Adriat 47: 195–206.

Zakaria, H.Y., Radwan, A.A. and Said, M.A. 2007. Influence of salinity variations on zooplankton community in El-Mex bay, Alexandria, Egypt. Egypt. J. Aquat. Res. 33(2): 52–67.

Zakaria, H.Y. 2015. Impact of climate variability and anthropogenic activities on zooplankton community in the neritic waters of alexandria, egypt. J. King Abdulaziz Univ Mar. Sci. 25: 2.

Zhang, C., Dang, H., Azam, F., Benner, R., Legendre, L., Passow, U. et al. 2018. Evolving paradigms in biological carbon cycling in the ocean. National Science Review 5(4): 481–499.

Zhang, R., Grimi, N., Marchal, L., Lebovka, N. and Vorobiev, E. 2019. Effect of ultrasonication, high pressure homogenization and their combination on efficiency of extraction of bio-molecules from microalgae parachlorella kessleri. Algal Res. 40: 101524.

Zizah, S., Ettahiri, O., Salah, S., Yahyaoui, A. and Ramdani, M. 2012. Evolution spatio-temporelle des abondances zooplanctoniques au large de la côte atlantique marocaine entre Cap Boujdour (26°30′N) et Cap Blanc (21°N). Bull. Inst. Sci. Sect. Sci. Vie 34(2): 79–94.

Index